# Synthese Library

Studies in Epistemology, Logic, Methodology, and Philosophy of Science

Volume 418

The aim of *Synthese Library* is to provide a forum for the best current work in the methodology and philosophy of science and in epistemology. A wide variety of different approaches have traditionally been represented in the Library, and every effort is made to maintain this variety, not for its own sake, but because we believe that there are many fruitful and illuminating approaches to the philosophy of science and related disciplines.

Special attention is paid to methodological studies which illustrate the interplay of empirical and philosophical viewpoints and to contributions to the formal (logical, set-theoretical, mathematical, information-theoretical, decision-theoretical, etc.) methodology of empirical sciences. Likewise, the applications of logical methods to epistemology as well as philosophically and methodologically relevant studies in logic are strongly encouraged. The emphasis on logic will be tempered by interest in the psychological, historical, and sociological aspects of science.

Besides monographs *Synthese Library* publishes thematically unified anthologies and edited volumes with a well-defined topical focus inside the aim and scope of the book series. The contributions in the volumes are expected to be focused and structurally organized in accordance with the central theme(s), and should be tied together by an extensive editorial introduction or set of introductions if the volume is divided into parts. An extensive bibliography and index are mandatory.

More information about this series at http://www.springer.com/series/6607

Hitoshi Omori • Heinrich Wansing
Editors

# New Essays on Belnap-Dunn Logic

Springer

*Editors*
Hitoshi Omori
Ruhr-University Bochum
Bochum, Germany

Heinrich Wansing
Ruhr-University Bochum
Bochum, Germany

Synthese Library
ISBN 978-3-030-31135-3    ISBN 978-3-030-31136-0 (eBook)
https://doi.org/10.1007/978-3-030-31136-0



This Springer imprint is published by the registered company Springer Nature Switzerland AG.
The registered company address is: Gewerbestrasse 11, 6330 Cham, Switzerland

**Prof. J. Michael Dunn and Prof. Nuel D. Belnap**
**Pittsburgh, April 2018**

(Photo: courtesy of Prof. Anil Gupta)

# Contents

# An invitation to *New Essays on Belnap-Dunn logic*

Hitoshi Omori and Heinrich Wansing

**Abstract** In this introductory note, we place the new essays on Belnap-Dunn logic, **FDE**, of the present volume against the background of the development of **FDE**. This note is an invitation to study the volume. It presents a chronological perspective on Belnap-Dunn logic and a slightly idiosyncratic list of further research topics.

**Keywords** Belnap-Dunn logic • First degree entailment • Tautological entailment • Dunn's semantics • Relevance logic • Routley star • American plan • Australian plan • Paraconsistent logic • Exactly true logic • Non-falsity logic • Bilattices • Trilattices • Constructible falsity • Catuṣkoṭi • Negation as a modal operator • Connexive logic • Dialetheism

## 1 Introduction

Among the continuum many systems of nonclassical logic, Belnap-Dunn logic seems to enjoy a very special status for a number of different reasons. With Belnap-Dunn logic at hand, we are able to offer different topics of interest to philosophers, computer scientists, and mathematicians, such as relevance of entailment, aboutness, negation, paraconsistency, inconsistency-tolerant information processing, definitional equivalence, etc. Moreover, at class rooms, we can show some of the basic ideas behind different techniques that are standard in philosophical logic by only focusing on Belnap-Dunn logic, including many-valued semantics, two-valued relational semantics, possible worlds semantics, algebraic semantics, and so on. When it comes to the latter pedagogical aspect, Belnap-Dunn logic seems to score better than the more popular classical logic.

The present edited volume is in fact the second volume by the same editors dedicated to Belnap-Dunn logic. The first volume, which appeared as a special issue of *Studia Logica*,[1] focused on the more technical developments related to Belnap-Dunn logic. This volume, in contrast, aims at including also contributions that touch some more philosophical issues related to Belnap-Dunn logic. Moreover, we are very happy and proud to be able to reprint the three seminal papers by Nuel D. Belnap and J. Michael Dunn [4,5,10], publish the famous manuscript "Natural Language versus Formal Language" by J. Michael Dunn for the first time, and include an interview with Nuel D. Belnap as well as a new essay by J. Michael Dunn. These form the first part of the volume, titled **Essays by the Founders**.

Hitoshi Omori
Institute of Philosophy I, Ruhr-Universität Bochum, Germany
e-mail: hitoshi.omori@rub.de

Heinrich Wansing
Institute of Philosophy I, Ruhr-Universität Bochum, Germany
e-mail: heinrich.wansing@rub.de

[1] Special issue "40-years of FDE," *Studia Logica* 105(6), 2017.



H. Omori, H. Wansing (eds.), *New Essays on Belnap-Dunn Logic*,
Synthese Library 418, https://doi.org/10.1007/978-3-030-31136-0_1

As one can see by reading them, these essays themselves deliver a lot of information regarding the early exciting developments of Belnap-Dunn logic. It even seems to us that they are already serving as an excellent introduction to the whole literature on Belnap-Dunn logic, and we therefore felt justified to decide to keep our introduction short and not to keep our readers wait too long before jumping into the whole volume.

Still, we believe that it is one of the editors' responsibilities to guide our readers into the new essays on Belnap-Dunn logic. We will, however, refrain from reviewing the basic technicalities of Belnap-Dunn logic since our introduction [25] to the first volume should serve well for that purpose. Therefore, for the rest of this introduction, we first offer a brief guide to the new essays, included in the second part of the volume under the title **New Essays**, by reviewing some major developments related to Belnap-Dunn logic, and putting each essay in context (but without giving a summarizing overview). We then turn to pointing out a few topics that are related to Belnap-Dunn logic and seem to be quite interesting from the editors' perspective.

## 2 Major developments and contributions to this volume

### *1959: Belnap on the first degree entailment of relevance logic* E

What we refer to as Belnap-Dunn logic is also known as first degree entailment, or first-degree entailment logic, **FDE**. This is so because of the origin of the logic under discussion. More specifically, first degree entailments are formulas of the form $A \to B$, where the formulas $A$ and $B$ contain at most conjunction, disjunction, and negation. Belnap, in his unpublished doctoral dissertation from 1959, presented an axiom system that captures the first degree entailment fragment of the Anderson-Belnap system **E** of relevant entailment. Moreover, Belnap, in an abstract published also in 1959, reported on a characterization of the provable first-degree entailments in an intuitive way as the tautological entailments.[2] These results can be found in [2]. Semantically, the valid first degree entailments of **E** were characterized by an eight-valued matrix. This was later improved by Timothy Smiley, who pointed out that a four-valued matrix will suffice for the characterization, but the four-values did not have their intuitive readings yet.

As one can see from the interview with Belnap, included in this volume, Belnap is still in favor of viewing Belnap-Dunn logic as first degree entailment in the language with implication rather than in the language without implication, implication being replaced by a (semantic or proof-theoretic) consequence relation. The latter seems to be the more popular and prevailing presentation nowadays, but it is also important to keep in mind the origin of Belnap-Dunn logic.

*A note on the papers in this volume (we will use this way of highlighting contributions to the present volume with a bar on the left).* Allen Hazen and Jeff Pelletier also take the more recent presentation of Belnap-Dunn logic, and in view of the result that we cannot define reasonable implication connectives in the language with negation, conjunction, and disjunction, their contribution discusses, among other things, some expansions of

[2] $A \to B$ is a *tautological entailment* iff it can be put into a provably equivalent normal form $A_1 \vee \ldots \vee A_m \to B_1 \wedge \ldots \wedge B_m$ and for every $A_j \to B_k$, the conjunction $A_j$ and the disjunction $B_k$ share a propositional variable (so that $A_j \to B_k$ is tautologically valid in this sense).

**FDE**, as well as Kleene's strong three-valued logic, **K3**, and the logic of paradox, **LP**, by different implications.

## *1966: Dunn's intuitive semantics*

So, it should be now clear why Belnap-Dunn logic is also known as **FDE**. We next turn to a contribution by Dunn, which substantially improved the semantic understanding of **FDE**. In brief, Dunn offered a semantics for **FDE** in terms of *two* truth values. This was made possible by using a non right unique valuation relation instead of a total valuation function. Roughly speaking, the truth and the falsity of a formula now come apart, and as a result, the given connectives will receive not only truth conditions but also falsity conditions in the semantics. In particular, the negation of $A$ is true iff $A$ is false, and the negation of $A$ is false iff $A$ is true.

Even though this new semantics was published only in 1976, in a paper which is reprinted in this volume, the results were already in Dunn's dissertation from 1966, and thus we decided to place Dunn's contribution here, not later.

*A note on the papers in this volume.* Jc Beall's contribution addresses the question whether Belnap-Dunn logic may serve as "The One True Logic." As Beall acknowledges explicitly at the beginning of his essay, this was not a question addressed by Belnap or Dunn, but Beall aims at giving some arguments in favor of the claim that **FDE** is The One True Logic.

## *1972: Routleys' star semantics*

Yet another *two*-valued semantics for **FDE** was presented by Richard Routley (later Sylvan) and Valerie Routley (later Plumwood). In contrast to Dunn's semantics, this was achieved by using a possible worlds semantics and, especially, by making use of the so-called star operation on worlds, which is an involutive operation. In particular, the negation of $A$ is true at a world $w$ iff $A$ is not true at the star world $w^*$ of $w$.

*A note on the papers in this volume.* Adam Přenosil considers two expanded languages of **FDE**, one obtained by adding two propositional constants, and the other obtained by adding intuitionistic implication besides the two constants. For the latter language, Přenosil makes an interesting use of star semantics to formulate a system that can be seen as a combination of classical logic and intuitionistic logic.

So, we now have two different but equivalent semantics for **FDE** with two truth values, namely Dunn's semantics and Routleys' star semantics. Even though these semantics are equivalent for the language of **FDE**, they turned out to be quite different when it comes to devising semantics for the full language of relevance logics. Two different approaches to the semantics of relevance logics are sometimes called in the literature *the American plan*, that does without the star operation, and *the Australian plan*, that makes use of the Routley-star.

*A note on the papers in this volume.* Takuro Onishi's contribution is concerned with the relation between the American plan and the Australian plan. Building on a paper by Richard Routley, [31], Onishi aims at showing that the Australian plan is obtained by developing the American plan.

## *1977: Belnap's seminal papers*

Belnap's four-valued semantics as presented in the two seminal papers [4,5] can be seen as obtained by a combination of Belnap's ideas together with Smiley's observation that for characterizing **FDE** it is sufficient to have four values, instead of eight values, and Dunn's semantics. More specifically, the four values can be represented as the elements of the powerset **4** of the set of classical truth values $\mathbf{2} = \{1, 0\}$. Thus, we have $\mathbf{t} = \{1\}$, $\mathbf{f} = \{0\}$, $\mathbf{b} = \{1, 0\}$, and $\mathbf{n} = \{\}$. These four-values were motivated and read by considering "How a computer should think" in terms of information passed to a question-answering computer confronted possibly with contradictory information or nor information at all concerning some given atomic formulas. If the value **t** is assigned to an atomic formula $p$, then $p$ is understood as *told only true* by at least one information source. For the remaining values, we obtain that the value **f** is read as *told only false*, **b** as *both told true and told false*, and **n** as *neither told true nor told false*. Moreover, the representation of the four values as the subsets of **2** led Belnap to distinguishing two partial orders on **4**, an "approximation" ordering with **b** as the top element and an ordering referred to as a truth ordering, with **t** as the top element. Belnap called the resulting lattices an approximation lattice and a logical lattice, respectively, and then introduced **FDE** as the logic of the four-element logical lattice, **L4**, with conjunction interpreted as lattice meet and disjunction as lattice join. Moreover, negation is uniquely determined by the requirement that the function interpreting it in addition to mapping **t** to **f** and **f** to **t**, is also monotonic, i.e., respecting the ordering $\leq$ of **L4**. Finally, Belnap defined semantic consequence as order preservation; $B$ follows from $A$ just in case for every four-valued valuation function $v$, $v(A) \leq v(B)$.

*A note on the papers in this volume.* Katalin Bimbó's contribution aims at exploring non-monotonic consequence relations based on Belnap-Dunn logic, instead of taking classical logic. More specifically, Bimbó defines several default rules, and considers some applications of them. As non-monotonic logic forms an important area in computer science, Bimbó's contribution can be seen as echoing the original intuition of Belnap that was brought in from computer-related issues.

By seeing how Dunn's idea nicely motivates a reading of the four values, it is also quite natural to define the consequence relation as is done in many-valued logic. In particular, if we assume the standard truth preservation account of the semantic consequence relation based on Dunn's semantics, then the equivalent way is to preserve the values that contain 1, namely **t** and **b**.

Belnap's two papers and the presentation of **FDE** as a *useful four-valued logic* for how a computer should process information meant a kind of break-through for Belnap-Dunn logic. Given a many-valued semantics, there are, however, also other choices for the set of designated values from **4**. Pursuing the idea that in some sense positive values are preserved in semantic consequence, the following choices are perhaps suggestive, namely to take **t** only, and all values except **f**. The resulting logics are known as **ETL** (exactly true logic) and **NFL** (non-falsity logic), respectively, and discussions on these logics can be found in [20,28] and [32,36], respectively.

> *A note on the papers in this volume.* Yaroslav Shramko offers various proof systems that are equivalent to the original formulation given by Belnap, but more flexible in considering various extensions of Belnap-Dunn logic. One of the implications of Shramko's results include sound and complete proof systems for **ETL** and **NFL**.

Note also that there is some room for discussing how we give intuitive readings to the four values. See, for example, the paper [9] by Didier Dubois and a reply [40] to it by Wansing and Belnap concerning whether the four values are to be understood in epistemic or rather in informational terms.

> *A note on the papers in this volume.* Andreas Kapsner focuses on the issue of possible readings of the four values with some constructive flavors. The term "constructive" is here understood along the line of thought given by Michael Dummett, who is also one of the main figures discussed earlier by Kapsner in [18].

The algebraic structure enjoyed by the four values in Belnap's semantics was later generalized in terms of the notion of a *bilattice*, introduced by Matthew Ginsberg in [12]. Briefly speaking, and as already pointed out, the four values form a lattice structure with respect to two different orderings, one measuring the amount of truth (in a sense), the other measuring the amount of information. Further interesting developments can be found in, e.g. [3,11].

> *A note on the papers in this volume.* Francesco Paoli's contribution is concerned with an expansion of bilattices by what is called *demi-negation*, introduced by Lyold Humberstone in [14]. In brief, demi-negation is a unary operation such that it behaves as a single classical negation (or possibly other negations one prefers) once they are iterated twice. Paoli explores both semantics and axiomatizations from the perspective of abstract algebraic logic.

Note also that the developments of bilattice logics were followed by the introduction of the notion of a *trilattice*, by adding an ordering to measure constructivity, a suggestion made by Yaroslav Shramko, J. Michael Dunn, and Tatsutoshi Takenaka in [33], or to measure falsity, a suggestion made by Yaroslav Shramko and Heinrich Wansing in [34]. For the latter approach, see also [35].

## *1984: Almukdad and Nelson's constructible falsity*

Quite independently of the developments we have seen so far, there is also another related development by David Nelson. Nelson's first work on the related topic of strong negation was already published in 1949, namely [21], but that was, seen from the current perspective, resulting in a paracomplete logic, but not in a logic that is both paracomplete and paraconsistent. It was in [1] that Nelson together with Ahmad Almukdad came up with the system nowadays called **N4**, which can be seen as an expansion of Belnap-Dunn logic by intuitionistic implication. The logic **N4** enjoys constructible falsity as a constructive feature in addition to the constructive features of intuitionistic logic, namely that $\sim(A \wedge B)$ is provable iff it holds that $\sim A$ is provable or $\sim B$ is provable. However, Almukdad and

Nelson did not refer to the literature on Belnap-Dunn logic, and it was only much later that the relationship between **FDE** and **N4** has been pointed out *explicitly*, for example in [16]. A discussion of **N4** in connection with paraconsistency, relevance logic, and information processing can already be found in the dissertation [38].

Note that there exists some intensive and systematic research on Nelson's logics. This includes investigations on algebraic semantics given by Sergei Odintsov in [22], as well as on proof systems studied by Norihiro Kamide and Heinrich Wansing in [16,17]. One of the more recent developments related to Nelson's logic includes the systematic development of modal logics based on the classical extension of Nelson's logic since [23] (cf. [24] for an overview).

*A note on the papers in this volume.* The joint paper by Igor Sedlár and Ondrej Majer takes the modal logic developed by Odintsov and Wansing in [23], and develops a framework which not only allows to deal with inconsistent bodies of information (this was already possible within the framework presented in [19]), but also to keep track of the sources of inconsistency. Sedlár and Majer also expand the semantic framework by adding a compatibility relation to allow more expressivity on the relation between sources, and presents some basic results.

## *2010: Priest's Logic of Catuṣkoṭi*

The application of Belnap-Dunn logic to philosophy is not restricted to the Western tradition, but also stretches out to the Eastern tradition. The idea of applying paraconsistent logic to buddhism has been in the air among dialetheists since the 1980s, but one of the first clear applications of paraconsistent logic was given by Graham Priest in [29].

*A note on the papers in this volume.* Jay Garfield's contribution offers a comparison between Belnap and Nāgārjuna by pointing to a few similarities and differences, with a special emphasis on the notion of truth. This is also related to the reading of the four truth values of Belnap-Dunn logic, already mentioned earlier.

Graham Priest's contribution is concerned with developing systems of natural deduction for what Priest calls the **FDE** family. This family obviously includes **FDE** itself, but also comprises **FDEe**, which is the system introduced and discussed in [29]. It is worth highlighting that the construction Priest uses to obtain **FDEe** from **FDE** is an application of a very general construction called *plurivalent semantics* which produces many-valued semantics out of another many-valued semantics with less truth values. The details of the construction is given in [30], and a comparison of Priest's technique with Dunn's semantics, with a focus on truth-value gaps, is briefly discussed in [37].

## 3 Some further topics: a small and idiosyncratic list

There are some other topics related to Belnap-Dunn logic, but not discussed as central topics in the contributed essays. We will pick three topics that are of special interest to the editors, and briefly discuss why we believe these topics are interesting.

## *3.1 Negation*

As we saw earlier, Belnap-Dunn logic enjoys at least two kinds of two-valued semantics, namely Dunn's relational semantics and the Routleys' star semantics. This has an implication on two different semantics for negation.[3] There are some recent discussions comparing negation as a modal operator with negation as a contradictory-forming operator (in the sense mentioned above that (i) the negation $\sim A$ of a sentence $A$ is true iff $A$ is false and (ii) $\sim A$ is false iff $A$ is true), see [6,7,8]. In the end, the discussion may boil down to a matter of taste. Still, it remains to be seen what we can learn about negation from the two different semantics. In particular, it seems to be worthwhile to set up some criteria and compare the two accounts of negation in more detail.

## *3.2 Connexive logic*

One of the charms of Belnap-Dunn logic, when formulated in terms of Dunn's semantics, is the possibility to formulate highly non-classical principles, such as Aristotle's theses and Boethius' theses that characterize connexive logic.[4] In fact, if we set the formulation of connexive principles as one of the criteria to compare two accounts of negation, then Dunn semantics seems to score far better than Routleys' semantics. Indeed, we only need to modify the falsity condition for the conditional in a very simple manner in the former, whereas it is extremely complicated to achieve that in the latter. In fact, it seems that we can reach a much broader family of contra-classical logics (cf. [15,26]) in a relatively simple manner. It again remains to be seen to which extent this is the case, and what is the exact picture we obtain of contra-classical logics.

## *3.3 Inconsistency and dialetheism*

Finally, there is the issue of interpreting the *both* value in Belnap's four-valued semantics, or allowing propositions to be related to *both* true and false in Dunn's relational semantics. As one can see from their writings, both Belnap and Dunn are strictly resisting the dialetheic reading, defended by Routley and Priest, among others. Still, there seems to be a reason why we still need to keep discussing this issue. Here is why. As a byproduct of the approach to connexive logics, mentioned above, the propositional logic will have a formula and its negation as both valid for certain formulas. Of course, one might see this as a bad result, and apply a *reductio* to conclude that the very approach to connexive logic should be rejected.[5] But if, for whatever reasons, may they be more technical or philosophical, the approach to connexive logic turns out to be favored, then we need to make sense of the inconsistency. How exactly that should be done, again, remains to be seen.

---

[3] Needless to say, these two semantics do not exhaust the options. For an up-to-date overview, see [13].

[4] For an overview of connexive logics, see [39], and for some current trends, see [27].

[5] Note that some connexive logics are indeed consistent.

## 4 Concluding remarks

Already 60 years have passed since the publication of Belnap's abstract on first degree entailment in 1959, and 50 years since the talk "Natural Language versus Formal Language" delivered by Dunn in 1969. Half a century is quite a long time, and we have already seen a lot of exciting developments related to **FDE**. Still, as the essays in this volume show, there are many directions for further explorations, and we hope that some readers will be motivated to join the continuing investigation and development of Belnap-Dunn logic.

**Acknowledgment** First of all, we would like to thank Nuel D. Belnap and J. Michael Dunn for agreeing to be part of this project and offering us numerous support. We would also like to thank the authors of the new essays for accepting our invitation and contributing excellent essays that shed light on a number of different formal as well as philosophical aspects of Belnap-Dunn logic. Moreover, we would like to thank Otávio Bueno for his enthusiastic support for our volume as the editor–in–chief of the Synthese Library series, Anil Gupta and Ties Nijssen for their kind help, which was necessary and substantial to deal with some of the challenges to reprint the seminal papers, and Tobias Koch for assisting us in the typesetting of the manuscript.

Hitoshi Omori's preparation of the final version of this introduction was partly supported by a Sofja Kovalevskaja Award of the Alexander von Humboldt-Foundation, funded by the German Ministry for Education and Research. Finally, but not the least, we would also like to thank the reviewers of the papers submitted to the volume.

## References

1. Ahmad Almukdad and David Nelson. Constructible falsity and inexact predicates. *The Journal of Symbolic Logic*, 49(1):231–233, 1984.
2. Alan Anderson and Nuel Belnap. Tautological entailments. *Philosophical Studies*, 13:9– 24, 1962.
3. Ofer Arieli and Arnon Avron. Reasoning with logical bilattices. *Journal of Logic, Language and Information*, 5:25–63, 1996.
4. Nuel Belnap. How a computer should think. In G. Ryle, editor, *Contemporary aspects of philosophy*, pages 30–55. Oriel Press, 1977.
5. Nuel Belnap. A useful four-valued logic. In J.M. Dunn and G. Epstein, editors, *Modern Uses of Multiple-Valued Logic*, pages 8–37. D. Reidel Publishing Co., 1977.
6. Francesco Berto. A modality called 'negation'. *Mind*, 124(495):761–793, 2015.
7. Francesco Berto and Greg Restall. Negation on the Australian plan. *Journal of Philosophical Logic*, 2019. https://doi.org/10.1007/s10992-019-09510-2.
8. Michael De and Hitoshi Omori. There is more to negation than modality. *Journal of Philosophical logic*, 47(2):281–299, 2018.
9. Didier Dubois. On ignorance and contradiction considered as truth-values. *Logic Journal of the IGPL*, 16(2):195–216, 2008.
10. J. Michael Dunn. Intuitive semantics for first-degree entailment and 'coupled trees'. *Philosophical Studies*, 29:149–168, 1976.
11. Melvin Fitting. Bilattices are nice things. In V. Hendricks and S.A. Pedersen, editors, *Self-Reference*, pages 53–77. CSLI-Publications, Stanford, 2004.
12. Matthew L. Ginsberg. Multivalued logics: A uniform approach to reasoning in AI. *Computer Intelligence*, 4:256–316, 1988.
13. Laurence R. Horn and Heinrich Wansing. Negation. In Edward N. Zalta, editor, *The Stanford Encyclopedia of Philosophy*. http://plato.stanford.edu/entries/negation/, Spring 2015 edition, 2015.
14. Lloyd Humberstone. Negation by iteration. *Theoria*, 61(1):1–24, 1995.

15. Lloyd Humberstone. Contra-classical logics. *Australasian Journal of Philosophy*, 78(4):438–474, 2000.
16. Norihiro Kamide and Heinrich Wansing. Proof theory of Nelson's paraconsistent logic: A uniform perspective. *Theoretical Computer Science*, 415:1–38, 2012.
17. Norihiro Kamide and Heinrich Wansing. *Proof Theory of N4-related Paraconsistent Logics*. Studies in Logic, Vol. 54. College Publications, London, 2015.
18. Andreas Kapsner. *Logics and Falsifications*, volume 40 of Trends in Logic. Springer, 2014.
19. Hector J Levesque. A logic of implicit and explicit belief. In *Proceedings of AAAI* 1984, pages 198–202, 1984.
20. João Marcos. The value of the two values. In J.-Y. Béziau and M.E. Coniglio, editors, *Logic without Frontiers: Festschrift for Walter Alexandre Carnielli on the occasion of his 60th birthday*, pages 277–294. College Publication, 2011.
21. David Nelson. Constructible falsity. *The Journal of Symbolic Logic*, 14(1):16–26, 1949.
22. Sergei P. Odintsov. *Constructive Negations and Paraconsistency*. Dordrecht: SpringerVerlag, 2008.
23. Sergei P. Odintsov and Heinrich Wansing. Modal logics with belnapian truth values. *Journal of Applied Non-Classical Logics*, 20:279–301, 2010.
24. Sergei P Odintsov and Heinrich Wansing. Disentangling FDE-based paraconsistent modal logics. *Studia Logica*, 105(6):1221–1254, 2017.
25. Hitoshi Omori and Heinrich Wansing. 40 years of FDE: an introductory overview. *Studia Logica*, 105(6):1021–1049, 2017.
26. Hitoshi Omori and Heinrich Wansing. On Contra-classical variants of Nelson logic N4 and its classical extension. *Review of Symbolic Logic*, 11(4):805–820, 2018.
27. Hitoshi Omori and Heinrich Wansing. Connexive logics. an overview and current trends. *Logic and Logical Philosophy*, pages 1–17, 2019.
28. Andreas Pietz and Umberto Rivieccio. Nothing but the truth. *Journal of Philosophical Logic*, 42:125–135, 2013.
29. Graham Priest. The Logic of the Catuskoti. *Comparative Philosophy*, 1(2):24–54, 2010.
30. Graham Priest. Plurivalent Logics. *The Australasian Journal of Logic*, 11(1):1–13, 2014.
31. Richard Routley. The American plan completed: Alternative classical-style semantics, without stars, for relevant and paraconsistent logics. *Studia Logica*, 43(1-2):131–158, 1984.
32. Yaroslav Shramko. Dual-Belnap logic and anything but falsehood. *IfCoLog Journal of Logics and their Applications*, 6(2):411–430, 2019.
33. Yaroslav Shramko, J. Michael Dunn, and Tatsutoshi Takenaka. The trilattice of constructive truth values. *Journal of Logic and Computation*, 11(6):761–788, 2001.
34. Yaroslav Shramko and Heinrich Wansing. Some useful 16-valued logics: How a computer network should think. *Journal of Philosophical Logic*, 34(2):121–153, 2005.
35. Yaroslav Shramko and Heinrich Wansing. *Truth and Falsehood. An Inquiry into Generalized Logical Values*. Trends in Logic. Vol. 36. Springer, Berlin, 2011.
36. Yaroslav Shramko, Dmitry Zaitsev, and Alexander Belikov. First-degree entailment and its relatives. *Studia Logica*, 105(6):1291–1317, 2017.
37. Damian Szmuc and Hitoshi Omori. A Note on Goddard and Routley's Significance Logic. *The Australasian Journal of Logic*, 15(2):431–448, 2018.
38. Heinrich Wansing. *The Logic of Information Structures*. Springer Lecture Notes in AI 681. Springer, Berlin, 1993.
39. Heinrich Wansing. Connexive logic. In Edward N. Zalta, editor, *The Stanford Encyclopedia of Philosophy*. http://plato.stanford.edu/archives/fall2014/entries/logicconnexive/, Fall 2014 edition, 2014.
40. Heinrich Wansing and Nuel Belnap. Generalized truth values. A reply to Dubois. *Logic Journal of IGPL*, 18(6):921–935, 2009.

# Part I
# Essays by the Founders

# Natural Language versus Formal Language*

J. Michael Dunn

The comparison of natural languages and formal languages has become quite popular of late. The topic was on the program of the last International Congress for Logic, Methodology and Philosophy of Science in Amsterdam, and also on the program of the 1968 New York University Institute of Philosophy. I have read the published results of both meetings [1], and I must say that I am not quite sure what all the fuss is about.

On both occasions it was pointed out that a natural language typically differs from a formal one in that a natural language is ambiguous, vague, context dependent, and generally untidy. I agree that one can typically point to these differences, but frankly my reaction is, so what? We all know that untidiness has both its good points and its bad.

The very title of this symposium, "Natural Language *versus* Formal Language", suggests a certain opposition that I think is inappropriate. It sounds rather like the opposition, "Ford cars versus General Motor cars", but it seems to me that the opposition is more like that of "Ford cars versus John Deere tractors". It could be that both are useful, for different purposes.

---

* *For presentation at the joint APA-ASL symposium, New York, Dec. 27, 1969.* I presented "Natural Language versus Formal Language" as an invited speaker (together with Frederic Fitch, Bas van Fraassen, and Richard Montague) in the joint symposium by that title of the Association for Symbolic Logic and the American Philosophical Association at their joint meeting in New York, December, 1969. While it covers a number of topics related to that symposium, it was also the first public presentation I gave of the 4-valued semantics for my Ph.D. supervisor Nuel Belnap's system FDE of First-Degree Entailments.
I prepared a typed manuscript just prior to that talk. Heinrich Wansing and Hitoshi Omori, working from a computer scan of a poor photo scan that I provided, prepared a much more readable transcript for this volume. My computer scan is at `http://www.philosophy.indiana.edu/people/papers/natvsformal.pdf`. Unfortunately the copy that I had did not contain the references, though it did contain their citations in the text. I was easily and unambiguously able to reconstruct all but one of the intended references. The only one I could not find is [4] on p. 1 when it mentions "the program of the 1968 New York University Institute of Philosophy," whatever that is. The manuscript builds on material from my dissertation *The Algebra of Intensional Logics*, Univ. of Pittsburgh, 1966 (Director: Nuel D. Belnap), which also does "An Intuitive Semantics for First Degree Relevant Implications," contributed paper, meeting of the Association for Symbolic Logic, Chicago, May, 1967, (Abstract) "An Intuitive Semantics for First Degree Relevant Implications," *The Journal of Symbolic Logic*, 36, 1971, pp. 362 363. And it is a precursor to my "Intuitive Semantics for First Degree Entailments and Coupled Trees," *Philosophical Studies*, 29, pp. 149 168.
For more information about the relationships of the items mentioned above, and to work by others (particularly, R. and V. Routley and N. Belnap) see my paper "Partiality and its Dual," *Partiality and Modality*, eds. E. Thijsse, F. Lepage & H. Wansing, special issue of *Studia Logica,* Vol. 66, 2000, pp. 5-40. Another place to look is my "Relevance Logic and Entailment," in *Handbook of Philosophical Logic*, vol. 3, eds. D. Gabbay and F. Guenthner, D. Reidel, Dordrecht, Holland, 1985, pp. 117-224, or the newer version "Relevance Logic" (with G. Restall), *Handbook of Philosophical Logic*, 2nd edition, vol. 6, eds. D. Gabbay and F. Guenthner, Kluwer Academic Publishers, pp. 1-128.

J. Michael Dunn
Department of Philosophy and School of Informatics, Computing, and Engineering, Indiana University – Bloomington, Bloomington, IN 47405-7000, USA
e-mail: dunn@indiana.edu



H. Omori, H. Wansing (eds.), *New Essays on Belnap-Dunn Logic*,
Synthese Library 418, https://doi.org/10.1007/978-3-030-31136-0_2

Thus if one is writing poetry, it seems desirable to have a language that is ambiguous; but not necessarily if one is writing mathematics. And if one is talking to one's wife, vagueness might be a convenience; but maybe not if one is programming a computer.

Now surely the purpose which is of most interest to the people here today is that of "doing philosophy", as we say. With this purpose in mind, one might think that there is a definite choice, at least in principle, between natural languages and formal languages. But I doubt that there would be any universal agreement on this choice, simply because I am sure that there would be no universal agreement about what doing philosophy involves. Indeed, if doing philosophy is an activity having something to do with the gaining of insights, these insights might just as well be produced by a Zen master's stick as by the use of either natural or formal language.

But if one believes that part of doing philosophy involves attempts at constructing valid arguments, then it seems to me that one should be concerned with making explicit why they are valid, at least in problematic cases, and the best way of doing that is by means of a formal language. I am not suggesting that arguments should actually be written completely in a formal language, nor that the steps in arguments be numbered and labeled according to, say, Copi's rule. Mathematicians do not do this either, but the formal structure of their arguments is usually (though not always) clear enough that it could be reconstructed in some appropriate formal language if one so desired.

When the classical logic of *Principia Mathematica* was the only brand of formal language on the market, it was understandable for some philosopher to feel that his argument lost something in translation (often its validity) when it was formalized. But nowadays, what with modal logics, free logics, tense logics, deontic logics, epistemic logics, entailment logics, *et al*, this feeling deserves less sympathy. Of course it is always possible that a philosopher with such a feeling has some genuine insight about what follows from what, and that this insight is not captured by any extant formal logic. But insights being rather rare, and logical errors being all too common, a little shopping around among at least the more well known formal logics would not hurt. And if none of these fit, one is always free to knit one's own.

I have pointed out that logicians seem to be getting away from the bugaboo, to paraphrase Ramsey out of context, "What we can't say in *PM* we can't say, and we can't whistle it either". Formal languages are becoming increasingly natural, and I believe that this undercuts part of the supposition behind the topic of this symposium.

There is a converse development which also tends to undercut the distinction between natural and formal languages. Roughly put, natural languages are lately appearing to be more formal. What with the work that Chomsky and others have done on generative grammar, it is no longer clear that the so-called natural languages such as English are not formal languages after all. It is now frequently conjectured that the grammatical sentences of, say, English are recursive. It is true that the transformational rules most often suggested for generating them are context-dependent, whereas the rules for generating the well-formed sentences of a formal logic are typically context-independent. This, however, seems to be but a mere difference in detail (which is not to say that detail cannot be extremely interesting).

Now it might be thought that the difference between a natural language and a formal language arises not at the level of syntax, but instead at the level of semantics. The most extreme view that might be taken here is that of the formalist: a formal language,by definition, is regarded as uninterpreted. I learned very early to avoid quarreling with definitions. But I cannot help pointing out that many of the languages created by logicians are regarded as interpreted. Indeed, Gödel's famous completeness and incompleteness theorems take full advantage of such interpretations.

Now it is quite true that the semantical theories proposed by logicians for their formal languages have typically differed from the semantical theories of natural languages proposed by linguists. Thus the semantics of formal languages has typically centered around a Tarski-type recursive definition of truth. This definition can be quite complex in the case of some of the more sophisticated formal languages whose semantics is some variant of the Kripke semantics for modal logic. (Professor Montague has a general semantical theory for such languages, calling them all *pragmatic languages*.) But still the basic objective is to define under what conditions a sentence is true (relativized to a model, a world, a history, a speaker, or what have you).

On the other hand, the semantical theories of natural languages proposed by linguists have typically avoided the notion of truth altogether. Instead they have tried to provide "readings" of sentences by some ideal representation of their semantic structures. Simple semantic components, for example Katz's semantic markers which are supposed to stand for simple ideas, are strung together in such a way so as to provide an unambiguous reading. These semantic theories have concentrated on the ways that readings are generated from sentences, and of central interest here has been the disambiguation of sentences into their different readings.

At the risk of oversimplifying, the quickest way to characterize the difference between the logicians' and the linguists' semantic theories is to mobilize Quine's division of semantics into the theory of reference and the theory of meaning. Logicians have talked as if they have been concerned with the former, and linguists as if they have been concerned with the latter.

Lately, however, there have been developments on the side of the logicians which challenge this easy dualism. Thus it has been argued that an account of meaning for both formal and natural languages alike should proceed via a Tarski-type truth definition. In defense of this, Donald Davidson says [2]:

> There is no need to suppress, of course, the obvious connection between a definition of truth of the kind that Tarski has shown how to construct, and the concept of meaning. It is this: the definition works by giving necessary and sufficient conditions for the truth of every sentence, and to give truth conditions is a way of giving the meaning of a sentence. To know the semantic concept of truth for a language is to know what it is for a sentence – any sentence – to be true, and this amounts, in one good sense that we can give to the phrase, to understanding the language.

Professor Montague has also argued on several occasions for handling the semantics of natural languages in the same general manner as the semantics of formal languages, and his English as a Formal Language I is an ingenious and penetrating application of this point of view.

Now let us leave aside the obvious problems caused for this approach by the fact that there are many sentences in natural languages that appear to be neither true nor false (questions, imperatives, etc.). These cases need not necessarily vitiate the Montague-Davidson approach. First, because there may be ways of handling the troublesome cases by some extension of the concept of truth (perhaps along the lines suggested in Michael Dummett's paper "Truth", and as actually carried out in the Belnap approach to the logic of questions and the Casteñada approach to the logic of imperatives). Second, because even though the Montague-Davidson approach might not be appropriate to all of English, it might still be applicable to that large fragment of English consisting of ordinary declarative sentences.

Indeed, both Montague (in [1], p. 276) and Davidson (in [2]) have claimed that the Tarski truth definition can be straightforwardly applied so as to provide a satisfactory semantics for that fragment of English that consists of the literal translations of the formulas of the classical predicate calculus.

Should we identify the meaning of a sentence with its truth conditions? I do not ask this as a metaphysical question like the question are numbers *really* classes of equinumerous classes? If classes of equinumerous classes behave enough like numbers, then at least we have some sort of isomorphism, and that is all we need for certain purposes dear to logicians. But the present question is whether truth conditions do behave enough like meanings. Clearly, meanings determine truth conditions, but I do not find the converse so obvious.

Now there is an obvious truism that we can take to justify the claim that truth conditions determine meaning. Thus consider the sentence 'snow is white'. A truth condition for this sentence is: 'Snow is white' is true iff snow is white. I readily agree that if one knew and understood this truth condition, then one would know the meaning of the sentence 'snow is white' (for the unexciting reason that one must understand this sentence in order to understand the truth condition, which is formulated by using this very sentence itself). If truisms such as this were the only things in the wind, I would not bother to turn my head.

But typically when people claim that truth conditions determine meaning, they go on to say some profound but ultimately silly things, such as that any two logically equivalent sentences have the same meaning since they have the same truth conditions. This leads quickly to the view that any two logically false sentences (or any two logically true sentences) are synonymous. We get the most striking application of this line of thought in Wittgenstein's *Tractatus*, where he says [8, 4.461]:

> Propositions show what they say: tautologies and contradictions show that they say nothing.
>
> A tautology has no truth conditions, since it is unconditionally true: and a contradiction is true on no condition,
>
> Tautologies and contradictions lack sense.

I simply do not believe that the sentence

$$\exists x \forall y (y \in x \equiv y \notin y)$$

has the same meaning as

$$1 \neq 1$$

nor do I believe that they are both meaningless, even though I grant that they are both logically false.

This Tractarian view survives today in the best logic texts. Jeffrey, in his *Formal Logic* [5] says a little more than most authors to justify that the truth table rules of valuation give meaning to the connectives. Thus he says (p. 15):

> The rules of valuation make no mention of the meanings of sentences; they are couched entirely in terms of truth-values. Nevertheless, the rules of valuation determine the meanings of compound sentences in terms of their ingredient sentence letters, for we know the meaning of a sentence (we know what statement the sentence makes) if we know what facts would make it true and what facts would make it false. Now if we have this information about the letters that occur in a sentence, the truth conditions supply the corresponding information about the whole sentence.

A little later (pp. 30-31) in discussing contradictions, Jeffrey says:

> The sentence
>
> It is and is not raining
>
> is only apparently about the weather, just as the sentence

$$2 + 2 = 4 \text{ and } 2 + 2 \neq 4$$

is only apparently about numbers. In fact the two sentences have exactly the same truth conditions: in all possible cases, both are false.

I think we can avoid the necessity of Jeffrey's conclusion while yet agreeing, in a trivial sense, that the meaning of a sentence is determined by its truth conditions. Thus, let $p$ be the sentence 'it is raining' and let $q$ be the sentence '$2 + 2 = 4$'. By standard truth table considerations it follows that $p \wedge \neg p$ is true iff $p$ is true and $\neg p$ is true, that is, iff $p$ is true and $p$ is false. Similarly $q \wedge \neg q$ iff $q$ is true and $q$ is false. The question bluntly then is whether the condition that $p$ is true and $p$ is false is the same condition as that $q$ is true and $q$ is false. I think it is not.

Notice that it is no argument against me to reply that the first is a contradiction meaning $p$ is true and $p$ is not true, while the second is also a contradiction meaning $q$ is true and $q$ is not true, and that of course any two contradictions have the same meaning. This only pushes the question with which we began up into the metalanguage.

Intuitively, $p \wedge \neg p$ and $q \wedge \neg q$ describe different situations, granted that neither situation is realizable. What we need is a semantics that is sensible to this intuition.

I may as well let any who do not know me in on a little secret at this point. I was a student of Belnap and Anderson's at the university of Pittsburgh, and I am one of those crazy people who think that there is something in their system E of entailment (and in the other similar relevant logics that have been developed). I believe that there is a sense of 'entails' (or 'implies') in which it simply is not true that a contradiction entails or implies any old sentence whatsoever. It thus becomes extremely critical that not just any two contradictions are synonymous. For if $p \wedge \neg p$ were synonymous with $q \wedge \neg q$, then since it is true that $q \wedge \neg q$ entails $q$, then by substitution of synonyms *salva veritatae*, it would be true that $p \wedge \neg p$ entails $q$. Having made a clean breast of my motivation, I hope that they will not be held against me as I continue.

I mention at this point that both Professor van Fraassen and myself have developed semantical ways of ruling out $p \wedge \neg p$'s entailing $q$. I, in my dissertation [3], in terms of $q$'s possibly being about some topic that $p \wedge \neg p$ is not about, van Fraassen [7] in terms of some fact forcing $p \wedge \neg p$ which does not force $q$. Both of these semantics lead to completeness proofs for a very narrow fragment of the system E, namely those sentences of the form A entails B, where A and B are purely truth functional (the so-called *first degree* entailments), and it is very difficult to see how these semantics might be generalized so as to take care of all of E (with entailments entailing entailments, etc.). Furthermore, both these semantics suffer from the defect that they are formulated in terms of concepts that are out of fashion in logic (*topics* that sentences are about, *facts* that force sentences to be true).

Let us then recall which concepts are in fashion, and I shall try my best to talk that language in trying to communicate the notion that not just any two contradictions are synonymous. The standard realization of a proposition as found in Montague, Kaplan, Scott and others is a mapping from possible worlds (or reference points, situations, call them what you will) to truth values. That corresponds to the principle that different meaning can be distinguished by different situations with different truth values, i.e., by different truth conditions. But it has the untoward consequence that (relative to a given set of situations) there is only one contradictory proposition, simply because there is only one constant false mapping.

However, we need modify this picture only slightly to provide a kind of extensional apparatus that allows us to distinguish contradictory propositions from one another. Starting from the intuition expressed above that a contradiction can be true in some situations (of course, unrealizable) in which some other contradiction is not true, we can identify a proposition with a relation from a set of situation into the set $\{T, F\}$, where every situation is related to at least one of $T$ and $F$. A contradictory proposition is then such a relation where $F$ is in the image of every situation. There can then be many different contradictory propositions. These can be distinguished by a situation such that one of the propositions has $T$ in its image while the other does not.

What this means as far as the modeling of truth functional logic is concerned is that a valuation is a relation from sentences into the set $\{T, F\}$, rather than a mapping, and of course it is required that every sentence be related to at least one of $T$ and $F$ (we shall eventually speculate upon what happens if we drop this last requirement). This relation is determined inductively in just the classical truth table way. Thus

i) $\neg A$ is $T$ iff $A$ is $F$,
$\neg A$ is $F$ iff $A$ is $T$;
ii) $A \wedge B$ is $T$ iff $A$ is $T$ and $B$ is $T$,
$A \wedge B$ is $F$ iff $A$ is $F$ or $B$ is $F$;
iii) $A \vee B$ is $T$ iff $A$ is $T$ or $B$ is $T$,
$A \vee B$ is $F$ iff $A$ is $F$ and $B$ is $F$.

Note that in each of i) – iii), we need two clauses, one giving truth conditions and the other giving falsity conditions. We cannot rely upon the standard intuition that a sentence which has been given the value $T$ is not $F$.

We can already give a semantical explication of one of the principal features of entailment, namely, that $p \wedge \neg p$ need not entail $q$. For there is a valuation in which $p \wedge \neg p$ receives the value $T$ and yet $q$ does not. This is a valuation in which $p$ receives both the values $T$ and $F$, while $q$ receives the single value $F$.

We can also give a semantical explication of perhaps the most controversial feature of entailment, namely, that $\neg p \wedge (p \vee q)$ need not entail $q$ (the failure of the so-called rule of disjunctive syllogism). Let me give this explication in the context of examining the supposed proof of Lewis's that a contradiction entails everything.

The proof starts out by supposing that $p \wedge \neg p$ is true. We then detach $p$ by the rule of simplification, and from $p$ we obtain $p \vee q$ by the rule of addition. Next we obtain $\neg p$ from our supposition of $p \wedge \neg p$ by another use of the rule of simplification. So far, O.K. But finally we claim that $q$ follows from $\neg p$ and $p \vee q$ by disjunctive syllogism. In producing this proof for a class, it used to be my habit to motivate this last step by telling the following story. "So on our assumption that $p \wedge \neg p$ is true, we have obtained that one of $p$ or $q$ is true. But we have also obtained $\neg p$, which says that $p$ is not the true one." When I was once telling this story, some wise guy yelled out, "But $p$ was the true one – look again at your assumption."

That wise guy was right. If we assume that $p \wedge \neg p$ is true, we are thereby assuming that $p$ is both true and false, and hence it should not be surprising that $p \wedge (\neg p \vee q)$ comes out true under that assumption, while $q$ might still be false.

Do not get me wrong; I am not claiming that there are sentences which are in fact both true and false. I am merely pointing out that there are plenty of situations where we suppose, assert, believe, etc., contradictory sentences to be true, and we therefore need a semantics which expresses the truth conditions of contradictions in terms of the truth values that the ingredient sentences would have to take for the contradictions to be true.

I must unfortunately remark that these particular insights have not as yet given the degree of illumination regarding entailment that might be expected. In particular, they have not led to completeness proofs for the system E of Entailment. But I have obtained reasonably intuitive completeness results based upon this framework for an extension of E called R-mingle. These I now announce for the first time, and they should not be confused with earlier purely algebraic completeness results obtained by both my colleague Robert K. Meyer [6] and myself [4]. The intuitive results are rather like those of Kripke for intuitionist logic, and even more like those of Thomason for the system of Professor Fitch's *Symbolic Logic*, though with assignments allowed to give both the values $T$ and $F$. Furthermore, the extension to R-mingle with quantifiers seems not a bother. The basic idea is that instead of doing the classical thing of interpreting n $n$-ary predicate as a propositional *function* (a mapping from the $n$-tuples of objects in the domain into $\{T, F\}$), we rather interpret the predicate as a propositional *relation* (a relation from the $n$-tuples of objects of the domain into $\{T, F\}$, with the requirement that every $n$-tuple be related to at least one of $T$ and $F$).

The reason why we do not obtain a semantics for the system E in this framework is that it is difficult to rule out $p \wedge \neg p$'s entailing $q \vee \neg q$, since $p \wedge \neg p$ is always false and $q \vee \neg q$ is always true. Thus whenever the antecedent is true, the consequent is true (it being always true); and whenever the consequent is false, the antecedent is false (it being always false). Thus we are stuck no matter how we try to falsify the entailment, and yet the entailment is not a theorem of the system E. One way out that suggests itself is to let $q \vee \neg q$ have no truth value, and we could naturally arrange this by allowing an assignment in which $q$ was related to neither $T$ nor $F$. Then $p \wedge \neg p$ could be given the value $T$ (by giving $p$ both the values $T$ and $F$), while $q \vee \neg q$ is not given the value $T$ (by giving $q$ no value whatsoever). This works for first degree entailments, but there are vast problems both of an intuitive and a technical sort in generalizing this approach to entailments nested in entailments.

In closing, let me urge that even it the particular approach I have suggested for distinguishing contradictions semantically is not your liking, still something should be done in this area. It may be good logic to say that any two contradictory sentences are logically equivalent, but I would think that it would be bad linguistics to say that any two contradictory sentences have the same meaning.

## References

1. *Abstracts of Papers: Third International Congress for Logic, Methodology, and Philosophy of Science*, Amsterdam, 1967.
2. Donald Davidson. Truth and Meaning. *Synthese*, 17(3): 304-323, 1967.
3. J. Michael Dunn. *The Algebra of Intensional Logics*. PhD thesis, University of Pittsburgh, Ann Arbor, 1966.
4. J. Michael Dunn. Algebraic Completeness results for R mingle and its extensions. *Journal of Symbolic Logic*, 1970.
5. Richard C. Jeffrey. *Formal Logic: Its Scope and Limits*. Hackett Publishing Co. Inc., Indianapolis, 1967.
6. Robert K. Meyer. R-mingle and relevant disjunction. *Journal of Symbolic Logic*, 36(2): 366, 1971.
7. Bas van Fraassen. Facts and tautological entailments. *The Journal of Philosophy*, 66(15): 477-487, 1969.
8. Ludwig Wittgenstein. *Tractatus Logico-Philosophicus*. Routledge & Kegan Paul Ltd., 1961. Translated by David Pears and Brian McGuinness, with an introduction by Bertrand Russell.

# Intuitive Semantics for First-Degree Entailment and 'Coupled Trees'

J. Michael Dunn

## 1 Introduction[1]

Classically, an argument $A$ therefore $B$ is 'valid' (or $A$ is said to 'entail' $B$) if and only if (iff) each situation (model) is such that either $A$ is false or $B$ is true. This fits well with so-called 'tableau' methods for showing that $A$ entails $B$ by working out the mutual inconsistency of $A$ and $\sim B$. But both the classical notion of validity and the corresponding tableau methods allow that $A$ may entail $B$ because of some feature of $A$ alone, irrespective of $B$, and vice versa. Thus if $A$ is a contradiction, then each situation is such that $A$ is false, and so *a fortiori* is such that $A$ is false or $B$ is true. And if $A$ is a contradiction, then a tableau construction will show that $A$ is inconsistent, and so *a fortiori* that $A$ and $\sim B$ are inconsistent. Of course, the same points can be made dually when $B$ is a logical truth.

A competing theory of 'entailment' developed by Anderson and Belnap requires that for $A$ to entail $B$ there must be some relation of real relevance between $A$ and $B$, e.g., they share some sentence letter. In this paper I shall develop a notion of 'relevant validity' and a corresponding tableau method that tie in with the Anderson-Belnap theory.

Jeffrey (1967) introduced 'coupled trees' as a modified tableau method for testing an argument for validity. In Section 2 I shall describe the formalism of the coupled tree method and explain how by pruning it of complications (needed by Jeffrey to get precisely the classically valid arguments) we get a well-motivated syntactical characterization of when an entailment holds relevantly between truth-functional sentences. In Section 3 an 'intuitive' semantical characterization is presented and motivated using inconsistent and incomplete 'situations'. In Section 4 these two characterizations are connected by completeness and soundness results. In Section 5 the semantical characterization is connected similarly with a well-known syntactical characterization ('tautological entailment') due to Anderson and Belnap (1962), and connected by them to the provable 'first-degree entailments' (formulas of the form $A \rightarrow B$, where $A$ and $B$ contain no occurrences of $\rightarrow$) of their system **E**. In Section 6 another semantical characterization using 'topics' and having an information-theoretic flavor is related to the semantics of Section 3. So we have the happy circumstance that all these characterizations coincide. In Section 7, I ruminate.

I must mention that there are in the literature by now at least two other semantical modelings of the first-degree entailments, one due to van Fraassen (1969) and the second

J. Michael Dunn
Department of Philosophy and School of Informatics, Computing, and Engineering, Indiana University – Bloomington, Bloomington, IN 47405-7000, USA
e-mail: dunn@indiana.edu

[1] This work was supported in part by NSF Grant GS-33708. I suppose the 'coupled trees' provide the excuse for the present publication, for the rest has seen 'semi-publication' (abstracts, mimeo, talks) some years previously, as specific references in the sequel will indicate. Basically it all stems from my dissertation (Dunn, 1966) (cf. Section 6 of the present paper) and so I must once more express indebtedness to all who helped there. I recall as being particularly 'relevant' here my teachers N. D. Belnap, Jr. (the director) and the late A. R. Anderson, and my then fellow students R. K. Meyer, B. van Fraassen, and P. Woodruff.



H. Omori, H. Wansing (eds.), *New Essays on Belnap-Dunn Logic*,
Synthese Library 418, https://doi.org/10.1007/978-3-030-31136-0_3

due to Routley and Routley (1972). Both of these have certain similarities to my own and to each other.[2] It would be too lengthy an excursion to compare them all, so I will content myself with just a quick flight over.

The major advantage I see of my semantics over van Fraassen's is that it is less complicated and more familiar both philosophically and mathematically. Van Fraassen's philosophical motivation rests on certain somewhat quaint ontological intuitions concerning complex facts. Mathematically, it is just plain hard to keep straight the various steps and typelevels of his clever set-theoretical reification of these intuitions. There need ultimately be nothing wrong in any of this, but since I have a fairly narrow partisan interest in boosting the Anderson-Belnap relevance program I would like to here present a semantics that is more accessible to the evergrowing number of people who have cut their semantical eye-teeth on the 'possible world' semantics for modal logic.

Now the Routleys' approach does relate directly to 'possible world' semantics. Their 'set ups' are advertised as liberalizations of the notion of 'possible world' so as to include impossible and incomplete 'worlds'. My semantics has the same end – it is the means over which we differ. The Routleys perform some magic with a 'star operation' in giving the truth condition for negation. By a feat of prestidigitation one 'set up' $H$ is switched with another set up $H^*$. Thus $\sim A$ is true in $H$ iff $A$ is not true in $H^*$ (instead of the usual plain $H$). But just what is this 'star operation' and why does it stick its nose into the truth condition for negation?[3] This seems to me to remain an ultimate mystery in the Routleys' semantics, and I count it as a philosophical virtue of my semantics that it does without the 'star operation'.

## 2 Relevantly Coupled Trees

Jeffrey's logic text (1967) provides an excellent introductory treatment of the method of 'analytic tableaux' of Smullyan (1968). Jeffrey (p. 92) compares the method of 'truth trees' (his suggestive name for analytic tableaux) with indirect proof, the essential point being that in order to test an argument, say, $A$ therefore $B$ (in symbols $A \vdash B$), for validity one uses the method of truth trees to test for the mutual inconsistency of $A$ and $\sim B$. The idea of a truth tree is that it diagrams in a branching treelike fashion (so as to keep track of various alternatives) all of the truth conditions of a set of sentences. Each path represents a way in which the given sentences might become true, and when testing for inconsistency we search to see whether all of these paths are 'closed' by virtue of containing both a sentence and its denial.

[2] The Routleys' and my modelings are basically isomorphic, in a mathematically precise sense. The only reason for the qualification 'basically' is that the Routleys take their 'set ups' to be certain sets of sentences, and so (assuming a denumerable language) crude considerations of cardinality prevent my larger models from having Routley correlates. However, taking 'set ups' more abstractly (as is in fact done in Routley and Meyer, 1973) so as to provide collections of 'set ups' of all cardinalities, each Routley model can be regarded as an assignment to each truth-functional sentence of an element in a 'quasi-field' of sets (this notion from Bialynicki-Birula and Rasiowa, 1957). In Dunn (1966) (cf. also Dunn, 1967,1971) my models (directly in their Section 6 version here) are similarly connected to de Morgan lattices of 'proposition surrogates' or the somewhat dual '2-products' of a field of sets, and these related by isomorphisms to quasi-fields of sets. It would be nice to have some similar connections to van Fraassen's modeling.

[3] Almost any inference could be 'invalidated' by some analogous device. Thus, e.g., one could invalidate the inference from $A \wedge B$ to $A$ by changing the truth condition for conjunction so that $A \wedge B$ is true in $H$ iff both $A$ and $B$ are true in $H^*$.

Jeffrey (p. 93) provides a modification of the basic method of truth trees, which modification he calls the method of 'coupled trees' and compares to direct proof. The basic idea is that in order to test the validity of an argument, $A \vdash B$, one constructs two truth trees, one for $A$ (coming down) and one for $B$ below (going up). Since each path in the tree for $A$ represents an alternative set of truth conditions for $A$, and similarly for $B$, it is natural to require that every path in the upper tree 'cover' some path in the lower tree in the sense that every sentence letter or denial of a sentence letter (the term *atom* will henceforth embrace both these) that appears in the covered path appears also in the covering path. Thus every way in which $A$ is true is also a way in which $B$ is true.

However, there are two technical complications that Jeffrey needs in order to get precisely the classically valid arguments. I shall explain these complications after I describe with more precision the formalism of the coupled tree method.

Jeffrey's formalism includes sentence letters (we shall suppose they are $p, q, r$, etc.), and connectives for negation, conjunction and disjunction (we suppose these are $\sim$, $\wedge$ and $\vee$) as well as for the truth-functional conditional and biconditional. We shall ignore these last two since they are not primitive in the standard formulations of the system **E** (though they can of course be introduced as abbreviatory devices *via* their ordinary contextual definitions). There are then schematically the following five rules:[4]

$$(\sim\sim)\ \frac{\sim\sim A}{A}$$

$$(\wedge)\ \frac{A \wedge B}{\begin{array}{c} A \\ B \end{array}} \qquad\qquad (\sim\wedge)\ \frac{\sim(A \wedge B)}{\sim A \quad \sim B}$$

$$(\vee)\ \frac{A \vee B}{A \quad B} \qquad\qquad (\sim\vee)\ \frac{\sim(A \vee B)}{\begin{array}{c} \sim A \\ \sim B \end{array}}$$

The rules are reasonably self-explanatory. Note that $(\sim\wedge)$ and $(\vee)$ have branching conclusions representing alternatives and are the source of the 'tree' structure.

The basic idea of Jeffrey's coupled tree method is illustrated by the argument $p \wedge q \vdash q \vee r$, for which we can construct the following coupled trees (the arrow indicates covering):

$$\begin{array}{c} p \wedge q \\ p \\ q \\ \swarrow \\ q \qquad r \\ q \vee r \end{array}$$

[4] Jeffrey actually avoids formal recognition of the first rule by a practice of erasing pairs of juxtaposed negation signs, but the rule is formally in Smullyan (1968).

The corresponding entailment $(p \wedge q) \rightarrow (q \vee r)$ is not only a theorem of **E** but it is basic to the motivation of **E** presented in Anderson and Belnap (1962), being paradigmatic of what they there call a *primitive entailment*.

The first of the two technical complications needed by Jeffrey is nicely illustrated by the argument $p \wedge {\sim}p \vdash q$. The coupled tree, if any, for this argument would be the following:

$$\begin{array}{c} p \wedge {\sim}p \\ p \\ {\sim}p \\ q \end{array}$$

But there is a conspicuous absence of covering. This fits nicely with the Anderson-Belnap intuition that there is no relevance between premiss and conclusion. Jeffrey though is concerned to capture this classically valid inference. Thus he complicates the basic idea that every path above must cover some path below by excepting the closed paths above.[5] Thus trivially the above diagram (with a cross written under ${\sim}p$ to indicate that the path is closed) counts as a coupled tree.

The second complication is nicely illustrated by the dual of the last argument, namely, $p \vdash q \vee {\sim}q$. It would seem that the following would represent a failed attempt at constructing an appropriate coupled tree:

$$\begin{array}{ccc} & p & \\ q & & {\sim}q \\ & q \vee {\sim}q & \end{array}$$

Again the lack of covering can be taken to be in accord with Anderson-Belnap intuitions about irrelevance.

Jeffrey's device to wash this one through is to permit in the construction of the tree coming down from the premiss a simultaneous branching with a sentence and its negation. Thus the following counts as a coupled tree:

$$\begin{array}{ccc} & p & \\ q & & {\sim}q \\ \downarrow & & \downarrow \\ q & & {\sim}q \\ & q \vee {\sim}q & \end{array}$$

This amounts to tacitly adding as a rule (in constructing upper trees only) the following, which we shall call 'punt':

$$\frac{A}{B \qquad {\sim}B}$$

Let us close this section with the definition it has been motivating: An argument $A \vdash B$ *passes the relevantly coupled tree test* iff in constructing truth trees for $A$ and $B$, every path in the tree for $A$ (including the closed paths) covers some path in the tree for $B$ (not allowing use of 'punt').

[5] Of course this appears as no complication at all in the (classical) context Jeffrey has working for him, where 'closed' paths are discountenanced in just the way their name suggests.

## 3 Intuitive Semantics[6]

Wittgenstein says in the *Tractatus* (Wittgenstein, 1961, translation):

> 4.461 Propositions show what they say: tautologies and contradictions show that they say nothing.
>
> A tautology has no truth-conditions, since it is unconditionally true: and a contradiction is true on no condition.
>
> Tautologies and contradictions lack sense.

This Tractarian view survives today in some of the best logic texts. Jeffrey (1967) says a little more than most authors to justify that the truth table rules of valuation give meaning to the connectives. Thus he says (p. 15):

> The rules of valuation make no mention of the meanings of sentences; they are couched entirely in terms of truth-values. Nevertheless, the rules of valuation determine the meanings of compound sentences in terms of their ingredient sentence letters, for we know the meaning of a sentence (we know what statement the sentence makes) if we know what facts would make it true and what facts would make it false. Now if we have this information about the letters that occur in a sentence, the truth conditions supply the corresponding information about the whole sentence.

A little later (pp. 30-31) in discussing contradictions, Jeffrey says:

> The sentence
>
> *It is and is not raining*
>
> is only apparently about the weather, just as the sentence
>
> $2 + 2 = 4$ and $2 + 2 \neq 4$
>
> is only apparently about numbers. In fact the two sentences have exactly the same truth conditions: in all possible cases, both are false.

I think that we can avoid the necessity of Jeffrey's conclusion while yet agreeing, in a trivial sense, that the meaning of a sentence is determined by its truth conditions. Thus let $p$ be the sentence 'It is raining' and let $q$ be the sentence '$2 + 2 = 4$'. By standard truth table considerations it follows that $p \wedge {\sim}p$ is true iff $p$ is true and ${\sim}p$ is true, that is, iff $p$ is true and $p$ is false. Similarly, $q \wedge {\sim}q$ is true iff $q$ is true and $q$ is false. The question bluntly then is whether the condition that $p$ is true and $p$ is false is the same condition as that $q$ is true and $q$ is false. I think it is not.

Notice that it is no argument against me to reply that the first is a contradiction meaning $p$ is true and $p$ is not true, while the second is also a contradiction meaning $q$ is true and $q$ is not true, and that of course any two contradictions have the same meaning. This only pushes the question with which we began up into the metalanguage.

Intuitively, $p \wedge {\sim}p$ and $q \wedge {\sim}q$ describe different situations, granted that neither situation is realizable. What we need is a semantics that is sensitive to this intuition.

The by now orthodox realization of a proposition is a function from possible worlds (or indices, reference points, situations, whatever) to truth values (cf. for explicitness Montague, 1972, who credits the idea to Kripke – cf. also the articles by Lewis and Stalnaker in the same volume as the Montague paper). This corresponds to the principle that different meanings can be distinguished by different situations with different truth values, i.e., by different truth conditions. It too has the untoward consequence that (relative to a given

[6] The bulk of this section is taken verbatim from the middle of Dunn (1969).

set of situations) there is only one contradictory proposition, simply because there is only one constant false function.

However, we need modify this picture only slightly to provide a kind of extensional apparatus that allows us to distinguish contradictory propositions from one another. Starting from the intuition expressed above that a contradiction can be true in some situations (of course, unrealizable) in which some other contradiction is not true, we can identify a proposition with a relation from a set of situations into the set {**T, F**}. A contradictory proposition is then such a relation where **F** is in the image of every situation. There can then be many different contradictory propositions. These can be distinguished by a situation such that one proposition has **T** in its image while the other does not.

We could go on to develop the notion of an 'interpretation' as an assignment of propositions to (truth-functional) sentences, setting down straightforward rules by which the propositions assigned to complex sentences are determined inductively from the propositions assigned their ingredient sentence letters. However, such an assignment of a proposition to a sentence is obviously interchangeable with a rule telling us whether the sentence is true or false for each situation. We allow of course the options that the rule tells us sometimes both as well as neither, so as to be faithful to our construal of a proposition as relational but not necessarily functional in character. So such a rule or *valuation* could be identified with a three-placed relation $\phi$ relating sentences, situations, and truth-values. A *situation model* then is an ordered pair $\langle \mathbf{K}, \phi \rangle$, where **K** is a non-empty set (its members being called 'situations') and $\phi$ is such a relation relating sentences, situations, and truth values in a natural recursive manner soon to be tied down.

As a matter of simplification in working up a semantics to assess the validity of *first-degree* entailments we can forget situations and deal just with two-placed relations simply relating sentences to truth values. This is for the familiar reason that in assessing the validity of an entailment every variety of situation whatsoever must be considered in which the antecedent is true to see if it is also a situation in which the consequent is true, and this is in effect done by looking at just the specifications of truth and falsity given the antecedent by situations and discarding the situations themselves.

We thus define a *relevance valuation* as a certain kind of inductively determined relation whose domain is the set of truth-functional sentences and whose range is a subset of {**T, F**}. Before going on to state the exact inductive clauses required to be satisfied, we introduce some conventions useful for their statement and also in the sequel. For a given such valuation $V$, we let $V^*(A)$ be the image of $A$ under $V$ (i.e., the set of truth values to which $A$ is related by $V$). We can then indicate that $V$ relates $A$ to **T[F]** by $\mathbf{T[F]} \in V^*(A)$. In context we simply say $A$ is **T[F]**, or $A$ is true [false]. We can adopt similar conventions regarding the more complicated situation-relative kind of valuation $\phi$ of a situation model described above, letting for a given situation $H$, $\phi^*(A, H)$ be the set of truth values **X** ('**X**' ranges over **T** and **F**) such that $\phi(A, H, \mathbf{X})$. Also in context we allow ourselves to say things like '$A$ is **T** in $H$', or when $H$ is fixed even '$A$ is **T**'.

With these understandings behind us, we require of a relevance valuation $V$, and also of a valuation $\phi$ on a situations model (relative to each situation $H$) the following:

(i) $\sim A$ is **T** iff $A$ is **F**,
$\sim A$ is **F** iff $A$ is **T**;
(ii) $A \wedge B$ is **T** iff $A$ is **T** and $B$ is **T**,
$A \wedge B$ is **F** iff $A$ is **F** or $B$ is **F**;
(iii) $A \vee B$ is **T** iff $A$ is **T** or $B$ is **T**,
$A \vee B$ iff $A$ is **F** and $B$ is **F**.

Note that in each of (i)-(iii), we need two clauses, one giving truth conditions and the other giving falsity conditions. We cannot rely upon the standard intuition that a sentence which has been given the value **T** is not **F**.[7]

We can already give a semantical explication of one of the principal features of entailment, namely, that $p \wedge {\sim}p$ need not entail $q$. For there is a valuation in which $p \wedge {\sim}p$ receives the value **T** and yet $q$ does not. This is a valuation in which $p$ receives both the values **T** and **F**, while $q$ receives the single value **F**.

We can also give a semantical explication of perhaps the most controversial feature of entailment, namely, that ${\sim}p \wedge (p \vee q)$ need not entail $q$ (the failure of the so-called rule of disjunctive syllogism). Let me give this explication in the context of examining the supposed proof of Lewis and Langford (1932) that a contradiction entails everything.

The proof starts out by supposing that $p \wedge {\sim}p$ is true. We then detach $p$ by the rule of simplification, and from $p$ we obtain $p \vee q$ by the rule of addition. Next we obtain ${\sim}p$ from our supposition of $p \wedge {\sim}p$ by another use of the rule of simplification. So far, O.K. But finally we claim that $q$ follows from ${\sim}p$ and $p \vee q$ by disjunctive syllogism. In producing this proof for a class, it used to be my habit to motivate this last step by telling the following story. "So on our assumption that $p \wedge {\sim}p$ is true, we have obtained that one of $p$ or $q$ is true. But we have also obtained ${\sim}p$, which says that $p$ is not the true one. So $q$ must be the true one." When I was once telling this story, some wise guy yelled out, "But $p$ was the true one – look again at your assumption."

That wise guy was right. If we assume that $p \wedge {\sim}p$ is true, we are thereby assuming that $p$ is both true and false, and hence it should not be surprising that $p \wedge ({\sim}p \vee q)$ comes out true under that assumption, while $q$ might still be false.

Do not get me wrong – I am not claiming that there are sentences which are in fact both true and false. I am merely pointing out that there are plenty of situations where we suppose, assert, believe, etc., contradictory sentences to be true, and we therefore need a semantics which expresses the truth conditions of contradictions in terms of the truth values that the ingredient sentences would have to take for the contradictions to be true.

The careful reader will by this time have noticed that in making a valuation a relation from sentences to truth values rather than a function we have thereby allowed a sentence be neither true nor false as well as both true and false. We have seen that the latter move is connected with invalidating inferences like $p \wedge {\sim}p \vdash q$ or $p \wedge {\sim}p \vdash q \wedge {\sim}q$. The former move is dually connected with invalidating inferences like $p \vdash q \vee {\sim}q$ or $p \vee {\sim}p \vdash q \vee {\sim}q$. And the two moves in concert succeed in invalidating inferences like $p \wedge {\sim}p \vdash q \vee {\sim}q$.[8]

How do we go about motivating allowing sentences to be assigned no truth value? The answer is, of course, 'dually' to our motivation for both truth values. Rather than thinking about the (*per impossible*) truth conditions for contradictions, we think about the (*per*

[7] There is no difficulty in extending all this to quantifiers. The basic idea is that instead of doing what is sometimes done classically, to wit, interpreting an $n$-ary predicate as a *function* from the ordered $n$-tuples of individuals in the domain into $\{\mathbf{T}, \mathbf{F}\}$, one rather allows a *relation* from the $n$-tuples to the truth values.

[8] Incidentally, one can capture semantically the first-degree implications of the system **RM** by working with relevance valuations that always relate a sentence to at least one of the two truth values. But one must then bring into the definition of 'validity' the additional requirement that whenever the conclusion is false so is the premiss. This, together with the usual requirement of truth preservation, washes out only the first four of the above five inferences, leaving the fifth, which is kind of characteristic of **RM**. Also these ideas can be extended to a Kripke-style semantics for all of **RM** using a *binary* accessibility relation, as was indicated in Dunn (1969) and set down in Dunn (1971b). In all frankness I have not had much luck in extending my framework to the whole of the systems **E** or **R**, the way the Routley and Routley (1972) ideas were extended in Routley and Meyer (1973) (even using their new *ternary* accessibility relation). The Routleys' star operation interacts with the accessibility relation in a seemingly essential way.

*impossible*) 'non-truth conditions' for tautologies. The classical truth table considerations of (i)-(iii) above tell us that the only way $p \vee \sim p$ could possibly (better, *impossibly?*) be non-true is for $p$ to have neither truth value.

Here again I am not arguing that there are sentences which are in fact neither true nor false.[9] It is just that in working out the truth conditions for a compound sentence it is very easy to overlook the condition that an ingredient sentence have some truth value. (To consider this as a 'background assumption' seems to amount to the same.) Once such conditions are made explicit by relevance valuations it no longer looks as if all tautologies have the same truth conditions, to wit, any.

By the way, I am painfully aware of the strangeness of some of my remarks motivating the semantics. Phrases like 'impossible conditions' strain, if not break. It is at least as difficult to speculate about the impossible without seeming to talk of it as possible as it is to speculate about the merely possible without seeming to talk of it as actual. Frege had a (very roughly) similar problem in telling us about his concepts and seeming to talk of them as objects. I am tempted to join with him and say (in the Geach and Black, 1960, translation): "By a kind of necessity of language, my expressions, taken literally, sometimes miss my thought. . . . I fully realize that in such cases I was relying upon a reader who would be ready to meet me halfway – who does not begrudge a pinch of salt."

Let us end this section by officially defining some notions implicit in the preceding motivations. An argument $A \vdash B$ is *relevantly valid* iff for every relevance valuation $V$, either $\mathbf{T} \notin V^*(A)$ or $\mathbf{T} \in V^*(B)$. This disjunction may be more naturally expressed (speaking in the material conditional tone of voice that seems characteristic of metatheoretical investigations of the relevance logics) as: if $\mathbf{T} \in V^*(A)$, then $\mathbf{T} \in V^*(B)$.

I next introduce some cognate notions using situation models. We shall say that *A entails B in a situation model* $\langle \mathbf{K}, \phi \rangle$ iff for all $H \in \mathbf{K}$, if $\phi(A, H, \mathbf{T})$ then $\phi(B, H, \mathbf{T})$. We shall say that *A universally entails B in a set of situations* $\mathbf{K}$ iff for all situation models $\langle \mathbf{K}, \phi \rangle$, *A entails B* in $\langle \mathbf{K}, \phi \rangle$. Finally, we shall say that *A logically entails B* iff $A$ universally entails $B$ in all sets of situations.

From my remark motivating the introduction of the relevance valuations $V$ as applying directly to sentences without the intervention of situations, it should be obvious to the reader that $A$ logically entails $B$ iff $A \vdash B$ is relevantly valid. This makes the notions involving situations somewhat otiose as far as providing a semantical modeling of the first-degree entailments, and indeed we shall make no further mention or use of such notions until Section 6.

## 4 Coupled Trees and the Semantics

In this section we establish the following and its converse:

*Soundness Theorem for Relevantly Coupled Trees.* If an argument $A \vdash B$ passes the relevantly coupled tree test, then it is relevantly valid.

*Proof.* The rules are *downward correct* in the sense that every relevance valuation that makes the premiss true also makes the conclusions in at least one branch true. Also they are upward correct since every relevance valuation that makes the conclusions in at least one branch true also makes the premiss true. Suppose then that $A \vdash B$ passes the relevantly

[9] I am not saying that there are not either. There may be 'truth value gaps' due for example to failure of denotations of singular terms, but such things seem to me to be 'irrelevant' to present concerns.

coupled tree test and that $V$ is a relevance valuation in which $A$ is true. It is easy to see on the basis of the downward correctness of the rules that in the tree constructed downward from $A$ there must be some path $\mathbf{P}$ so that every sentence in $\mathbf{P}$ is true in $V$. But the tree constructed upward from $B$ must have some path $\mathbf{Q}$ so that $\mathbf{P}$ covers $\mathbf{Q}$. All of the atoms that appear in $\mathbf{Q}$ must then be true in $V$, and it is easy to see on the basis of the upward correctness of the rules that $B$ must be true in $V$ as well.

We next attack the converse:

*Completeness Theorem for Relevantly Coupled Trees.* If an argument $A \vdash B$ is relevantly valid, then it passes the relevantly coupled tree test.

*Proof.* Suppose the argument fails to pass the relevantly coupled tree test. Then there exists a tree constructed downward from $A$ and one constructed upward from $B$ so that the tree for $A$ has a path $\mathbf{P}$ that fails to cover each path in the one for $B$. Let the paths in the tree for $B$ be $\mathbf{Q}_1,\ldots,\mathbf{Q}_n$, and let $\alpha_1,\ldots,a_n$ be, respectively, atoms that keep $\mathbf{P}$ from covering $\mathbf{Q}_1,\ldots,\mathbf{Q}_n$. Define the relevance valuation $V$ so that for each sentence letter $p$,

(i) $\mathbf{T} \in V^*(p)$ iff $p$ is in $\mathbf{P}$,
(ii) $\mathbf{F} \in V^*(p)$ iff $\sim p$ is in $\mathbf{P}$.

It is easy to see because of upward correctness that every sentence in $\mathbf{P}$ is true in $V$, and hence in particular that $A$ is true in $V$. Also, it is easy to see because of downward correctness that if $B$ is true in $V$ then every sentence in at least one of the paths $\mathbf{Q}_i$ must be true. But it is easy to see that $\alpha_i$, a member of $\mathbf{Q}_i$, is not true in $V$. So $A$ is true in $V$ while $B$ is not, and so the argument fails to be relevantly valid.

## 5 Tautological Entailments and the Semantics

I submit that the relevantly coupled tree test provides a plausible proof-theoretic characterization of when two truth-functional sentences $A$ and $B$ relevantly entail one another. In this section I shall briefly rehearse an earlier such characterization, the so-called 'tautological entailments', due to Anderson and Belnap (1962), and then provide soundness and completeness results for the tautological entailments with respect to the intuitive semantics. Since Anderson and Belnap (1962) have shown that a first-degree entailment $A \to B$ is a theorem of their system $\mathbf{E}$ iff it is a tautological entailment, the results of this section and the last provide the hookup *via* the semantics between coupled trees and entailment promised in Section 1.

According to Anderson and Belnap, a *primitive conjunction (disjunction)* is a conjunction (disjunction) of atoms, and a *primitive entailment* is of the form $A \to B$ where $A$ is a primitive conjunction and $B$ is a primitive disjunction. A primitive entailment $A \to B$ is *explicitly tautological* iff some (conjoined) atom of $A$ is the same as some (disjoined) atom of $B$. A first degree entailment $A \to B$ is a *tautological entailment* iff when $A$ is put into disjunctive normal form $A_1 \vee \cdots \vee A_m$ and $B$ is put into conjunctive normal form $B_1 \wedge \cdots \wedge B_n$ it turns out that each $A_i \to B_j$ is an explicitly tautological entailment.[10]

[10] Once we have shown tautological entailment and relevantly coupled trees go well together, we will have in effect another normal form test. This is because, as is rather obvious (cf. Smullyan, 1968, exercise 2, p. 30), a finished truth-tree can be thought of as a disjunctive normal form (each path the conjunction of its atoms, with the paths then the disjuncts). So one can put *both* $A$ and $B$ into disjunctive normal form, $A_1 \vee \cdots \vee A_m$ and $B_1 \vee \cdots \vee B_n$ respectively, and require that each $A_i$ 'cover' some $B_j$.

The replacement rules which are permitted in reducing sentences to the normal forms are the familiar ones (we use the notation $\rightleftarrows$ to indicate mutual replaceability):

*Commutativity:* $A \wedge B \rightleftarrows B \wedge A$,
$A \vee B \rightleftarrows B \vee A$;
*Associativity:* $A \wedge (B \wedge C) \rightleftarrows (A \wedge B) \wedge C$,
$A \vee (B \vee C) \rightleftarrows (A \vee B) \vee C$;
*Distributivity:* $A \wedge (B \vee C) \rightleftarrows (A \wedge B) \vee (A \wedge C)$,
$A \vee (B \wedge C) \rightleftarrows (A \vee B) \wedge (A \vee C)$;
*Double Negation:* $\sim\sim A \rightleftarrows A$;
*De Morgan:* $\sim(A \wedge B) \rightleftarrows \sim A \vee \sim B$,
$\sim(A \vee B) \rightleftarrows \sim A \wedge \sim B$

We now prove in one swat the

*Soundness and Completeness Theorem for Tautological Entailments.* A sentence $A \to B$ is a tautological entailment if (completeness), and only if (soundness) $A \vdash B$ is relevantly valid.

*Proof.* Now $A \to B$ is a tautological entailment iff there is a disjunctive normal form of $A$, $A_1 \vee \cdots \vee A_m$, and a conjunctive normal form of $B$, $B_1 \wedge \cdots \wedge B_n$, so that each $A_i \to B_j$ is explicitly tautological. Since the rules of replacement are truth-preserving, the question of whether $A \vdash B$ is relevantly valid amounts to the question of whether $A_1 \vee \cdots \vee A_m \vdash B_1 \wedge \cdots \wedge B_n$, is relevantly valid. We next observe that it may routinely be argued that this last is relevantly valid iff $A_i \vdash B_j$ is relevantly valid for each $A_i$ and $B_j$. So the whole theorem reduces by chains of equivalences to the question as to whether for a primitive entailment $A_i \to B_j$, $A_i \to B_j$ is explicitly tautological iff $A_i \vdash B_j$ is relevantly valid.

Pursuing this, let $A_i$ be the conjunction of atoms, $\alpha_1 \wedge \cdots \wedge \alpha_k$ and let $B_j$ be the disjunction of atoms, $\beta_1 \vee \cdots \vee \beta_l$. If $A_i \to B_j$ is explicitly tautological, then some $\alpha_s$ is the same as some $\beta_t$. It is obvious for a given relevance valuation that if a conjunction is true then so is each conjunct, and that if a disjunct is true then so is the disjunction. So if a given relevance valuation makes $A_i$ true it will also make $\alpha_s$, i.e. $\beta_t$ true, and hence $B_j$ true. So $A_i \vdash B_j$ is relevantly valid. On the other hand if $A_i \to B_j$ fails to be explicitly tautological it is easy to construct a relevance valuation that invalidates $A_i \to B_j$. For a given sentence letter $p$, let $\mathbf{T} \in V^*(p)$ iff $p$ is a conjunct of $A_i$, and let $\mathbf{F} \in V^*(p)$ iff $\sim p$ is a conjunct of $A_i$. Note that this has the effect of making an atom true iff it is a conjunct of $A_i$. So since each conjunct of $A_i$ is true in $V$, then obviously $A_i$ is true in $V$. But also clearly $B_j$ is not true in $V$, for if it were true then some disjunct $\beta_t$ would be true and hence would be a conjunct of $A_i$, contrary to our assumption that $A_i$ and $B_j$ fail to share an atom.

## 6 An Earlier Semantical Gloss of Essentially the Same Mathematics

This was contained in my dissertation (Dunn, 1966) (and also reported in Dunn, 1971). There a *proposition surrogate* was defined to be an ordered-pair $\langle X, Y\rangle$, where $X$ and $Y$ are sets. The members of $X$ are to be thought of as the 'topics' that a given proposition gives definite information about, and the members of $Y$ as the 'topics' that the negated proposition gives definite information about. These 'topics' are members of some arbitrary but fixed set **U** called 'the universe of discourse', and $\langle X, Y\rangle$ is said to be a proposition surrogate 'in' **U**. An *interpretation* in **U** is then a function $I$ assigning a proposition surrogate in **U** to each truth-functional sentence $A$ in accord with the rules (i)–(iii) next given (in our statement of these rules we adopt the notational convention that if $I(A) = \langle X, Y\rangle$, then $I^+(A) = X$ and $I^-(A) = Y$):

$$\begin{aligned} &(I\text{i})\ \ I(\sim A) = \langle I^-(A), I^+(A)\rangle; \\ &(I\text{ii})\ \ I(A \wedge B) = \langle I^+(A) \cup I^+(B), I^-(A) \cap I^-(B)\rangle; \\ &(I\text{iii})\ \ I(A \vee B) = \langle I^+(A) \cap I^+(B), I^-(A) \cup I^-(B)\rangle. \end{aligned}$$

It is natural to say that $I(A)$ *implies* $I(B)$ iff both $I^+(B) \subseteq I^+(A)$ and $I^-(A) \subseteq I^-(B)$, and to regard a first-degree entailment $A \rightarrow B$ to be *valid in a universe of discourse* **U** iff for every interpretation $I$ in **U**, $I(A)$ implies $I(B)$, and to be *valid* simpliciter iff it is valid in every universe of discourse. The basic result about all this, proved in my dissertation, is that a first-degree entailment is a theorem of **E** iff it is valid, indeed iff it is valid in a universe of discourse with but a single topic.

The above semantical treatment of first-degree entailments was explicitly designed as a generalization of the Carnap and Bar-Hillel (1952) theory of semantic information. In that theory to each sentence $A$ is assigned a certain set representing the informational 'content' of $A$ (in symbols, $C(A)$). This assignment turns out to satisfy the following rules:

$$\begin{aligned} &(C\text{i})\ \ C(\sim A) = -C(A); \\ &(C\text{ii})\ \ C(A \wedge B) = C(A) \cup C(B); \\ &(C\text{iii})\ \ C(A \vee B) = C(A) \cap C(B). \end{aligned}$$

The basic nature of the generalization may be seen by comparing ($I$i) and ($C$i). Clause ($C$i) says that the information given by the negation of a sentence may be determined in a very simple way, i.e., set-theoretical complement, from the information given directly by the sentence. Whereas clause ($I$i) leaves open the relationship between the information given by a sentence and the information given by its negation. They may overlap, they may not be exhaustive; hence the need for the 'double-entry bookkeeping' done by proposition surrogates.

Carnap and Bar-Hillel are quite explicit about the particular nature of the elements of $C(A)$. They first define a *content-element* as the negation of a state description, and then define $C(A)$ as the set of all content-elements **L**-implied by $A$. They might just as well (cf. Bar-Hillel, 1964, Ch. 17) have defined $C(A)$ as the set of all state descriptions in which $A$ is false, i.e., which **L**-imply $\sim A$, and it will be convenient for our purposes to suppose this is what they did. It is also useful to mention the dual concept of the *range* of $A$, in symbols, $R(A)$, defined by Carnap and Bar-Hillel as the set of state-descriptions in which $A$ is true, i.e., which **L**-imply $A$.

Now I could go on to give a similarly explicit syntactical account of $I(A)$. I would define a 'situation description' as a conjunction of atoms (formed from some fixed finite set of atomic sentences, as is typical with state descriptions), *not* requiring (as does a state description) that precisely one of every atomic sentence and its negate occur in the conjunction. Thus, e.g., if the atomic sentences are $p$, $q$, and $r$, the following would all be situation descriptions, although only the first would be a state description:

$$p \wedge \sim q \wedge r,\ p \wedge \sim p \wedge \sim q \wedge r,\ p \wedge \sim q,\ p \wedge \sim p \wedge \sim q.$$

Since situation descriptions differ from state descriptions in that they may be inconsistent and/or incomplete, there are in general many more of them. Thus from $n$ atomic sentences only $2^n$ state descriptions may be formed, whereas there are $2^{2n}$ situation descriptions (counting the void situation description).

Relative to a fixed finite set of atomic sentences including all the atomic sentences ingredient in a given truth-functional sentence $A$, we can analogously define notions of the 'relevant content' of $A$ ($C_{rel}(A)$) and the 'relevant range' of $A$( $R_{rel}(A)$), defining $C_{rel}(A)$ as the set of situation descriptions that tautologically entail $\sim A$ and $R_{rel}(A)$ as the set of state descriptions that tautologically entail $A$. We can then set $I(A) = \langle C_{rel}(A), R_{rel}(A)\rangle$. I leave to that unsung hero of logical-mathematical researches, 'the interested reader', the task of working out the detail of all this. I content myself with observing that in general $C_{rel}(A) \neq -R_{rel}(A)$, since sometimes $C_{rel}(A) \cap R_{rel}(A) \neq \Lambda$, e.g., $p \wedge \sim p$ is a member of both $C_{rel}(p \wedge \sim p)$ and $R_{rel}(p \wedge \sim p)$, and also since sometimes $C_{rel}(A) \cup R_{rel}(A) \neq \mathrm{V}$, e.g., $q$ is not a member of either $C_{rel}(p \wedge \sim p)$ or $R_{rel}(p \wedge \sim p)$.

In semantical treatments of logic these days, possible worlds are much more used than their syntactical alter-egos, the state descriptions. We have already introduced in Section 4 the fashionable identification of a proposition with the set of worlds in which it is true. This corresponds to $R(A)$, and there is also the option of introducing a possible world analogue of $C(A)$, namely, the set of worlds in which $A$ is false. Similarly, we can give situation analogues of $C_{rel}(A)$ and $R_{rel}(A)$, respectively the set of situations in which $A$ is false and the set of situations in which $A$ is true.

It should by now be obvious to the reader that the situation models of Section 4 and the interpretations in universes of discourse come very close to being mere stylistic variants of one another.[11] Through the moves motivated in the last paragraph as natural and even 'familiar', it is possible to construct from a given situation model $(\mathbf{K}, \phi)$ a corresponding interpretation $I$ in a universe of discourse $\mathbf{U}$, setting $\mathbf{U} = \mathbf{K}$, and $I(A) = \langle\{H \in \mathbf{K} : \phi(A, H, \mathbf{F})\}, \{H \in \mathbf{K} : \phi(A, H, \mathbf{T})\}\rangle$. It is easy to see that $I$ so defined really is an interpretation, i.e., satisfies clauses ($I$i)–($I$iii) above. And, of course, one can also go the other way. Thus given an interpretation $I$ in a universe of discourse $\mathbf{U}$, one can set $\mathbf{K} = \mathbf{U}$, defining $\phi(A, H, \mathbf{T})$ to hold iff $H \in I^-(A)$, and $\phi(A, H, \mathbf{F})$ to hold iff $H \in I^+(A)$. The fact that we were able to pull off completeness and soundness results in Section 5 without the full apparatus of situation models (but only needing the relevance valuations $V$) can be regarded as a direct analogue of the result of Dunn (1966) already and that a universe of discourse with but a single member is all that is needed to determine the valid first-degree entailments. For, as has already been explained in Section 4, doing things with the relevance valuations $V$ is in effect to work with, in turn, various single situations.

There is one subtle defect in this story of stylistic variance, and that has to do with the fact that I required for $A \to B$ to be valid in a universe of discourse $\mathbf{U}$ that for every interpretation $I$ in $\mathbf{U}$ both $I(B) \subseteq I(A)$ and $I(A) \subseteq I(B)$. The last conjunct corresponds nicely to my requirement that for $A$ to imply $B$ in a situation model $\langle \mathbf{K}, \phi\rangle$ that for all $H \in \mathbf{K}$, if $\phi(A, H, \mathbf{T})$ then $\phi(B, H, \mathbf{T})$. However, the first conjunct corresponds to a requirement that I *never* made, to wit, that for all $H \in \mathbf{K}$, if $\phi(B, H, \mathbf{F})$ then $\phi(A, H, \mathbf{F})$. Putting the same basic point another way, it was not required for the relevant validity of $A \vdash B$ that whenever $F \in V^*(B)$ then $F \in V^*(A)$, but only that whenever $T \in V^*(A)$ then $T \in V^*(B)$.

[11] I was aware of this as a 'formal' move when I wrote my dissertation using the 'topics' semantics, and in conversation then already talked of sentences being both true and false, and also neither, but I lacked the philosophical nerve to embrace this as a serious way of talking for about another year. This was because I somehow thought that it required that it should be *possible* for sentences to *really* be *both* true and false, or *really* be *neither*, and this seemed plain mad. I hope my presentation in Section 3 avoids that.

I think it is in fact odd that no requirements were made about $B$'s passing **F** to $A$, but only requirements about $A$'s passing **T** to $B$. In a rough and ready fashion it would seem that implication, entailment, validity, etc. should have as much to do with falsity preservation as with truth preservation – it is just that the direction is reversed. But for now I think basically what needs to be said is that when it comes to assessing relevant validity and logical entailment that a situation where a sentence is both true and false symmetrizes nicely with a situation where it is neither true nor false. Thus, fixing ideas to the case of relevant validity, corresponding to every relevant valuation $V$ is its 'dual' $\overline{V}$ such that $\mathbf{T} \in \overline{V}^*(A)$ iff $\mathbf{F} \notin V^*(B)$, and also $\mathbf{F} \in \overline{V}^*(A)$ iff $\mathbf{T} \notin \overline{V}^*(B)$. (Technically it needs to be argued inductively that the dual of a relevance valuation is also indeed a relevance valuation, but this is fairly obvious.) What happens then is that it just so happens that if there is a relevance valuation $V$ so that $\mathbf{F} \in V^*(B)$ and yet $\mathbf{F} \notin V^*(A)$, then there is (in general) *another* relevance valuation (its dual) $\overline{V}$, so that $\mathbf{T} \in \overline{V}^*(A)$ and yet $\mathbf{T} \notin \overline{V}^*(B)$. Thus in assessing relevant validity, where we consider *all* relevance valuations, we can by merely requiring truth preservation also get as a 'spin off' falsity preservation in the converse direction.

## 7 Ruminations

I suppose the dominant trend of this paper has been to suggest that co-entailment in the system **E** might serve as the foundation of synonymy. I am not sure how much I myself am willing to go along with that trend in virtue of the fact that $p \leftrightarrow p \wedge (p \vee q)$ is such a provable co-entailment. As the reader can easily see, both entailments are tautological, and the one from left-to-right holds 'mainly' because of the primitive entailment $p \rightarrow p \vee q$.

I find myself in an inconsistent triad of intuitions here. I do believe that $p$ entails $p \vee q$.[12] I also believe that co-entailment (*properly understood*) should suffice for synonymy (again *properly understood*). But I find it difficult to accept that an atomic sentence that talks just about the rain is synonymous with some compound sentence that seems to talk about the number two as well.

I am not going to solve this problem here, but it seems worth saying that unless one is willing to give up on the relation of synonymy altogether (except perhaps as holding only between each sentence and itself), one should be prepared to buy sentences as synonymous which may differ from one another in their grammatical visage. Perhaps this is just an extreme such case. More moderate cases, e.g., that $p$ is synonymous with $p \wedge p$, might even raise some eyebrows ("the last talks about conjunction, whereas the first does not").

## References

1. Anderson, A. R., and Belnap, Jr., N. D., 'Tautological Entailments', *Philosophical Studies*, 13, (1961), 9-24.
2. Bar-Hillel, Y., *Language and Information*, Reading, Mass., 1964.
3. Carnap, R., and Bar-Hillel, Y., 'An Outline of a Theory of Semantic Information', MIT Research Lab. of Electronics Tech. Report, No. 247, 1952. Also reprinted as Ch. 15 of Bar-Hillel (1964).

[12] If I were not so convinced of this, I would take the opportunity to push Parry's 'analytic implication' or some version thereof (cf. Dunn (1972)).

4. Dunn, J. M., *The Algebra of Intensional Logics*, unpublished dissertation, University of Pittsburgh, 1966.
5. Dunn, J. M., 'The Effective Equivalence of Certain Propositions about de Morgan Lattices', (Abstract) *Journal of Symbolic Logic*, 32, 1967.
6. Dunn, J. M., 'Natural Language versus Formal Language', a mimeo of a talk in the joint APA-ASL Symposium by that title, New York, Dec. 27, 1969.
7. Dunn, J. M., 'An Intuitive Semantics for First Degree Relevant Implications', (Abstract) *Journal of Symbolic Logic*, 36, (1971a), pp. 362-363.
8. Dunn, J. M., 'A Kripke-Style Semantics for R-Mingle', contributed paper, Tarski Symposium, Univ. of California, Berkeley, June, (mimeo abstracts of contributed papers distributed to participants), 1971b.
9. Dunn, J. M., 'A modification of Parry's analytic implication', *Notre Dame Journal of Formal Logic*, **XIII**, (1972), pp.195–205.
10. Geach, P. T., and Black, M., *Translations from the Philosophical Writings of Gottlob Frege*, Oxford, 1960.
11. Jeffrey, R. C., *Formal Logic: Its Scope and Limits*, New York, 1967.
12. Lewis, C. I., and Langford, C. H., *Symbolic Logic*, New York, 1932.
13. Montague, R., 'Pragmatics and Intensional Logic'. in *Semantics of Natural Language*, ed. by D. Davidson and G. Harman, (Dordrecht, Holland), 1972, pp. 142-168.
14. Routley, R., and Routley, V., 'The Semantics of First Degree Entailment', 1972, *Noüs*, 6, (1972), pp. 335-359.
15. Routley, R., and Meyer, R. K., 'The Semantics of Entailment', in *Truth, Syntax and Modality*, ed. by H. Leblanc, Amsterdam 1973, pp. 198-243.
16. Smullyan, R., *First-Order Logic*, Berlin, 1968.
17. van Fraassen, B., 'Facts and Tautological Entailments', *The Journal of Philosophy*, 66, (1969), pp. 477-487.
18. Wittgenstein, L., *Tractatus Logico-Philosophicus*. Translated by D. F. Pears and B. F. McGuinness. London, 1961.

# How a Computer Should Think*

Nuel D. Belnap

INTRODUCTION; *the computer*. I propose that a certain four-valued logic should sometimes be used. It is to be understood that I use "logic" in a narrow sense, the old sense: a logic as an *organon*, a tool, a canon of inference. And it is also to be understood that I use "should" in a straightforward normative sense.

My suggestion for the utility of a four-valued logic is a local one. It is not the Big Claim that we all ought always use this logic (this paper does not comment on that claim), but the Small Claim that there are circumstances in which someone – not you – ought to abandon the familiar two-valued logic and use another instead. It will be important to delineate these circumstances with some care.

The situation I have in mind may be described as follows. In the first place, the reasoner who is to use this logic is an artificial information processor; that is, a (programmed) computer. This already has an important consequence. People sometimes give as an argument for staying with classical two-valued logic that it is tried and true, which is to say that it is imbued with the warmth of familiarity. This is a good (though not conclusive) argument for anyone who is interested, as I am, in practicality; it is kin to Quine's principle of "minimal mutilation," though I specifically want the emotional tone surrounding familiarity to be kept firmly in mind. But given that in the situation I envisage the reasoner is a computer, this argument has no application. The notion of "familiarity to the computer" makes no sense, and surely the computer does not care what logic is familiar to *us*. Nor is it any trouble for a programmer to program an unfamiliar logic into the computer. So much for emotional liberation from two-valued logic.

In the second place, the computer is to be some kind of sophisticated question-answering system, where by "sophisticated" I mean that it does not confine itself, in answering questions, to just the data it has explicitly in its memory banks, but can also answer questions on the basis of *deductions* which it makes from its explicit information. Such sophisticated devices barely exist today, but they are in the forefront of everyone's hopes. In any event, the point is clear: unless there is some need for reasoning, there is hardly a need for logic.

Thirdly, the computer is to be envisioned as obtaining the data on which it is to base its inferences from a variety of sources, all of which indeed may be supposed to be on the whole trustworthy, but none of which can be assumed to be that paragon of paragons, a universal truth-teller. There are at least two possible pictures here. One puts the computer in the context of a lot of fallible humans telling it what is so and what is not; or with rough

---

* This paper and Belnap (1976) are complements. There the introductory paragraphs and Part I are heavily abbreviated, while the more technical Parts 2 and 3 are given in full instead of, as here, barely summarized. Thanks are due to the National Science Foundation for partial support through Grant SOC71 03594 A02.

Nuel D. Belnap
Department of Philosophy, University of Pittsburgh, Pittsburgh, USA
e-mail: belnap@pitt.edu



H. Omori, H. Wansing (eds.), *New Essays on Belnap-Dunn Logic*,
Synthese Library 418, https://doi.org/10.1007/978-3-030-31136-0_4

equivalency, a single human feeding it information over a stretch of time. The other picture paints the computer as part of a network of artificial intelligences with whom it exchanges information. In any event, the essential feature is that there is no single, monolithic, infallible source of the computer's data, but that inputs come from several independent sources. In such circumstances the crucial feature of the situation emerges: *inconsistency threatens*. Elizabeth tells the computer that the Pirates won the Series in 1971, while Sam tells it otherwise. What is the computer to do? If it is a classical two-valued logician, it must give up altogether talking about anything to anybody or, equivalently, it must say everything to everybody. We all know all about the fecundity of contradictions in two-valued logic: contradictions are never isolated, infecting as they do the whole system. Of course the computer could refuse to entertain inconsistent information. But in the first place that is unfair either to Elizabeth or to Sam, each of whose credentials are, by hypothesis, nearly impeccable. And in the second place, as we know all too well, contradictions may not lie on the surface. There may be in the system an *undetected* contradiction, or what is just as bad, a contradiction which is not detected until long after the input which generated it has been blended in with the general information of the computer and has lost its separate identity. But still we want the computer to use its head to reason to just conclusions yielding sensible answers to our questions.

Of course we want the computer to report any contradictions it finds, and in that sense we by no means want the computer to ignore contradictions. It is just that in these cases in which there is a possibility of inconsistency, we want to set things up so that the computer can continue reasoning in a sensible manner even if there is such an inconsistency, discovered or not. With respect to the latter, even if the computer has discovered and reported an inconsistency in its baseball information such as that the Pirates both won and did not win the Series in 1971, we would not want that to affect how it answered questions about airline schedules; but if the computer is a two-valued logician, the baseball contradiction will lead it to report that there is no way to get from Bloomington to Chicago. And also of course that there are exactly 3,000 flights per day. In an elegant phrase, Shapiro calls this "polluting the data." What I am proposing is to Keep Our Data Clean.

So we have a *practical* motive to deal with situations in which the computer may be told both that a thing is true and that also it is false (at the same time, in the same place, in the same respect, etc., etc., etc.).

There is a fourth aspect of the situation, concerning the significance of which I remain uncertain, but which nevertheless needs mentioning for a just appreciation of developments below: my computer is *not* a complete reasoner, who should be able to do something better in the face of contradiction than just report. The complete reasoner should, presumably, have some strategy for *giving up* part of what it believes when it finds its beliefs inconsistent. Since I have never heard of a practical, reasonable, mechanizable strategy for revision of belief in the presence of contradiction, I can hardly be faulted for not providing my computer with such. In the meantime, while others work on this extremely important problem, my computer can only accept and report contradictions without divesting itself of them.

This aspect is bound up with a fifth: in answering its questions, the computer is to reply strictly in terms of what it has been told, not in terms of what it could be programmed to believe. For example, if it has been told that the Pirates won and did not win in 1971, it is to so report, even though we could of course program the computer to recognize the falsity of such a report. The point here is both subtle and obvious: if the computer would not report our contradictions in answer to our questions, we would have no way of knowing that its data-base harbored contradictory information. (We could, if we wished, ask it to give a *supplementary* report, e.g., as follows: "I've been told that the Pirates won and did not win; but of course it ain't so"; but would that be useful?)

*Approximation Lattices.* Always in the background and sometimes in the foreground of what I shall be working out is the notion of an *approximation lattice*, due to Scott (see e.g., Scott 1970, 1971, 1972). Let me say a word about this concept before getting on. You are going to be disappointed at the mathematical definition of an approximation lattice: mathematically it is just a complete lattice. That is, we have a set $A$ on which there is a partial ordering $\sqsubseteq$, and for arbitrary subsets $X$ of $A$ there always exist least upper bounds $\sqcup X \in A$ and greatest lower bounds $\sqcap X \in A$ (two-element ones written $x \sqcup y$ and $x \sqcap y$). But I don't call a complete lattice an approximation lattice unless it satisfies a further, non-mathematical, condition: it is appropriate to read $x \sqsubseteq y$ as "$x$ approximates $y$." Examples worked out by Scott include the lattice of "approximate and overdetermined real numbers," where we are identifying approximate real number with an interval, and where $x \sqsubseteq y$ just in case $x \subseteq y$. The (only) overdetermined real number is the empty set. As a further example Scott offers the lattice of "approximate and overdetermined functions" from $A$ to $B$, identified as subsets of $A \times B$. Here we want $f \sqsubseteq g$ just in case $f \subseteq g$.

In such lattices the *directed* sets are important: those sets such that every pair of members $x$ and $y$ of the set have an upper bound $z$ also in the set. For such a set can be thought of as approximating by a limiting process to its union $\sqcup X$. That is, if $X$ is directed, it makes sense to think of $\sqcup X$ as the limit of $X$. (An ascending sequence $x_1 \sqsubseteq \cdots \sqsubseteq x_i \sqsubseteq \ldots$ is a special case of a directed set.) And now when we pass to the family of functions from one approximation lattice into another (or of course the same) approximation lattice, Scott has demonstrated that what are important are the *continuous* functions: those that preserve non-trivial directed unions (i .e., $f(\sqcup X) = \sqcup\{fx : x \in X\}$, for nonempty directed $X$). These are the only functions which respect the lattices qua *approximation* lattices. This idea is so fundamental to developments below that I choose to catch it in a Thesis to be thought of as analogous to Church's Thesis:

> *Scott's Thesis.* In the presence of complete lattices $A$ and $B$ naturally thought of as approximation lattices, pay attention only to the continuous functions from $A$ into $B$, resolutely ignoring all other functions as violating the nature of $A$ and $B$ as approximation lattices.

(Though honesty compels me to attribute the Thesis to Scott, the same policy bids me note that the formulation is mine, and that as stated he may not want it, or may think that some other Thesis in the neighborhood is more important; for example, that every approximation lattice (intuitive sense) is a continuous lattice (sense of Scott 1972).)

You will see how I rely on Scott's Thesis in what follows.

*Program.* The rest of this paper is divided into three parts. Part 1 considers the case in which the computer accepts only *atomic* information. This is a heavy limitation, but provides a relatively simple context in which to develop some of the key ideas. Section 3 allows the computer to accept also information conveyed by *truth-functionally compounded* sentences; and in this context I offer a new kind of meaning for formulas as certain mappings from epistemic states into epistemic states. In Part 3 the computer is allowed also to accept *implications construed as rules* for improving its data base.

*Part* 1. *Atomic inputs*

*Atomic Sentences and the approximation lattice* **A4**. Now and throughout this paper you must keep firmly fixed in mind the circumstances in which the computer finds itself, and especially that it must be prepared to receive and reason about inconsistent information. I want to suggest a natural technique for employment in such cases: when an item comes in as asserted, mark it with a "told True" sign, and when an item comes in denied, mark with a "told False" sign, treating these two kinds of tellings as altogether on a par. It is easy to see that this leads to four possibilities. For each item in its basic data file, the computer is going to have it marked in one of the following four ways: (1) just the "told True" sign, indicating that that item has been asserted to the computer without ever having been denied. (2) just the value "told False", which indicates that the item has been denied but never asserted. (3) no "told" values at all, which means the computer is in ignorance, has been told nothing. (4) the interesting case: the item might be marked with both "told True" and "told False". (Recall that allowing this case is a *practical necessity* because of the fallibility of man.)

These four possibilities are precisely the four values of the many-valued logic I am offering as a practical guide to reasoning by the computer. Let us give them names:

***T:*** just told True
***F:*** just told False
***None:*** told neither True nor False
***Both:*** told both True and False

So these are our four values, and we baptize: **4** = {***T, F, None, Both***}. Of course four values do not a logic make, but let us nevertheless pause a minute to see what we have so far.

The suggestion requires that a system using this logic code each of the atomic statements representing its data base in some manner indicating which of the four values it has (at the present stage). It follows that the computer cannot represent a class merely by listing certain elements, with the assumption that those not listed are not in the class. For just as there are four values, so there are four possible states of each element: the computer might have been told none, one, or both of "in the class" and "not in the class." Two procedures suggest themselves. The first is to list each item with one of the values ***T, F***, or ***Both***, for these are the elements about which the computer has been told something; and to let an absence of a listing signify ***None***, i.e., that there is no information about that element. The second procedure would be to list each element with one *or both* of the "told" values, "told True" and "told False," not listing elements lacking both "told" values. Obviously the procedures are equivalent, and we shall not in our discourse distinguish between them, using one or the other as seems convenient.

The same procedure works for relations, except that it is ordered pairs that get marked. For example, a part of the correct table for Series winners, conceived as a relation between teams and years, might look like this:

⟨Pirates, 1971⟩ ***T*** and ⟨Orioles, 1971⟩ ***F***

or

⟨Pirates, 1971⟩ True and ⟨Orioles, 1971⟩ False

But if Sam slipped up and gave the wrong information after Elizabeth had previously entered the above, the first entry would become

⟨Pirates, 1971⟩ ***Both***

*or*

⟨Pirates, 1971⟩ True, False

To be specific, we envision (in this Part of the paper) the epistemic state of the computer to be maintained in terms of a table giving one of four values to each atomic sentence. We call such a table a *set-up* (following an isomorphic use of Routley and Routley, 1972); i.e., a set-up is, mathematically, a mapping from atomic sentences into the set **4** = {***T, F, None, Both***}. When an atomic formula is entered into the computer as either affirmed or denied, the computer modifies its current set-up by adding a "told True" or "told False" according as the formula was affirmed or denied; it does not subtract any information it already has, for that is the whole point of what we are up to. In other words, if $p$ is affirmed, it marks $p$ with ***T*** if $p$ were previously marked with ***None***; with ***Both*** if $p$ were previously marked with ***False***; and of course leaves things alone if $p$ was already marked either ***T*** or ***Both***. So much for $p$ as input.

The computer not only accepts input, but answers questions. We consider only the basic question as to $p$; this it answers in one of four ways: Yes, No, Yes and No, or I don't know, depending on the value of $p$ in its current set-up as ***T, F, Both***, or ***None***. (It would be wrong to suppose that these four answers are either dictated by the four-valued logic or excluded by the two-valued logic; it is just that they are made more useful in the four-valued context. See Belnap, 1963, and Belnap and Steel, 1975.)

Warning – or, as N. Bourbaki says, *tournant dangereux*: "told True" is not equivalent to ***T***. The relationships are rather as follows. In the first place, the computer is told True about some sentence $A$ just in case it has *either* marked $A$ with ***T*** or with ***Both***. Secondly, the computer marks $A$ with ***T*** just in case it has been told True of $A$ *and* has not been told False of $A$. And similarly for the relation between ***F*** and "told False." These relationships are certainly obvious, but also in practice confusing. It might help always to read "told True" as "told *at least* True," and ***T*** as "told *exactly* True."

I now make the observation which constitutes the foundation of what follows: these four values naturally form a lattice under the lattice-ordering "approximates the information in", and indeed an *approximation-lattice* in the sense we described above:

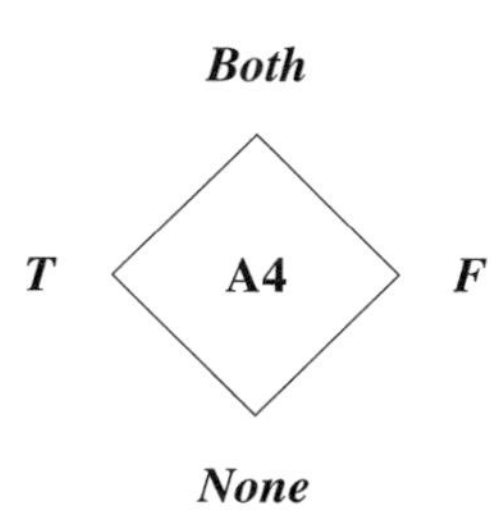


(In this Hasse diagram joins (⊔) and meets (⊓) are least upper bounds and greatest lower bounds respectively, and ⊑ goes uphill. ***None*** is at the bottom because it gives no information at all; and ***Both*** is at the top because it gives too much (inconsistent) information, saying as it does that the statement so marked is held both True and False. As we mentioned above, Scott has studied approximation-lattices in detail and in a much richer setting than we have before us; yet still this little four-element lattice is important for much of his work. We remarked above that according to Scott's Thesis the important functions in the neighborhood of an approximation lattice like **A4** are the continuous ones. We do not, fortunately, have to deal with continuity for a while, since in the finite case it turns out that for a function to be continuous is just for it to be *monotonic*, i.e., for it to preserve the lattice ordering: $a \sqsubseteq b$ implies $fa \sqsubseteq fb$.

For example, suppose a function *g* on **A4** is such that it takes ***T*** into ***F*** and ***F*** into ***T***: *g*(***T***) = ***F***, *g*(***F***) = ***T***. Then given that *g* is monotonic, since ***T***⊑***Both*** we must have ***F***⊑ *g*(***Both***) and similarly ***T***⊑ *g*(***Both***). So we must have *g*(***Both***) = ***Both***. In a similar way, it is easy to calculate that *g*(***None***) = ***None*** – if *g* is to be monotonic, as all good functions should be.

*Compound sentences and the logical lattice* **L4**. Now this function, *g*, is no *mere* example of a monotonic function on the lattice **A4** of approximate and inconsistent truth-values. In fact we are in the very presence of *negation*, which some have called the original sin of logic but which we clearly need in a sufficiently rich language for our computer to use – just to be able to answer simple yes-no questions. To see that *g* really *is* negation, consider first that the values ***T*** and ***F***, representing as they do the pure case, should act like the ordinary truth values the True and the False, so that obviously we want ~***T*** = ***F***, and ~***F*** = ***T***. And then Scott's Thesis now imposes on us a *unique* solution to the problem of extending negation to the values of our foursome; we *must* have ~***None*** = ***None*** and ~***Both*** = ***Both*** if negation is to be an acceptably monotonic function on the approximation-lattice **A4**.

We can summarize the argument in a small table for negation.

| | ***None*** | ***F*** | ***T*** | ***Both*** |
|---|---|---|---|---|
| ~ | ***None*** (m) | ***T*** (tt) | ***F*** (tt) | ***Both*** (m) |

Here "tt" in the upper right hand corner means that the value was given by truth table considerations while "m" indicates that monotonicity was invoked.

Having put negation in our pocket, let's turn to conjunction and disjunction. We start with truth table considerations for ***T–F*** portions of the tables, and then invoke monotony (in each argument place) and easy considerations to extend them as indicated.

| & | ***None*** | ***F*** | ***T*** | ***Both*** |
|---|---|---|---|---|
| ***None*** | ***None*** (m) | | ***None*** (m) | |
| ***F*** | | ***F*** (tt) | ***F*** (tt) | |
| ***T*** | ***None*** (m) | ***F*** (tt) | ***T*** (tt) | ***Both*** (m) |
| ***Both*** | | | ***Both*** (m) | ***Both*** (m) |

| ∨ | ***None*** | ***F*** | ***T*** | ***Both*** |
|---|---|---|---|---|
| ***None*** | ***None*** m | ***None*** m | | |
| ***F*** | ***None*** m | ***F*** tt | ***T*** tt | ***Both*** m |
| ***T*** | | ***T*** tt | ***T*** tt | |
| ***Both*** | | ***Both*** m | | ***Both*** m |

With just ordinary truth tables and monotonicity, it would appear we have to stop with these partial tables; on this basis neither conjunction nor disjunction – unlike negation – are uniquely determined. Of course we might make some guesses on the basis of intuition, but this part of the argument is founded on a desire not to do that; rather, we are trying to see how far we can go on a purely theoretical basis.

It turns out that if we ask only that conjunction and disjunction have some minimal relation to each other, then every other box is uniquely determined. There are several approaches possible here, but perhaps as illuminating as any is to insist that the orderings determined by the two in the standard way be the same; which is to say that the following equivalence holds:

$$a\&b = a \quad \text{iff} \quad a \vee b = b$$
$$a\&b = b \quad \text{iff} \quad a \vee b = a$$

For look at the partial table for conjunction. One can see that ***T*** is an identity element: $a\&\boldsymbol{T} = a$, all $a$. So if conjunction and disjunction fit together as they ought, we must have $a \vee \boldsymbol{T} = \boldsymbol{T}$, all $a$, which fills in two boxes of the ∨-table. And similar arguments fill in all except the corners.

For the corners we must invoke monotony (*after* the above lattice argument). For example, since ***F*** ⊑ ***Both***, by monotony (***F&None***) ⊑ (***Both&None***) so ***F*** ⊑ (***Both&None***). Similarly, ***None*** ⊑ ***F*** leads to (***Both&None***) ⊑ (***Both&F***), i.e. (***Both&None***) ⊑ ***F***. So by antisymmetry in **A4**, (***Both&None***) = ***F***. These additional results are brought together in the following tables, where "f" indicates use of the above fit between & and ∨, and "m" again indicates monotony.

| & | ***None*** | ***F*** | ***T*** | ***Both*** |
|---|---|---|---|---|
| ***None*** | ***None*** | ***F*** f | ***None*** | ***F*** f |
| ***F*** | ***F*** f | ***F*** | ***F*** | ***F*** f |
| ***T*** | ***None*** | ***F*** | ***T*** | ***Both*** |
| ***Both*** | ***F*** m | ***F*** f | ***Both*** | ***Both*** |

| ∨ | ***None*** | ***F*** | ***T*** | ***Both*** |
|---|---|---|---|---|
| ***None*** | ***None*** | ***None*** | ***T*** f | ***T*** m |
| ***F*** | ***None*** | ***F*** | ***T*** | ***Both*** |
| ***T*** | ***T*** f | ***T*** | ***T*** | ***T*** f |
| ***Both*** | ***T*** m | ***Both*** | ***T*** f | ***Both*** |

I don't know if we should be surprised or not, but in fact these tables do constitute a lattice, with conjunction as meet and disjunction as join; a lattice which can be pictured as follows:

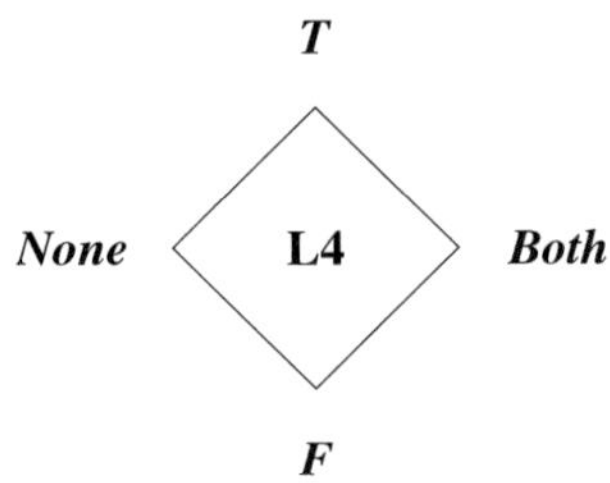


Let us agree to call this the *logical* lattice, **L4**, to distinguish it from the *approximation* lattice, **A4**. The ordering on **L4** we write as $a \leq b$; we write meets as $a\&b$, and joins as $a \vee b$. I note that in the logical lattice, both of the values ***None*** and ***Both*** are intermediate between ***F*** and ***T***; and this is as it should be, for the worst thing is to be told something is false, simpliciter. You are better off (it is one of our hopes) in either being told nothing about it, or in being told both that it is true and also that it is false; while of course best of all is to be told it is true, with no muddying of the waters. Nevertheless, surely most of you must be puzzled, if you are thinking about it, concerning the rules for computing the conjunction and disjunction of ***None*** and ***Both***: ***None&Both*** = ***F***, while ***None*** ∨ ***Both*** = ***T***. I ask you for now only to observe that we were driven to these equations by only three considerations: ordinary truth tables, monotony, and fit between & and ∨. But I shall have more to say about this.

We can now use these logical operations on **L4** to induce a semantics for a language involving &, ∨, and ~, in just the usual way. Given an arbitrary set-up $s$ – a mapping, you will recall of atomic formulas into **4** – we can extend $s$ to a mapping of all formulas into **4** in the standard inductive way:

$$\begin{aligned} s(A\&B) &= s(A)\&s(B) \\ s(A \vee B) &= s(A) \vee s(B) \\ s({\sim}A) &= {\sim}s(A) \end{aligned}$$

And this tells us how the computer should answer questions about complex formulas based on a set-up representing its epistemic state (what it has been told): just as for answering questions about atomic formulas, it should answer a question as to $A$ by Yes, No, Yes and No, or I don't know, according as the value of $A$ in $s$ (i.e., $s(A)$) is ***T***, ***F***, ***Both***, or ***None***.

The preceding discussion will have struck you as abstractly theoretical; I should next like to take up negation, conjunction, and disjunction from an altogether different and more intuitive point of view. The question to which we are going to address ourselves is this: given our intuitive understanding of the meaning of the four truth-values as indicating markings of sentences with either or both of the True and the False, what is a plausible way to extend these values to compound sentences when we know the values of the ingredient sentences? Let us take up negation first.

The inevitable thing to say seems to be that ~*A* should be marked with the True just in case *A* is marked with the False, and should be marked with the False just in case *A* is marked with the True. In other words, ~*A* should be marked with at least the True just in case *A* is marked with at least the False, and marked with at least the False just in case *A* is marked with at least the True. But then consider the correspondences

***None:*** marked with neither,
***F:*** marked with just the False,
***T:*** marked with just the True,
***Both:*** marked with both.

It immediately comes out that we should mark ~*A* with ***Both*** if *A* is, with ***None*** if *A* is, and with ***T*** or ***F*** if *A* is ***F*** or ***T***. For example, if *A* is marked ***None***, i.e., with neither the True nor the False, then ~*A* should also be marked with neither. If you know nothing about *A*, then you know nothing about ~*A*. And the same reasoning works for ***Both***: if you know too much about *A*, then you also know too much about ~*A*.

In a similar way, we can give intuitive clauses for evaluation of conjunctions and disjunctions, as follows:

- Mark (*A*&*B*) with at least True just in case both *A* and *B* have been marked with at least the True.
- Mark (*A*&*B*) with at least the False just in case at least one of *A* and *B* have been marked with at least the False.

This completely determines how to mark conjunctions.

- Mark ($A \vee B$) with at least the True just in case at least one of *A* and *B* have been marked with at least the True.
- Mark ($A \vee B$) with at least the False just in case both *A* and *B* have been marked with at least the False.

And this similarly uniquely determines disjunction, given our intuitive correspondence between our four values ***None, F, T, Both*** on the one hand, and markings with neither, one, or both of the True and the False on the other. And furthermore, this intuitive account of the connectives exactly agrees with the theoretically based account deriving from Scott's approximation lattices. For example, consider one of the odd corners, ***Both*&*None*** = ***F***. Well, suppose *A* has been marked with both the True and the False, and *B* with neither (corresponding to ***Both*** and ***None***, respectively). Then we must mark (*A*&*B*) with at least the value the False, since one of its components has at least the value the False; and we must not mark it with at least the True, since not both of its components are so marked. So we must mark it with exactly the False: So ***Both*&*None*** = ***F***. In other even more informal words, in this circumstance the computer has a reason to suppose (*A*&*B*) told false, but none to suppose it told true. So, although the oddity of ***Both*&*None*** = ***F*** doesn't go away, it anyhow gets explained.

*Entailment and inference: the four-valued logic.* Where are we? Well, we haven't got a logic, i.e., rules for generating and evaluating inferences. (In our case we really want the former; we want some rules for the computer to use in generating what it implicitly knows from what it explicitly knows.) What we do have is four interesting values, with indications as to how those are to be used by friend computer, and three splendid connectives, with complete and wellmotivated tables for each. And as we all know lots of other connectives can be defined in terms of these, so that for our purposes three is enough.

Suppose we have an argument involving these connectives. The question is, when is it a good one? Again I want to give an abstractly theoretical answer, and then an intuitive answer. (And then several more answers, too, were there time enough. For the question is fascinating.)

The abstract answer relies on the *logical* lattice we took so much time to develop. It is: entailment goes up hill. That is, given any sentences $A$ and $B$ (compounded from variables by negation, conjunction, and disjunction), we will say that $A$ *entails* or *implies* $B$ just in case for each assignment of one of the four values to the variables, the value of $A$ does not exceed (is less-than-or-equal-to) the value of $B$. In symbols: $s(A) \leq s(B)$ for each set-up $s$. This is a plausible definition of entailment whenever we have a lattice of values which we can think of as somehow being graded from bottom to top; and as I suggested when first presenting you with the logical lattice, we can indeed think of ***None*** and ***Both*** as being intermediate between awful ***F*** and wonderful ***T***.

Now for an account which is close to the informal considerations underlying our understanding of the four values as keeping track of markings with the True and the False: say that the inference from $A$ to $B$ is valid, or that $A$ entails $B$, if the inference never leads us from the True to the absence of the True (preserves Truth), *and also* never leads us from the absence of the False to the False (preserves non-Falsity) . Given our system of markings, to ask this is hardly to ask too much.

(I note that Dunn, 1975, has shown that it suffices to mention truth-preservation, since if some inference form fails to always preserve non-Falsity, then it can be shown by a technical argument that it *also* fails to preserve Truth. Just take the assignment to the propositional variables which *switches* ***Both*** and ***None***, but leaves ***T*** and ***F*** alone, and show that the value of any compound has the same feature. But I agree with the spirit of a remark of Dunn's, which suggests that the False really is on all fours with the True, so that it is profoundly natural to state our account of "valid" or "acceptable" inference in a way which is neutral with respect to the two.)

Finally we have a logic, that is a canon of inference for our computer to use in making inferences involving conjunction, negation, and disjunction as well of course as whatever can be defined in terms thereof. I note that this logic has two key features. In the first and most important place it is rooted in reality. We gave reason why it would be good for our computer to think in terms of our four values and why the logic of the four values should be as it is. In the second place, though we have not thrown around many hen-scratches it is clear that our account of validity is mathematically rigorous. And obviously the computer can decide by running through a truth-tabular computation whether or not a proposed inference is valid. But there is another idea to the logician's job, which is codifying inferences in some axiomatic or semi-axiomatic way which is transparent and accordingly usable. If this sounds mysterious, it is not; I just mean that a logician, given a semantics, ordinarily tries to come up with proof-theory for it; a proof-theory which is consistent and complete relative to the semantics.

This job has been done any number of times, though in the beginning the proof-theory came first and the semantics only later. The history of the matter is about as follows. A long time ago Alan Anderson and I in 1962 proposed a group of inferences, which we called *tautological entailments*, as comprising all of the *sane* inferences (involving &, ∨, ~) which anyone not a psychotic, or not badly trained, would want to make. We had various proof-theoretical formalizations for these, and showed that an eight-valued matrix was sufficient to characterize them semantically.

Later T. J. Smiley showed (in a letter to us) that a certain four-valued matrix would do exactly the same work; namely, though with numbers instead of fancy names, precisely the four values I am offering you here. So that's where I learned them. Smiley of course intended his result as merely technical, without logical point, in the sense of "logical" of this paper.

Still later Dunn in 1966 gave a variety of semantics for tautological entailments, some of them highly intuitive, some closely related to the four-valued matrix of Smiley; and to him is due one of the central ideas with which I have been working, namely, the identification of the four values with the four subsets of {the True, the False}. Dunn in 1975 (belatedly) presents much of both the intuitive and the technical significance of this idea. Other semantics for tautological entailments, again with intuitive considerations, are due to van Fraassen (1966b) and Routley and Routley (1972). The algebraic structure corresponding to this logic has been investigated in detail by Dunn and others; all this is reported by Dunn in Chapter III of Anderson and Belnap (1975), wherein also will be found Gentzen calculuses and the like.

My own deeper interest and the thought that there might be a computer application, came after overlapping with Dana Scott in Oxford in 1970 as guests of Christopher Strachey's – to whom thanks, now tragically posthumous, are due. The four values emerged as an important approximation lattice in Scott's work and the connection with the Smiley four values was not hard to see. The realization of the importance of the epistemic interpretation is more recent.

Stuart Shapiro has independently argued for the utility of "relevance logics" for question-answering systems, and has suggested implementation in a research proposal; see also Shapiro and Wand 1975.

So much for history. Let me briefly present a group of principles which are semantically valid, and taken together semantically complete. They will also be redundant, but recall that the byword for this exercise is usefulness; I am offering a set of principles for the computer to use in making its inferences.

Let $A$, $B$, etc., be formulas in &, ∨, and ~. Let $A \rightarrow B$ signify that the inference from $A$ to $B$ is valid in our four values, i.e., that $A$ entails $B$. Also let $A \leftrightarrows B$ signify that $A$ and $B$ are semantically equivalent, and can be intersubstituted in any context. Then the following have proved to be a useful (complete) set of principles.

$A_1 \& \ldots \& A_m \rightarrow B_1 \vee \ldots \vee B_n$
provided some $A_i$ is some $B_j$ (sharing)
$(A \vee B) \rightarrow C$ iff (if and only if) $A \rightarrow C$ and $B \rightarrow C$
$A \rightarrow B \& C$ iff $A \rightarrow B$ and $A \rightarrow C$

$A \rightarrow B$ iff $\sim B \rightarrow \sim A$[1]
$A \vee B \leftrightarrows B \vee A$ — $A \& B \leftrightarrows B \& A$
$A \vee (B \vee C) \leftrightarrows (A \vee B) \vee C$ — $(A \& B) \& C \leftrightarrows A \& (B \& C)$
$A \& (B \vee C) \leftrightarrows (A \& B) \vee (A \& C)$ — $A \vee (B \& C) \leftrightarrows (A \vee B) \& (A \vee C)$
$(B \vee C) \& A \leftrightarrows (B \& A) \vee (C \& A)$ — $(B \& C) \vee A \leftrightarrows (B \vee A) \& (C \vee A)$

[1] Editors' note: The original publication contains a misprint: $A \rightarrow B$ iff $\sim A \rightarrow \sim B$.

$\sim\sim A \leftrightarrows A$
$\sim(A\&B) \leftrightarrows \sim A \vee \sim B$ — $\sim(A \vee B) \leftrightarrows \sim A\&\sim B$
if $A \to B$ and $B \to C$, then $A \to C$
if $A \leftrightarrows B$ and $B \leftrightarrows C$, then $A \leftrightarrows C$
$A \to B$ iff $A \leftrightarrows (A\&B)$ iff $(A \vee B) \leftrightarrows B$

*Observations.* Some observations now need to be made before pushing further. First, I note that not derivable from these principles, and not semantically valid, are the paradoxes of "implication" $A\&\sim A \to B$ and $A \to B \vee \sim B$. In context the failure of the principles is evident. The failure of the first simply means that just because we have been told both that $A$ is True, and that $A$ is False, we cannot conclude: everything. Indeed, we may have been told nothing about $B$, or just that it is False. And the failure of the second is equally evident: from the fact that we have been told that $A$ is True, we cannot conclude that we know something about $B$. Of course $B$ is *ontologically* either True or False, and such ontological truth-values will receive their due; but for $B \vee \sim B$ to be marked with the True is either for $B$ to be marked with the True or for $B$ to be marked with the False; and it may have neither mark. Or, for a different way of counterexampling $A \to B \vee \sim B$, $A$ may have just the True while $(B \vee \sim B)$ has *both* values because $B$ does.

These inferences are not wanted in a scheme which is designed not to break down in the presence of "contradictions"; and since contradictions really do threaten in the circumstances we describe, their absence is welcome.

I would be less than open, however, if I failed to point out the absence of what at first sight looks like a more harmless principle: $(A \vee B)\&\sim A \to B$. Surely, one would think, our computer should be able to argue that if one of $A$ and $B$ is true, and it's not $A$, then it must be $B$. That's true; unless – and of course this is a critical "unless," – there is an inconsistency around. In fact the inference the canon allows is just exactly

$$(A \vee B)\&\sim A \to (A\&\sim A) \vee B$$

That is, having determined that the antecedent is at least True, we allow the computer to conclude: either $B$ is at least True, or something funny is going on; i.e., it's been told that $A$ is both True and False. And this, you will see, is right on target. If the *reason* that $(A \vee B)\&\sim A$ is getting thought of as a Truth is because $A$ has been labeled as both True and False, then we certainly do *not* want to go around inferring $B$. The inference is wholly inappropriate in a context where inconsistency is a live possibility.

The second observation is that our four values are proposed only in connection with *inferences*, and are definitely not supposed to be used for determining which formulas in &, $\vee$, and $\sim$ count as so-called logical truths. In fact no formula takes always the value ***T***, so *that* property surely won't do as semantic account of logical truth. There are, on the other hand, formulas which never take the value ***F***, e.g., $A \vee \sim A$; but this set is not even closed under conjunction and do not contain $(A \vee \sim A)\&(B \vee \sim B)$, which can take ***F*** when $A$ takes ***None*** and $B$ takes ***Both***. So just don't try to base logical truth on these values.

Thirdly, let us consider ontology vs. epistemology. One of the difficulties which often arises in relating many-valued logics to real concerns is that one tends to vacillate between reading the various values as epistemic on the one hand, or ontological on the other. Does Łukasiewicz' middle value, 1/2, mean "doesn't have a proper truth-value" or does it mean "truth-value unknown"? In informal explanations of what is going on, logicians sometimes move from one of these readings to the other in order to save the interest of the enterprise.

My four values are unabashedly epistemic. According to my instructions, sentences are to be marked with either a ***T*** or an ***F*** a ***None*** or a ***Both***, according as to what the computer has been told; or, with only a slight metaphor, according to what it believes or knows.[2] Does this somehow make the enterprise wrong headed? Or not logic? No. Of course these sentences *have* truth-values independently of what the computer has been told; but who can gainsay that the computer cannot *use* the actual truth-value of the sentences in which it is interested? All it can possibly *use* as a basis for inference is what it knows or believes, i.e., what it has been told.

But we can do better than this. Let us get the ontology into the act by *splitting* our four epistemological values into two, one representing the case in which the sentence is ontologically true, the other the case in which it is false. Obviously we then get eight values instead of four, each of which we may visualize as an ordered pair, the left entry of which is an epistemic value ***T***, ***F***, ***None***, or ***Both***, while the right entry is one of Frege's ontological values the True and the False. Giving the usual classical two-valued tables to the connectives, and also and equivalently, interpreting the implicative connective in the usual way, we are led to the following lattice picture (this is *not* an approximation lattice).

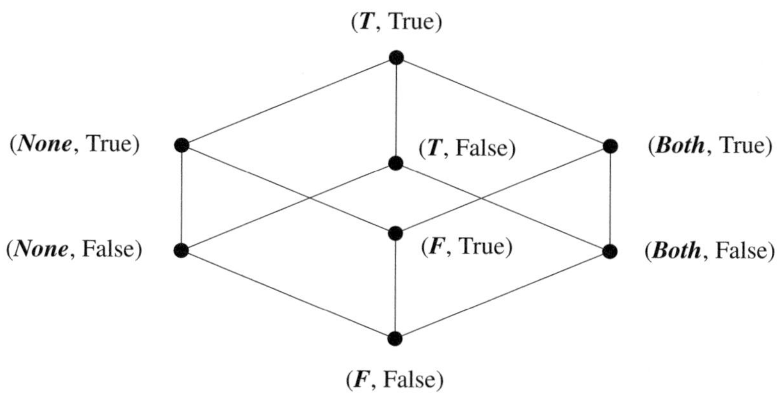


The &'s and ∨'s can be computed respectively as g.l.b.s and l.u.b.s, while negation-pairs are: two left, two center, two right, and top-bottom (*not* the boolean way). The values of this new many-valued logic have a mixed status: they are in part epistemological and in part ontological. Should we then move to this logic? It is entertaining to observe that there is no need to do so for inferences; for *exactly* the same inferences are valid with this as with our four-valued canon of inference. Nor for two reasons should this be surprising.

In the first place, as we already observed, the only thing we can actually *use* in inference are the epistemic values ***T, F, None*** and ***Both***, representing what we know, believe, or in any event have been told by authority we by and large trust. Secondly, and more prosaically, observe that all the inferences sanctioned by the four-valued canon are already approved in two-valued logic; so that adding as a condition that ontological truth is to be preserved is to add a condition that is already satisfied and yields no new constraints. So for practical reasons there is no need to move from four to eight values for judging inferences. In the words of a famous philosopher, "Do not multiply many values beyond necessity."

[2] Editors' note: On p. 521 of Alan Ross Anderson, Nuel D. Belnap, and J. Michael Dunn (1992), *Entailment: The Logic of Relevance and Necessity, Vol. II*, Princeton University Press, we find the modification "or with only a slight (but dangerous) metaphor, according to what it believes or knows."

If, however, for some reason (I do not just now know what) someone wanted an account of logical truth in &, $\vee$, and $\sim$, then one could invoke as a criterion: being always True (by the right entry of the pair) regardless of what you've been told (according to the left entry). Then, not surprisingly, one finds out that the two-valued tautologies are precisely the logical truths on this account. Not surprisingly because we invoke values only ontological, throwing away (in the eight-valued case) all the information of the epistemic values.

Let me say explicitly, if it is not obvious, that I think this codification of truth-functional logical truths not very important to the computer; for what was *wanted* was a way of *reasoning* from and to truth-functional compounds, not a sorting of these compounds.

My fourth "observation" is not so much that as it is an inconclusive discussion of the role of ***Both*** and ***None***. The problem is that one is inclined to the view that they should be identified, that the computer is in the same state having been told that $A$ is both True and False as it is having been told nothing about $A$. A. Kenny and S. Haack have each, in quite different ways, suggested something like this. If you will be satisfied with a dialectical flourish I can supply one of the form "wrong, but understandable." It goes like this. In the first place, it is somehow magnificently obvious that ***Both*** and ***None*** should not be identified, as H. S. Harris noted in conversation, just because we want the computer to distinguish *for* us when it has been told a contradiction from when it has been told nothing. This is surely essential on anyone's view. In the second, our developments can be taken as explaining the feeling that they should be identified, for just look at the logical lattice **L4**: there ***Both*** and ***None*** occupy (distinct but) absolutely symmetrical positions between ***F*** and ***T***, and in this sense are "identified." For instance, we allow the inference from neither to ***F***, nor from ***T*** to either, and thus treat them alike.

Still, though this response may be helpful, I am not altogether happy with it. And I much prefer to leave the discussion as at this stage incomplete.

My penultimate observation concerns the suggestion that the computer keep more information than I have allowed it to keep. Perhaps it should count the number of times it has been told True or told False, or perhaps it should keep track of its sources by always marking, for example, "told True by Sam at 22:03 on 4 August 1973." I do not see why these ideas should not be explored, but two comments are in order. The first is that it is by no means self-evident how this extra information is to be utilized in answering questions, in inference, and in the input of complex sentences. That is, do not be misled by the transparency of the idea in the case of atomic sentences. The consequence of the first comment is merely that the exploration lies ahead. The second is the practical remark that there are severe costs in carrying extra information, costs which may or may not be worth incurring. And if there are circumstances in which they are not worth incurring, we are back to the situation I originally described.

Lastly, I want to mention some alternatives without (much) discussing them. A. Gupta has noted that one could define the value of $A$ in $s$ not directly as we have done, but rather by reference to all the consistent sub-set-ups of $s$. Definitions: $s'$ is a sub-set-up of $s$ if it approximates it: $s' \sqsubseteq s$. $s'$ is consistent if it never awards ***Both***. Finally, let $s(A)$ be defined, à la Gupta, by $s(A) = \{s'(A) : s'$ is a consistent sub-set-up of $s\}$, where $s'(A)$ is as already defined. The idea is clearly dual to van Fraassen's definition of true-in-a-valuation by reference to all the complete (i.e. all truth value gaps filled) supervaluations of a given valuation. One note that if $s(p) =$ ***Both***, then the question as to $p$ (on $s$) will be answered "Yes and No" as before, while the question as to $p\&{\sim}p$ will be answered just "No," instead of "Yes and No."

This idea could be used, incidentally, to improve on Rescher's suggestion in *Hypothetical Reasoning* (North-Holland, 1964). There he suggests reasoning from an inconcsistent set of premisses by looking at all the consistent subsets. The difficulty with that idea, in my judgment is that it is too sensitive to the way the set of premisses is broken down into separate sentences. Gupta's notion could be used, I think to present an idea having the initial flavor of Rescher's without its bitter aftertaste.

A related idea is to follow van Fraassen directly by looking at all the complete super-set-ups of a given set-up; this would give always "Yes" to $p \vee \sim p$. And carrying this idea to its logical conclusion would combine the two ideas (carefully).

All of these things are possible. One would hope, however, that the discussion of the alternatives would circulate around the question, How in fact do we want the computer to answer our questions? In this way they would not be *mere* possibilities.

*Quantifiers.* Quantifiers introduce a number of subleties to which I shall merely tip my hat, while recognizing that treating them in detail is quite essential to my enterprise.

There is in the first place the question of whether "the" domain is finite or infinite. Both cases can plausibly arise. In the latter case, there is the question of how the computer is to represent infinite information with its finite resources, but one should not infer from the existence of this problem that the computer can't or shouldn't involve itself with quantification over infinite domains. Surely it should be allowed to answer "is there a number such that . . . ?" queries (if it can).

In the second place, there is the question of whether the computer has a name for everything in "the" domain so that we can employ the substitutional interpretation of the quantifiers, or on the other hand does not have a name for each entity in "the" domain so that the domain-and-values interpretation is forced. Again: both cases can plausibly arise, though attending to standard examples like baseball queries or airline flights might have made one think that in the computer situation everything always has a name. But, for example, in some of Isner's work the computer is told "there is something between $a$ and $b$" in a context in which it hasn't got a complete list of either the names or the entities against which to interpret this statement. And still it must work out the consequences, and answer the questions it is given. (Of course it is O.K. for the computer to make up its own name for the "something" between $a$ and $b$; but that is both an important and an entirely different matter.)

In any event, the semantics given for the connectives extend to universal and existential quantifiers in an obvious way, and I suppose the job done. And the various alternatives mentioned above turn out not to make any difference to the *logic* (with the obvious exception of the finite everything-has-a-name case): the valid "first degree entailments" of Anderson and Belnap (1965) do admirably (supplemented, in the finite case, with the principle that a conjunction which runs through the domain implies the appropriate universal statement).

*Part 2. Compound truth and functional inputs.*

My aim in this part and the next is barely to give the flavor of the developments I suggest. They are based much more heavily than are those of Part 1 on approximation lattices, and in general tend to be more technical. But I think something can be said to give you, as I say, the flavor. Details are to be found in Belnap (1976).

*Epistemic States.* If we allow the computer to receive as inputs not only atomic formulas but also complexes such as $p \vee q$, a single set-up will no longer suffice to represent its epistemic state. A well-known solution to this kind of problem, going back at least to Carnap (1942), exploited by Hintikka (1962) in epistemic and doxastic logic, and worked out for computers in Isner (1972, 1975), is to employ instead a *collection* of set-ups as the representation of the epistemic state of the computer. Let it be done. Further, it is possible to use approximation ideas to motivate and define how we want the computer to answer our questions about $A$ in each such state; which is to say, where $E$ is such a state, we can compute $E(A)$, the value of $A$ in $E$, as one of the four values. Let this also be done. (Skippable definition: $E(A) = \sqcap\{s(A) : s \in E\}$.) I note two special epistemic states for use further on: $Tset(A)$ and $Fset(A)$ are defined in such a way that they represent what the computer has been told when precisely $A$ has been, respectively, affirmed or denied. (Skippable: $Tset(A) = \{s : \boldsymbol{T} \sqsubseteq s(A)\}$, and $Fset(A) = \{s : \boldsymbol{F} \sqsubseteq s(A)\}$.) *More Approximation*

*Lattices.* One of the principal lessons to be drawn from Scott's work is that wherever there is one approximation lattice there are lots of them. In particular, the family of all set-ups forms a natural approximation lattice, **AS**; and then the family of all epistemic states constitutes (or almost constitutes; subtleties omitted) still another approximation lattice **AE**. (Skippable: in **AS**, $s \sqsubseteq s'$ iff $s(p) \sqsubseteq s'(p)$ for all $p$; in **ES**, $E \sqsubseteq E'$ iff for every $s' \in E'$ there is an $s \in E$ such that $s \sqsubseteq s'$.)

*Formulas as mappings; a new kind of meaning.* Now I turn to a question of considerable interest and a question on which our various approximation lattices can shed considerable light: How is the computer to interpret a truth-functional formula, $A$, as input? Clearly it is going to use $A$ to modify its present epistemic state; and indeed it is not too much to say that defining how the computer uses the formula $A$ to transform its present epistemic state into a new epistemic state is a way, and a *good* way, of giving $A$ a meaning. Consequently we want to associate with each formula $A$ a transformation, a mapping from epistemic states into new epistemic states. Furthermore, we also want to know what the computer is to do when the formula $A$ is *denied* to the computer; so actually we associate with a formula $A$ *two* functions, one representing the transformation of epistemic state when $A$ is affirmed, the other the transformation when $A$ is denied. Let us call these two functions $A^+$ and $A^-$. How to define them?

Recall that $A^+$ is to map states into states: $A^+(E) = E'$. The key ideas in defining what we want $E'$ to be come from the approximation lattice. First, in our context we are assuming that the computer uses its input always to increase its information, or at least it never uses input to throw information away. (That would just be a different enterprise; it would be nice to know how to handle it in a theory, but I don't.) And we can say this accurately in the language of approximation: $E \sqsubseteq A^+(E)$. Second, $A^+(E)$ should certainly say no less than the affirmation of $A$: $Tset(A) \sqsubseteq A^+(E)$. Third and lastly, we clearly want $A^+(E)$ to be the *minimum mutilation* of $E$ that renders $A$ at least True. "Minimum mutilation" is Quine's fine phrase, but in the approximation lattice we can give a sense that is *no longer merely metaphorical*: namely, we want the *least* of those epistemic states satisfying our first two conditions. That is, we should define

$$A^+(E) = E \sqcup Tset(A),$$

for that is precisely the minimum mutilation of $E$ which makes $A$ at least True. (Recall that in any lattice, $x \sqcup y$ is the "least (minimum) upper bound".) Having agreed on this as the definition of $A^+$, it is easy to see that $A^-(E)$ should be the minimum mutilation of $E$ which makes $A$ at least False:

$$A^{-}(E) = E \sqcup Fset(A)$$

The above definitions accurately represent the meaning of $A$ as input, but they do involve a drawback: the $Tset$s and $Fset$s may be infinite, or at least large, and so do not represent something the computer can really work with. For this reason, and also for its intrinsic interest, another explication of $A^+$ and $A^-$ is offered in Belnap (1976), this time inductive, but still very much involving the idea of minimum mutilation. Here we omit it.

*Part* 3. *Implicational inputs and rules.*
In Part 1 we pretended that all information fed into the computer was atomic, so we could get along with set-ups. In Part 2 we generalized to allow information in the form of more complex truth-functional formulas, a generalization which required moving to epistemic states. Now we must recognize that it is *practically* important that sometimes we give information to the computer in the form of rules which allow it to modify its own representation of its epistemic state in directions we want. In other words, we want to be able to instruct the computer to make inferential moves which are not mere tautological entailments. For example, instead of physically handing the computer the whole list of Series winners and non-winners for 1971, it is obviously cheaper to tell the computer: "the Pirates won; and further, if you've got a winner and a team not identical to it, that team must be a non-winner" (i.e., $\forall x \forall y(Wx \& x \neq y \rightarrow {\sim}Wy)$. In the presence of an obviously needed table for identity and distinctness, or else in the presence of a convention that different names denote different entities (not a bad convention for *practical* use in many a computer setting), one could then *infer* that "The Orioles won" is to be marked false.

Your first thought might be that you could get the effect of "given $A$ and $B$, infer $C$," or "if $A$ and $B$, then $C$," by feeding the computer "${\sim}A \vee {\sim}B \vee C$." But that won't work: the latter formula will tend to *split* the set-up you've got into three, in one of which $A$ is marked False, etc.; while what is wanted is (roughly) just to improve the single set-up you've got by adding True to $C$ provided $A$ and $B$ are marked True (and otherwise to leave things alone). It is (roughly) this idea we want to catch.

*Implicational inputs.* Let us introduce "$A \rightarrow B$" as representing the implication of $A$ to $B$; so what we have is notation in search of a meaning. But we have in the previous section found just the right way of giving meaning to an expression construed as an input: the computer is to improve its epistemic state in the minimum possible way so as to make the expression True. So let us look forward to treating $(A \rightarrow B)^+$ as signifying some mapping from states into states such that $A \rightarrow B$ is true in the resultant state.

Without going into details (to be found in Belnap, 1976), the procedure is this. First we atomize the problem by concentrating on a given set-up $s$. Then we divide the problem by recalling that implication has two parts: $B$ must be at least $\boldsymbol{T}$ if $A$ is, and $A$ must be at least $\boldsymbol{F}$ if $B$ is. So we define two functions, $(A \rightarrow_T B)^+$ and $(A \rightarrow_F B)^+$ the first *making* $B$ True if $A$ is, the second *making* $A$ True if $B$ is – and in each case by minimum mutilation. Lastly, we put these functions together, in a certain way (omitted here), to come up with $(A \rightarrow B)^+$ as a function.

Just a bit more on the part $(A \rightarrow B)^+$ which is supposed to minimally mutilate to make $B$ True if $A$ is. Clearly if $A$ is not True in $s$, we know what to do: nothing. No mutilation is minimum mutilation. Suppose now that $A$ is True in $s$: then what is wanted is the minimum mutilation which makes $B$ True – namely, the already defined function, $B^+$. So putting these clauses together we evidently have a proper definition for $(A \rightarrow_T B)^+$ as the minimum mutilation which makes $B$ True if $A$ is.

*Rules and information states.* This last section of this paper is going to be altogether tentative, and altogether abstract, with just one concrete thought that needs remembering, which I learned from Isner: probably the best way to handle sophisticated information states in a computer is by a judicious *combination* of tables (like our epistemic states) and rules (like our $A \rightarrow B$, or a truth-functional formula the computer prefers to remember, or a quantificational formula which it must remember). For this reason, as well as for the quite different reason that some rules, since they may have to be used again (are not permanent, must be remembered), we can no longer be satisfied by representing what the computer knows by means of an epistemic state. Rather, it must be represented by a *pair* consisting of an epistemic state *and* a set of rules:

$$\langle R, E\rangle$$

$E$ is supposed to represent what the computer explicitly knows, and is subject to increase by application of the rules in the set $R$. For many purposes we should suppose that $E$ is finite, but for some not.

Let us dub such a pair an *information state*, just so we don't have to retract our previous definition of "epistemic state." But what is a rule? Of what is $R$ a set? A good thing to mean by *rule*, or *ampliative* rule in this context might be: *any* continuous and ampliative mapping from epistemic states into epistemic states. As I mentioned above the set of *all* continuous functions from an approximation lattice into itself has been studied by Scott; it forms itself a natural approximation lattice. It is furthermore easy to see that the ampliative continuous functions form a natural approximation lattice, and one which is an almost complete sublattice of the space of all the continuous functions: all meets and joins agree, except that the join of the empty set is the identity function $I$ instead of the totally undefined function. Intuitively: the effect of an empty set of rules is to leave the epistemic state the way it was. The notion of an information state is considered a bit in Belnap (1976), but the explorations there are so tentative that we omit them here.

*Closure.* Lest it have been lost, let me restate the principal aim of this paper: to propose the *usefulness* of the scheme of tautological entailments as a guide to inference in a certain setting; namely, that of a reasoning, question-answering computer threatened with contradictory information. The reader is not to suppose that Larger Applications have not occurred to me; e.g., of some of the ideas to a logic of imperatives, or to doxastic logic, or even to the development of The One True Logic. But because of my fundamental conviction that logic is (all in all) practical, I did not want these possibilities to loom so large as to shut out the light required for dispassionate consideration of my far more modest proposal.

*Professor Nuel D. Belnap*
*University of Pittsburgh*

## References


1. Anderson, A. R. and Belnap, N. D. Jr. (1962). 'Tautological Entailments', *Philosophical Studies*, 13(1), 9-24. (See also Ch. III of Anderson and Belnap (1975).)
2. Anderson, A. R. and Belnap, N. D. Jr. (1965). 'First-Degree Entailments', *Mathematische Annalen*, 149(4), 302-319.
3. Anderson, A. R. and Belnap, N. D. Jr. (1975). *Entailment: The Logic of Relevance and Necessity, vol. I.* Princeton: Princeton University Press.
4. Belnap, N. D. Jr. (1963). *An Analysis of Questions: Preliminary Report.* Santa Monica: System Development Corporation.
5. Belnap, N. D. Jr. (1976). 'A Useful Four-Valued Logic'. *Modern Uses of Multiple-Valued Logic – Proceedings of the 1975 International Symposium on Multiple-Valued Logic*, G. Epstein and J. M. Dunn (eds.). Dordrecht: Reidel.
6. Belnap, N. D. Jr. and Steel, T. B. Jr. (1976). *Erotetic Logic: An Introduction to the Logic of Questions and Answers*. New Haven: Yale University Press.
7. Carnap, R. (1942). *Introduction to Semantics*. Cambridge: Harvard University Press.
8. Dunn, J. M. (1966). *The Algebra of Intensional Logics*, dissertation, University of Pittsburgh.
9. Dunn, J. M. (1976). 'Intuitive Semantics for First-Degree Entailments and "Coupled Trees"', *Philosophical Studies*, 29(3), 149-168.
10. Dunn, J. M. and Belnap, N. D. Jr. (1968). 'The Substitution Interpretation of the Quantifiers', *Noûs*, 2(2), 177-185.
11. Hintikka, J. (1962). *Knowledge and Belief.* Ithaca: Cornell University Press.
12. Isner, D. W. (1975). 'An Inferential Processor for Interacting with Biomedical Data Using Restricted Natural Language', *Proceedings of Spring Joint Computer Conference*, 1107-1124.
13. Isner, D. W. (1975). *'Understanding "Understanding" Through Representation and Reasoning'*, dissertation, University of Pittsburgh.
14. Kripke, S. (1963). 'Semantical Analysis of Modal Logic I', *Zeitschrift für mathematische Logik und Grundlagen der Mathematik*, 9(5-6), 67-96.
15. Routley, R. and Routley, V. (1972). 'Semantics of First-Degree Entailment', *Noûs*, 6(4), 335-359.
16. Scott, D. (1970). 'Outline of a Mathematical Theory of Computation', *Proceedings of the Fourth Annual Princeton Conference on Information Sciences and Systems*, 169-176.
17. Scott, D. (1972). 'Continuous Lattices: *Toposes, Algebraic Geometry and Logic*', *Springer Lecture Notes in Mathematics*, 274, 97-136.
18. Scott, D. (1973). 'Models for Various Type-Free Calculi', *Logic, Methodology, and Philosophy of Science IV*. Proceedings of the Fourth International Congress for Logic, Methodology and the Philosophy of Science, Bucharest, 1971. Suppes, Henkin, Juja, Moisil (eds.), Elsevier: North-Holland.
19. Shapiro, S. and Wand, M. (1976). 'The Relevance of Relevance', Technical Report, Indiana University, Computer Science Department.
20. van Fraassen, B. (1969a). 'Presuppositions, Supervaluations, and Free Logic', *The Logical Way of Doing Things*, K.Lambert (ed.), New Haven: Yale University Press.
21. van Fraassen, B. (1969b). 'Facts and Tautological Entailments', *The Journal of Philosophy*, 66(15), 477-487.

# A Useful Four-Valued Logic

Nuel D. Belnap

**Abstract** It is argued that a sophisticated question-answering machine that has the capability of making inferences from its data base should employ a certain four-valued logic, the motivating consideration being that minor inconsistencies in its data should not be allowed to lead (as in classical logic) to irrelevant conclusions. The actual form of the four-valued logic is 'deduced' from an interplay of this motivating consideration with certain ideas of Dana Scott concerning 'approximation lattices'.

## I. INTRODUCTION

A lot of work has been done recently on applying many-valued logics to the design of computer circuitry and thus giving them application; so what, you may ask, is special about offering a four-valued logic as 'useful'? In fact I think I am indeed involved in an odd sort of enterprise, for in the present context I want to use 'logic' in a narrow sense, the old sense: 'logic' in the sense of an *organon*, a tool, a canon of inference. And it is my impression that hardly any of what individual many-valued logicians have done is directly concerned with developing logics to use as practical tools for inference. Hence the peculiarity of my task, which is to suggest that a certain four-valued logic ought to be used in certain circumstances as an actual guide to reasoning.

## II. THE COMPUTER

The situation I have in mind may be described as follows. In the first place, the reasoner who is to use this logic is an artificial information processor; that is, a (programmed) computer. In the second place, the computer is to be some kind of sophisticated question-answering system, where by 'sophisticated' I mean that it does not confine itself, in answering questions, to just the data it has explicitly in its memory, but can also answer questions on the basis of *deductions* which it makes from its explicit information. Thirdly, the computer is to be envisioned as obtaining the data on which it is to base its inferences from a variety of sources, all of which indeed may be supposed to be on the whole trustworthy, but none of which can be assumed to be that paragon of paragons, a universal truth-teller. In such circumstances the crucial feature of the situation emerges: inconsistency threatens. Elizabeth tells the computer that the Pirates won the Series in 1971, while Sam tells it otherwise. What is the computer to do? If it is a classical two-valued logician, it must give up altogether talking about anything to anybody or, equivalently, it must say everything to everybody. We all know all about the fecundity of contradictions in two-valued logic: contradictions are never isolated, infecting as they do the whole system. Of course the computer could refuse to entertain inconsistent information. But in the first place that is unfair either to Elizabeth or to Sam, each of whose credentials are, by hypothesis, nearly

Nuel D. Belnap
Department of Philosophy, University of Pittsburgh, Pittsburgh, USA
e-mail: belnap@pitt.edu

 
H. Omori, H. Wansing (eds.), *New Essays on Belnap-Dunn Logic*,
Synthese Library 418, https://doi.org/10.1007/978-3-030-31136-0_5

impeccable. And in the second place, as we know all too well contradictions may not lie on the surface.

There is a fourth aspect of the situation, concerning the significance of which I remain uncertain, but which nevertheless needs mentioning for a just appreciation of developments below: my computer is *not* a complete reasoner, who should be able to do something better in the face of contradiction than just report. The complete reasoner should, presumably, have some strategy for *giving up* part of what it believes when it finds its beliefs inconsistent. Since I have never heard of a practical, reasonable, mechanizable strategy for revision of belief in the presence of contradiction, I can hardly be faulted for not providing my computer with such. Furthermore, it seems to me plausible that a *part* of each complete reasoner should be a capability such as I describe. So in the meantime, while others work on this extremely important problem, my computer can only accept and report information without divesting itself of it. This aspect is bound up with a fifth: in answering its questions, the computer is to reply strictly in terms of what it has been told, *not* in terms of what it could be programmed to believe. For example, if it has been told that the Pirates won and did not win in 1971, it is so to report, even though we could of course program the computer to recognize the falsity of such a report. The point here is both subtle and obvious: if the computer would not report out contradictions in answer to our questions, we would have no way of knowing that its data-base harbored contradictory information. (We could, if we wished, ask it to give a *supplementary* report, e.g., as follows: "I've been told that the Pirates won and did not win; but of course it ain't so"; but would that be useful?)

## III. APPROXIMATION LATTICES

Always in the background and sometimes in the foreground of what I shall be working out is the notion of an *approximation lattice*, due in all but terminology to Scott (see e.g., Scott, 1970, 1972, 1973). Let me say a word about this concept before getting on. You are going to be disappointed at the mathematical definition of an approximation lattice: mathematically it is just a complete lattice. That is, we have a set $A$ on which there is a partial ordering $\sqsubseteq$, and for arbitrary subsets $X$ of $A$ there always exist least upper bounds $\sqcup X \in A$ and greatest lower bounds $\sqcap X \in A$ (finite ones written $x \sqcup y$ and $x \sqcap y$). But I don't call a complete lattice an approximation lattice unless it satisfies a further, non-mathematical condition: it is appropriate to read $x \sqsubseteq y$ as 'x approximates y'. Examples worked out by Scott include the lattice of 'approximate and overdetermined real numbers', where we identify an approximate real number with an interval and where $x \sqsubseteq y$ just in case $y \subseteq x$. The (only) overdetermined real number is the empty set. As a further example Scott offers the lattice of 'approximate and overdetermined functions' from $A$ to $B$, identified as subsets of $A \times B$. Here we want $f \sqsubseteq g$ just in case $f \subseteq g$.

In such lattices the directed sets are important: those sets such that every pair of members $x$ and $y$ of the set have an upper bound $z$ also in the set. For such a set can be thought of as approximating, by a limiting process, its union $\sqcup X$. That is, if $X$ is directed, it makes sense to think of $\sqcup X$ as the limit of $X$. (An ascending sequence $x_1 \sqsubseteq \cdots \sqsubseteq x_i \sqsubseteq \ldots$ is a special case of a directed set.) And now when we pass to the family of functions from one approximation lattice into another (or of course the same) approximation lattice, Scott has demonstrated that what are important are the *continuous* functions: those that preserve non-trivial directed unions (i.e. $f(\sqcup X) = \sqcup\{fx : x \in X\}$, for nonempty directed $X$). These are the only functions which respect the lattices qua *approximation* lattices. This idea is so fundamental to developments below that I choose to catch it in a Thesis to be thought of an analogous to Church's Thesis:

**Scott's Thesis:** In the presence of complete lattices $A$ and $B$ naturally thought of as approximation lattices, pay attention only to the continuous functions from $A$ into $B$, resolutely ignoring all other functions as violating the nature of $A$ and $B$ as approximation lattices.

(Though honesty compels me to attribute the Thesis to Scott, the same policy bids me note that the formulation is mine, and that as stated he may not want it, or may think that some other Thesis in the neighborhood is more important; for example, that every approximation lattice (intuitive sense) is a continuous lattice (sense of Scott, 1972).)

You will see how I rely on Scott's thesis in what follows.

## IV. PROGRAM

The rest of this paper is divided into three parts. Section 2[1] considers the case in which the computer accepts only *atomic* information. This is a heavy limitation, but provides a relatively simple context in which to develop some of the key ideas. (In this paper this part is abbreviated; see Belnap (1976) for a filling out.) Section 3 allows the computer to accept also information conveyed by *truth-functionally* compounded sentences; and in this context I offer a new kind of meaning for formulas as certain mappings from epistemic states into epistemic states. In section 4 the computer is allowed also to accept *implications construed as rules* for improving its data base.

## *Part I. Atomic Inputs*

### I. ATOMIC SENTENCES AND THE APPROXIMATION LATTICE **A4**

Now and throughout this paper you must keep firmly fixed in mind the circumstances in which the computer finds itself, and especially that it must be prepared to receive and reason about inconsistent information I want to suggest a natural technique for employment in such cases: when an item comes in as asserted, mark it with a 'told True' sign, and when an item comes in denied, mark with a 'told False' sign, treating these two kinds of tellings as altogether on a par. It is easy to see that this leads to four possibilities. For each item in its basic data file, the computer is going to have it marked in one of the following four ways: (1) just the 'told True' sign, indicating that that item has been asserted to the computer without ever having been denied (2) just the value 'told False', which indicates that the item has been denied but never asserted. (3) no 'told' values at all, which means the computer is in ignorance, has been told nothing. (4) the interesting case: the item might be marked with both 'told True' and 'told False'.

These four possibilities are precisely the four values of the many-valued logic I am offering as a practical guide to reasoning by the computer. Let us give them names:

***T:*** just told True
***F:*** just told False
***None:*** told neither True nor False
***Both:*** told both True and False

[1] Editor's note: references to parts of the paper had to be adjusted for this reprint.

So these are our four values, and we baptize: **4** = {***T, F, None, Both***}. Of course four values do not a logic make, but let us nevertheless pause a minute to see what we have so far.

The suggestion requires that a system using this logic code each of the atomic statements representing its data base in some manner indicating which of the four values it has (at the present stage). Two procedures suggest themselves. The first is to list each item with one of the values ***T***,***F***, or ***Both***, for these are the elements about which the computer has been told something; and to let an absence of a listing signify ***None***, i.e., that there is no information about that element. The second procedure would be to list each element with one or both of the 'told' values, 'told True' and 'told False', not listing elements lacking both 'told' values. Obviously the procedures are equivalent, and we shall not in our discourse distinguish between them.

The same procedure works for relations, except that it is ordered pairs that get marked. For example, apart of the correct table for Series winners, conceived as a relation between teams and years, might look like this:

⟨Pirates, 1971⟩ ***T*** and ⟨Orioles, 1971⟩ ***F***

or

⟨Pirates, 1971⟩ True and ⟨Orioles, 1971⟩ False

But if Sam slipped up and gave the wrong information after Elizabeth had previously entered the above, the first entry would become

⟨Pirates,1971⟩ ***Both***

or

⟨Pirates, 1971⟩ True, False

To be specific, we envision (in this part of the paper) the epistemic state of the computer to be maintained in terms of a table giving one of four values to each atomic sentence. We call such a table an *(epistemic) set-up* (following an isomorphic use of Routley and Routley (1972)); i.e., a set-up is, mathematically, a mapping from atomic sentences into the set **4** = {***T, F, None, Both***}. When an atomic formula is entered into the computer as either affirmed or denied, the computer modifies its current set-up by adding a 'told True' or 'told False' according as the formula was affirmed or denied; it does not subtract any information it already has, for that is the whole point of what we are up to. In other words, if $p$ is affirmed, it marks $p$ with ***T*** if $p$ were previously marked with ***None***, with ***Both*** if $p$ were previously marked with ***False***; and of course leaves things alone if $p$ was already marked either ***T*** or ***Both***. So much for $p$ as input.

The computer not only accepts input, but answers questions. We consider only the basic question as to $p$; this it answers in one of four ways: Yes, No, Yes and No, or I don't know, depending on the value of $p$ in its current set-up as ***T, F, Both***, or ***None***.

I now make the observation which constitutes the foundation of what follows: these four values naturally form a lattice under the lattice ordering 'approximates the information in', and indeed an *approximation-lattice* in the sense described above:

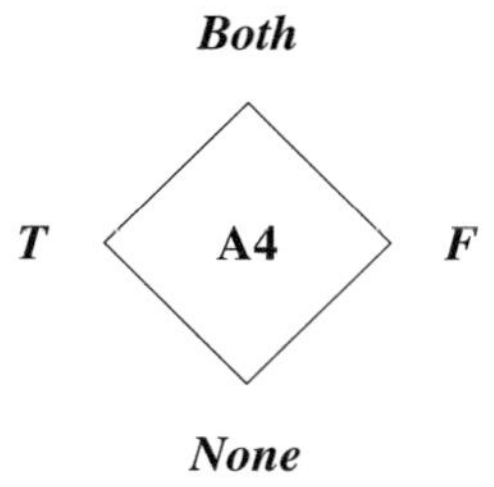


(In this Hasse diagram joins (⊔) and meets (⊓) are least upper bounds and greatest lower bounds respectively, and ⊑ goes uphill.) I remarked above that according to Scott's Thesis the important functions in the neighborhood of an approximation lattice like **A4** are the continuous ones. We do not, fortunately, have to deal with continuity for a while, since in the finite case it turns out that for a function to be continuous is just for it to be *monotonic*, i.e., for it to preserve the lattice ordering: $a \sqsubseteq b$ implies $fa \sqsubseteq fb$.

### II. COMPOUND SENTENCES AND THE LOGICAL LATTICE L4

The pay-off from the approximation idea is immediate: negation, if it is to be classical on ***T*** and ***F***, is uniquely determined to have the table[2]

| | ***None*** | ***F*** | ***T*** | ***Both*** |
|---|---|---|---|---|
| ~ | ***None*** | ***T*** | ***F*** | ***Both*** |

by the condition that it be monotonic! And although conjunction and disjunction are not uniquely determined by classical considerations alone, they must be as given below if in addition we postulate that they yield the same ordering in the standard way: $a \& b = a$ iff $a \vee b = b$, and $a \& b = b$ iff $a \vee b = a$ (in these tables it is convenient to use ***B*** for ***Both*** and ***N*** for ***None***):

| & | ***N*** | ***F*** | ***T*** | ***B*** |
|---|---|---|---|---|
| ***N*** | ***N*** | ***F*** | ***N*** | ***F*** |
| ***F*** | ***F*** | ***F*** | ***F*** | ***F*** |
| ***T*** | ***N*** | ***F*** | ***T*** | ***B*** |
| ***B*** | ***F*** | ***F*** | ***B*** | ***B*** |

| ∨ | ***N*** | ***F*** | ***T*** | ***B*** |
|---|---|---|---|---|
| ***N*** | ***N*** | ***N*** | ***T*** | ***T*** |
| ***F*** | ***N*** | ***F*** | ***T*** | ***B*** |
| ***T*** | ***T*** | ***T*** | ***T*** | ***T*** |
| ***B*** | ***T*** | ***B*** | ***T*** | ***B*** |

These tables constitute a lattice, with conjunction as meet and disjunction as join, which can be pictured as follows:

[2] Editors' note: The original publication contains a misprint presenting ***Both*** and ***None*** as the negation of each other.

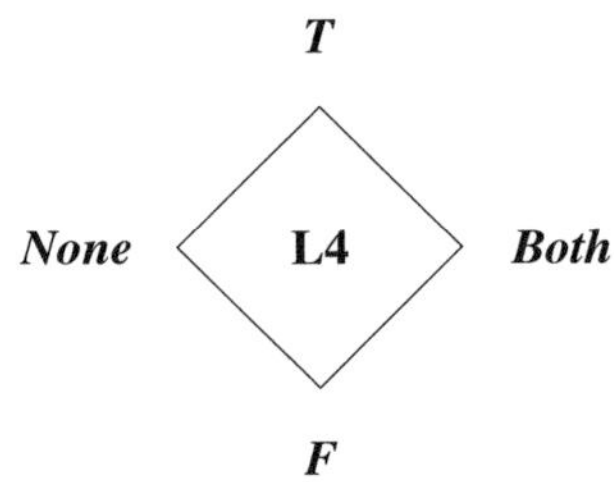


Let us agree to call this the *logical* lattice, **L4**, to distinguish it from the *approximation* lattice, **A4**. The ordering on **L4** we write as $a \leq b$; we write meets as $a\&b$, and joins as $a \vee b$.

We can now use these logical operations on **L4** to induce a semantics for a language involving &, ∨, and ∼, in just the usual way. Given an arbitrary set-up $s$ – a mapping, you will recall, of atomic formulas into **4** – we can extend $s$ to a mapping of *all* formulas into **4** in the standard inductive way:

$$\begin{aligned} s(A\&B) &= s(A)\&s(B) \\ s(A \vee B) &= s(A) \vee s(B) \\ s(\sim A) &= \sim s(A) \end{aligned}$$

And *this* tells us how the computer should answer questions about complex formulas based on its epistemic set-up (what it has been told): just as for answering questions about atomic formulas, it should answer a question as to $A$ by Yes, No, Yes and No, or I don't know, according as the value of $A$ in $s$ (i.e., $s(A)$) is ***T***, ***F***, ***Both***, or ***None***.

### III. ENTAILMENT AND INFERENCE: THE FOUR-VALUED LOGIC

Where are we? Well, we still haven't got a logic, i.e., rules for generating and evaluating inferences. (In our case we really want the former; we want some rules for the computer to use in generating what it implicitly knows from what it explicitly knows.) What we do have is four values and three connectives. Suppose we have an argument involving these connectives. The question is, when is it a good one? I rely on the *logical* lattice **L4**: entailment goes up hill. That is, given any sentences $A$ and $B$ (compounded from variables by negation, conjunction, and disjunction), we will say that $A$ *entails* or *implies* $B$ just in case for each assignment of one of the four values to variables, the value of $A$ does not exceed (in **L4**) the value of $B$. In symbols: $s(a) \leq s(b)$ for each set-up $s$.

Finally we have a logic, that is a canon of inference for our computers to use in making inferences involving conjunction, negation, and disjunction, as well of course as whatever can be defined in terms thereof. Let me briefly present a group of principles which are semantically valid, and taken together semantically complete. They will also be redundant, but recall that the byword for this exercise is usefulness; I am offering a set of principles for the computer to use in making its inferences.

Let $A$, $B$, etc., be formulas in &, ∨, and ∼. Let $A \rightarrow B$ signify that the inference from $A$ to $B$ is valid in our four values, i.e., that $A$ entails $B$. Also let $A \rightleftarrows B$ signify that $A$ and $B$ are semantically equivalent, and can be intersubstituted in any context. Then the following have proved to be a useful (complete) set of principles.

$A_1\&\ldots\&A_m \rightarrow B_1 \vee \ldots \vee B_n$
provided some $A_i$ is some $B_j$ (sharing)

$(A \vee B) \to C$ iff (if and only if) $A \to C$ and $B \to C$
$A \to B\&C$ iff $A \to B$ and $A \to C$
$A \to B$ iff $\sim B \to \sim A$

$A \vee B \leftrightarrows B \vee A$ — $A\&B \leftrightarrows B\&A$
$A \vee (B \vee C) \leftrightarrows (A \vee B) \vee C$ — $(A\&B)\&C \leftrightarrows A\&(B\&C)$
$A\&(B \vee C) \leftrightarrows (A\&B) \vee (A\&C)$ — $A \vee (B\&C) \leftrightarrows (A \vee B)\&(A \vee C)$
$(B \vee C)\&A \leftrightarrows (B\&A) \vee (C\&A)$ — $(B\&C) \vee A \leftrightarrows (B \vee A)\&(C \vee A)$
$\sim\sim A \leftrightarrows A$
$\sim(A\&B) \leftrightarrows \sim A \vee \sim B$ — $\sim(A \vee B) \leftrightarrows \sim A\&\sim B$
if $A \to B$ and $B \to C$, then $A \to C$
if $A \leftrightarrows B$ and $B \leftrightarrows C$, then $A \leftrightarrows C$
$A \to B$ iff $A \leftrightarrows (A\&B)$ iff $(A \vee B) \leftrightarrows B$

## IV. OBSERVATIONS

Not derivable from these principles, and not semantically valid, are the paradoxes of 'implication' $A\&\sim A \to B$ and $A \to B \vee \sim B$. In context, the failure of these principles is evident. The failure of the first simply means that just because we have been told both that $A$ is True, and that $A$ is False, we cannot conclude: everything. These inferences are not wanted in a scheme which is designed not to break down in the presence of 'contradictions'; and since contradictions really do threaten in the circumstances we describe, their absence is welcome.

## V. CONNECTIONS

The logical system developed is precisely the system of 'tautological entailments' of Anderson and Belnap (1975), Chapter III. That the system is characterizable by four values was noted, in correspondence, by T. J. Smiley. The connection of these values with subsets of {True, False} is due to Dunn, arising out of Dunn (1966), reported in Dunn (1969), and fully presented in Dunn (1976). I overlapped with Scott at Oxford – we were both guests of Christopher Strachey, to whom thanks (now tragically posthumous) are due – in 1970, whence emerged the connections with approximation lattices. The epistemic interpretation came later. Stuart Shapiro has independently argued the utility of 'relevance logics' for question-answering systems, and suggested implementation in a research proposal; see also Shapiro and Wand (1975).

## VI. QUANTIFIERS

Quantifiers introduce a number of subtleties to which I shall merely tip my hat, while recognizing that treating them in detail is quite essential to my enterprise.

There is in the first place the question of whether 'the' domain is finite or infinite. Both cases can plausibly arise. In the latter case, there is the question of how the computer is to represent infinite information with its finite resources, but one should not infer from the existence of this problem that the computer can't or shouldn't involve itself with quantification over infinite domains. Surely it should be allowed to answer "Is there a number such that. . . ?" queries (if it can).

In the second place, there is the question of whether the computer has a name for everything in 'the' domain so that we can employ the substitutional interpretation of the quantifiers, or on the other hand does not have a name for each entity in 'the' domain so that the domain-and-values interpretation is forced (see Dunn and Belnap, 1968). Again: both cases can plausibly arise, though attending to standard examples like baseball queries or airline flights might have made one think that in the computer situation everything always has a name. But, for example, in some of Isner's work the computer is told "there is something between *a* and *b*" in a context in which it hasn't got a complete list of either the names or the entities against which to interpret this statement. And still it must work out the consequences, and answer the questions it is given. (Of course it is O.K. for the computer to make up its own name for the 'something' between *a* and *b*; but that is both an important and an entirely different matter.)

In any event, the semantics given for the connectives extend to universal and existential quantifiers in an obvious way, and I suppose the job done. And the various alternatives mentioned above turn out not to make any difference to the *logic* (with the obvious exception of the finite everything-has-a-name case): the valid 'first degree entailments' of Anderson and Belnap (1965) do admirably (supplemented, in the finite case, with the principle that a conjunction which runs through the domain implies the appropriate universal statement).

## *Part II. Compound Truth Functional Inputs*

### I. EPISTEMIC STATES

I can pause now if I like with regard to the title of the paper, for it would be possible to do so and still claim the title appropriate: I really have presented a four-valued logic and argued that it is useful. But there is a fair bit more to do, some of it of theoretical interest, some of it practical. I begin with considerations closer to the practical.

So far, in Part I, we have been considering the situation in which the epistemic state of the computer could be represented by tables specifying for the various atomic formulas which of the four values in **4** each is to take. We called the mathematical equivalent of such a table a *set-up*; that is, a set-up, *s*, is a mapping from all atomic formulas into **4**: $s(p) \in \mathbf{4}$. Let **S** be the set of all set-ups, and recall that each $s \in \mathbf{S}$ extends uniquely to map all formulas into **4**: $s(A) \in \mathbf{4}$.

Each set-up *s*, represents (not what is true but) what the computer has been told. But can every epistemic state of the computer be represented by a set-up? If in fact, as in Part I, only atomic sentences are affirmed or denied to the computer, of course; but not in general otherwise. For example, no single set-up can represent the state the computer should be in if it is told that either the Pirates or the Orioles won in 1971, but it isn't told which. Set-ups can, by judicious use of ***None***, represent some kinds of incomplete information, but not this kind. For any single set-up in which 'either P or O' (with obvious meaning) is marked True is a set up in which either P or O is *also* marked True, and hence has too much information. Any such set-up would lead the computer to answer 'Yes' either to the question, Did the Pirates win?, or to the question, Did the Orioles win? And the computer should not be able to answer either of these questions having been told only that either the Pirates or the Orioles won.

The solution to the problem is well known in the logical literature, going back to Carnap (1942) at least. It has been used in epistemic and doxastic logic by Hintikka (1962), and has also been worked out for computers by Isner (1972, 1975): one uses a *collection* of set-ups to represent a single epistemic state, the rough and partial idea being that the computer takes a formula as something it has been told if it comes out True on *each* of the set-ups forming its current epistemic state. For example, when told that either the Pirates or the Orioles won, the computer would represent this information by building *two* set-ups, one in which the Pirates get ***T*** and the Orioles ***None***, and the other in which the Orioles get ***T*** and the Pirates ***None***. Later, when it is asked if the Pirates won, it will say that it doesn't know (since the Pirates are not marked ***T*** in every state, nor ***F*** in every state); and similarly if it is asked about the Orioles. But if it is asked "Did either the Pirates or the Orioles win?", then it will answer affirmatively, since that sentence is told True in both of the set-ups in its epistemic state.

Let us therefore, at least for the duration of this Part, define an epistemic state as a nonempty collection of set-ups, a (nonempty) subset, that is, of **S**. (If we later omit 'nonempty,' please supply it; or identify the empty set with the unit set of the set-up which marks everything in sight with ***Both***.) We let **E** be the set of all epistemic states, and use '$E$' as ranging over **E**. Let $E$ be an epistemic state. Then the 'meaning' of $E$ is that the computer has been told that the world is accurately (but perhaps incompletely) described by at least one of the set-ups in $E$. As from the beginning, the possibility exists that such a description is inconsistent.

$E$ represents the basis on which we want the computer to answer our questions. And let me now state more completely and more accurately how we want our questions answered by defining the value of a sentence in an epistemic state; in symbols, $E(A)$ for $E \in \mathbf{E}$ and $A$ a formula. Note how the key idea of approximation is mobilized to give insight into what is going on: the value of a sentence in an epistemic state is to be determined by taking the *meet* of all its values in the separate states – the meet not to be taken in the logical lattice **L4** but in the approximation lattice, **A4**. In notation:

$$E(A) = \sqcap\{s(A) : s \in E\}$$

The idea of this definition is straightforward and intuitively appealing. In the first place, we noted that set-ups individually tend to give us more information than we've got about a formula; or in the language of approximation,

$$E(A) \sqsubseteq s(A), \text{ all } s \in E$$

Now what I am saying is that $E(A)$ should be defined so as to be *maximal* while retaining this relationship; i.e., $E(A)$ – the value of $A$ in $E$ – should be the greatest lower bound of all the $s(A)$ for $s \in E$.

Example. Let $E = \{s, s'\}$, where

| | | | |
|---|---|---|---|
| $s(P) = \boldsymbol{T}$, | $s'(P) = \boldsymbol{None}$ | $\Rightarrow$ | $E(P) = \boldsymbol{None}$ |
| $s(O) = \boldsymbol{None}$, | $s'(O) = \boldsymbol{T}$ | $\Rightarrow$ | $E(O) = \boldsymbol{None}$ |
| $s(B) = \boldsymbol{T}$, | $s'(B) = \boldsymbol{F}$ | $\Rightarrow$ | $E(B) = \boldsymbol{None}$ |
| $s(M) = \boldsymbol{T}$, | $s'(M) = \boldsymbol{Both}$ | $\Rightarrow$ | $E(M) = \boldsymbol{T}$ |

Further, though $E(P) = E(O) = \boldsymbol{None}$, clearly $E(P \vee O) = \boldsymbol{T}$.

Let me, as usual, relate this to marking with the True and the False: it all amounts to saying that we should mark $A$ with True in $E$ if it is marked True in all set-ups in $E$, and mark it with False if it is marked False in all set-ups in $E$; recognizing, as always, that this recipe allows marking $A$ with neither or both. On this account, the similarities emerge to van Fraassen (1969a)'s super-valuations, to definitions of necessity and impossibility in modal logic (e.g., Kripke, 1963), and to evaluation of epistemic operators in Hintikka (1962). But of course in all of those cases set-ups are restricted to those that are consistent, nor is there any sense in which any of those logics are four-valued, or even threevalued. (van Fraassen's formulas can take three values, but the third does not have a *logical* relation to the other two, nor are his semantics truth-functional.)

In now extending the account of question-answering, again I treat only the simple question as to $A$ in the context of an epistemic state, $E$. It goes just as before: the computer answers Yes, No, Yes and No, or I don't know, according as the value $E(A)$ of $A$ in $E$ is ***T***, ***F***, ***Both***, or ***None***. If, for example, the value of $A$ in its current state is ***Both***, the computer answers "Yes and No," i.e., $A\&{\sim}A$. Of course in this case the asker of the question will know that the answer is based on an inconsistency – and so will the computer. Indeed, this is how the computer would naturally *report* an inconsistency in an epistemic state; recall that the answer does not have the ontological force, "That's the way the world is," but rather the epistemic force, "That's what I've been told (by people I trust to get it generally right)."

There are at least three situations in which the computer has to deal with formulas: when asked a question, as I just discussed; when calculating or inferring, which I have discussed some and to which I shall return, and when a formula is input. It is this last which is now up for discussion, but further developments are going to be easier if some additional approximation-lattices are introduced at this point.

## II. MORE APPROXIMATION LATTICES

Note first that the family **S** of all set-ups constitutes a natural approximation lattice **AS**, where the order is pointwise:

$s \sqsubseteq s'$ iff for each atomic sentence $p$, $s(p) \sqsubseteq s'(p)$ (in **A4**). That is, one set-up approximates another if for each atomic formula $p$, the information the first set-up gives about $p$ approximates the information the other set-up gives about $p$. My point is not only that **AS** is a complete lattice (we need that mathematically), but that it is *natural* to interpret its ordering as an approximation: if one increases the information on one of the atomic formulas, one increases the information in the set-up.

Since **AS** is infinite, for the first time in the course of these deliberations the approximation-lattice ideas of limit and of continuity now come into their own. I won't dwell on this, but do point out one application. Let us say a set-up is *finite* if it gives values other than ***None*** to only finitely many atomic formulas. Then every set-up, $s$, is the limit of a set of finite set-ups; $s = \sqcup X$ for some $X$ a directed set of finite set-ups. This is important if the computer can only directly represent finite set-ups.

Moving up a level, we can also define a natural approximation-lattice ordering on the set **E** of epistemic states. Naturally we want

$$E \subseteq E' \quad \text{implies} \quad E' \sqsubseteq E$$

since the smaller epistemic state $E$ gives more definite information; but the converse won't do (unless both $E$ and $E'$ are closed upward; see below). The right definition, yielding the above as a special case, is as follows:

$$E \sqsubseteq E' \text{ iff every } s' \in E' \text{ is approximated by some } s \in E$$

For example let

| | | |
|---|---|---|
| $s(p) = \boldsymbol{T}$ | $s'(p) = \boldsymbol{T}$ | $s''(p) = \boldsymbol{T}$ |
| $s(q) = \boldsymbol{None}$ | $s'(q) = \boldsymbol{T}$ | $s''(q) = \boldsymbol{None}$ |
| $s(r) = \boldsymbol{None}$ | $s'(r) = \boldsymbol{None}$ | $s''(r) = \boldsymbol{T}$ |

Then $E = \{s\}$ approximates $E' = \{s', s''\}$. Note that neither $E$ nor $E'$ give any information about $q$ or $r$, but $E'$ tells us $q \vee r$ is $\boldsymbol{T}$.

It is not true that this ordering of **E** yields a lattice; anti-symmetry fails. There are two ways to make it a lattice, both of which we mention and neither of which we employ. The first is to first define an equivalence relation by

*E is equivalent to E′* iff each approximates the other. Then 'divide through' by this relation: take equivalence classes. This easily turns out to be a complete lattice, and a natural approximation lattice (partly since the equivalence is natural).

The second uses the method of 'representatives' instead of equivalence classes. Define a state $E$ as *closed upward* if $s \sqsubseteq s'$ and $s \in E$ imply $s' \in E$. Where **CE** is the set of all nonempty closed upward states, it constitutes a natural approximation lattice **ACE** under the ordering defined above. Indeed in this case it is obvious that the ordering we defined above does in fact agree with the superset relation, so that obviously we have a complete lattice. One might worry, however, that we have cut out some interesting states. Not so: define the *upward closure of* E by:

$C(E)$ = the family of set-ups approximated by some set-up in $E$. Clearly $C(E)$ is both upward closed, and equivalent to $E$; so if we liked we could use $C(E)$ as the 'representative' of $E$. (Note also that $E$ and $E'$ are equivalent just in case $C(E) = C(E')$; everything fits.)

But we choose to stay with **E** and its ordering even though it is not a lattice, for although both the lattice of equivalence classes and the lattice **ACE** are mathematically convenient (indeed we constantly rely on the convenience of the latter), they depart from practicality: the computer cannot work with the elements of these lattices since these elements are grossly infinite.

Let us then define **AE** as **E** supplied with the ordering above, and also with a couple of lattice-like operations which (1) give results equivalent to those obtained by passing through ACE and (2) *preserve finiteness.*

The most natural meet operation is obviously just union:

$$E \sqcap E' = E \cup E'$$

And the join:

$$E \sqcup E' = \{s \sqcup s' : s \in E, s' \in E'\}$$

Also analogously for the general meet$\sqcap X$ and general join$\sqcup X$, $X$ a subset of **E**.

It is important that our valuation function, $E(A)$, is in the argument $E$ not only monotonic but, in an appropriate sense, continuous in **AE**; in spite of the fact that meets, used in the definition of $E(A)$, are notoriously badly behaved in approximation-lattices.

Certain elements of **E** are of particular interest, namely those which characterize and are characterized by formulas. For each $A$, define *the truth-set of A* and *the falsity-set of A* as follows:

$$T\mathit{set}(A) = \{s : \boldsymbol{T} \sqsubseteq s(A)\}$$

$$F\mathit{set}(A) = \{s : \boldsymbol{F} \sqsubseteq s(A)\}$$

$A$ is marked told True by all and only members of $T\mathit{set}(A)$, and told False by the members of $F\mathit{set}(A)$. Both of these sets are closed upwards, hence in **CE**. Dunn (1976) has investigated some of their properties.

### III. FORMULAS AS MAPPINGS; A NEW KIND OF MEANING

Now I turn to a question of considerable interest, and a question on which our various approximation lattices can shed considerable light: How is the computer to interpret a truth-functional formula, $A$, as input? Clearly it is going to use $A$ to modify its present epistemic state; and indeed it is not too much to say that defining how the computer uses the formula $A$ to transform its present epistemic state into a new epistemic state is a way, and a *good* way, of giving $A$ a meaning. Consequently we want to associate with each formula $A$ a transformation, a mapping from epistemic states into new epistemic states. Furthermore, we also want to know what the computer is to do when the formula $A$ is *denied* to the computer; so actually we associate with a formula $A$ *two* functions, one representing the transformation of epistemic state when $A$ is affirmed, the other the transformation when $A$ is denied. Let us call these two functions $A^+$ and $A^-$. How to define them?

Recall that $A^+$ is to map states into states: $A^+(E) = E'$. The key ideas in defining what we want $E'$ to be come from the approximation lattice. First, in our context we are assuming that the computer uses its input always to increase its information, or at least it never uses input to throw information away. (That would just be a different enterprise; it would be nice to know how to handle it in a theory, but I don't.) And we can say this accurately in the language of approximation: $E \sqsubseteq A^+(E)$. Second, $A^+(E)$ should certainly say no less than the affirmation of $A$: $T set(A) \sqsubseteq A^+(E)$. Third and lastly, we clearly want $A^+(E)$ to be the *minimum mutilation* of $E$ that renders $A$ at least True. 'Minimum mutilation' is Quine's fine phrase, but in the approximation lattice we can give a sense that is *no longer merely metaphorical*: namely, we want the *least* of those epistemic states satisfying our first two conditions. That is, we should define

$$A^+(E) = E \sqcup T set(A)$$

for that is precisely the minimum mutilation of $E$ which makes $A$ at least told True. (Recall that in any lattice, $x \sqcup y$ is the 'least (minimum) upper bound'.) Having agreed on this as the definition of $A^+$, it is easy to see that $A^+(E)$ should be the minimum mutilation of $E$ which makes $A$ at least told False:

$$A^-(E) = E \sqcup F set(A)$$

The above definitions accurately represent the meaning of $A$ as input, but they do involve a drawback: the Tsets and Fsets may be infinite, or at least large, and so do not represent something the computer can really work with. For this reason, and also for its intrinsic interest, we offer another explication of $A^+$ and $A^-$, this time inductive, but still very much involving the idea of minimum mutilation.

First, what is the computer to do to its present epistemic state, when an atomic formula, $p$, is affirmed? Recalling that $p$ must be marked at least told True in the result, and that it will not be such unless it is such in each member of $E$, it is clear that what the computer must do is run through each set-up in $E$ and add a told True to $p$. This will make $p$ ***T*** if it was ***None*** before, it will leave it alone if it was already either ***T*** or ***Both***, and will make it ***Both*** if it was ***F***. And this is obviously the *minimum* thing the computer can do. Defining $p_T$ as that set-up in which $p$ has ***T*** and all other atoms have ***None***, we can say this technically as follows (note where minimum mutilation comes in):

$$p^+E = \{s \sqcup p_T : s \in E\}$$

And with $p_F$ defined simlarly,

$$p^{-}E = \{s \sqcup p_F : s \in E\}$$

The union is in the approximation-lattice **AS** of all set-ups.

The recursive clauses, which represent a way of giving meaning to the connectives (different from – though of course related to – the usual 'truth-conditions' account), now come easily.

$$(A\&B)^{+} = A^{+} \circ B^{+}$$

That is, to make $A\&B$ true by minimum, first minimally mutilate to get $B$ true, and then mutilate the result to get A true as well. (The '$\circ$' is for composition of functions.) It had better turn out, and it does, that $(A\&B)^{+} = (B\&A)^{+}$ – i.e., that the order of minimal mutilation makes no difference. Next,

$$(\sim A)^{+} = A^{-},$$

obviously. And

$$(A \vee B)^{+} = \lambda E(A^{+}(E) \sqcap B^{+}(E))$$

That is, one makes the minimum mutilations for $A$ and $B$ separately, and then finds the best (maximum) among all states which approximate both of these – which is just their set-theoretical union. For example, if $E$ is a singleton $\{s\}$ in which $p \vee q$ has ***None***, $(p \vee q)^{+}E$ is obtained by 'splitting' $s$ into two new states, in one of which $p$ has ***T*** while $q$ and everything else stays the same, and in the second of which $q$ has ***T*** while $p$ and everything else remains fixed.

I give the clauses for $A^{-}$ without comment:

$$\begin{aligned}(\sim A)^{-} &= A^{+}\\ (A\&B)^{-} &= \lambda E(A^{-}(E) \sqcap B^{-}(E))\\ (A \vee B)^{-} &= A^{-} \circ B^{-}\end{aligned}$$

I have given two separate accounts of the meaning of $A$ as input (affirmed or denied), so I had better observe that they agree. A third account of some merit begins by defining $A^{+}$ and $A^{-}$ as functions from set-ups $s$ into states $E$:

$$A^{+}s = \{s \sqcup s' : s' \in T\,set(A)\}$$

$$A^{-}s = \{s \sqcup s' : s' \in F\,set(A)\}$$

Then

$$A^{+}E = \bigcup\{A^{+}s : s \in E\}$$

$$A^{-}E = \bigcup\{A^{-}s : s \in E\}$$

And there are a number of other variations, e.g.,

$$(A\&B)^{+}E = A^{+}E \sqcup B^{+}E$$

## IV. More Observations

What we have done is to use the approximation lattices not only to spell out in reasonably concrete terms what the computer is to do when it receives a formula as affirmed or denied, but further, to give a new theoretical account of the meaning of formulas as certain sorts of mappings from epistemic states into epistemic states. It is clear that there remains work to be done here in finding the right abstract characterizations and general principles, unless it has already been done somewhere or other; but I make a few comments.

To set the stage, I recall that Scott has observed that the family of all *continuous* functions from an approximation-lattice into itself (or indeed another) naturally forms a new approximation-lattice, and it is important that our $A^+$ and $A^-$ functions are members in good standing. But the $A^+$ functions (we may drop reference now to the $A^-$ functions since $A^- = (\sim A)^+$) form but a limited subset of all of these functions, and it would be desirable to characterize an appropriate subset, without however leaning too heavily on linguistic considerations. One feature they all have in common is that they are one and all *ampliative*:

$$E \sqsubseteq A^+E$$

Or where $\boldsymbol{I}$ is the identity function on $E$,

$$\boldsymbol{I} \sqsubset A^+.$$

And this feature is a hallmark of our entire treatment: the computer is never to throw away information, only to soak it up. It is easy to see that the family of all continuous functions 'above' $\boldsymbol{I}$ themselves form an approximation-lattice – the lattice of all ampliative and continuous functions – which is closed under such pleasant operations as composition. ($\boldsymbol{I}$ is the bottom of this lattice.)

Another feature of the $A^+$ is that they are permanent: once $A^+$ is done to a state $E$, it stays done, and does not have to be done again, no matter how much the computer later learns. In symbols:

$$A^+E \sqsubseteq E' \text{ implies } A^+E' = E'$$

These three features taken together can very likely be taken as a proper intrinsic characterization of the 'kind' of functions represented by our truth-functional formulas. For a function, $f$, is continuous, ampliative, and permanent just in case $f$ can be characterized as improving the situation by some fixed amount. That is, just in case there is some fixed element $E_0$ such that $f(E) = E \sqcup E_0$, all $E$. And that sounds right.

The interested reader can verify that from these principles one can deduce what are perhaps the most amusing of the properties of the $A^+$ functions: composition is the same as join, hence commutative and idempotent:

$$A^+ \circ B^+ = B^+ \circ A^+$$
$$A^+ \circ A^+ = A^+$$

I do not like to leave this discussion on such an abstract note, and so I conclude with a more practical remark. What 'permanence' in the above sense *means* for the computer is that it has a choice when it receives $A$ as an input: it can if it likes 'remember' the formula $A$ in some convenient storage; or, if it prefers, it can 'do' $A$ to its epistemic state, and then *forget* about it. Since the meaning of $A$ is a permanent function, $A$ will be permanently built into the computer's present *and future* epistemic states. In the next part there will emerge important contrasts with this situation.

## V. QUANTIFIERS AGAIN

Quantifiers again introduce problems which *must* be worked through but which I do *not* work through. The chief difficulty comes from the fact that we must keep our set-ups and epistemic states finite for the sake of the computer, whereas a quantified statement contains infinitely much information if the domain is infinite. I am going to offer only some murky comments.

In the first place, I will stay with the substitutional interpretation of the quantifiers so as not to have to modify the definition of set-up. So quantification is always with respect to a family of constants suitable for substitution: $(x)Ax$ is the generalized conjunction of all its instances, and $\exists xAx$ the generalized disjunction of its instances. So given a substitutional range, the reader can supply the right definitions for $s((x)Ax)$ and $s(\exists xAx)$. Second, I am going to suppose that the substitution range is infinite; otherwise there is no problem. Third, with considerable hesitation, I am going to attach the substitution-range to the entire epistemic state $E$, rather than permitting the various set-ups in $E$ to come with different substitutional ranges.

The problem is not really how to answer questions about quantified formulas (though there may be difficulty in practice), but in how to treat them as input. Perhaps it is obvious what we want for the existential quantifier: given $\exists xAx$ as input, add a new constant, $c$, to the substitutional range, and then make the minimum mutilation making $Ac$ True. But I am not yet clear how to justify this procedure in approximation terms.

The universal quantifier as input is where the real problem lies: it can lead from a finite state $E$ (i.e., a finite collection of finite set-ups) to an infinite state $E'$. What is probably best is to apply the universal quantifier (minimally mutilate to make an instance true) only for a while; which will force the computer to *remember* the universally quantified formula so that it can be applied again later, if necessary. (What counts as 'necessary' is: as much as is needed to answer the questions asked.) The various finite states obtained by repeatedly applying $(x)Ax$ in this way clearly have as a limit the minimum mutilation in which $(x)Ax$ is True.

Some of what is needed can be better appreciated from the point of view of section 4, and we drop the matter for now.

## *Part III. Implicational Inputs and Rules*

In Part 1 we pretended that all information fed into the computer was atomic, so we could get along with set-ups. In Part II we generalized to allow information in the form of more complex truth-functional formulas, a generalization which required moving to epistemic states. Now we must recognize that it is *practically* important that sometimes we give information to the computer in the form of rules which allow it to modify its own representation of its epistemic state in directions we want. In other words, we want to be able to instruct the computer to make inferential moves which are not mere tautological entailments. For example, instead of physically handing the computer the whole list of Series winners and non-winners for 1971, it is obviously cheaper to tell the computer: "the Pirates won; and further, if you've got a winner and a team not identical to it, that team must be a non-winner" (i.e., $(x)(y)(Wx\&x \neq y \rightarrow {\sim}Wy)$). In the presence of an obviously needed table for identity and distinctness, or else in the presence of a convention that different names denote different entities (not a bad convention for *practical* use in many a computer setting), one could then *infer* that 'The Orioles won' is to be marked told False.

Your first thought might be that you could get the effect of 'given $A$ and $B$, infer $C$,' or 'if $A$ and $B$, then $C$'; by feeding the computer '${\sim}A \vee {\sim}B \vee C$'. But that won't work: the latter formula will tend to *split* the set-up you've got into three, in one of which $A$ is marked told False, etc.; while what is wanted is (roughly) just to improve the single set-up you've got by adding told True to $C$ provided $A$ and $B$ are marked told True (and otherwise to leave things alone). It is (roughly) this idea we want to catch.

## I. IMPLICATIONAL INPUTS

Let us introduce '$A \to B$' as representing the implication of $A$ to $B$; so what we have is notation in search of a meaning. But we have in the previous section found just the right way of giving meaning to an expression construed as an input: the computer is to improve its epistemic state in the minimum possible way so as to make the expression true. So let us look forward to treating $A \to B$ as signifying some mapping from states into states such that $A \to B$ is true in the resultant state.

Obviously if we are to pursue this line, we must know what it is for $A \to B$ to be true in a state. This is a delicate matter. One definition which suggests itself is making $A \to B$ true in a state $E$ just in case $E(A) \leq E(B)$ (in the logical lattice **L4**); but while I don't have any knock-down arguments against the fruitfulness of this definition, I am pretty sure it is wrong. I think it will be more fruitful to define $A \to B$ closer to the following: modify *every* set-up you are considering to make $A \to B$ true in it. So let us first define what it is for $A \to B$ to be true in a set-up; and naturally for this we turn to the logical lattice **L4**, specifying that $A \to B$ is true in $s$ just in case $s(A) \leq s(B)$. (Note that we do *not* give $A \to B$ values in **4**; $A \to B$ is just true or false in $s$, never both nor neither.)

It might be tempting now to define $A \to B$ as true in state $E$ if true in every set-up $s$ in $E$, and false otherwise, but that would be wrong. The reason is that the truth of $A \to B$ is *not* closed upward: $s \sqsubseteq s'$ and $A \to B$ true in $s$ do not together guarantee $A \to B$ true in $s'$. But epistemic states are supposed to be equivalent to their upward closures. The next thing to try is to look just at the *minimal members* $M(E)$ of each state $E$; i.e., those set-ups in $E$ which are minimal with respect to the approximation ordering between set-ups. For in any state $E$ in which every set-up is approximated by some minimal set-up, non-minimal set-ups (those not in $M(E)$) can be thought of as redundant. In particular, they do not contribute to the value of any formula and should not contribute to the value of implications. So it would be plausible to define $A \to B$ as true in a state if true in every minimal member. And indeed this will work if $E$ is finite, or if every $s$ in $E$ is finite, for then every descending sequence

$$s_1 \sqsupseteq \ldots s_i \sqsupseteq \ldots,$$

in $E$ is finite, so that in fact $M(E)$ is equivalent to $E$. Of course for real applications on the computer this will always be so. But let us nevertheless give a definition which will work in the more general case: $A \to B$ is true in $E$ if for every $s \in E$, there is some $s' \in E$ such that $s' \sqsubseteq s$, and $A \to B$ is true in $s'$.

We claim for this definition the merit of passing over equivalent states: given $E$ and $E'$ equivalent, $A \to B$ will have the same truth-value in each. The reason that, if $A \to B$ is true in the closure of $E$ then it is true as well in $E$ (the hard part), is that there cannot be in the closure of $E$ an infinitely descending chain of set-ups in which the truth-value of $A \to B$ changes infinitely often. Sooner or later as you pass down the chain, the truth-value of $A \to B$ will have to stabilize as either true or false. And under the hypothesis that $A \to B$ is true in the closure of $E$, in each chain it will have to stabilize as true; which is enough to get it true in $E$. And the reason that there cannot be an infinitely descending chain of set-ups in which the truth-value of $A \to B$ flickers is that any such flicker must be caused by a change in either $s(A)$ or $s(B)$; but since this function is monotonic in $s$, once having changed the value of $A$ or $B$ in the only permitted (downward) direction, one can never change it back up again. So at most the value of $A$ can change twice, and similarly for $B$; which means $A \to B$ can change at most four times.

One more note of profound caution: the notion of the truth of $A \to B$ in $E$ is dramatically different from the notion of $A$'s being told True in $E$, in that the former is *not* monotonic in $E$, whereas the latter is: $E \sqsubseteq E'$ guarantees that if $A$ is at least $\boldsymbol{T}$ in $E$, then it is so in $E'$; but does not guarantee that if $A \to B$ is true in $E$, it is so in $E'$. (The falsity of $A \to B$ fares no better.) We shall see later how this influences the computer to manipulate $A \to B$ and $A$ quite differently; now, however, I remark that this fact is not in conflict with Scott's Thesis, since we have not got something which can be represented as a function from one approximation lattice into another. In particular, the usual characteristic function representing the set of $E$ in which $A \to B$ is true will not work, since the two truth values True and False do not constitute an approximation lattice.

Now back to our enterprise of defining $A \to B$ in such a way as to make it a mapping from epistemic states $E$ to states $E'$ in such a way as to represent minimum mutilation yielding Truth – in exact analogy with our results in the previous section.

Since we know what it is for $A \to B$ to be true in $s$, we know it has a *Tset*: $Tset(A \to B) = \{s : A \to B \text{ true in } s\}$. So one might try just defining

$$(A \to B)^{+}E = E \sqcup Tset(A \to B)$$

as before. There may be something in the vicinity which works, but this doesn't; since one of the set-ups in which $A \to B$ is true is that in which every atomic formula has ***None***, this $(A \to B)^{+}$ is just the identity function (up to equivalence). It is also worth noting that $Tset(A \to B)$ is not closed upward and so not well-behaved. Nor will it do to try to close it upward – by the remark above, that would yield the family of *all* set-ups. In any event, I take a different and I think intuitively plausible path to defining $(A \to B)^{+}$ as a function which minimally mutilates $E$ to make $A \to B$ true.

I propose first to define $A \to B$ on set-ups $s$, looking forward to the following extension to states:

$$(A \to B)^{+}E = \bigcup\{(A \to B)^{+}s : s \in E\}$$

So we are up to defining $(A \to B)^{+}$ on a set-up $s$ – with the presumption doubtless that the value will be some state $E'$ (we may have to 'split' $s$). The idea is, as always, that we want to increase the information in $s$ as little as possible so as to make $A \to B$ true. If we keep firmly in mind that 'increase of information' is no mere metaphor, but is relative to an approximation lattice, it turns out that we are guided as if by the hand of the Great Logician.

One case is easy. If $A \to B$ is *already* true in $s$, the minimum thing to do is to just leave $s$ alone. Now in order to motivate the definition to come, consider all the ways that $p \to q$, for example, could be false in $s$. The possibilities, which refer to the *logical* lattice **L4**, are layed out under '$p$' and '$q$' below (ignore for now the right column).

| $p$ | $q$ | Essential to make $p \to q$ true |
|---|---|---|
| ***T*** | ***None*** | Raise $q$ to ***T*** |
| ***T*** | ***Both*** | Raise $p$ to ***Both*** |
| ***T*** | ***F*** | Raise $p$ to ***Both*** and $q$ to ***Both*** |
| ***None*** | ***F*** | Raise $p$ to ***F*** |
| ***Both*** | ***F*** | Raise $q$ to ***Both*** |

Now try the following. Keep one eye on the logical lattice **L4** and the other on the approximation lattice **A4**, and use your third to verify the claims made in the right column. For example, the first entry says, in effect, that if $p \rightarrow q$ is false because $p$ is $\boldsymbol{T}$ and $q$ is ***None***, then *it does no good* to raise $p$ (in the approximation lattice **A4**), for the only place to which to raise it is to ***Both***; and (in the logical lattice **L4**) that still doesn't imply $q$ (make $p \rightarrow q$ true). So $q$ must be raised. (An important presupposition of these remarks is that we may only speak of 'raising' (in the approximation lattice **A4**), never of 'lowering'; the computer is never to treat an input as *reducing* its information, never to treat it as a cause to 'forget' something. And you will recall that this constraint is a local one, certainly not part of what I think is essential to the Complete Reasoner.)

Next note the following analysis of the table, where $s$ is the current set-up and $E'$ is the new state. All the raisings of $q$ occur when $\boldsymbol{T} \sqsubseteq s(p)$, and all the raisings of $p$ occur when $\boldsymbol{F} \sqsubseteq s(q)$. Further, the raising of $q$ consists of always making $\boldsymbol{T} \sqsubseteq E'(q)$ and the raising of $p$ consists in making $\boldsymbol{F} \sqsubseteq E'(p)$. That is, as might have been expected, making $p \rightarrow q$ true consists in making $q$ have at least $\boldsymbol{T}$ when $p$ does, and of making $p$ have at least $\boldsymbol{F}$ when $q$ does.

Let us divide the problem (and abandon the special case of atomic formulas). One thing we must do is make $B$ have at least $\boldsymbol{T}$ when $A$ does. Let us call the corresponding statement: $A \rightarrow_T B$. We want to make $B$ true if $A$ is, and in a minimal way. But we *already* know the minimal way of making $B$ true. So the following definition of $(A \rightarrow_T B)^+$ is pretty well forced:

$$\begin{aligned}(A \rightarrow_T B)^+ s &= B^+\{s\} \text{ if } s \in Tset(A), \text{ i.e., if } \boldsymbol{T} \sqsubseteq s(A)\\ &= \{s\} \quad\ \text{if } s \notin Tset(A), \text{ i.e., if } \boldsymbol{T} \not\sqsubseteq s(A)\end{aligned}$$

This account of $(A \rightarrow_T B)^+$ matches very well the intuitions leading Ryle to say that if-then's are inference tickets. For $(A \rightarrow_T B)^+$ is exactly a license to the computer to infer the conclusion whenever it has got the premiss in hand. For example, if it finds that 'The Pirates won' is marked $\boldsymbol{T}$, then 'The Pirates won $\rightarrow_T$ the Orioles didn't' will direct it to make the minimum mutilation which marks 'the Orioles didn't' with at least $\boldsymbol{T}$. (Recall from the previous section that $B^+$ is the minimum mutilation making $B$ at least $\boldsymbol{T}$.)

There is already much food for thought here, and a host of unanswered questions. I do note that Scott's thesis is not violated: $(A \rightarrow_T B)^+$ is indeed a continuous function from the space of set-ups to that of states – and, with the previous extension, from states into states. That it is depends on the fact that *Tsets* are (1) always closed upward and (2) 'open': if $\bigsqcup X \in Tset(A)$ for directed $X$, then $x \in Tset(A)$ for some $x \in X$. (The topological language fits the situation: it means that no point in $X$ can be approached as the limit of a family of points lying entirely outside of $X$.) The point of this remark is to draw the consequence that we cannot sensibly use $(A \rightarrow_T B)$ in the absence of these conditions; hence, since the *Tset* for $A \rightarrow_T B$ is *not* closed upward, we cannot make sense of $(A \rightarrow_T B) \rightarrow_T C$. In contrast, all we need from $B$ is the continuity of $B^+$; so $A \rightarrow_T (B \rightarrow_T C)$ is acceptable. (Note how the approximation idea and Scott's Thesis guide us through the thicket.)

For its intrinsic interest, note that in the lattice of all ampliative functions, we have

$$((A \rightarrow_T B)^+ \circ A^+) \sqsupseteq B^+$$

but not

$$(A^+ \circ (A \rightarrow_T B^+)^+) \sqsupseteq B^+$$

Maybe this has something to do with some of the nonpermutative logics; and maybe not.

Turning back now to our principal task, the defining of $(A \to B)^+$ we have completed part of our task by defining $(A \to_T B)^+$, which makes $B$ true if $A$ is. The other part is by way of the function

$$\begin{aligned}(A \to_F B)^+ s &= A^- \text{ if } s \in Fset(B), \text{ i.e., if } \boldsymbol{F} \sqsubseteq s(B) \\ &= s \quad \text{if } s \notin Fset(B), \text{ i.e., if } \boldsymbol{F} \not\sqsubseteq s(B)\end{aligned}$$

This is the function which makes $A$ told False, minimally, if $B$ is.

Before pushing on to define $(A \to B)^+$, let us pause to note just a thing or two about $(A \to_T B)^+$. This family of functions has in common with the $A^+$ that each is *ampliative*:

$$\boldsymbol{I} \sqsubseteq (A \to_T B); \text{ i.e., } E \sqsubseteq (A \to_T B)^+ E$$

In contrast, however, these new functions are *not* 'permanent' in the sense defined at the end of section 3. That means that once the computer has 'done' $(A \to_T B)^+$, it may have to do it again; a consequence of the fact that the $Tset$ of $A \to_T B$ is not upward closed; adding new information can falsify $A \to_T B$. But there is one property in the vicinity that $(A \to_T B)^+$ shares with $A^+$: at least one doesn't have to do it twice in a row:

$$f \circ f = f$$

for $f = (A \to_T B)^+$.

Closely related to the permanence-impermanence distinction between the two sorts of ampliative functions is the way they behave under composition: all the truth-functional ampliative functions permute with each other ($A^+ \circ B^+ = B^+ \circ A^+$), but the $\to_T$ functions permute neither with each other nor with the truth-functions. The clearest example of the latter is that

$$(p^+ \circ (p \to_T p)^+) \neq ((p \to_T q)^+ \circ p^+)$$

Applying the right hand side to an $s$ in which $p$ and $q$ each have ***None*** yields a state in which first $p$ is made told True, and then as a consequence of this, $q$ is made told True, too. But applying the left hand side to $s$ does not fare as well: $p \to_T q$ does no work since $p$ is not at least $\boldsymbol{T}$ in $s$, so the outcome is only the marking of $p$ as told True *without* changing $q$.

By noting that $(A \to_F B)^+ = (\sim B \to_T \sim A)^+$, we can be sure that this function has both the virtues and shortcomings of $(A \to_T B)^+$. Except that it has the additional shortcoming that not only is $(A \to_F B) \to_F C$ impossible (since the $Fset$ of $(A \to_F B)^+$ is not closed upwards), but so is $A \to_F (B \to_F C)$ (since $(B \to_F C)^-$ is not defined). (We can if we like have $A \to_T (B \to_F C)$.)

The shortcomings of the arrow functions make us see that we cannot define $(A \to B)^+$ as simply the composition of $(A \to_T B)^+$ and $(A \to_F B)^+$. For $A \to B$ might not be true in the result. Intuitively, $(A \to_F B)^+$ might cause nothing to happen since $B$ is ***None*** in the set-up $s$ in question, while $(A \to_T B)^+$ causes $B$ to be marked not only True (since $A$ is) but False as well. This can happen if $B$ is a formula like $p\&{\sim}p$ which cannot be made True without being made False *as well*. Then if $A$ still has the value $\boldsymbol{T}$, $A \to B$ will be false. So the composition of $(A \to_T B)^+$ with $(A \to_F B)^+$ (in either order) is *not* the minimum increase making $A \to B$ true in the result. As a special solution to this problem, one finds that $(A \to_F B)^+ \circ (A \to_T B)^+ \circ (A \to_F B)^+$ works admirably: first make $A$ False if $B$ is; then $B$ True if $A$ is; then, once more, $A$ False if $B$ is. Since $A \to B$ is true in the result, one need do nothing else; one has indeed found the minimum. In particular,

$$((A \to B)^+ \circ (A \to B)^+ s = (A \to B)^+ s$$

$$(A \to B)^+ = (A \to_T B)^+ \circ (A \to_F B)^+ \circ (A \to_T B)^+$$

So we take this as a definition of what $A \to B$ *means* as a mapping of epistemic states into epistemic states.

I conclude this section with two remarks. First, I have offered no *logic* for rules $(A \to B)^+$; there is just much work to be done.

Second, $A \to B$ has been construed as a rule, and has been given 'input' meaning. It has been given no output meaning, and it is not intended that the computer answer questions about it. In particular, we have given no meaning to *denying* $A \to B$; $(A \to B)^-$ has not been given a sense. I am not sure if this is a limitation to be overcome, or just a consequence of my presenting $A \to B$ as a rule; for I do not know what it would mean to tell the computer *not* to use the rule $(A \to B)^+$. One might try to give sense to $(A \to B)^-$ by instructing the computer to make $E(A) \nleq E(B)$; but this is an instruction which it is not always possible for the computer to carry out.

## II. Rules and Information States

This last section of this paper is going to be altogether tentative, and altogether abstract, with just one concrete thought that needs remembering, which I learned from Isner: probably the best way to handle sophisticated information states in a computer is by a judicious *combination* of tables (like our epistemic states) and rules (like our $A \to B$, or a truthfunctional formula the computer prefers to remember, or a quantificational formula which it must remember). For this reason, as well as for the quite different reason that some rules, since they may have to be used again (are not permanent, must be remembered), we can no longer be satisfied by representing what the computer knows by means of an epistemic state. Rather, it must be represented by a *pair* consisting of an epistemic state *and* a set of rules:

$$\langle R, E \rangle.$$

$E$ is supposed to represent what the computer explicitly knows, and is subject to increase by application of the rules in the set $R$. For many purposes we should suppose that $E$ is finite, but for some not.

Let us dub such a pair an *information state*, just so we don't have to retract our previous definition of 'epistemic state.' But what is a rule? Of what is $R$ a set? A good thing to mean by *rule*, or *ampliative* rule, in this context, might be: *any* continuous and ampliative mapping from epistemic states into epistemic states. As I mentioned above, the set of *all* continuous functions from an approximation lattice into itself has been studied by Scott; it forms itself a natural approximation lattice. It is furthermore easy to see that the ampliative continuous functions form a natural approximation lattice, and one which is an almost complete sublattice of the space of all the continuous functions: all meets and joins agree, except that the join of the empty set is the identity function $\boldsymbol{I}$ instead of the totally undefined function. Intuitively: the effect of an empty set of rules is to leave the epistemic state the way it was.

So much for the general concept of rule, of which the various functions $A^+$, $A^-$, $(A \to_T B)^+$, $(A \to_F B)^+$, and $(A \to B)^+$ are all special cases. We now have to say what a set, $R$, of rules *means*. Of course we want to express it as a mapping from epistemic states into epistemic states. Let us begin by saying that a rule

$\rho$ is *satisfied in* a state $E$ if applying it to $E$ does not increase information: $\rho(E) = E$

and also that a set $R$ of rules is satisfied in $E$ if all its members are. Then what we want a set of rules to do is to make the minimum mutilation of $E$ that will render all its members satisfied. Even if $R$ is a unit set, simple application of its member might not work to satisfy it. And even if $R$ is a finite set of rules, each of which is satisfied after its own application, the simple composition of $R$ might not be adequate; all of this can be derived from considerations I adduced in defining $(A \rightarrow B)^+$. But there is a general construction which is bound to work.

Let $R$ be a set of rules. Let $R^\circ$ be the closure of $R$ under composition. This is a directed set; the composition of $f$ and $g$ will always provide an upper bound for both $f$ and $g$ if they are monotonic and ampliative. Now take the limit: $\bigsqcup R^\circ$.

Claim: take any $E$, and any set, $R$, of rules. Then $\bigsqcup R^\circ(E)$ is the minimum mutilation of $E$ in which all rules in the set $R$ are satisfied (Below we write $R(E)$ for $\bigsqcup R^\circ(E)$.)

In this way we give meaning to the pair consisting of an epistemic state $E$ and a set of rules $R$. There is that state $\bigsqcup R^\circ E$ consisting of 'doing' the rules in all possible ways to $E$; and it is in regard to this state that we want our questions answered in the presence of $E$ and $R$. Of course $R(E)$ can be infinitely far off from $E$. This will certainly happen if the computer is dealing with infinitely many distinct objects and some rule involves universal quantification; so in practice, "I don't know" might have to mean either: "I haven't computed long enough", or "I have positive evidence that I haven't been told."

Because of the importance of computers which maintain both (1) sets of rules and (2) tables (epistemic states), the idea of information states $\langle R, E\rangle$ should be studied in detail. I close this section with just a few idle definitions in the area which might or might not turn out to be fruitful.

When are two states equivalent? There seem to be at least two ideas: $\langle R_1, E_1\rangle$ is *currently equivalent* to $\langle R_2, E_2\rangle$ just in case $R_1(E_1) = R_2(E_2)$; which is to say, they give the same answers to the same questions. And they are *strongly equivalent* if adding the same information to each always produces currently equivalent results: $\langle R_1, E_1 \sqcup E\rangle$ is currently equivalent to $\langle R_2, E_2 \sqcup E\rangle$, all $E$. Such information states would not only answer all present questions alike, but also all future questions asked after the addition of the same information to each.

We defined a rule, $\rho$, as satisfied in an epistemic state $E$ if $\rho(E) = E$. We could similarly define a rule as satisfied in an information state $\langle R, E\rangle$ in one of two ways: *currently* satisfied if satisfied in $E$, and *ultimately satisfied* if satisfied in $R(E)$. A third notion brings in just the set $R$: perhaps saying a rule, $\rho$, is *in force* in $R$ might be defined by: $\rho \sqsubseteq R$; i.e., $\rho$ approximates $R$. (This is not a 'relevant' idea – see Anderson and Belnap (1975) – of being in force; e.g., for each $A$ the rule $(A \rightarrow A)^+$ is in force in every $R$. Problem: what is a relevant idea?)

## III. CLOSURE

Lest it have been lost, let me restate the principal aim of this paper: to propose *usefulness* of the scheme of tautological entailments as a guide to inference in a certain setting; namely, that of a reasoning, question-answering computer threatened with contradictory information. The reader is not to suppose that Larger Applications have not occurred to me; e.g., of some of the ideas to a logic of imperatives, or to doxastic logic, or even to the development of The One True Logic. But because of my fundamental conviction that logic is (all in all) practical, I did not want these possibilities to loom so large as to shut out the light required for dispassionate consideration of my far more modest proposal.

*University of Pittsburgh,*
*Pittsburgh, PA 15260, U.S.A.*

## References

1. Anderson, A. R. and Belnap, N. D. Jr. (1962). 'Tautological Entailments', *Philosophical Studies*, 13(1), 9-24. (See also Ch. III of Anderson and Belnap (1975).)
2. Anderson, A. R. and Belnap, N. D. Jr. (1965). 'First-Degree Entailments', *Mathematische Annalen*, 149(4), 302-319.
3. Anderson, A. R. and Belnap, N. D. Jr. (1975). '*Entailment: The Logic of Relevance and Necessity, vol. I*'. Princeton: Princeton University Press.
4. Belnap, N. D. Jr. (1963). *An Analysis of Questions: Preliminary Report.* Santa Monica: System Development Corporation.
5. Belnap, N. D. Jr. (1976). 'How a Computer Should Think', *Contemporary Aspects of Philosophy*, Proceedings of the Oxford International Symposium.
6. Belnap, N. D. Jr. and Steel, T. B. Jr. (1976). *Erotetic Logic: An Introduction to the Logic of Questions and Answers.* New Haven: Yale University Press.
7. Carnap, R. (1942). *Introduction to Semantics.* Cambridge: Harvard University Press.
8. Dunn, J. M. (1966). *The Algebra of Intensional Logics*, dissertation, University of Pittsburgh.
9. Dunn, J. M. (1969). 'Natural Language versus Formal Language', Mimeo of a talk in the joint A.P.A./A.S.L. symposium of that title, New York, Dec. 27. Privately Circulated.
10. Dunn, J. M. (1976). 'Intuitive Semantics for First-Degree Entailments and "Coupled Trees"'. *Philosophical Studies*, 29, 149.
11. Dunn, J. M. and Belnap, N. D. Jr. (1968). 'The Substitution Interpretation of the Quantifiers', *Noûs*, 2, 177-185.
12. Hintikka, J. (1962). *Knowledge and Belief.* Ithaca: Cornell University Press.
13. Isner, D. W. (1975). 'An Inferential Processor for Interacting with Biomedical Data Using Restricted Natural Language', *Proceedings of Spring Joint Computer Conference*, pp. 1107-1124.
14. Isner, D. W. (1975). '*Understanding "Understanding" Through Representation and Reasoning*', dissertation, University of Pittsburgh.
15. Kripke, S. (1963). 'Semantical Analysis of Modal Logic I', *Zeitschrift für mathematische Logik und Grundlagen der Mathematik*, 9, 67-96.
16. Routley, R. and Routley, V. (1972). 'Semantics of First-Degree Entailment'. *Noûs*, 6, 335-359.
17. Scott, D. (1970). 'Outline of a Mathematical Theory of Computation'. *Proceedings of the Fourth Annual Princeton Conference on Information Sciences and Systems*, pp. 169-176.
18. Scott, D. (1972). 'Continuous Lattices: *Toposes, Algebraic Geometry and Logic*'. *Springer Lecture Notes in Mathematics*, 274, 97-136.
19. Scott, D. (1973). 'Models for Various Type-Free Calculi'. *Logic, Methodology, and Philosophy of Science IV.* Proceedings of the Fourth International Congress for Logic, Methodology and the Philosophy of Science, Bucharest, 1971. Suppes, Henkin, Juja, Moisil (eds.), Elsevier: North-Holland.
20. Shapiro, S. and Wand, M. (1976). 'The Relevance of Relevance'. Technical Report, Indiana University, Computer Science Department.
21. van Fraassen, B. (1969a). 'Presuppositions, Supervaluations, and Free Logic', *The Logical Way of Doing Things*, K.Lambert (ed.), New Haven: Yale University Press.
22. van Fraassen, B. (1969b). 'Facts and Tautological Entailments', *The Journal of Philosophy*, 66(15), 477-487.

# Two, Three, Four, Infinity: The Path to the Four-valued Logic and Beyond*

J. Michael Dunn

**Abstract** I give a kind of intellectual history of the so-called "Belnap-Dunn Four-valued Logic," examining its evolution: the 4-element De Morgan lattice of Antonio Monteiro, and related work of Bialynicki – Birula and Helena Rasiowa, and John Kalman; Timothy Smiley's 4-element matrix for Belnap's Tautological Entailment; Dunn's interpretation in terms of "aboutness;" Bas van Fraassen's semantics for Tautological Entailment using "facts;" and Dunn's interpretation in terms of how a sentence can be assigned both true and false, or neither true nor false, as well as the usual two values simply true, or simply false. Of course I discuss Nuel Belnap's viewing the four values as elements in a "bi-lattice" and his famous use of this interpretation for "How a Computer Should think." I also examine relationships to Richard Routley and Valerie Routley's "star semantics." Moreover, I discuss extension of the 4-valued semantics to the whole system R (allowing nested relevant implications), focusing especially on Edwin Mares' work.
I then examine later adaptations and extensions of the Four-valued Logic, including work by Yaroslav Shramko, Tatsutoshi Takenaka, Dunn, Heinrich Wansing, and Hitoshi Omori on "trilattices." I also explain my recent extension to an infinite valued "Opinion Tetrahedron" (extending Audun Jøsang's Opinion Triangle) which has the four values as its apexes.
I end by acknowledging that ideas involving the 4-values date back to classical Indian logic (Sanjay's "Four Corners"), prior to the 6th century B.C.E.

**Keywords** 4-valued logic • First degree entailment • Tautological entailment • Relevance logic • Aboutness • Bilattice • Opinion tetrahedron

* We of course borrow our title from George Gamow's (1947), *One, Two, Three...Infinity.* This books was a great influence on my intellectual development in high school and likely had a lot to do with developing my interest in logic.

J. Michael Dunn
Department of Philosophy and School of Informatics, Computing, and Engineering, Indiana University – Bloomington, Bloomington, IN 47405-7000, USA
e-mail: dunn@indiana.edu

 
H. Omori, H. Wansing (eds.), *New Essays on Belnap-Dunn Logic*,
Synthese Library 418, https://doi.org/10.1007/978-3-030-31136-0_6

# 1 Introduction

What has often been referred to as the "Belnap-Dunn[1] 4-valued Logic" was first presented by me in "Natural Language versus Formal Language," as an invited speaker in the joint symposium by that title of the Association for Symbolic Logic and the American Philosophical Association at their joint meeting, New York, December, 1969.[2] *No, that is not quite right.* Belnap first presented an axiomatization of the logic known as "First Degree Entailments" (FDE). I was the first to give a 4-valued semantics for it. *No, that is not quite right either.* Timothy Smiley was the first to provide a 4-valued semantics for it, but the values were just the numbers 1, 2, 3, 4. I was the first one to give an intuitive interpretation of the values. *Wait a minute!* Ideas suggesting the 4-values True, False, Both, None date back to classical Indian logic (Sanjaya's "Four Corners"), prior to the 6th century B.C.E.[3] *Anyway!* I did not actually have a valuation as a function that gave values, but rather a relation to the values **t** and **f**. But the image of the relation can be viewed as a valuation giving the values {**t**}, {**f**}, {**t**, **f**}, { }. *Etc., etc.* Below I shall describe more fully the "Who, What, When, Where, and Why." I will be repeating parts of the story that I have told earlier.[4]

# 2 Historical Background and Overview

Once upon a time there was two-valued logic with Aristotle and his syllogism, and Philo and his conditional. It took well over 2 millennia for this to fully develop, and finally the two were merged through the work of Frege, and then Whitehead and Russell, and others in the late 19th and early 20th centuries. Then three valued logic was developed by Łukasiewicz (and others) starting around 1920, with values true, false and some indeterminate third value (neither, undefined, etc.). And Grigore Moisil extended this to $n$-valued logic. And there have been a number of infinite-valued logics. To name just one, I mention the Łukasiewicz logic defined on the unit interval [0, ..., 1] of real numbers between 0 and 1.

Two valued logic was, and still is, the standard logic. Richard M. Martin wrote a book in 1978 defending classical logic, in what can be viewed as a "rear guard action." He wrote "Strictly there is only one logic, which, however, can be extended variously for specific purposes as needed. 'One God, one country, one logic,' in the stunning phrase of

[1] I am going to often refer to Nuel Dinsmore Belnap, Jr. in this paper as "Belnap" so as to make clear his connection to the "Belnap-Dunn Four-valued Logic." Also because of our close friendship I will often refer to him as just "Nuel." I remember a fellow graduate student Richard Schuldenfrei referring to him as "Nuel Call-Me-Nuel," because of Nuel's practice of wanting students to call him by his first name. I am going to refer to myself as "Dunn," or "I" or some grammatical synonym, such as "me" shortly after the next comma.

[2] The text of a typed manuscript prepared just prior to that talk is included in this book. It builds on material from my dissertation (Dunn 1966) and was in effect a preliminary draft of my published paper (Dunn, 1976).

[3] See Dunn (1999) and for 3 more values see Pragati Jain (2000).

[4] See for example Dunn (1999, 2000, 1985), and Anderson, Belnap, and Dunn (2002). For an external point of view see Shramko and Wansing (2011), p. 44 ff.

Whitehead." Martin has a stunning phrase of his own when he goes on to say "God made first-order logic and all the rest is the handiwork of man."[5]

I personally think that at least many logics were created by the handiwork of humans to serve as intellectual tools,[6] and this includes one of my favorites, Anderson and Belnap's system E of Entailment. Alan Ross Anderson, (somewhat) jokingly in conversation, referred to the system E of Entailment as "The One True Logic."

I was fortunate to be in the first wave of graduate students to attend the University of Pittsburgh's Philosophy Department just as it was being "renovated," arriving the same year as Nuel Belnap and just two years before Alan Ross Anderson. This was an exciting time to be there, amidst many first-rate students who would become first-rate philosophers. I hesitate to name them, because I will undoubtedly leave someone out. But just to name those with a logic orientation, they included Louis Goble, Robert K. Meyer, Bas van Fraassen, and Peter Woodruff. If you are interested in learning more I refer you to an autobiography: Dunn (2016). I repeat here some of what I said there about the wonderful environment in which it was to be a graduate student in the Philosophy Department at "Pitt" back then.

> At that time Philosophy was located in a building that was a remodeling of the old Schenley Hotel. I shared an office with fellow graduate student Tryg Ager, and we had our own bathroom since our office was once a hotel room. The office was right down the hall from Nuel's. He would often stop by, at least several times a week, and ask me to prove something for him. Also I felt free to look in on him and see what he was working on, and he would often stop me as I walked by his door.

I went on to say:

> Nuel Belnap was a huge positive influence when I was a student, and still to this day. To give credit where credit is due, this was also a practice of Nuel's own teacher Alan Ross Anderson. Alan moved from Yale to Pitt in my last year there, and shortly after his arrival invited me to lunch. He told me during lunch that there were two things he didn't like in a student: first, if the student didn't call him by his first name, and second, if the student didn't tell him when he was being stupid. I said something as we parted like "Thanks for lunch Professor Anderson." About a week later he asked me to lunch again, and told me the same thing. I said "Which would you prefer Professor Anderson, that I call you Alan or that I call you stupid." He replied, "Please call me Alan," which I did.

I wrote my dissertation *The Algebra of Intensional Logics* with Nuel as my director. This was a perfect match, at least as far as I was concerned (and I feel confident Nuel would agree). Nuel, working with his own dissertation director Alan Ross Anderson, had done significant early work on relevance logic. Perhaps the most "relevant" to this essay is his axiomatization and characterization of the first-degree entailment fragment of the system E of entailment. *First degree entailments* are formulas of the form $A \rightarrow B$, where the formulas $A$ and $B$ themselves contain no implications, but only conjunction, disjunction, and negation.

Belnap, in his own dissertation (1959), had formulated an axiom system that exactly captured the provable first degree entailments of the Anderson-Belnap system E of Entailment. And in an abstract the same year as his dissertation he also characterized the provable first-degree entailments in an intuitive way as the "Tautological Entailments." Put quickly, a first degree entailment is "tautological" if when $A$ is put in disjunctive normal form $A_1 \vee \cdots \vee A_m$ and $B$ is put in conjunctive normal form $B_1 \wedge \cdots \wedge B_n$, each $A_i$ shares an "atom," i.e., and atomic sentence or its negation, with each $B_j$.

[5] This is of course a clever takeoff on the remark attributed to Leopold Kronecker: "God made the integers; all else is the work of man."

[6] See Dunn (2017).

Both of these can be found in Anderson and Belnap (1975), and this is often cited for these. But actually both were first published by Belnap alone. The axiom system for First Degree Entailments was in Belnap's 1959 dissertation, and the intuitive characterization of them as called "Tautological Entailments" was published as an abstract by Belnap in 1959.[7]

My dissertation was largely technical and was mostly about the relation of De Morgan lattices[8] to relevance logics, focusing on their representations. The title in retrospect was somewhat unintentionally misleading – a better choice would have been "relevance logics." Just google "intensional logic" and you will find in *Wikipedia*, the *Stanford Encyclopedia of Philosophy*, the *Routledge Encyclopedia of Philosophy*, etc. a lot about Frege sense and reference, Carnap's sense and reference, Kripke's possible world semantics for modal logic and Montague's similar semantics for natural languages. The title was Nuel's idea, as I acknowledged in my dissertation (p. 1): "Following a suggestion by Nuel D. Belnap, Jr., we call these systems "intensional logics" because they may be viewed as attempts to explicate a species of implication that in some sense explicates the relation of the *meanings* of the antecedent and consequent."

It had become common by then, stemming from Boole, to view a proposition as the set of times (or occasions, states, state descriptions, situations, worlds, whatever) in which it is true (see Dunn 2008). In Chapter IX of my dissertation I introduced a different way of looking at a proposition (I called it a "proposition surrogate"), namely as a pair of sets, rather than as a single set. As I noted in motivating the idea of a proposition surrogate, Bar-Hillel and Carnap (1952) already had the idea of two sets – the first set they called "range" and it was the set of state descriptions that make a sentence true, and the second they called "content" and it was the set of state descriptions that make the sentence false. They did not really need two sets, for in classical logic these would be the set-theoretical complements of each other relative to the universe of all state descriptions. But I wanted the sets to possibly overlap and also to possibly not be exhaustive. I was too timid though to speak of the proposition as being "both true and false" or "neither true nor false," so I invented a more conservative way to look at this.

Rather than talking of "state descriptions" I talked of "topics. " And rather than saying that a sentence was true in one of these I instead said that the sentence gave definite positive information about the topic, and instead of saying that a sentence was false in one of these I said the sentence gave definite negative information about the topic. That way I could say that the sentence *Florida is Democratic* gives definite positive information about Florida, and that the sentence *Florida is not Democratic* gives us definite negative information about Florida.

Why did I qualify the information as "definite"? I took off from Nelson Goodman's (1961) paper "About," which had a corresponding notion of "absolutely about" (which was related but different). He considered the fact that *Florida is Democratic* entails *Florida is Democratic or Maine is Democratic.* The consequent apparently gives some information about Maine, and given that, so should the antecedent. But *Florida is Democratic* seems to give no information about Maine whatsoever. If one somehow thought it did, then if

[7] The characterization of the probable first-degree entailments in the system E as "tautological entailments" first appeared in a very public way in Anderson and Belnap (1962). But the characterization was first publicly mentioned by Belnap in an abstract (Belnap, 1959a). Belnap's 1959 dissertation was "semi-published" in 1960 as a technical report by the Office of Naval Research.

[8] De Morgan lattices were perfectly matched to first-degree entailments, but I also created the notion of a De Morgan monoid to correspond to the full relevance logic R with its nested implications.

one accepts the relevant entailment $A \to (A \vee B)$ (a standard theorem in classical logic and in the Anderson-Belnap relevance logics), then since $B$ could be any sentence mentioning anything whatsoever, $A$ would be about anything whatsoever. This is too much, so I brought in the notion of "definite information" to protect this entailment because it would seem that *Florida is Democratic or Maine is Democratic* gives no "definite information" about Maine.[9]

## 3 An Interlude on van Fraassen's Interpretation on Tautological Entailment

Bas van Fraassen was a fellow graduate student when I was at "Pitt," and we were good friends. Bas wrote a paper "Facts and Tautological Entailments" while he was still a student, that was published in 1969.[10] Bas builds on Bertrand Russell's 1918 lectures on Logical Atomism, and constructs a model for tautological entailments using "facts." The idea is that a model M consists of a non-empty set $D$ (domain) together with relations $R_1, R_2, \ldots$ on that domain. A "complex" is an ordered $n+1$-tuple consisting on an $n$-placed relation $R_i$ followed by $n$ individuals from $D$: $\langle R_i, d_1, \ldots, d_n\rangle$. A *fact* $e$ is then defined as any non-empty set of complexes. It is required that facts be closed under union, which van Fraassen symbolizes as $e \cdot e'$. At first glance union might seem the wrong notion, but what this in effect does is to allow conjunctive facts. And going a step higher, where $X$ and $Y$ are two sets of facts we define $X \bullet Y = \{e \cdot e' : e \in X \text{ and } e' \in Y\}$

Let $d$ be an assignment of elements of the domain to the individual variables, and let us think of each predicate $P_i$ as being interpreted by the relation $R_i$. Given an atomic sentence $P_i x_1 \ldots x_n$, there is just one fact that makes it true, namely $\{\langle R_i, d(x_1), \ldots, d(x_n)\rangle\}$, and just one fact that makes it false, namely $\{\langle \overline{R}_i, d(x_1), \ldots, d(x_n)\rangle\}$. Thus each atomic sentence $P_i x_1 \ldots x_n$ can be assigned a set $T(A)$ of facts that make it true, and a set $F(A)$ of facts hat make it false, and then these are then extended to compound sentences as follows:

$$T({\sim}A) = F(A);$$
$$F({\sim}A) = T(A);$$
$$T(A \wedge B) = T(A) \cdot T(B);‘$$
$$F(A \wedge B) = F(A) \cup F(B).$$

Bas took disjunction as implicit, defining $A \vee B = {\sim}({\sim}A \wedge {\sim}B)$. This corresponds to the valuation clauses:

$$T(A \vee B) = T(A) \cup T(B);$$
$$F(A \vee B) = F(A) \cdot F(B).$$

Since facts are sets, Bas can define *e forces e′* iff $e \subseteq e'$. He then defines a kind of closure of $T(A)$, defining $T^*(A)$ = the set of all facts that force some fact in $T(A)$. Bas defines a kind of entailment relation $A \Vdash B$ iff $T^*(A)$ is included in $T^*(B)$ for every model

[9] As I said in my dissertation, footnote 3 of Ch, IX, "Goodman has a goal complementary to ours; he wants to explicate *aboutness* in terms of entailment. ... Some of the difficulties that arise for Goodman are due to his meaning by ‘entailment’ (provable) material implication. If he were to mean by ‘entailment,’ entailment (in the system E), many of these difficulties would be abviated. Thus, for example, Goodman is forced to maintain that a tautology is not about anything (roughly because it is implied by any statement)."

[10] Bas also published his results as a contributing author to Anderson and Belnap (1975), sec. 20.3. His paper van Fraassen (1973) is also related.

M, and shows it is equivalent to $A$ tautologically entails $B$. . He points out that without the closure, one could not even show $A \wedge B \Vdash A$.

I think that my topics came before Bas's facts, but that his facts were before my 4 values - this at least in publication. In fn 12 of van Fraassen (1969) Bas says: "J. Michael Dunn showed in his doctoral dissertation (Pittsburgh, 1966) that tautological entailment could be explicated in terms of the topics that sentences are "about," but this relation of *being about* was not further explicated."

Bas is right that I did not do much to explicate "aboutness." For me I think it was just a handy tool for avoiding talk of sentences being *both true and false* (I was not so scared of their being *neither true nor false*). I did not have the foresight to see that Stephen Yablo (2014) would write a whole book titled *Aboutness* or I might have made more of it. Instead I abandoned talk of aboutness and screwed up my courage to move to the 4-valued semantics, where sentences could be just true, just false, neither true nor false, or both true and false. More on this below.[11]

## 4 My Dissertation: *The Algebra of Intensional Logics*

My dissertation contained a number of representations for De Morgan lattices,[12] some of these in analogy to representations of Boolean algebras. De Morgan lattices were well-known to Nuel and I am sure he pointed me towards them.[13] Stone (1936) had famously shown that every Boolean algebra is isomorphic to a *field of sets*., i.e., a collection of subsets of a set $U$ closed under intersection, union, and complement set-theoretic relative to $U$. Not all of these representations of De Morgan lattices were original with me.

The basic idea of a De Morgan lattice was introduced by Białynicki-Birula and Rasiowa (1957), Kalman (1958), and Monteiro (1960). These were all independent and about the same time.[14] But they used different names, respectively *quasi-Boolean algebra*, *distributive lattice with involution* (*distributive i-lattice*), *De Morgan lattice*, and sometimes were required to have a top/bottom element (no big deal since these can always be added). These names all had their own advantages. *Quasi-Boolean algebras* suggests that they are a weakening of Boolean algebras. *Distributive lattices with involution* defines them in a single phrase. They are structures $(L, \leq, \sim)$, where $(L, \leq)$ is a partially ordered set which is a lattice, i.e., for each $a, b \in L$, there exists a greatest lower bound $a \wedge b$ and a least upper bound $a \vee b$. And they satisfy the Distributive Law $a \wedge (b \vee c) = (a \wedge b) \vee (a \wedge c)$.[15] Further $\sim$ is an *involution*, i.e., a unary operation on $A$ satisfying

[11] In case you are wondering, Yablo's book does not mention my early work on aboutness, which I find quite understandable given what I just said about how I did not very much develop or promote it, and also I spoke of *positive* and *negative definite* aboutness – not aboutness more generally.

[12] Most of these can be found in Dunn (1967a).

[13] Before I started working with him at Pitt, Nuel had already written a paper with Joel Spencer at Yale on "intensionally complemented distributive" lattices" (*icdls*) – Belnap and Spencer (1966). An intensionally complemented distributive lattice is a De Morgan lattice where for every element $x$, $x \neq \sim x$, and they showed that a De Morgan lattice has a "truth filter (i.e., a proper filter containing for each element $x$, exactly one of $x$ or $\sim x$) iff it is an icdl. And Nuel and I wrote a paper together on the homomorphisms of icdls – Dunn and Belnap (1967).

[14] Though I have recently read that Moisil (1935) had already introduced them.

[15] The logic FDE has the corresponding logical axiom of distribution:
$A \wedge (B \vee C) \to (A \wedge B) \vee (A \wedge C)$.

$\sim\sim a = a$ Period Two
$a \leq b$ implies $\sim b \leq \sim a$ Order Inverting.

These last two are equivalent to the familiar *De Morgan Laws*

$\sim(a \vee b) = \sim a \wedge \sim b$
$\sim(a \wedge b) = \sim a \vee \sim b,$

and so not only is George Boole's contemporary Augustus De Morgan recognized, but we have a more saleable trademark.

Each of these papers had a bellwether result. Thus Andrzej Białynicki-Birula and Helena Rasiowa (1957) showed that every De Morgan lattice[16] is isomorphic to what they called a *quasi-field of sets*. A quasi-field of sets is closed under intersection, union, and a "quasi-complement." The quasi-complement is obtained by assuming that there is an involution, i.e., a function $g$ of period 2 ( i.e., $g(g(x)) = x$) on $U$, and for a set $A \subseteq U$, defining its quasi-complement $\sim A = U - g_*(A)$, where $g_*(A)$ is the image of $A$ under $g$.

And John Kalman (1958) had shown that every De Morgan lattice is embeddable in a direct product of the quintessential De Morgan lattice DM4.

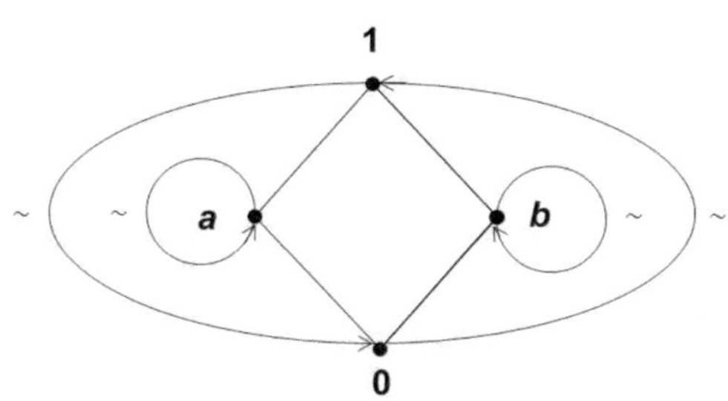


**Fig. 1** 4-element De Morgan Lattice DM4

Note that the order $\leq$ is pictured by putting one node below another in the common Hasse diagram way. The De Morgan complement is picture by the curved arrows and

It inherits this as the first-degree fragment of the relevance logics E and R. But there are many non-classical logics that lack distribution, e.g., the orthomodular logic of Birkhoff and von Neumann and the linear logic of Girard. Bimbó and Dunn (2001) develop a "non-distributive" semantics that keeps the spirit of the 4-valued logic but makes it more complicated, invoking in effect Birkhoff's notion of a "polarity." We will not go into the details here.

[16] They did not use the name "De Morgan lattice" but instead "quasi-Boolean algebras."

amounts to turning the diagram upside down.[17] We will not bother to have the curved arrows in future pictures of DM4 – you can just imagine them.

Monteiro (1960) introduced De Morgan lattices to describe a class of logical matrices that are characteristic for classical propositional logic (ccpl), showing that a matrix $(M, D)$, where $M$ is a De Morgan lattice and $D$ is a proper filter of $M$ is a ccpl iff every element of the form $x \vee {\sim}x \in D$.

I added my own representation to those above, one that would prove critical in the development path of the "4-valued logic." I defined a *2-product of a Boolean algebra* $B$ as an ordered quadruple $(B^2, \wedge, \vee, \sim)$, where $B^2 = \{(a_1, a_2) : a_1, a_2 \in B\}$, $(a_1, a_2) \wedge (b_1, b_2) = (a_1 \wedge b_1, a_2 \wedge b_2)$, ${\sim}(a_1, a_2) = (a_2, a_1)$. This would be simply a direct product of the Boolean algebra with itself except for the definition of ${\sim}(a_1, a_2) = (a_2, a_1)$. The direct product would have this be $-(a_1, a_2) = (-a_1, -a_2)$ (where $-$ is Boolean complement), not the "switcheroo." Let $U$ be a given set. Then for sets $A^+, A^-, B^+, B^- \subseteq U$, we have:

$$
\begin{aligned}
&{\sim}(A^+, A^-) = (A^-, A^+)\\
&(A^+, A^-) \wedge (B^+, B^-) = ((A^+ \cap B^+), (A^- \cap B^-))\\
&(A^+, A^-) \vee (B^+, B^-) = ((A^+ \cup B^+), (A^- \cup B^-)).
\end{aligned}
$$

I showed that every De Morgan lattice is isomorphic to a 2-product of a Boolean algebra of sets (often called a *field* of sets). This was intended to mimic Stone's famous representation theorem for Boolean algebras: Every Boolean algebra is isomorphic to a field of sets.

I was able to show that all of these representations are "effectively equivalent" to each other, i.e., one can show that any one of these implies any other, without using the axiom of choice.[18]

I came to realize that one logician's "and" is another logician's "or" (and vice versa), and so $\wedge$ and $\vee$ could be represented in a kind of mixed way as:

$$
\begin{aligned}
&(A^+, A^-) \wedge (B^+, B^-) = ((A^+ \cup B^+), (A^- \cap B^-))\\
&(A^+, A^-) \vee (B^+, B^-) = ((A^+ \cap B^+), (A^- \cup B^-)).
\end{aligned}
$$

This is what actually led to my talk of "proposition surrogates," and I was pleased that the left-hand components behaved like Carnap and Bar-Hillel's "content" and the right-hand components behaved like their "range."

A proposition surrogate $(A^+, A^-)$ *entails* a proposition surrogate $(B^+, B^-)$ when both $B^+ \subseteq A^+$ and $A^- \subseteq B^-$. Here is DM4 interpreted as proposition surrogates with just a single topic x.

Before we leave the subject of proposition surrogates, there is one more and perhaps the philosophically most obvious way to interpret them and to define the "truth-functional" operations on them:

$$
\begin{aligned}
&{\sim}(A^+, A^-) = (A^-, A^+)\\
&(A^+, A^-) \wedge (B^+, B^-) = ((A^+ \cap B^+), (A^- \cup B^-))\\
&(A^+, A^-) \vee (B^+, B^-) = ((A^+ \cup B^+), (A^- \cap B^-)).
\end{aligned}
$$
[19]

[17] There is also another 4 element De Morgan lattice, which is at the same time a Boolean algebra, where $a$ and $b$ are complements of each other. This amounts to rotating the diagram both horizontally and vertically.

[18] It would be nice to fit van Fraassen's fact semantics under this same umbrella with various representations of De Morgan lattices.

[19] If one does not look too closely, there is an obvious similarity of these clauses to van Fraassen's clauses about $T(A)$ and $F(A)$, but his clauses are in fact more complicated because of the way the $\cdot$ operator enters in.

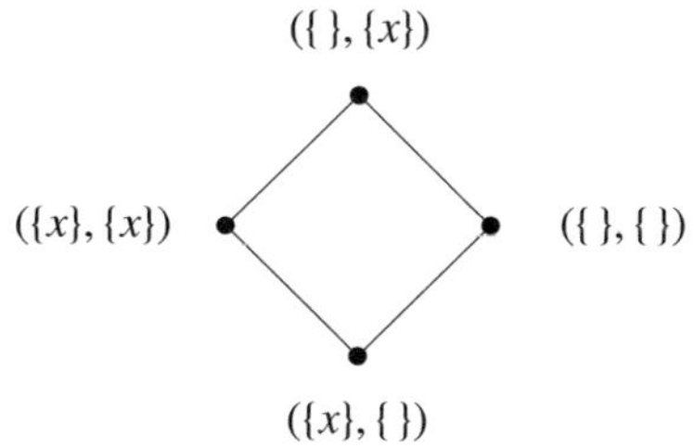


**Fig. 2** DM4 as Proposition Surrogates with Topic x

The obvious interpretation is to think of the elements of the set $U$ not as topics, but rather as say "situations." And then to think of $(A^+, A^-)$ so that $A^+$ is the set of situations where the proposition $A$ is true, and $A^-$ is the set of situations where the proposition $A$ is false. The the clauses above make perfect sense, except for two small details. Since it is allowed that $A^+ \cup A^- \neq U$ there can be situations in which $A$ is neither true nor false, and since it is allowed that $A^+ \cap A^- = \varnothing$ there can be situations in which $A$ is both true and false. I was not ready for these, particularly the last.

## 5 From Topics and Aboutness to Truth/Falsity/ Both/Neither

I do not remember why I finally got the nerve to talk about a sentence being both true and false as opposed to merely providing both definite positive information and definite negative information about a topic. As I said in Dunn (1976, fn 10):

> But I was aware of it at the time I was writing, and as I said I was aware of this as a 'formal' move when I wrote my dissertation using the 'topics' semantics, and in conversation then already talked of sentences being both true and false, and also neither, but I lacked the philosophical nerve to embrace this as a serious way of talking for about another year. This was because I somehow thought that it required that it should be possible for sentences to really be both true and false, or really be neither, and this seemed plain mad. I hope my presentation avoids that in Section 3.

The following paragraph was in Section 3 of Dunn (1976, p. 157):[20]:

> Do not get me wrong - I am not claiming that there are sentences which are in fact both true and false. I am merely pointing out that there are plenty of situations where we suppose, assert, believe, etc., contradictory sentences to be true, and we therefore need a semantics which expresses the truth conditions of contradictions in terms of the truth values that the ingredient sentences would have to take for the contradictions to be true.

The idea was then to allow sentences to take either the values **t**, **f**, or neither (this last was familiar from say the Łukasiewicz 3-valued logic), but also to take both of the values **t** and **f**. Again I was a bit cautious about this so I introduced the idea of a *valuation* $V$ as being a relation from sentences to the truth values **t** and **f**, and not just the usual function. This allows sentences to be related to just the value **t**, just the value **f**, both the values **t** and **f**, or to neither. I did use the image of $V_*(A) = \{v : AVv\}$, and of course $V_*$ is a function from the set of sentences to $\wp(\{\mathbf{t}, \mathbf{f}\})$,and can be viewed as a valuation satisfying:

[20] And also in Dunn (1969, p. 12).

$\mathbf{t} \in V_*(\sim A)$ iff $\mathbf{f} \in V_*(A)$;
$\mathbf{f} \in V_*(\sim A)$ iff $\mathbf{t} \in V_*(A)$;
$\mathbf{t} \in V_*(A \wedge B)$ iff $\mathbf{t} \in V_*(A)$ and $\mathbf{t} \in V_*(B)$;
$\mathbf{f} \in V_*(A \wedge B)$ iff $\mathbf{f} \in V_*(A)$ or $\mathbf{f} \in V_*(B)$;
$\mathbf{t} \in V_*(A \vee B)$ iff $\mathbf{t} \in V_*(A)$ or $\mathbf{t} \in V_*(B)$;
$\mathbf{f} \in V_*(A \vee B)$ iff $\mathbf{f} \in V_*(A)$ and $\mathbf{f} \in V_*(B)$.

I was fortunate to have the opportunity to present this semantics as an invited speaker in the 1969 joint APA-ASL symposium on Natural Language vs. Formal Language, and it gave me the opportunity to write it up (see Dunn (1969)) and circulate it. But I waited to publish this paper longer than I should. I was taken aback by the Routleys' (1972) publication of their "star-semantics" (see sec. 6 below), and felt I needed an excuse to publish my 4-valued semantics. My excuse was to link it to Richard Jeffrey's (1967) coupled truth trees. I will not go into the details here, but Jeffrey's idea (building on Raymond Smullyan's "Analytic Tableaux") was to validate $A \rightarrow B$ by working out the branching ways that $A$ could be true, working out the branching ways that $B$ could be true, and requiring that every branch of the first "cover" some branch of the second. Jeffrey got classical logic by two "tricks." The first trick, coming down from $A$, was to disregard those branches where an inconsistency $C$ and $\sim C$ occurs. The second trick, coming up from $B$, was to allow one to branch out at any time, to $C$ on the one hand and $\sim C$ on the other. This is how Jeffrey accommodated the classical irrelevancies $(A \wedge \sim A) \rightarrow B$, and $A \rightarrow (B \vee \sim B)$. I showed that if we did not allow these "work arounds" (and restricted $A$ and $B$ to truth functional formulas, i.e., formulas containing only negation, conjunction, and disjunction) we obtain exactly the first-degree relevant entailments of R. So with this addition, finally my paper was published as Dunn (1976).[21]

## 6 Belnap's Four Values

Nuel in 1977 published two very influential papers, "A Useful Four-valued Logic (Belnap 1977) and "How a Computer Should Think" (Belnap 1977a). The second of these was largely reprinted as sec. 81 of Anderson, Belnap, and Dunn (1992). As suggested by their titles, the papers were extremely well-written and well motivated. Nuel used the terminology "told values," and the metaphor of two different people populating an unstructured database, sometimes with incomplete and other times with contradictory information. And this was before the World Wide Web and its over 3 billion users!

In both papers he motivated the four-valued logic with a simple example about a computer database entry regarding whether the Pittsburgh Pirates or the Baltimore Orioles won the 1971 World Series. In both papers he puts the same set of constraints on the computer. I will not repeat all five of these here, but simply quote the fifth constraint as stated by Belnap (1977, p. 9; see also 1977a, p. 32):

> ... in answering its questions, the computer is to reply strictly in terms of what it has been told, *not* in terms of what it could be programmed to believe. For example, if it has been told that the Pirates won and din not win in 1971, it is to so report, even though we could of course program the computer to recognize the falsity of such a report.

[21] I feel I should mention that Dunn (1971) is not an abstract for this paper. It might be taken to be so because of an overlap in their titles ("Intuitive Semantics"), but it is instead a report on a 1967 ASL talk on the aboutness semantics and proposition surrogates from my dissertation.

This motivated Nuel's introducing what he called "told values":

> For each item in its basic data file, the computer is going to have it marked in one of the following four ways: (1) just the 'told True' sign, indicating that item has been asserted to the computer without ever having been denied.(2) Just the value 'told False', which indicates that the item has been denied but never asserted. (3) No 'told' values at all, which means the computer is in ignorance, has been told nothing. (4) The interesting case: the item might be marked with both 'told True' and 'told False'.

Nuel says that "These four possibilities are precisely the four values of the many-valued logic I am offering as a practical guide to reasoning by a computer," and he gives them the names **T**, **F**, **None**, and **Both**. He notes that these form a lattice:

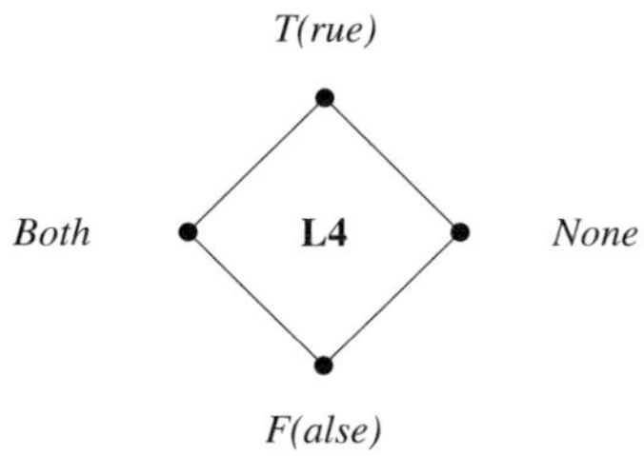


**Fig. 3** Belnap's Lattice L4

He gives "truth tables" for negation, conjunction, and disjunction, using these four values, and also observes that negation just rotates the diagram vertically, leaving *Both* and *True* as fixed points. These tables correspond to those of Smiley that we shall discuss below in section 7. Nuel Belnap observes that conjunction is greatest lower bound ($a \wedge b = glb\{a, b\}$) and that disjunction is least upper bound ($a \vee b = lub\{a, b\}$).

Nuel also examined a number of other topics in these two papers, including "formulas as mappings, "truth-functional inputs," "implicational inputs," quantifiers, and last but not least "approximation lattices." Except for the last, I think these have not gotten the attention they deserve. But I will guiltily follow suit and ignore all of these except for the last, which I will discuss in Section 9 below.

The relationship of Belnap's version of the "4-valued semantics" to the version I presented in Dunn (1969, 1976)[22] was perhaps not so obvious. This is frankly why I have taken the time to explain it here. Because of the nature of the 1969 ASL-APA symposium ("Natural Language versus Formal Language") in which I first presented these ideas, I gave my talk and wrote the accompanying paper (unpublished, available in this volume) in a very philosophical style with almost no symbols and certainly no mention of the 4-valued lattice that was so much a centerpiece of my dissertation. I thought this was a good way to help "sell" the ideas. But ironically it got in the way of connecting those ideas to the algebraic way of looking at things using the 4-element lattice made popular by Nuel. For no sooner had I done this than Nuel published his own version of the 4-valued semantics in the already mentioned two separate venues: Belnap (1977) and Belnap (1977a).

To be fair, in each of these Nuel carefully cited my work. Thus in Belnap (1977), p. 16, he says:

[22] Dunn (1976) was essentially reprinted in Anderson, Belnap, and Dunn (1976).

The logical system developed is precisely the system of "tautological entailments" of Anderson and Belnap (1975), Chapter III. That the system is characterizable by four values was noted, in correspondence, by T. J. Smiley. The connection of these values with subsets of {True, False} is due to Dunn, arising out of Dunn (1966), reported in Dunn (1969), and fully published in Dunn (1976).

And Belnap (1977a) says, after talking about Smiley's matrix:

Still later Dunn in 1966 gave a variety of semantics for tautological entailments, some of them highly intuitive, some closely related to the four-valued matrix of Smiley; and to him is due one of the central ideas with which I have been working, namely, the identification of the four values with the four subsets of {the True, the False}. Dunn in 1975 (belatedly) presents much of both the intuitive and the technical significance of this idea."[23]

## 7 Smiley's Four Values

These are mentioned in the two quotes from Nuel just above, but how did Tim Smiley enter the story? Anderson and Belnap in their investigations of the first-degree implications developed an 8 element matrix they called $M_0$, and Smiley (in correspondence) told them that it could be simplified to a 4-element matrix (cf. Anderson and Belnap (1975)) and still capture the same first-degree entailments.[24]

Smiley's Four-valued Matrix

| $\sim$ | |
|---|---|
| 1 | 4 |
| 2 | 2 |
| 3 | 3 |
| 4 | 1 |

| $\wedge$ | 1 | 2 | 3 | 4 |
|---|---|---|---|---|
| 1 | 1 | 2 | 3 | 4 |
| 2 | 2 | 2 | 4 | 4 |
| 3 | 3 | 4 | 3 | 4 |
| 4 | 4 | 4 | 4 | 4 |

| $\vee$ | 1 | 2 | 3 | 4 |
|---|---|---|---|---|
| 1 | 1 | 1 | 1 | 1 |
| 2 | 1 | 2 | 1 | 2 |
| 3 | 1 | 1 | 3 | 3 |
| 4 | 1 | 2 | 3 | 4 |

| $\rightarrow$ | 1 | 2 | 3 | 4 |
|---|---|---|---|---|
| 1 | 1 | 4 | 4 | 4 |
| 2 | 1 | 1 | 4 | 4 |
| 3 | 1 | 4 | 1 | 4 |
| 4 | 1 | 1 | 1 | 1 |

Notice that the matrix for $\rightarrow$ is essentially coding and on-off relation, with 1 (designated = on), and 4 (an undesignated = off). This can be seen as determining an ordering relation reflected by the following diagram:

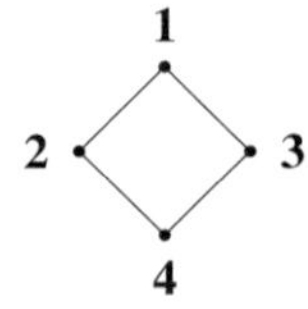


**Fig. 4** Lattice Order of Smiley's Matrix

[23] Two points: First, "Dunn 1975" is in Belnap's references as "forthcoming" and was actually published as Dunn (1976). Second, Belnap also goes on to cite van Fraassen (1966), and Routley and Routley (1972).

[24] It should be noted that $M_0$ is still of some importance since even though it is not needed to capture the provable first-degree entailments it still plays a useful role in capturing the provable "first-degree formulas." These are 'truth functions' of first-degree entailments and/or formulas containing no $\rightarrow$ at all (the 'zero-degree formulas'). See Belnap (1960).

This is of course again the famous De Morgan lattice DM4. Note that Smiley seemed to give no special meaning to the 4 elements (nor did Anderson nor Belnap), but merely labeled them as 1, 2, 3, 4, and made 1 the designated value.

## 8 Relation to the Routleys' * Semantics

Richard and Valerie Routley[25] (1972) published a semantics for first-degree entailments, often referred to as the "star semantics."[26] They call a set $H$ of formulas a "set up," saying (note, to match the usage to the rest of this article I use $\wedge$ where they used &):

> A *set-up* is a class of sentences or wff. An *&-normal set-up* $H$ is a set-up such that (i) $A \wedge B$ is in $H$ iff $A$ is in $H$ and $B$ is in $H$, for every $A$ and $B$. &-normal set-ups are vastly more liberal structures than the state-descriptions of Carnap or the worlds required by modal logicians for strict implication. For worlds and state-descriptions are constrained by the classical negation requirement:
> (ii) $\sim A$ is in $H$ iff $A$ is not in $H$. But this requirement is, we shall argue, very much too restrictive. (p. 337)

To remove the restrictions they introduce a function $^*$ from set-ups into set-ups to satisfies $H^{**}(A) = A$. They provide some motivations which we shall not go through here. In essence they require the following for what they call a a *normal set-up*:

(1) $A \wedge B \in H$ iff $A \in H$ and $B \in H$,
(2) $A \vee B \in H$ iff $A \in H$ or $B \in H$,
(3) $\sim A \in H$ iff $A \notin H^*$.

The important point is that while "set-ups" are analogous to "worlds" in say the Kripke semantics for modal logics, they are also vast generalizations, allowing for incompleteness and inconsistency. In Val and Richard's words:

> A set-up $H$ is a consistent and complete set-up, a world, iff $H^* = H$. For worlds, and only for worlds, non-membership of a statement universally coincides with membership of its negation. Worlds are those set-ups whose logical behaviour is indistinguishable from that of $G$, the actual world. ... Strict implication considers only worlds: but worlds by no means exhaust the class of set-ups needed for the assessment of entailment, as well as for many other intensional functors. (p. 339)

As Richard and Val say, set-ups are also analogous to Carnap's state descriptions, and this analogy plays a role in their formal definition, which starts with an *atomic setup*, which is simply a set of atomic formulas. They then use the clauses (1) - (3) above to inductively build up what they call a "normal set-up" from an atomic setup.

In a later publication (Routley and Meyer, 1973) the semantics for negation was used as a component for a full semantics for relevance logic, including nested implications. But the semantics took "set-ups" as primitive, not as sets of formulas. This amounts to taking the semantics of negation to be based on a frame $(U, ^*)$, where $U$ is a non-empty set of "set-ups," and $*$ is a function from $U$ into $U$ that is of period two, i.e., $x^{**} = x$. (Look

[25] They later changed their last names to Richard Sylvan and Valerie Plumwood.

[26] It also has often been referred to as "Australian Plan" in contrast to the four-valued "American Plan." I believe the first occurrence of this terminology occurred in a semi-published paper Meyer (1979) and first appeared in fully published form in Meyer and Martin (1986).

familiar? If not please review what was said above about Białynicki-Birula and Rasiowa's representation of De Morgan lattices as quasi-fields of sets.)

Val and Richard show familiarity with my dissertation.[27] They say:

> By combining range, defined as usual $r(A) =_{def} \{H : A \text{ is in } H\}$, with content we can derive as an associated modelling, and thereby provide an interpretation for, Dunn's modelling [they cite my dissertation) for f.d. entailments. An *rc model* (range content model) associates, with every atomic wff $P$ a pair $\langle r(P), c(P)\rangle = rc(P)$. $rc(P)$ may be extended recursively to all truth-functional wff by the following connexions - all of which we can of course prove
>
> $$rc(\sim A) = \langle r(\sim A), c(\sim A)\rangle = \langle c(A), r(A)\rangle;$$
> $$rc(A \wedge B) = \langle r(A) \cap r(B), c(A) \cup (B)\rangle;$$
> $$rc(A \vee B) = \langle r(A) \cup r(B), c(A) \cap (B)\rangle;$$
>
> Finally, for the given model,
>
> $A \to B$ iff $c(A) \supseteq c(B)$ & $r(B) \supseteq r(A)$, *i.e.* iff $c(A) \supseteq (B)$;
>
> *i.e.* entailment is content inclusion.

There was controversy about the Routleys' star-semantics, especially just after it first appeared. Jack Copeland (1979) wrote about the Routleys' semantics for first degree entailments, and also about Richard Routley and Robert K. Meyer's semantics for the systems R, E, etc. which incorporated it. But most of his negative comments were about the star-semantics and its lack of an intuitive interpretation. And Bob Meyer himself made fun of the star-semantics, at least at first. And in Dunn (1976) I described the switching of $a$ with $a^*$ as "a feat of prestidigitation."

I felt that way myself, which is what influenced me to develop first the topics semantics and then the 4-valued semantics. But many years later (Dunn 1993) I actually produced a somewhat intuitive way to think of the star-semantics. It starts with a (binary) incompatibility relation $\perp$ ("perp") on set-ups. I next introduced an order relation on set-ups, and saw that $a^*$ could be understood as the least upper bound of the set-ups compatible with $a$: $a^* = lub\{x : \text{not } x \perp a\}$. Thus, as I put it, $a^*$ is the strongest information state compatible with the information state $a$. I said "somewhat intuitive" because I presented this largely as a technical result, but Restall (1999) made a big deal of the intuitive character.

## 9 Bilattices, Trilattices, ...

Belnap (1977, 1977a) made clear the connection of the lattice DM4 with Dana Scott's work on continuous lattices, where he used these in his study of models for lambda calculi and the denotational semantics of computer programs. The connection is that the lattice DM4 could be viewed as having either a "logical order" or an "(information) approximation order." The first starts with F and goes through B, or N, to T (getting "more true"). The second starts with N and goes through T, or F, to B (getting "more information."). This gave rise to Ginsberg's (1988) concept of a "bilattice" and, in turn, Fitting's (1988)

[27] I should note that it is not clear that when the Routleys wrote their paper that they had actually seen a copy of my dissertation. I am not surprised by this since I first met Richard at the 3rd International Congress for Logic, Methodology, and Philosophy of Science, Amsterdam, 1967 (where I presented an algebraic completeness theorem for FDE relative to the icdl $M_0$ – see Dunn (1967)). I spent some hours explaining my dissertation to Richard while he took notes, as we sat on the stairs of the Grand Hotel Krasnapolsky. I cannot remember whether I explained the Białynicki-Birurula and Rasiowa representation theorem.

work on 4-valued logic programming.[28] Put quickly DM4 can be viewed as a lattice either vertically or horizontally. Belnap's two lattices are as follows:

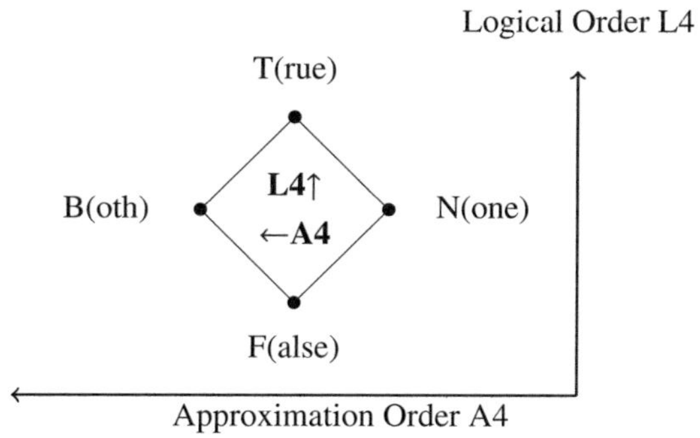


**Fig. 5** Logical Lattice L4 and Approximation Lattice A4 Combined

I have put these into a single diagram but Belnap had two separate diagrams. The main point is that the lattice diagram can be rotated $90^o$ and still be a lattice, but with a different order. The Logical Order has to do with preserving truth, whereas the Information Order has to do with increasing information.

Nuel and Dana spent some time together visiting at Oxford in early 1970. I have a letter from Nuel dated March 9, 1970, thanking me for my "elegant symposium paper," and saying he would be returning to Pittsburgh in a week.) Nuel says that my paper is "Beautifully put together, and beautifully written. It sings." But despite his praise, I think that Dana's work on continuous lattices was the driving force to Nuel's developing his 4-valued lattice.

I have a copy of an intensive and lengthy correspondence between Nuel and Dana Scott, with 11 typed letters or inserts dating from May 20 1970, to August 20, 1970. The correspondence was paper-clipped together, with also a handwritten 3 page note on top. Page 1 of the note contains the words "SCOTT CORRESPONDENCE" written as an addition near the top in what seems to be Nuel's handwriting. The note is titled: "Operations on $T$ – Belnap" and is dated January 30, 1970. I believe Nuel gave me the copy of the note sometime shortly after this correspondence ended.

The first paragraph of the note reads: "If we wish $\wedge$ and $\vee$ on $T = \{\bot, \mathbf{t}, \mathbf{f}, \top\}$ to form a lattice and to satisfy $\mathbf{t} \wedge \mathbf{f} = \mathbf{f}$ and $\mathbf{t} \vee \mathbf{f} = \mathbf{t}$ and to be continuous, then

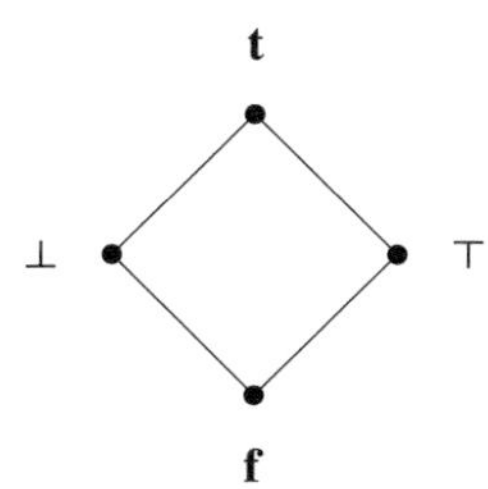


[28] Ofer Arieli, and Arnon Avron are two others that should be mentioned for their notable work on bilattices, starting in the 1990's. See Shramko and Wansing (2011) for references.

is the *only* possibility. And there is only one continuous function ~ such that ~**t** = **f**, ~**f** = **t**: namely, $\sim\top = \top$ and $\sim\bot = \bot$." Belnap (1977, 1977a) actually contain a formal proof of these statements using Scott's notion of a continuous lattice.[29]

Later, when Yaroslav Shramko and Tatsutoshi Takenaka were visitors at IU, I worked with them on 8 and 16 valued structures that we called "trilattices," because they could be viewed as lattices in 3 different ways. Shramko then went on with Wansing to develop even more trilattices. See their book Shramko and Wansing (2011)[30] and/or their *SEP* article Shramko and Wansing (2018).

Recent work by Jc Beall and Caleb Camrud shows that after combining the two classical values to get four values that characterize FDE, then the process of continuing to combine values 'all the way up' to many values, for any ordinal , results in the same system of logical consequence, i.e., FDE.

## 10 4-valued Semantics for the Full System R of Relevant Implication

Routley and Meyer (1973) gave a semantics that characterizes the whole of the system R (and also the system E). It deftly combined the Routleys' star-semantics for negation with a 3-placed accessibility relation to model implication. It is then an obvious and important question whether the four-valued logic can be similarly extended to nested implications.

The earliest attempt at this seems to be in Meyer (1979) – but as is explained in Anderson, Belnap, and Dunn (1982, p. 171) while the attempt seems successful, "it is at the price of great technical complexity," and also the star "can still be said to sneak in through the back door."

The best result along these lines is due to Ed Mares (2004). As he explains (p. 328) in motivating his result:

> To extend this [four-valued] semantics to full systems of relevant logic, one needs truth and falsity conditions for implicational formulae. A truth condition was given in Richard Routley and Robert Meyer's two-valued semantics. The short story is that in Routley (1984) the Routley–Meyer semantics is combined with Dunn's four-valued semantics to give a model theory for very weak systems of relevant logic. Routley's theory is ingenious but very complicated. The two-valued semantics uses a ternary relation on worlds to give a truth condition for implication. Routley adds a second ternary relation to give a falsity condition for implication. In addition, Routley's soundness proof requires that we show both that for every implicational theorem, $A \to B$, in every world in which $A$ holds, $B$ obtains as well and that in every world in which $B$ is false, $A$ must be false also. In the standard Routley–Meyer semantics, we only need to show that the first of these conditions obtains. This addition makes soundness proofs difficult, especially with the more complicated implicational axioms of stronger relevant logics.

Ed goes on to say that "The situation was improved greatly by Greg Restall (1995)," since "Restall's model theory utilizes only one ternary accessibility relation," and the natural falsity condition. "Restall provides semantics for systems weaker than the strong relevant logic, R, but not for R itself."

[29] Nuel's letter of June 12, where he copied me, says he is sending to Scott an abstract by me which points out all the senses in which the 2-element Boolean algebra's relation to the lattice above is like the general relation of Boolean algebras to De Morgan algebras. Scott replies in his letter of June 22 "Dunn's lattice is sideways. Thanks for the paper, though. I will think about it more. All these studies are connected."

[30] For obvious reasons having to do with the title of my present essay, I love the names of Ch. 3 (Generalized Truth Values: From $FOUR_2$ to $SIXTEEN_3$) and Ch. 4 ((Generalized Truth Values: $SIXTEEN_3$ and Beyond). But where is Infinity? :)

Ed does provide a 4-valued semantics for all of R (also a new two-valued semantics motivated by 4-valued considerations), saying (p. 328):

> But there is a catch. The present semantics are not a pure frame theories. They include a frame (a structure based on a set of worlds) and a set of propositions, together with an operator on propositions. More concisely, the present semantics is a neighbourhood semantics for R.

Ed adds:

> The reader is justified in being disappointed upon finding out that the elegance of the theory is gained by using neighbourhoods, because neighbourhood semantics are rather easily produced by, in effect, reproducing the axioms of the logic in the semantics. But the present semantics is not like this. The conditions on the operator, which is used to characterize negation, are all in effect taken from the falsity conditions for the various connectives, creating a more elegant class of structures.

So maybe this is as good as it gets.

## 11 A 3-valued Cousin of the 4-valued Logic

Let us take a step backward on our path to Infinity, and consider the 3-element De Morgan lattice DM3, which arises naturally in both the Łukasiewicz logics (as his 3-valued logic – see Łukasiewicz(1920)) and in the extensions of the semi-relevant logic R-Mingle (see Dunn 1970):

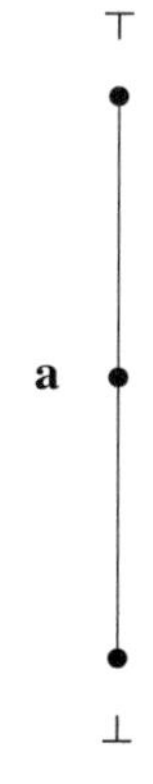


**Fig. 6** 3-element De Morgan Lattice DM3

DM3 may be viewed as a sublattice of DM4 in two difference ways: either as the left-hand side or as the right-hand side. There are several different ways to define a consequence lattice using DM3 (see Dunn (2000)), The most "relevant" of these is to view DM3 as the 3-element Sugihara lattice $S_1 + 0$, where $\top = +1, \bot = -1$, and $\mathbf{a} = 0$. **a** is viewed (along with $\top$) as a "designated element." This is in effect to interpret **a** as $\{\mathbf{t}, \mathbf{f}\}$, $\top$ as $\{T\}$, and $\bot$ as $\{\mathbf{f}\}$.

This has been an important lattice in the study of (what Arnon Avron has dubbed) the "semi-relevant" logic R-Mingle (RM), as the smallest "Sugihara algebra" that is not a

Boolean algebra (cf. Dunn (1970)). In Dunn (1976a) I showed how to provide a "Kripke semantics" for RM using valuations into DM3 defined relative to "situations" on which there was a linear accessibility relation. I later began to use the term "information state" (because, as I have joked in more than one lecture, "it provided more funding opportunities").

## 12 Infinity: The Opinion Tetrahedron

Having started with Two, let us end then with Infinity. Dunn (2010)[31] embedded the four-valued lattice DM4 into the apexes of the "Opinion Tetrahedron." In the diagram below we picture how the 4 element De Morgan lattice DM4 can be expanded to an infinite valued tetrahedron.

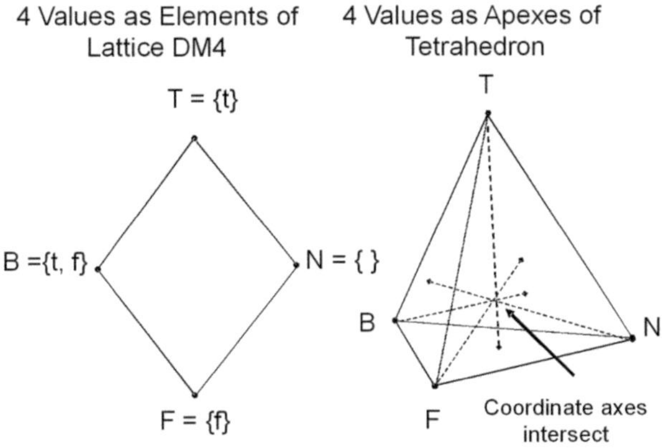


**Fig. 7** DM4 and the Opinion Tetrahedron

The Opinion Tetrahedron was created by me as an extension of Audun Jøsang's (1997) "Opinion Triangle," which cleverly accounted for degrees of uncertainty, using "Barycentric coordinates, but in just one sense, that of degrees of ignorance (lack of evidence). The Opinion Tetrahedron allows also for degrees of conflict (contradictory evidence). Each point in the Opinion Tetrahedron is an ordered quadruple of real numbers $(b, d, u, c)$, each of which lies in the unit interval $[0, 1]$. These are akin to subjective probabilities, and "normalized" to require that they sum to 1. Degree of belief is $b$, degree of conflict is $d$, degree of ignorance is $u$, and degree of conflict is $c$. Who would have guessed in the 1960's that this could have come out of my dissertation.

[31] Or maybe we should have strictly followed Gamow, and started with One, the "values" for the universal dialethic logic. :)

## 13 Dunn-Belnap, or Belnap-Dunn Four-valued Logic, or . . . ?

Clearly I have a stake in this, but I do not think I am biased. I in fact think Belnap-Dunn should be preferred to Dunn-Belnap. There is a kind of relevant parallel in the case of what is often referred to as the *Anderson* and Belnap (italics mine) system of first-degree entailment. I think it should simply be the Belnap system of first-degree entailment. Let me explain. First what are the first-degree entailments? They came into existence with Anderson and Belnap's system E of Entailment, where they are the provable formulas of the form $\varphi \to \psi$ where $\varphi$ and $\psi$ contain no occurrences of $\to$, i.e., they have only the truth functional connectives $\wedge$, $\vee$, and $\sim$ – put simply, there are no arrows nested within arrows. That is a good definition, but that does not actually make it a system of logic. It was first presented as a system of logic in Belnap's 1959 dissertation – see fn 8 above.

So Belnap first created the formal logic of first-degree entailments. Smiley provided the interpretation of it in terms of an abstract 4-valued matrix. And Dunn first provided the intuitive interpretation of it. Therefore it should be called the "Belnap-Smiley-Dunn Four-valued Logic."

But we should not just stop here without acknowledging that ideas suggesting the 4-values (True, False, Both, None) date back to classical Indian logic (Sanjaya's "Four Corners"), prior to the 6th century B.C.E. See Dunn (1999, 2000) for references. So maybe it should be called the "Sanjaya-Belnap-Smiley-Dunn Four-valued Logic." :)

## References


1. Alan Ross Anderson and Nuel D. Belnap (1962), "Tautological Entailments," *Philosophical Studies*, vol. 13, pp. 9-24.
2. Alan Ross Anderson and Nuel D. Belnap (1975), *Entailment: The Logic of Relevance and Necessity, Vol. I*, Princeton University Press.
3. Alan Ross Anderson, Nuel D. Belnap, and J. Michael Dunn (1992), *Entailment: The Logic of Relevance and Necessity, Vol. II*, Princeton University Press.
4. Yehoshua Bar-Hillel. and Rudolph Carnap (1952), "An Outline of a Theory of Semantic Information," Technical report 247, Massachusetts Institute of Technology, Research Laboratory of Electronics. http://hdl.handle.net/1721.1/4821.
5. Jc Beall and Caleb Camrud (2009), "A Note on FDE 'All the Way Up'," draft, http://entailments.net/papers/beall-camrud-fde.pdf.
6. Nuel D. Belnap (1959), *A Formalization of Entailment*, Yale University Ph. D. dissertation.
7. Nuel D. Belnap (1959a), (Abstract) "Tautological Entailments," *Journal of Symbolic Logic*, vol. 24, p. 316.
8. Nuel D. Belnap (1960), *A Formal Analysis of Entailment*. Technical Report No. 7, Office of Naval Research, Group Psychology Branch, Contract No. SAR/Nonr-609(16), New Haven.
9. Nuel D. Belnap (1967), "Intensional Mocels for First Degree Formulas," *The Journal of Symbolic Logic*, vol. 32, pp. 1-22.
10. Nuel D. Belnap (1977), "A Useful Four-Valued Logic," *Modern Uses of Multiple-Valued Logic*, Invited Papers from the Fifth International Symposium on Multiple-valued Logic, Held at Indiana University Bloomington, May 11-16, 1975, ed. J. M. Dunn and G. Epstein,with a Bibliography on Many-valued by Robert G. Wolf, D. Reidel, 1977, pp. 8-37.
11. Nuel D. Belnap (1977a), "How a Computer Should Think," *Contemporary Aspects of Philosophy*, ed. G. Ryle, Oriel Press, pp. 30-55.
12. Nuel D. Belnap and Joel H. Spencer (1966), "Intensionally Complemented Distributive Lattices," *Portugaliae Mathematica*, vol. 25, pp. 99-104.
13. Andrzej Białynickii-Birula and Helena Rasiowa (1957), "On the Representation of Quasi-Boolean Algebras," *Bulletin de l' Académie Polonaise des Sciences*, vol. 5, pp. 259-261.

14. Katalin Bimbó and J. Michael Dunn (2001), "Four-valued Logic," *Notre Dame Journal of Formal Logic*, vol. 42, pp. 171-192.
15. B. J. Copeland (1979), "On When a Semantics Is Not a Semantics: Some Reasons for Disliking the Routley-Meyer Semantics for Relevance Logic," *Journal of Philosophical Logic*, vol. 8, pp.399-413.
16. J. Michael Dunn (1966), *The Algebra of Intensional Logics*, Ph. D. Dissertation, University. of Pittsburgh, 1966. University Microfilms.
17. J. Michael Dunn (1967), (Abstract) "An Algebraic Completeness Proof for the First Degree Fragment of Entailment," *Abstracts of Papers: Third International Congress for Logic, Methodology and Philosophy of Science*, Amsterdam, p. 9.
18. J. Michael Dunn (1967a), (Abstract) "The Effective Equivalence of Certain Propositions about de Morgan Lattices," *Journal of Symbolic Logic*, vol. 32, pp. 433-434.
19. J. Michael Dunn (1969), "Natural Language vs. Formal Language," unpublished manuscript written for presentation as an invited speaker (together with Frederic Fitch, Bas van Fraassen, and Richard Montague) in the joint symposium by that title of the Association for Symbolic Logic and the American Philosophical Association at their joint meeting, December, 1969, 14 pp. double-spaced typewritten. A transcript and additional information is available in the present volume.
20. J. Michael Dunn (1970), "Algebraic Completeness Results for R Mingle and its Extensions," *Journal of Symbolic Logic*, vol. 35, pp. 1-13.
21. J. Michael Dunn (1971),(Abstract) "An Intuitive Semantics for First Degree Relevant Implications," *Journal of Symbolic Logic*, 36, pp. 362-363.
22. J. Michael Dunn (1976), "Intuitive Semantics for First Degree Entailments and Coupled Trees," *Philosophical Studies*, 29, pp. 149-168.
23. J. Michael Dunn (1976a), "A Kripke Style Semantics for R-Mingle Using a Binary Accessibility Relation," *Studia Logica*, 35, pp. 163 172.
24. J. Michael Dunn (1985), "Relevance Logic and Entailment," in *Handbook of Philosophical Logic*, vol. 3, eds. D. Gabbay and F. Guenthner, D. Reidel, Dordrecht, Holland, 1985, pp. 117-224. Revised with Greg Restall (2002) as "Relevance Logic," *Handbook of Philosophical Logic*, 2nd edition, vol. 6, eds. D. Gabbay and F. Guenthner, Kluwer Academic Publishers, pp. 1-128.
25. J. Michael Dunn (1993), "Star and Perp: Two Treatments of Negation," in *Philosophical Perspectives* vol.7: *Language and Logic*, ed. James Tomberlin, pp. 331-357.
26. J. Michael Dunn (1999), "A Comparative Study of Various Semantical Treatments of Negation: A History of Formal Negation," in *What is Negation?*, eds. D. Gabbay and H. Wansing, Kluwer Academic Publishers, pp. 23-51.
27. J. Michael Dunn (2000), "Partiality and its Dual," *Partiality and Modality*, eds. E. Thijsse, F. Lepage & H. Wansing, special issue of *Studia Logica*, vol. 66, pp. 5-40.
28. J. Michael Dunn (2008), "Information in Computer Science," *Handbook of the Philosophy of Information*, eds. J. van Benthem and P. Adriaans, Elsievier. pp. 581-608.
29. J. Michael Dunn (2010), "Contradictory Information: Too Much of a Good Thing," *Journal of Philosophical Logic*, vol. 39, pp. 425-452.
30. J. Michael Dunn (2016), "An Engineer in Philosopher's Clothing," autobiography in In *J. Michael Dunn on Information Based Logics*, ed. Katalin Bimbó, Springer, Outstanding Contributions to Logic, vol. 8., pp. xvii-xxxiii.
31. J. Michael Dunn (2017), "Humans as Rational Toolmaking Animals," in *Modern Logic: Its Subject Matter, Foundations and Prospects*, in Russian, ed. D. Zaitsev. Moscow: Forum, pp. 128-160.
32. J. Michael Dunn and Nuel D. Belnap (1967), "Homomorphisms of Intensionally Complemented Distributive Lattices, *Mathematische Annalen*, vol. 126, pp. 28-38.
33. Bas C. van Fraassen (1969), "Facts and Tautological Entailments," *The Journal of Philosophy*, vol. 66, pp. 472-487.
34. Bas C. van Fraassen (1973), "Extension, Intension and Comprehension," in *Logic and Ontology*, ed. M. Munitz, New York University Press, pp. 101-131.
35. George Gamow (1947), *One, Two, Three...Infinity: Facts and Speculations of Science* (1947, revised 1961), Viking Press (copyright renewed by Barbara Gamow, 1974), reprinted by Dover Publications, eBook edition, Dover, 2012.
36. Matthew L. Ginsberg (1992), "Multivalued logics: A Uniform Approach to Inference in Artificial Intelligence," *Computational Intelligence*, vol. 4, pp. 256-316.
37. Nelson Goodman (1961), "About," *Mind*, vol. 70, pp. 1-24.
38. Pragati Jain (2000), "Saptabhaṅgī: The Jaina Theory of Sevenfold Predication: A Logical Analysis," *Philosophy East and West*, vol. 50, The Philosophy of Jainism, pp. 385-399.
39. Richard Jeffrey, (1967), *Formal Logic: Its Scope and Limits*, McGraw-Hill.

40. Audun Jøsang, A. (1997). "Artificial Reasoning with Subjective Logic,".in *Proceedings of the Second Australian Workshop on Commonsense Reasoning*. Perth.
41. Łukasiewicz, J. (1920), "O logice trojwartosciowej," *Ruch Filozoficny*, vol. 5, pp. 170-171. English translation in: *Łukasiewicz, Selected Works*, L. Borkowski (ed.), Amsterdam: North-Holland and Warsaw: PWN, 1970.
42. Edwin D. Mares (2004), "'Four-Valued' Semantics for the Relevant Logic R," *Journal of Philosophical Logic*, vol. 33, pp. 327-341
43. Richard M. Martin (1976), *Semiotics and Linguistic Structure: A Primer of Philosophic Logic*, SUNY Press, 1978.
44. Robert K. Meyer (1979), "A Boolean-valued Semantics for R," Research Paper No. 4, Logic Group, Department of Philosophy, Research School of Social Sciences, Australian national University.
45. Robert K. Meyer and Errol P. Martin (1986), "Logic on the Australian Plan," *Journal of Philosophical Logic*, vol. 15, pp. 305-332
46. A. Monteiro (1960), "Matrices De Morgan caratérisquiues pour le Calcul Propositionnel Classique," *Anais da Academia Brasileira de Ciencias*, vol. 32, pp. 1-7.
47. G.C. Moisil (1935), "Recherches sur l'algèbre de la logique," Ann. Sci. Univ. Jassy, vol. 22, pp. 1-117.
48. Greg Restall (1999), "Negation in Relevant Logics (How I Stopped Worrying and Learned to Love the Routley Star)," in *What is Negation?*, eds. D. Gabbay and H. Wansing, Kluwer Academic Publishers, pp. 53-76.
49. Richard Routley (1984), "The American Plan Completed: Alternative, Classical-style Semantics, without Stars, for Relevant and Paraconsistent Logics, *Studia Logica*, vol. 43, pp. 327-376.
50. Richard Routley and Valerie Routley (1972), "Semantics of First Degree Entailment,". *Noûs*, vol. 6, pp. 335-359.
51. Yaroslav Shramko and Heinrich Wansing (2005), "Some Useful 16-valued Logics: How a Computer Network Should Think," *Journal of Philosophical Logic*, vol. 34, pp. 121-153.
52. Yaroslav Shramko and Heinrich Wansing (2011), *Truth and Falsehood: An Inquiry into Generalized Logical Values*, Trends in Logic, vol. 36, Studia Logica Library, Springer
53. Yaroslav Shramko and Heinrich Wansing (2018), "Truth Values", *The Stanford Encyclopedia of Philosophy* (Spring 2018 Edition), Edward N. Zalta (ed.), https://plato.stanford.edu/archives/spr2018/entries/truth-values/. See also supplement "Generalized Truth Values and Multi Lattices," linked from above URL or at https://plato.stanford.edu/archives/spr2018/entries/truth-values/generalized-truth-values.html.
54. Marshal H. Stone (1936), "The Theory of Representations for Boolean Algebras," *Transactions of the American Mathematical Society,* vol. 40, pp. 37-111.
55. Stephen Yablo (2014), *Aboutness*, Princeton University Press.

# Interview with Prof. Nuel D. Belnap

Nuel D. Belnap and Heinrich Wansing

**Abstract** The interview between Nuel D. Belnap and Heinrich Wansing took place in Pittsburgh on November the 5th, 2015. The text below is a slightly edited version of the transcript based on the recording.[1]

Heinrich Wansing (H.W.): It's a great pleasure to see you again, Nuel, and to have the opportunity to talk to you about logic. My co-editor of the planned volume on FDE, Hitoshi Omori, and I are very grateful for your readiness to give us an interview.

There is a certain basic relevance logic called 'system of tautological entailments', 'Belnap's useful four-valued logic', 'Belnap and Dunn's four-valued logic', '$\mathbf{E}_{fde}$', 'first degree entailment logic',[2] or just 'FDE', and our interview will focus on that system. In your papers with Alan Ross Anderson, 'Tautological entailments' (*Philosophical Studies* 13 (1962), 9–24) and 'First degree entailments' (*Mathematische Annalen* 149 (1963), 302–319) you give a presentation of FDE as a system of tautological entailments and a sequent calculus for FDE, respectively, and there is also your abstract 'Tautological entailments' in the *Journal of Symbolic Logic* in 1959. What can you say about the origins of FDE? Does FDE already appear in your Yale University doctoral dissertation from 1959?

Nuel Belnap (N.B.): I think so; I think that this was part of my dissertation, isolating first degree entailments.

H.W.: Yes, as a subsystem of system **E**.

N.B.: Yes, of system **E** of entailment, that was in my dissertation.

H.W.: And do you remember where and when the term 'first degree entailments' first came up?

N.B.: Yes, I think in my dissertation. And also 'tautological entailments' – that phrase was in there.

H.W.: The dissertation remained unpublished, right?

N.B.: Correct.[3]

H.W.: Then very prominent places where FDE appeared are Chapter III of *Entailment. Volume I* (Princeton UP, 1975), Michael Dunn's paper 'Intuitive Semantics for First Degree

Nuel D. Belnap
Department of Philosophy, University of Pittsburgh, Pittsburgh, USA
e-mail: belnap@pitt.edu

Heinrich Wansing
Insitute of Philosophy I, Ruhr-Universität Bochum, Bochum, Germany
e-mail: heinrich.wansing@rub.de

[1] We are grateful to Mrs Claudia Smart for the careful transcription and to Nuel Belnap for his approval of it.

[2] or 'first-degree entailment logic' (with a hyphen).

[3] The doctoral dissertation, *The Formalization of Entailment*, Yale University, 1959, was, however, essentially "semi-published" as a Technical Report, no.7, by the Office of Naval Research, New Haven, Contract No. SAR/Nonr-609(16) in 1960.

 
H. Omori, H. Wansing (eds.), *New Essays on Belnap-Dunn Logic*,
Synthese Library 418, https://doi.org/10.1007/978-3-030-31136-0_7

Entailments and "coupled trees"' (*Philosophical Studies* 29 (1976), 149–168), and then, of course, your seminal papers 'How a computer should think' and 'A useful four-valued logic', both published in 1977.[4] In the latter paper, you mention correspondence with Timothy Smiley, where you say that he noticed that FDE is characterizable by four values. How did this exchange, this correspondence with Smiley, come about?

N.B.: To the best of my memory, Smiley was a friend of Alan Anderson's and that would have come from Alan's stay in England and I don't know what year it was, but he spent some time with Smiley at that time and that was the connection.

H.W.: Do you still have that correspondence?

N.B.: No, I don't have any correspondence whatsoever.

H.W.: What a pity.

N.B.: It is, yes, it was thrown all out a long time ago.

H.W.: In 'A useful four-valued logic' you also make some other comments on connections to your work. You point out that the representation of the four values as subsets of the sets of classical truth values, *the True* and *the False*, is due to Michael Dunn, arising out of his dissertation in 1966, and reported in a manuscript of a talk on 'Natural Language versus formal language', which Michael presented at a joint APA/ASL meeting in New York in December 1969. Both Michael Dunn's dissertation and the 1969 paper have remained unpublished so far,[5] and it is only in his 1976 paper 'Intuitive Semantics for First Degree Entailments and "coupled trees"', where he finally published the semantics using a relation between formulas and the classical semantical values **T** and **F**, as well as another semantics in terms of "proposition surrogates". Michael Dunn completed his doctoral dissertation at the University of Pittsburgh under your supervision. How did Michael and you interact in research related to FDE and relevance logic?

N.B.: I don't know how to answer that question. We spent a lot of time together, but I don't have any way to characterize the nature of our work together. I just don't remember.

H.W.: The supervision of a PhD student can mean quite different things.

N.B.: It was more collaboration. No doubt about that.

H.W.: Some people have weekly meetings, some people have meetings every now and then. Was it a very close collaboration?

N.B.: We spent *a lot of time* together. It was much more than a weekly meeting, but it is so long ago that I don't remember the details.

H.W.: Mike's dissertation is on algebraic logic, so it's very mathematical work, but he nevertheless did this dissertation in the philosophy department, right?

N.B.: Yeah. He brought in the algebra, taught it to me.

H.W.: Does it mean that you and maybe other philosophers in Pittsburgh at that time were very open-minded towards formal, mathematical work?

N.B.: Surely so. We were very happy about the mathematical connections of logic. I think that's your question.

---

[4] N.D. Belnap, 'How a computer should think', in: G. Ryle (ed.), *Contemporary Aspects of Philosophy*, 1977, Stocksfield, Oriel Press, 30–56. N.D. Belnap, 'A useful four-valued logic', in: J. M. Dunn and G. Epstein (eds.), *Modern Uses of Multiple-Valued Logic*, Dordrecht, Reidel, 1977, 5–37.

[5] The paper is available on the Web:
http://www.philosophy.indiana.edu/people/papers/natvsformal.pdf, and it appears in the present volume.

H.W.: Yes, that was my question. As I said before, you point out several connections to your work explicitly in 'A useful four-valued logic'; you also mention an exchange with Dana Scott during your stay in Oxford in 1970. How did this interaction or cooperation go and look like? Did you see Dana Scott quite frequently? Did you have a lot of discussions with him?

N.B: Yes, we overlapped for about a month at Oxford, and we had a number of sessions together. My memory of the details is gone, so I don't know any proper answer to "how did this interaction go or look like?". I just remember spending time with him.

H.W.: But if you overlapped only for a month, ...

N.B.: Yes, he was finishing a stay, and I was beginning a stay.

H.W.: Yes, but if it was just for a month that you overlapped there, then I would say that it must have been a very fruitful exchange.

N.B.: It was for me, yes. It meant a great deal to me.

H.W.: In the "Editor's prefatory" to 'A useful four-valued logic', Michael Dunn and George Epstein refer to some correspondence between Dana Scott and yourself in which Scott remarks that his approximation lattice A4, as you then called it, is your logical lattice L4 "on its side". Do you still have that correspondence with Dana Scott, or is it—like the other correspondence—gone?

N.B.: No, it was in the same filing cabinet that got thrown out.

H.W.: Yes, that happens. And there is still another connection you draw attention to in 'A useful four-valued logic'. You say that Stewart Shapiro had "independently argued the utility of 'relevance logic' for question-answering systems, and suggested implementation in a research proposal," and you refer to a forthcoming paper by Shapiro and Mitchel Wand ('The Relevance of relevance') that is available as technical report no. 46 at the Computer Science Department of Indiana University in Bloomington. Does that mean you had some co-operation with Stewart Shapiro at that time?

N.B.: Yes, but I don't remember. I remember writing what you just said, but I don't remember the connection with Stewart.

H.W.: We could ask Stewart as well, and see what he remembers.

N.B.: He might know.

H.W.: Ok, so these were some questions related to connections to your work you mentioned in 'A useful four-valued logic'. Let us now also talk about the four values and the particular interpretation you gave to them in 'How a computer should think' and 'A useful four-valued logic.' There you explain that you are thinking of a reasoner who is supposed to use the four-valued logic as an "artificial information processor; that is, a (programmed) computer". Moreover, you suppose the information processor to be a question-answering system being able to answer questions not only based on information given to or stored by the system, but also on the basis of deductions. Another assumption is that the information processor may or may not receive information concerning the truth or falsity of atomic propositions from different, in general trustworthy sources, none of which, however, can be assumed to be a universal truth-teller. Then, of course, the problem of processing inconsistent information and absence of information arises. Now I come to my question: How did you come up with this scenario? Was the idea more or less "in the air" due to the emerging computer or information age?

N.B.: No, I've been working on questions, that was with Thomas Steel, or under his supervision. We didn't work together much, but we talked a lot, and that was at the System

Development Corporation, where I spent a number of weeks I guess in the summer, when he was employed there.

H.W.: Was this a private company?

N.B.: I think I'd want to say more like "semi", it wasn't like a grand corporation, it was publicly funded, so in that sense ... but it was a separate company. It wasn't a division of the government, but it was funded by the government.[6]

H.W.: Does it mean you took some time off from university to visit him there?

N.B.: No, it was in the summer. It was summer employment really. I took a month out there in California with my family.

H.W.: Ah, great. So was it indeed this work on the logic of questions that resulted in the book with Thomas Steel (*The Logic of Questions and Answers*, Yale UP, 1976) that inspired you when you proposed this particular motivation for the four-valued logic?

N.B.: That's right.

H.W.: Ok, I said that we should also talk about the four values, so there are these four values, **T** ("just told True"), **F** ("just told False"), **None** ("told neither True nor False") and **Both** ("told both True and False"), and it is quite clear that they are not intended as ontic values.

N.B.: Correct.

H.W.: So in particular it is clear that the values **None** and **Both** are not meant to indicate that a sentence or proposition is indeed neither true nor false, or both true and false.

N.B.: Correct.

H.W.: Then it is quite natural to ask whether this rejection of an ontic reading of the four values reflects your attitude towards genuine truth value gaps and towards the so-called dialetheism, the view that there are sentences which are both true and false. Do you disbelieve that there are sentences that are neither true nor false and do you think that there exist dialethia?

N.B.: I still don't really know what to make of this. There's not a problem about sentences which are neither true nor false, but sentences that are both true and false. I don't really know what to make of that in an ontological way. I think that's an answer to your question.

H.W.: It certainly is an answer to my question. It's often claimed that one can appreciate paraconsistent logic, or relevance logic, where *ex contradictione quodlibet* fails, without necessarily then being committed to dialethia, and as far as I understood your answer, you are critical or cannot make much sense of dialethia, is this correct?

N.B.: That's correct. I've never been tempted in that direction.

H.W.: Interesting. Since the four values are certainly not intended as ontic values, it is perhaps natural to see them as epistemic values. In the "Editors' prefatory" to 'A useful four-valued logic' in *Modern Uses of Multiple-Valued Logic* (p. 6), Michael Dunn and George Epstein explain that 'Belnap's values are intended as epistemic, rather than ontic, in character," and you yourself speak of their "epistemic interpretation" (p. 16 of 'A useful four-valued logic') and say in *Entailment. Volume II* (Princeton UP, 1992, p. 521) that the "four values are unabashedly epistemic". But then you maybe recall that we both have a little note on generalized truth values (*Logic Journal of the Interest Group in Pure*

[6] The reader might want to consult this link:
https://en.wikipedia.org/wiki/System_Development_Corporation.

*and Applied Logics* 18 (2010), 921–935), where we emphasize that in *Entailment. Volume II* you explain that "sentences are to be marked with either a **T** or an **F**, a **None** or **Both**, according to what the computer has been told; or with only a slight (but dangerous) metaphor, according to what it believes or knows." Thus there you say that this is a slight but dangerous metaphor to say "according to what it believes or knows", so then strictly speaking, the epistemic reading is a metaphor and the proper reading is not epistemic but informational. Would you still agree?

N.B.: I'm not sure what the difference is.

H.W.: Well, the difference we point out in our note is that you can have some information, but you don't necessarily believe it, so that one would have to draw a distinction between the information one has and the doxastic attitude one has towards the propositions in question, so that although it's tempting to contrast this ontic reading, which is not intended, with an epistemic reading, strictly speaking, the scenario is about information, and the information concerning the truth or falsity of given atomic propositions is passed on to the computer.

N.B.: I guess I never wrestled with the distinction between informational and epistemic, so I never—probably wrongly—sort of run those together, so I don't have anything useful to say about that, I think.

H.W.: But would you think that it makes sense to draw a distinction between belief as a doxastic attitude and just information that one has available, or not? I don't mean to push you into a confession right now.

N.B.: I think I don't know. My own thoughts on the topic are just not good enough to make a sharp difference between informational values and epistemic values. Maybe I just haven't thought enough about it.

H.W.: But of course we can detach the notion of information from contrasting it with the notion of belief, and actually that brings me to my next question. Some logicians try to give a semantics for relevance logics in terms of information states that may support the truth of a proposition, for instance, so they explicitly talk about information. Do you endorse a particular notion of information?

N.B.: I haven't thought that much about it, so the answer to that question is "no", but it's not clear to me what it is to have information, I guess, in distinction from believing information. I like to have information without attaching some kind of belief state to it, either positive or negative.

H.W.: I can try to make that question a bit more concrete: There's a controversy about whether information is or should be understood as factive, as truthful or not, whether there can be false information or not. So there's a notion that's called "factive semantic information" and one of its defenders is Luciano Floridi.[7] According to him it's helpful to think of information as factive or truthful. And there is for instance also Michael Dunn, who thinks differently about information, and actually I agree with him on that, and I have a quote here from a paper Michael Dunn published in a volume that appeared in a collection of essays with the German title: 'Zwischen traditioneller und moderner Logik. Nichtklassische Ansätze' ('Between traditional and modern logic. Non-classical approaches', Mentis Verlag, Paderborn, 423–447), where he says "Information is what is left of knowledge

[7] The reader might want to consult Floridi's entry on information in the Stanford Encyclopedia of Philosophy, http://plato.stanford.edu/entries/information-semantic/.

when one takes away belief, justification, and truth ... Information is ... a kind of semantic content—the kind of thing that can be expressed by language." And elsewhere he says it's something like a Fregean thought. Are you sympathetic towards this?

N.B.: It doesn't strike me as making a whole lot of sense. 'Information' must have many different uses, but if I get having information, then I'm not quite sure how to distinguish if you have information that's true or false, but information that is not believed or not true, what that means.

H.W.: If we really think of Fregean thoughts here, then of course Frege had this idea of grasping a thought, so it's just the pure propositional content which you have or possess if you grasp it.

N.B: That's ok, I don't mind that.

H.W.: But then it seems to me it makes sense to say that I can grasp a thought, but need not have the attitude of belief towards it.

N.B.: Surely so. I'm just uncomfortable with 'information' as a synonym if you like for 'thought'. And it's not something I thought about deeply. Potential information, that's easier to digest than some written stuff, you don't have but you could have.

H.W.: One could also think of some so far not grasped thought that you could grasp, that may not be written, but could be really more abstract. And the factivity of information, do you have an opinion on the alleged factivity of information?

N.B.: Again, if information comes with the phrase 'having information', that seems to be factive, and I'm not sure what to do with information that isn't had.

H.W.: Yes, but would one also have to say that one can have false information, does that sound reasonable to you?

N.B.: Yes it does.

H.W.: Good, so now we have been talking a bit about the status of the four values, whether they are ontic, epistemic, or informational, whatever that would mean exactly. I think we should also talk about what Alan Ross Anderson and you refer to as "the heart of logic", in the very first paragraph of *Entailment. Volume I*, namely the notion 'if, ... then'. So, if First Degree Entailment Logic is based on the language with negation, conjunction, and disjunction, it has no primitive implication connective. The idea of defining implication as Boolean, i.e., defining 'if *A*, then *B*' as 'not *A* or *B*' is presumably not completely exotic, although adherents of relevance logic would consider it – I guess – as completely misguided. Material implication as a defined connective in FDE has a remarkable property; modus ponens fails, and in *Entailment Volume I* (on p. 259) we are told that "of course $\overline{A} \vee B$ is no kind of conditional, since *modus ponens* fails for it." In *Entailment. Volume I* we also find a lengthy discussion of the deduction theorem, where it is claimed that "no kind of deduction theorem ... is sufficient to make us want to call" the connective highlighted in it "an 'implication connective'." What is called for is a relevant deduction theorem, an entailment theorem, which is proved for system **E** in §23.6 of *Entailment Volume I*. This observation leads me to another question: The logic FDE is the implication-free fragment of David Nelson's constructive paraconsistent logic with strong negation ...

N.B.: I'm a little confused.

H.W.: Yes?

N.B.: The first degree entailment has a single arrow, so that's intended to mean "implication", so I'm not sure what to make of saying First Degree Entailment is implication-free. There aren't any nested implications. I'm not being argumentative, I'm just not clear.

H.W.: The implication there can be seen as the representing the entailment relation, right? So if you see it as that, then you have only conjunction, disjunction, and negation.

N.B.: It's a funny way to put it, suppose you don't have conjunction, then you don't have conjunction. You are saying something different.

H.W.: Yes. Ok, but it happens that the valid first degree implications coincide with the valid entailments, where you replace the arrow with a symbol for a semantic entailment relation. So in that sense, you have just conjunction, disjunction, and negation. That's the way first degree entailment logic is represented for instance in several publications, for instance in Graham Priest's textbook on non-classical logic. He presents FDE as a system with conjunction, disjunction, and negation, and then you have the entailment relation. It's not the arrow, in the object language, but the metalanguage notion of entailment. Would you find that a misrepresentation of first degree entailment?

N.B.: Yes, I guess so. That's a perfectly coherent language, just the truth-functional connectives. I think that's not the same language as when I say, now I'm going to put an arrow in between formulas and give it a meaning, and not allow nesting.

H.W.: But of course to have the entailment relation symbol nested would be very unusual. There are also approaches where the sequent arrow is nested, but to nest the entailment relation symbol is – I don't know – it's not something people think about. Ok, but then you could say that if you do not nest the arrow, the 'if, ... then', and you look at the valid first degree implications, it is as if you just treat it as an entailment relation symbol, a non-nested arrow, but you seem not to be very comfortable with this perspective.

N.B.: Say that perspective over again?

H.W.: In the object language you have non-nested 'if, ... then' and if you look at the valid first degree entailment formulas, then they coincide with the valid inferences, where you just have the semantical consequence symbol. Would you still say it's not justified to think of first degree entailment as being a system in this conjunction, disjunction, negation language?

N.B.: Well, it's certainly not what I intended.

H.W.: I see.

N.B.: I intended a language with a non-nestable arrow.

H.W.: So you saw it as a fragment of system **E**.

N.B.: Right. I mean, that somebody could think of it differently ...

H.W.: In Nelson's logic **N4**,[8] we add intuitionistic implications to this conjunction, disjunction, negation fragment, and it satisfies of course then the official non-relevant deduction theorem, and there's a simple and principled way of turning the intuitionistic implication into a relevant one. In a sequent calculus presentation one would just give up the thinning (or weakening or monotonicity) rule. Such a step was taken in various places, for instance also in my dissertation from 1993 (*The Logic of Information Structures*, Springer, 1993). But then we end up living in what you have called "the undistributed middle." If we also drop out contraction we get linear logic; we have relevance logic, but we do not have distribution of conjunction over disjunction and disjunction over conjunction. So we have this paper 'Life in the undistributed middle' in the volume on substructural logics, edited by Kosta Došen and Peter Schroeder-Heister (*Substructural Logics*, OUP, 1993,

[8] A. Almukdad and D. Nelson, "Constructible falsity and inexact predicates", *Journal of Symbolic Logic* 49 (1984), 231–233.

31–41). At that time, in 1993, distributionless relevance logics were referred to by you as non-standard. In particular you emphasized (p. 34) that "[t]he context of first degree entailments (tautological entailments) makes distribution seem almost indispensable: This logic has distributive normal forms in its very conception." Did your opinion on distributionless relevance logics change over the past 22 years? I mean, when you wrote this paper, this was in the early nineties, and linear logic just became very popular at the time, but many years have passed since then, so did you maybe change your attitude towards distributionless relevance logics, like linear logic since then?

N.B.: No, I don't think so.

H.W.: You don't think so, okay.

N.B.: I'm a fan of distribution, and I see that doesn't come cheap, so I have both of those thoughts, but I think a logic without distribution is hamstrung. It can't really do what we want for us when formalizing some reasoning on the one hand, but on the other hand I think if it doesn't arise out of some other principles which are very natural, it has to be added if you wanted it to be added. Is that an answer to your question?

H.W.: Yes, I think so, so you would still opt for adding it?

N.B.: I would.

H.W.: This brings me to my next question. Are your favourite relevance logics still the systems **E** and **R** you investigated together with Alan Ross Anderson?

N.B.: Well, you know, paternity ...

H.W.: Yes, you cannot escape paternity. I have many more questions, so let's carry on. I think as I already mentioned, FDE has reached textbook status a long time ago. One textbook I have in mind is Graham Priest's *An Introduction to Non-classical Logic: From If to Is* (Cambridge UP, 2nd edition 2008), where three different semantics are presented: Michael Dunn's semantics using a relation between formulas and truth values, your many-valued truth table semantics, and also the Routleys' star semantics, in which the interpretation of negated formulas makes use of a function that sends the world of evaluation to another semantical index. Graham Priest also mentions the algebraic semantics due to Michael Dunn, which is presented in detail in Chapter III, §18 of *Entailment. Volume I.* Some authors seem to have clear preferences with respect to these equivalent semantics, and here I have an example, where authors present a certain preference for one type of semantics. In a paper on a natural deduction proof system for FDE by Allard Tamminga and Koji Tanaka, that appeared in the *Notre Dame Journal of Formal Logic* in 1999, we find the following statement (p. 258):

> Anderson and Belnap provide a Hilbert-style system and a Gentzen-style system for FDE. Although they give characteristic matrices, Anderson and Belnap do not provide any formal semantics for FDE. For this, we had to wait for Routley and Routley ... and Dunn .... Routley and Routley provide a two-valued semantics for FDE. Although their semantics may be philosophically contentious, it serves as a basis for the semantics for various relevant logics. However, ... we are concerned only with Dunn's semantics, which is somewhat more intuitive. Together with a tableau system, Dunn presents an "intuitive" formal semantics for FDE.

Now my brief question: There are these different semantics for FDE. Do you have a preference for a certain semantics?

N.B.: I don't think I do. Could you give the choices again?

H.W.: The choices on offer are: the relational semantics due to Michael Dunn, the four-valued truth table semantics, the Routley star semantics, and the algebraic semantics.

N.B.: They all seem to me to shed light on formal systems, so in that sense I don't have any preference. I think the one I can understand mostly is the truth-table semantics. I don't know how much ice that cuts.

H.W.: So you would say they all give a different insight into the system?

N.B.: Yes.

H.W.: I have another, sort of related question: In §18 of *Entailment. Volume I* (p. 205), Michael Dunn writes that "We take the value of algebraic proofs to lie partly in the fact that they enable us to compare different systems." Some philosophers are critical about algebraic completeness proofs. Johan van Benthem in his chapter 'Correspondence theory' in the *Handbook of Philosophical Logic. Volume II* (1984) writes on p. 205, that to some the uses of modal algebra "show that the algebraic approach is merely 'syntax in disguise' ", and I think this term 'syntax in disguise' has been frequently used over the years and it certainly stresses the critical attitude towards algebraic semantics and algebraic completeness proofs. What is your opinion on algebraic semantics?

N.B.: I think it's very helpful. I don't know what to make of the phrase 'syntax in disguise'. I suppose in some sense all semantics is based on syntax, it has to be, it's part of the semantics. I don't know what the disguise is.

H.W.: Ok, so this is the sort of critical attitude one can find towards algebraic semantics, and another kind of semantics already mentioned is this Routley star semantics, and it has also been criticized. For instance Jack Copeland is well known for having drawn the distinction between pure and applied semantics in the paper 'On when a semantics is not a semantics: Some reasons for disliking the Routley-Meyer semantics for relevance logics', *Journal of Philosophical Logic* 8 (1979), 399–413), and according to him the ternary frame semantics for relevant logics and, in particular also the Routley star semantics fail to be an applied semantics. Since the logic we are focusing on has a Routley star semantics, my question is: what is your opinion on the Routley star semantics? Do you have a specific opinion concerning it?

N.B.: I think if I would put it together with the algebraic semantics that it is a formal device that, as far as it goes, it hasn't that much to do with the way we reason.

H.W.: So is the Routley star semantics for you in the first place a formal tool that one can use to obtain some results about the logic that is characterized?

N.B.: It can be an attractive formal enterprise, but I don't find it satisfying and it can't give the meaning of the language.

H.W.: So are you more critical towards the Routley star semantics than towards the algebraic semantics?

N.B.: I think I don't want to be critical of either of them, they are both formal devices and they only get to be informal if you are doing the work, and I myself don't know how to do the work of the star semantics. I'm not being entirely coherent, I know, but I'm just muddling around the topic, but that's the way I feel at the moment. Two-valued semantics is the one to contrast with, it seems to be closely rooted to the way we think in the application of logic, and there certainly can be applications for many-valued logic, but they have to tell me what the applications are.

H.W.: One prominent application is, of course, due to you. The application where you think of information passed to a computer, a model which has found many applications also in knowledge presentation, and people try to model inconsistency tolerant reasoning by basing description logics on paraconsistent logics. My next question is about a particular feature of First Degree Entailment Logic: FDE has the basic contraposition rule 'if $A$ entails $B$, then $\sim B$ entails $\sim A$' as an admissible rule; in the axiom system for $\mathbf{E}_{fde}$ in §15.2 of *Entailment. Volume I* it is a primitive rule. Basic contraposition does not extend to FDE with multiple-premises inferences.

N.B.: You mean flipping the conclusion with one of the premises?

H.W. : Yes, we have, say, two premises, $A$ and $B$, and they entail $C$, and then you keep $A$ and have $\sim C$, and you get the entailment to $\sim B$.

N.B.: That's what I meant. That's what I thought you meant, that you just interchange one of the conclusions with one of the premises.

H.W.: That's indeed what I meant by 'contraposition for multiple-premises inferences'. So then this extended notion of contraposition fails. Even basic contraposition fails for strong negation in Nelson's 3–[9] and 4–valued logics, and Nelson also considers a variant of his 3-valued logic with a contraposable strong negation, but in this system, which he called **S**,[10] the contraction axiom is given up in order to avoid collapse into classical logic, but that's perhaps not so bad at all because the contraction axiom gives rise to Curry's paradox. Do you have any particular view about the desirability of the basic contraposition rule? Some logics have it, some logics don't have it.

N.B.: Now you are talking about multiple premises?

H.W.: No, but we can talk about both, but let us start with single premises.

N.B.: I like contraposition. Is that an answer to your question?

H.W.: Yes, certainly it is an answer, but what does it mean for logics which do not have contraposition? Would you consider them as defective or problematic in some sense?

N.B.: Logics can serve a variety of purposes, I expect, but I'd want to see what the application was, but if it's supposed to represent some generalization of the way we ought to reason, then I wouldn't want to do without contraposition, because I think we ought to use it.

H.W.: This single premise basic contraposition rule is an admissible rule in FDE, and I talked for instance to Graham Priest about this, and he said it's sort of surprising that once you add intuitionistic implication to the system, as in Nelson's logic, then you lose it. And when I think of it, once you draw a clear distinction between truth and falsity, and you do not have the classical picture of falsity as a non-truth, so as in the four-valued logic, then why should one suppose that if $A$ entails $B$, it should also be the case that if $B$ is false, then $A$ is false? You'd have a clear separation between truth and falsity if they are two independent concepts. So one might be then surprised to find that contraposition is an admissible rule of FDE, and if one is surprised, then one is perhaps not so concerned about the failure of the basic contraposition rule, say in Nelson's logics. So you think, if I understood you correctly, that basic contraposition is useful.

N.B.: Yes. I think we use it, and I'd hate to do without it. I'd hate to be restricted to Nelson's logics for my own reasoning, and for your reasoning, too. But certainly I'm not going to object to a formal system that lacks contraposition. That might be interesting and useful.

H.W.: Another reason why Belnap and Dunn's useful four-valued logic has become so important and popular is that we have these two lattices, the logical lattice and what you

[9] D. Nelson, "Constructible falsity", *Journal of Symbolic Logic* 14 (1949), 16–26.

[10] D. Nelson, "Negation and separation of concepts in constructive systems", in: A. Heyting (ed.), *Constructivity in Mathematics*, Amsterdam, North-Holland, 1959, 208–225.

call the approximation lattice, and together they form the smallest non-trivial logical bilattice, which is now sometimes called $FOUR_2$ because we have the four values and the two lattice orderings, and the four values of this smallest logical bilattice, they are generalized truth values in the sense of being representable as elements of the power set of the given set of classical semantical values. Take the set of classical truth values and form the powerset, and then we have a set of generalized truth values, which are just the values of your four-valued logic or representations of the four values. Now, as you know, I had a co-operation with Yaroslav Shramko in a project on generalized truth values, which we to a large extent summarized in the book *Truth and Falsehood. An Inquiry into Generalized Logical Values* (Springer, 2011), and the starting point for that project was a quote from Bob Meyer, where he answers to a question he himself posed, namely, the question was: "If we take seriously both true and false and neither true nor false separately, what is to prevent our taking them seriously conjunctively?", and the answer is "This way, in the end, lies madness."

N.B.: That's what Bob said?

H.W.: That's what Bob said, yes. Shramko and I suggested that instead of ending up in madness, we obtain something very useful, namely a useful 16-valued logic that gives rise to another structure, not a bilattice but a trilattice, which we called $SIXTEEN_3$, a trilattice of generalized truth values. We claimed that one advantage of $SIXTEEN_3$, as compared to the bilattice $FOUR_2$, is that whereas the logical ordering of $FOUR_2$ (or of the logical lattice L4) is both a truth and falsity ordering, in $SIXTEEN_3$ we have two separate logical orderings, a truth ordering and a falsity ordering. The definition of the truth ordering refers only to the presence or absence of the classical value *the True* in a generalized truth value, and the falsity ordering refers only to the presence or absence of the classical value *the False* in a generalized truth value, so we can really separate these two orderings. This approach then also leads to a generalized view of logics, namely to thinking of logics as multiple consequence systems, as tuples consisting of a language and more than just one entailment relation. We have this logical falsity ordering and this logical truth ordering, and they both give rise to separate entailment relations. What is your opinion on such a conception of a logical system, a system that has more than one entailment relation or more than one consequence relation?

N.B.: I like it, it seems to me that it could be useful.

H.W.: I have another question that is related to both, to the bilattice $FOUR_2$ and to the trilattice $SIXTEEN_3$. The top-element of L4 is the value **T** ("just told True"), and the top element of the truth order of $SIXTEEN_3$ is $\mathbf{TB} = \{\mathbf{T}, \mathbf{B}\}$. This position of **TB** is justified by the definition of the truth ordering in terms "truth-containment", i.e., in terms of containing or not containing the classical value *the True* as an element of a generalized value. Nevertheless to some it seems tempting to consider the value $\{\mathbf{T}\}$ as "more true" than **TB**. There's a related question concerning FDE. In the truth-table semantics for FDE both the values **T** and **B** are designated values. Recently, in a paper by Andreas Pietz (Kapsner) and Umberto Rivieccio ('Nothing but the truth', *Journal of Philosophical Logic* 42 (2013), 125–135), it is suggested to consider a variant of FDE, which they call "exactly true logic", ETL, where one only has **T** as a designated value. So we take the same truth tables, but designate only **T** ("just told True"). But the logic is not paraconsistent, it's not a relevance logic. Would you say that this feature overrides the intuitive appeal of designating just **T** instead of both **T** and **B**?

N.B.: I just don't have a feeling about that. I never thought about that in that way at all, and I can't come up with a comment.

H.W. Ok, but I would guess that the fact that the logic is not paraconsistent and that it's also not a relevance logic is a drawback in your view, or isn't it?

N.B.: It surely is.

H.W. : Okay, now I'll come to another topic, but I try to establish a relation to our main topic, FDE. In Chapter V of *Entailment Volume I*, we have some paragraphs on connexive logic, written by Storrs McCall (§29.8 – §29.8.4). McCall refers to the systems of connexive logic presented there as neighbors of **E**, and expresses the hope that they "still fall within the embrace of **E**'s Good Neighbor Policy" (p. 435), so maybe they also fall within the embrace of $\mathbf{E}_{fde}$'s Good Neighbor Policy, and now we are back to tautological entailments. Connexive logic has recently received some new interest, as witnessed, for example, by an entry in *The Stanford Encyclopedia of Philosophy*, and a forthcoming special issue on connexive logics in the *The IfColog Journal of Logics and their Applications*. What is your opinion on connexive logics and, in particular, on their characteristic principles usually called Aristotle's Theses: $\sim(\sim A \rightarrow A)$, $\sim(A \rightarrow \sim A)$ and Boethius' Theses: $(A \rightarrow B) \rightarrow \sim(A \rightarrow \sim B)$, $(A \rightarrow \sim B) \rightarrow \sim(A \rightarrow B)$?

N.B.: I don't really have an opinion, I mean, what I love is Storrs, and that's why that was included in our book, but the system itself is not something I ever developed a fondness for.

H.W.: The early systems of connexive logic have been much criticized because of their unintuitive semantics or their complicated semantics, or having some not so convincing properties and features. But what in my view connexive logic is to a large extent about is how we think of the falsity conditions of implications. So the orthodox view is that '*A* implies *B*' is false if '*A*' is true and '*B*' is false, and the connexive view is that '*A* implies *B*' is false iff '*A* implies not *B*' is true. And it seems that people who have not been affected by logic courses, when they are asked when is '*A* implies *B*' false, they often say, well, if '*A* implies not *B*' is true. Do you think that this could be an interesting conception of negated implications, to say that '*A* implies *B*' is false iff '*A* implies not *B*' is true?

N.B.: I just want to say that whether it's interesting or not doesn't depend on the beginning, I think, but how it works out. That's not a short answer to your question, but I don't have a short answer to that.

H.W.: Now I have a much more general question: As one of the founding fathers of relevance logic and one of the most influential relevance logicians, what is your opinion of classical logic, or say intuitionistic logic? Are those rival or maybe just different logics, which have played, and are still playing, an important role in many fields? How would you think about them and their usefulness in comparison to the useful four-valued logic?

N.B.: I guess I think they are both more useful and descriptively accurate, especially when I think of the useful four-valued logic. It is at best an acquired taste, but we know there can be other uses, they can be properly characterized in terms of intuitionistic logic, and I think that's also true of classical logic, but I don't think there probably is a community properly characterized for the four-valued logic. There should be I guess, and there are some strands, of course. There's the rejection of the *ex falso quodlibet* that seems to be descriptively accurate, more or less.

H.W.: But if so, then classical and intuitionistic logic are descriptively inadequate because they validate *ex falso quodlibet*.

N.B.: Yes, but I don't take that very seriously because there don't seem to be any formal logics that are not going to be adequate, to be able to characterize somebody who was raised that way.

H.W.: The next question, as it appears, is perhaps my final question, and it's also related to classical logic. There is some discussion about what kind of logic one should use as a meta-logic for non-classical logics. There are papers where people said that they have presented an intuitionistically acceptable completeness proof for intuitionistic logic, for instance. Would you agree that classical logic is appropriate as a meta-logic for non-classical logics such as FDE?

N.B.: I would.

H.W.: Why?

N.B.: It's because I want to use it. I don't want to be limited how I talk about a particular logic. I mean even classical logic has got a limitation to it, but I'd rather not think about the kind of limitation that leads to intuitionistic logic, four-valued logic etc. when I'm thinking.

H.W.: But would you find it interesting to, say, prove the completeness of a certain non-classical logic only by means that are acceptable according to that non-classical logic?

N.B.: Absolutely. An entrancing prospect.

H.W.: There's actually a very closely related question: When it comes to contraction, the contraction rule in the sequent calculus or the contraction (or absorption) axiom in axiomatic proof systems, it's known that these are problematic principles because they give rise to the Curry paradox. So if one is serious about this, and wants to avoid the Curry paradox by giving up the contraction rule in the sequent calculus or the contraction axiom in a Hibert-style system, then one may wonder how much mathematics one can do if one uses a contraction-free logic. Would you think that this is also an interesting enterprise to explore?

N.B.: I want to say something like 'yes' and 'no'. I think it's an interesting prospect but in my own experience it doesn't lead anywhere, and one can do so little in one's own thinking without relying on some form of contraction, but using the same premise twice—I would not want to be limited in that way.

H.W.: Let me put the question somewhat differently: Suppose you would have to evaluate a research proposal for a project investigating how much mathematics can be done in contraction-less logic, would you support such a project?

N.B.: Sure.

H.W.: You would, okay.

N.B.: Oh yes, I think it would be interesting.

H.W.: Now that I have worked through my list of questions, let me thank you very much, Nuel, for your patience and your answers to all these questions.

N.B.: You are more than welcome.

# Part II
# New Essays

# FDE as the One True Logic

Jc Beall

**Abstract** The principal aim of this paper is to very briefly discuss one way in which FDE may be 'the one true logic' – an idea which neither Belnap nor Dunn have ever been keen to promote, but an idea which, properly understood, is worth having plainly on the table. After presenting the simple sense in which FDE may be plausibly thought of as the one true logic, I briefly rehearse a simple argument for thinking as much.[1] There are two other aims of this paper, one of which is to keep the paper very short. The last but not least aim is to honor the pioneering work of Belnap and Dunn on FDE and its philosophical applications. Let me say from the start that the paper necessarily fails to achieve its last aim, not for lack of aiming or even for poorly aiming; rather, the honor that both Belnap and Dunn deserve for their work is not something that this paper – perhaps any paper – can reflect. By my lights, their work on the topic (both mathematical and philosophical) is what we should all aspire to do. Both logic and its philosophy are vastly better for their efforts.[2]

**Keywords** Logical consequence • FDE • Paraconsistent • Paracomplete • Consequence relations • Universal closure • Universal consequence • Subclassical logic • Philosophy

## 1 FDE as the one true logic?

In what sense, if any, might it be plausible that FDE is The One True Logic? This is a question for which neither Belnap nor Dunn (nor a past proper part of myself!) have had much use. But I've come to think that there is some value in the question, at least in philosophy (if not in logic qua mathematics or even philosophically driven mathematics). But the first thing to do is to clarify the question. After doing that, I sketch an answer, pointing to other papers to fill out some of the details for anyone wishing to pursue them. I leave debate to decide the fate of the answer.[3]

Jc Beall
Department of Philosophy, University of Connecticut, Storrs, CT, USA
e-mail: jcbeall@gmail.com

[1] The argument is one I have advanced elsewhere [7] but it seems especially appropriate to repeat in this volume.

[2] I also want to explicitly record a small personal note to both Professor Belnap and Professor Dunn. Some people make a lot of their 'intellectual heritage' or 'family line' in PhDs and so forth. I'm not one of those people. Still, if one looks carefully beyond the exact letters of my 'lineage', one sees that I learned a lot of my logical background – though perhaps not logical proclivities – from Gary Hardegree, who gained much of his own logical background from Professor Dunn, who in turn gained valuable logical background from Professor Belnap. As a personal note to Nuel and Mike: thank you both. You remain, in so many ways, key sources in my intellectual development. (I'm sorry for the deviant turn that my own views have taken!)

[3] Important background note(s):

 
H. Omori, H. Wansing (eds.), *New Essays on Belnap-Dunn Logic*,
Synthese Library 418, https://doi.org/10.1007/978-3-030-31136-0_8

## 1.1 The role of logic: closure

'Logic' means many things to many users of the term.[4] From a mathematical perspective, anything that looks sufficiently like a formal entailment or consequence relation is a logic. And that's fine, as far as it goes. But there are sensible – even important – debates in philosophy on whether logical consequence (henceforth, logic) is (non-) classical. To make the discussion easier (and slightly more biased), focus on current debates over whether logic is *sub*classical – that is, properly weaker than classical logic. If just any old formal entailment-ish or consequence-ish relation is logical consequence then such apparently sensible (even important) debates are less than fruitful; such 'debates' would be just cross-talk. But they clearly aren't.

What, then, is being debated when philosophers debate whether logic (i.e., logical consequence) is subclassical? The question demands a prior answer: what is it that logic is taken to be in the debate? And here I think that the most fruitful (and, in fact, true) answer points to a familiar and traditional role: namely, the universal, basement-level closure relation involved in all of our true theories. Let me elaborate.

## 1.2 Theories and the twofold task of theorists

The account of logical consequence as the universal, basement-level closure (consequence) relation makes most sense when we think of theories along standard lines as *closed* theories – theories closed under some closure/consequence relation.[5]

The concern throughout is with truth-seeking disciplines wherein theorists aim to give the true and complete-as-possible theory of their target phenomenon. The task is twofold:

1. Formulate the truths about the target phenomenon and put them into the theory.
2. Construct the closure/consequence relation for the theory.

---

a. I assume throughout that the FDE-related work (including philosophical work) of Belnap and Dunn are sufficiently known to readers of this volume that rehearsal or citation is mostly unnecessary, especially so because (I presume) the editors present an overview for the volume.
b. Relatedly: I assume that FDE itself is presented in various ways in the introduction to this collection. Throughout this particular discussion I shall think of FDE largely from a so-called semantic or model-theoretic perspective, and in particular thinking of the relation as a formal entailment (i.e., lack-of-counterexample) relation over some space of 'logical possibilities' (viz., true, false, both, neither), where the relevant 'forms' are defined via standard first-order vocabulary without identity (and without function signs). (So, think of the boolean quartet – viz., truth (null) operator, falsity (negation) operator, conjunction, disjunction – together with first-order universal and existential quantifiers, and the only conditional involved in logic's said vocabulary is the material conditional defined via logic's falsity and disjunction connectives, namely, $A \supset B$ is defined as $\neg A \vee B$.)

[4] And, of course, nowadays there are five largely independent branches of the field of logic: model theory, proof theory, recursion theory, set theory, and philosophical logic – each branch having many, many numerous subbranches [8]. My concern throughout is with a small pocket of philosophical logic.

[5] For convenience I conflate closure relations (operators) and their corresponding consequence relations, but I presuppose the standard Tarskian idea of a closure operator defined via a 'prior' consequence relation: $Cn(X) = \{A : X \vdash A\}$ where $\vdash$ is the prior consequence relation. This conflation is safe given that I assume all standard properties (including so-called structural properties) of consequence that are required to induce a closure operator standardly understood (namely, idempotent/'transitive', increasing/'monotonic', extensive/'reflexive').

Here, the first task is as usual: after identifying the target phenomenon we start to record the truths about the phenomenon, recording them in a 'seed theory' (so to speak) to be 'completed' by the theory's consequence relation.

Where is logic (-al consequence) in this picture? The answer is clear. We are thinking of our true theories as closed theories, each with its own closure relation:

$$\langle T_1, \vdash_{\mathrm{T_1}} \rangle, \langle T_2, \vdash_{\mathrm{T_2}} \rangle, \langle T_3, \vdash_{\mathrm{T_3}} \rangle \ldots, \langle T_n, \vdash_{\mathrm{T_n}} \rangle$$

*Logical* consequence is in each and every one of these theories; it's at the bottom level, the foundational (nonempty) closure relation on top of which the theorist's second task begins.

But why does the theorist need to construct a consequence/closure relation for her theory if the theory already contains one (viz., logic)? The answer is plain in every direction, but consider a very simple and common example, namely, the true theory of (say) metaphysical necessity (or the true theory of which truths are metaphysically necessary). This theory relies on extra-logical, theory-specific vocabulary that's peculiar to its target phenomenon. One such bit of extra-logical vocabulary is the unary operator *it's metaphysically necessary that...* (say, $\mathbb{L}$). Now, let $p$ be logically atomic (i.e., contains no logical vocabulary). Logic itself – the universal closure relation – says about $\mathbb{L}p$ what it says about all claims: namely, that it interacts with logical vocabulary just so (where the *just so* spells out logically valid forms defined only over logical vocabulary). In particular, logic itself sees many, many counterexamples to the form $\mathbb{L}p \therefore p$. And that's the common rub. Our true theory of metaphysical necessity requires that $p$ be entailed by $\mathbb{L}p$, that $p$ be in our theory given that $\mathbb{L}p$ is in the theory. And that's why the theorist has the second task to complete. The theory of metaphysical necessity – similarly, of salamanders, tractors, quarks, triune gods, Pittsburgh logicians, Indiana logicians, and of anything else in reality (if any more there be) – will be vastly incomplete with respect to truths about its target phenomenon if the theory is closed only under logical consequence. Clearly.

## *1.3 Logic, its vocabulary, and the trinity of virtues*

On this very familiar picture logic is involved in every (true and complete-as-possible) theory. Two features of this familiar picture of logic jump out: a plausible (though not neutral or profound) account of logical vocabulary, and the trinity of traditional virtues of logical consequence. Each in turn.

### 1.3.1 Logical vocabulary

It is notoriously difficult to give an illuminating account of 'logicality' or 'logical vocabulary' that both gets enough vocabulary (e.g., standard first-order, shy of identity) while not getting too much (e.g., second-order vocabulary or the like). Alas, I have no solution, and no new tweaks to 'invariance' or the like that does the trick. What I do suggest is that the picture of logic qua universal closure (on all of our true theories) at least motivates the thought that standard first-order vocabulary (viz., boolean connectives and first-order quantifiers) is the logical vocabulary.

On the current picture, all true (and complete-as-possible) theories are closed under logic; it's just the basement-level closure relation on all such theories. But, then, logical vocabulary appears in all such (true, complete-as-possible) theories. And while there's no proof or even empirical study that suggests as much, it's highly plausible that the standard stock of first-order logical vocabulary is the vocabulary appearing in all true theories. One can't say as much about (for example) higher-order vocabulary, or notions of truth/falsity, or even notions of entailment itself (including, of course, logical entailment). These are one and all important topics that enjoy (still highly contested) rival candidates for true theories about them; but these are not pieces of vocabulary that figure in all of our true theories.

Let me be clear that I am not arguing by stipulation that since logical vocabulary is (stipulated to be) the vocabulary in all true theories, the standard stock (sans identity, function signs, etc.) is logical vocabulary. The claim is simply that on the picture of logic qua basement-level closure involved in all true theories, there's a natural account of what counts as logical vocabulary – and that that account suggests that the standard first-order vocabulary gets things roughly right.

### 1.3.2 Trinity of traditional virtues

In addition to carrying a natural account of logical vocabulary, the picture of logic qua basement-level closure sits very nicely with a trinity of traditional virtues of logic.

#### Universal

That logic is universal on the given picture is clear: logic is involved in each and every true theory, and hence there's no element of reality beyond the reach or demands of logic.

#### Topic-neutral

Similar to its universality, that logic is topic-neutral on the given picture is clear: regardless of the true theory's topic, logic is involved; it doesn't discriminate between topics; what's logically (in-) valid in theory $T$ is equally logically (in-) valid in theory $T'$, for all true theories $T$ and $T'$.[6]

#### Intransgressible

Finally, that logic is intransgressible is also obvious: your true theory cannot transgress logic because logic is part of the theory – full stop.

[6] Along these lines, a better slogan than Carnap's famous one about morals is that in logic there is no discrimination (of topics).

## 2 FDE as the one true logic!

The sense in which we might think of FDE as ('the one true') logic invokes the universal, basement-level closure role of logical consequence. But is there any degree of plausibility to the idea that FDE plays that role – at least versus the other standard candidate, namely, classical logic? This is a big question, and this paper is too short to answer it properly; however, there is a simple argument for taking the FDE account of logic to be the right account over the standard candidate. The simple argument runs as follows.[7]

The argument, in a nutshell, is that – between FDE and Classical – we lose no true theories by accepting the FDE account, and we gain live options for true theories. Each claim in turn.

* *Small parenthetical remark.* One might wonder not just about FDE versus Classical but about FDE versus salient and popular subclassical extensions of FDE, namely, so-called LP [1, 15] and Strong Kleene (K3) [12]. But the answer here is fairly clear: each of these 'stronger candidates' for an account of logic (understood as above) is very clearly imbalanced and awkward in ways that FDE (similarly, Classical) isn't. Each is an account according to which logic refuses to 'allow' one sort of theory (say, negation-inconsistent) but is 'happy' to allow exact duals of such theories (say, negation-complete). This sort of imbalance might be philosophically acceptable if the only aim were to have the strongest subclassical account of logic compatible with such and so theoretical goals; but generally, when doing philosophy, we don't just want 'strength' but also want naturalness, symmetry and balance. In this way, FDE clearly stands at the front versus the other leading subclassical candidates. Accordingly, my remarks throughout focus on the mainstream (so-called Classical) candidate versus FDE. *End remark.* *

### *2.1 Keeping true classically closed theories*

Perhaps the main worry that confronts the FDE account of logic (qua universal closure) is the ubiquity of true classically closed theories. So many of our true theories are classically closed (i.e., closed under what classical logic says is valid). Do we somehow lose these many true theories by accepting the FDE account of logic? To any who've thought a bit about FDE (or any other subclassical account of logic) the answer is plain: no.

The question is: what is going on in these theories if the basement-level closure relation – if logic as universal among all true theories – is FDE? How should we think about this? Here too the answer is fairly plain, and fairly traditional; but a snapshot of the answer is worth brief review.[8]

Logic, we're supposing, is FDE. This is the basement-level closure relation on all true theories; it's the one governing the very sparse set of logical vocabulary. This relation (I'm supposing) is an absence-of-counterexample relation over the broadest space of possibilities, namely, the logical possibilities – the ones recognized by logic (i.e., the ones that logic considers to be candidate counterexamples). But the task of theorists is in part to

[7] This simple argument is in the spirit – maybe even in slogans of lore – of early (perhaps especially Australasian) relevance and/or paraconsistent logicians. I spell out the argument in slightly more detail in [7]; and in that work I also give a bonus argument for the view that the FDE account of logic (qua universal closure) should be accepted over the standard (classical) account. I leave the bonus argument aside here.

[8] For a somewhat fuller discussion of some of the given answer see [3, 4].

build extra-logical consequence relations for our true (and complete-as-possible) theories. The space of possibilities recognized by *that* relation – namely, the extra-logical, theory-specific consequence relation of the theory – is just the space of theoretical possibilities (for the theory). This space of theoretical possibilities is also a space of logical possibilities but the converse rarely holds; the former space is generally a proper space. The logical possibilities that are beyond the theory-specific space of (theoretical) possibilities are theoretical *im*possibilities; they are logical possibilities that, according to the theory (and its consequence relation), are simply beyond what's possible for the target phenomenon.

Easy example: think of the true theory of arithmetic (say, PA). This is classically closed. As such, no predicate in the language of the theory is possibly glutty or gappy with respect to the theory's domain of objects, at least as far as the theory – its consequence relation – is concerned. Such logical possibilities are theoretically impossible.

Many other examples may be given but the idea is clear enough. The question is how such logical possibilities are ruled out by the theories.[9] The answer is familiar: the theory-specific consequence relations are built in ways that rule out the would-be possibilities. And this is done via 'shrieking' and 'shrugging'.[10]

### 2.1.1 Shrieking theories

To *shriek a predicate* $P$ in the language of theory $T$ is to impose the following condition on $T$'s closure relation $\vdash_{\mathrm{T}}$, where $\bot$ is true in no models of $T$:[11]

$$\exists x(Px \wedge \neg Px) \vdash_{\mathrm{T}} \bot$$

The effect of imposing this condition on $\vdash_{\mathrm{T}}$ is a reduction of logical possibilities to just those that are recognized by $T$ to be (thereby) theoretically possible. When $P$ is shrieked in theory $T$ we immediately get the $P$-driven instances of material detachment (which is logically invalid), where $t$ is any (say) name in the language of $T$:

$$Pt, Pt \supset B \vdash_{\mathrm{T}} B$$

For the same reason, the dual of excluded middle (viz., 'explosion' or 'EFQ') involving only shrieked predicates is now valid according to the theory (though, of course, still logically invalid):

$$Pt \wedge \neg Pt \vdash_{\mathrm{T}} B$$

If *all* predicates in the language of theory $T$ are shrieked – and this is called a *shrieked theory* (versus partially shrieked theory) – then such logically invalid patterns are valid

---

[9] This is not an epistemological question. The epistemology of constructing/discovering true theories is messy and still sorely unknown; it's governed by many default assumptions that are grounded by (empirical) induction and so on – including assumptions about rejecting glutty and/or gappy theories [5]. I have nothing new or helpful to say on the epistemology of finding the truth. (I remain an endorser of the old metaphor of rebuilding our raft at sea. It's a messy and difficult business even when the seas are calm.)

[10] The 'shriek' terminology comes from the not non-standard abbreviation of $A \wedge \neg A$ as $!A$ (pronounced 'shriek A'); the 'shrug' terminology comes from a graduate student (viz., Colin McCullough-Benner) who suggested that imposing the effect of excluded middle is like shrugging about each claim – either it is or isn't (said with a shrug).

[11] Without loss of generality I focus on unary predicates.

according to the theory's consequence relation. And that's precisely (part of) what's going on in the many true theories that are classically closed.

### 2.1.2 Shrugging theories

Shrieking a theory (i.e., shrieking all predicates of a theory) is insufficient for classical closure; shrugging is also required. But here the idea is very familiar.

To *shrug a predicate* $P$ in the language of a theory $T$ is to impose the following condition on $T$'s closure relation, where $\top$ is true in all models of the theory:

$$\top \vdash_{T} \forall x(Px \vee \neg Px)$$

The effect of imposing this condition on $\vdash_{T}$ is a reduction of logical possibilities to just those that are recognized by $T$ to be (thereby) theoretically possible. When $P$ is shrugged in theory $T$ we immediately get the $P$-instances of material identity (which is logically invalid), where $t$ is any (say) name in the language of $T$:

$$B \vdash_{T} Pt \supset Pt$$

For the same reason, the dual of 'explosion' or 'EFQ' involving only shrieked predicates is now valid according to the theory (though, of course, still logically invalid):

$$\top \vdash_{T} \neg Pt \vee Pt$$

If *all* predicates in the language of theory $T$ are shrugged – and this is called a *shrugged theory* (versus partially shrugged theory) – then such logically invalid patterns are valid according to the theory's consequence relation. And that's precisely (part of) what's going on in the many true theories that are classically closed.

### 2.1.3 Shrieking and shrugging a theory

And when a theory is both shrieked and shrugged we have a classically closed theory (i.e., closed under classical logic). In such theories the logical vocabulary behaves differently from the way it behaves in other theories; but that's just because of the theory-specific constraints imposed on those theories, constraints driven by the true account of the target phenomenon at the heart of the theory.

Accordingly, there is no loss of true (classically closed) theories – none whatsoever. This is obvious on reflection but worth more frequent reflection than the point has received in debates over whether logic is (non-) classical, and in particular debates over whether we should accept that logic (qua universal closure) is FDE.

## *2.2 Gaining live options for true theories*

As per §2.1 there is no loss of true theories on the FDE account of logic. But is there any gain? Here, I think that the answer is clearly affirmative. In particular, we gain very natural – and currently live-option – candidates for true theories of various extraordinary phenomena that appear, prima facie, to be 'gluts' or 'gaps'. The obvious examples from common paradoxical realms remain important.

### 2.2.1 Apparent gluts

Liar sentences (viz., 'It's false that I'm true') scream out as gluts.[12] This is not a forced or awkward position; it's a natural one, and one well-enough understood to make a review of the idea unnecessary.[13]

There are also other (admittedly extraordinary) phenomena that appear to demand a glut-theoretic approach. From a common religious tradition, consider the god-human figure in various Chalcedon-inspired Christian theories of the world. If there's truth to any such theory the truth is prima facie glutty – that the god-human figure is both divine and human (with all of the apparent contradictions entailed therein).

I am not hereby arguing for the truth of any such glut-theoretic approach to such phenomena; the point is simply that such an approach is very natural. These options are not mere graffiti on the surfaces of conceptual space; they are live options. And the FDE account of logic allows (though doesn't, of course, demand) that the options be genuine candidates for truth.

### 2.2.2 Apparent gaps

Just as there are strange phenomena that scream out for a glutty treatment, so too are there strange phenomena that, prima facie, scream out for a dual, gappy treatment. Sticking to the familiar truth-theoretic paradoxes, the various truth tellers (e.g., 'I am true') are very, very naturally treated as gaps. Ruling out such an option based on the mainstream (Classical) account of logic (qua universal closure) is prima facie unmotivated.

One needn't stick just to the peculiar paradoxical phenomena to find prima facie gaps; religious theories and/or other theories of strange metaphysical phenomena present familiar examples.

Again, I am not hereby arguing for the truth of any such gap-theoretic approach to such phenomena; the point, as above, is that such an approach is very natural, and to rule out such candidate theories without very good reason is unmotivated.

## 3 Closing remarks

While its most famous explorers (viz., Belnap and Dunn) have not exhibited great interest in whether FDE can plausibly be thought to be 'the one true logic', I hope that the brief account above sufficiently suggests that the answer is affirmative. My own view is that

[12] I grant that not all theorists – including Dunn or Belnap – have heard the screams as loudly as I have; but I trust that they clearly see the approach to be a genuine candidate.

[13] For some (not all) of the literature see [1, 2, 6, 9, 14, 15, 16, 17], and for further discussion [10, 11].

debates over whether logic is (sub-) classical should finally come to an end in favor of a subclassical account – and in particular, in favor of FDE.[14]

Of course, even if there's little truth in any of that (I think that there is), one can still enjoy FDE as a useful model of just such a conclusion. Either way, my hope is that this paper reflects my genuine admiration for FDE and its two most famous explorers.[15]

## An Unrelated Add-On: an 'informal reading' of Dunn-Belnap semantics

This appendix, which is simply an add-on to the volume, has nothing to do with the principal aim or argument of the foregoing paper; however, this volume specifically on FDE is a good place, if there is one, to record the following (simple and very, very minor) thought.

## 4 There's truth in that

There's a saying in American English, and probably many other variations of English (and quite possibly many other natural languages): namely, *there's truth in that*. A similarly natural though less common phrase is 'there's falsity in that'.

Philosophical logicians are driven by constructing precise models of different philosophical ideas or philosophically interesting fragments of language. One blindingly obvious way of modeling the idea that *there's truth in that* is to invoke some sort of degree-theoretic notion of truth. Any standard model off the block will do for a rough idea; the key is that truth (whatever else it might be) comes in degrees, and so saying *there's truth in that* is just to say that there's a degree of truth in that (where the *that* in question is some claim or other). Details will be important in the account, as always; and details in one direction will be more interesting than in other directions. But I raise the degree-theoretic direction only to set it aside.

### *4.1 There's truth in that: FDE*

There are undoubtedly many things that the phrase 'there's truth in that' is used to express. What I want to highlight is that the phrase can be understood along a combined Dunn-Belnap idea. First, think of FDE's four values along the Dunn-style way, where the full

[14] Well, there might be argument for going slightly lower to what Urquhart called *Ockham Logic* [18]; but I wouldn't fuss if that wound up being the right account.

[15] Thanks to the audience at the philosophical-logic plenary session of the 2018 Annual North American Association for Symbolic Logic Meeting where the gist of these ideas were presented, and likewise to audiences and research seminars at Yonsei University (Seoul), Melbourne University (Melbourne), George Washington University (Washington), University of North Carolina (Chapel Hill), West Virginia University (Morgantown), University of Notre Dame (Notre Dame IN), Syracuse University (Syracuse), St Andrews University (St Andrews), and Princeton University (Princeton). And – finally, and again – with utmost respect and deepest sincerity, my thanks to Nuel Belnap and Mike Dunn for their work: we are all better for it. Really.

space of values is the powerset of $\{1,0\}$, where we label the elements as follows:

$$\mathsf{t} = \{1\}$$

$$\mathsf{f} = \{0\}$$

$$\mathsf{b} = \{1,0\}$$

$$\mathsf{n} = \{0\}$$

Now, the famous Belnap way to think of these values, at least in the 'organon' context of a data base that Belnap was discussing, avoids the idea of the (let me say) extra-classical values (viz., b and n) as involving truth or falsity. Instead, Belnap's picture is one in which b represents 'too much information', and n 'not enough information'. And this way of thinking about those values remains a natural and powerful heuristic for FDE, as is well-known.

What I want to suggest is another (not a competitor) way of thinking of the extra-classical values but one that does involve truth and falsity. One way, of course, is simply the standard 'glut' and 'gap' terminology, where the value b represents (or models or etc.) truth-value gluts, sentences that are both true and false – 'true contradictions' as they're sometimes called (where 'contradiction' is a term for any logical conjunction of a sentence and its logical negation) – and n represents the dual (viz., truth-value gaps). While I do in fact endorse the existence of gluts and gaps among our true theories, my proposal here is to avoid thinking of b and n just so, and rather think of them along lines inspired by the Belnap and Dunn picture(s). In particular, return to the target idea: *there's truth in that* and its natural companion *there's falsity in that*. Combining the Dunn-style semantics with a Belnap-inspired take on the values, it's very natural to think of the extra-classical values along the following lines (pretty much at face value):

- *There's truth in A and there's falsity in A* is modeled by b.
- *There's no truth in A and no falsity in A* is modeled by n.

Note that the 'reading' of value n explicitly uses the 'no' terminology (quantifier), which is very natural. A natural way to think of the value b uses the corresponding 'some' terminology: b represents cases where $A$ 'has some truth to it' but (alas) $A$ also 'has some falsity to it'.

For the standard values there are natural ways to emphasize that that's that (so to speak): $A$'s having t represents (models, etc.) cases where $A$ 'has truth and nothing but the truth' – that is, that there's truth to $A$ and, as the English say, that's an end on't. Similarly for f: $A$'s having f represents (models, etc.) cases where $A$ 'has some falsity and nothing but falsity' – that is, that there's falsity to $A$ and that an end on't.

## *4.2 The FDE truth/falsity conditions*

Worth noting is the ease with which the *there's truth (falsity) in that* 'reading' motivates the FDE truth/falsity conditions (or, equivalently in this context, lattice conditions). For example, suppose that someone advances claim $A$ about the world. Suppose further that 'there's truth in that', that is, that there's some truth in what has been advanced (viz., in $A$).

Now suppose, yet further, that someone advances the (logical) negation of $A$,[16] namely, in English, *it's false that A*. But there's falsity in *it's false that A* if there's falsity anywhere; after all, there's truth in $A$. What all of this suggests is that the proposed 'reading' of the values motivates the condition on (logical) negation $\neg$ that if there's truth in $A$ then there's falsity in $\neg A$. The converse is equally motivated, and so too are the corresponding truth/falsity conditions – namely, that there's some truth in $\neg A$ just if there's some falsity in $A$. Likewise, the conjunction, disjunction and, in turn, quantifiers are naturally motivated by the target *there's truth in that* 'reading' of the values.

Let me be clear that the foregoing 'reading' of FDE's values, terribly sketchy as it remains, is only offered as an additional 'reading' somewhere between the famous Belnap one and the equally (in-) famous 'glut/gap' one. I believe that the informal *there's-truth-in-that* idea(s) can motivate the FDE account of consequence; but I leave that to any (potentially merely possible) interested reader to verify.

## References


1. F. G. Asenjo. A calculus of antinomies. *Notre Dame Journal of Formal Logic*, 7(1):103–105, 1966.
2. Jc Beall. *Spandrels of Truth*. Oxford University Press, Oxford, 2009.
3. Jc Beall. LP$^+$, K3$^+$, FDE$^+$ and their classical collapse. *Review of Symbolic Logic*, 6(4):742–754, 2013.
4. Jc Beall. A simple approach towards recapturing consistent theories in paraconsistent settings. *Review of Symbolic Logic*, 6(4):755–764, 2013.
5. Jc Beall. Free of detachment: Logic, rationality, and gluts. *Noûs*, 49(2):410–423, 2015.
6. Jc Beall. There is no logical negation: true, false, both, and neither. *Australasian Journal of Logic*, 14(1), 2017.
7. Jc Beall. The simple argument for subclassical logic. In Cory Juhl and Joshua Schechter, editors, *Philosophical Issues: Philosophy of Logic and Inference (Annual Supplement to Noûs)*. Wiley, 2018. Forthcoming.
8. Jc Beall and John P. Burgess. Logic. In Duncan Pritchard, editor, *Oxford Bibliographies in Philosophy*. Oxford University Press, 2017. DOI: 10.1093/obo/9780195396577-0336.
9. Bradley Dowden. Accepting inconsistencies from the paradoxes. *Journal of Philosophical Logic*, 13(2):125–130, 1984.
10. Hartry Field. *Saving Truth from Paradox*. Oxford University Press, Oxford, 2008.
11. Anil Gupta and Nuel Belnap. *The Revision Theory of Truth*. MIT Press, Cambridge, MA, 1993.
12. S. C. Kleene. *Introduction to Metamathematics*. North-Holland, 1952.
13. Robert L. Martin, editor. *Recent Essays on Truth and the Liar Paradox*. Oxford University Press, New York, 1984.
14. Robert L. Martin and Peter W. Woodruff. On representing 'true-in-L' in L. *Philosophia*, 5(3):217–221, 1975. Reprinted in [13].
15. Graham Priest. The logic of paradox. *Journal of Philosophical Logic*, 8(1):219–241, 1979.
16. Graham Priest. *In Contradiction*. Oxford University Press, Oxford, second edition, 2006. First printed by Martinus Nijhoff in 1987.
17. Richard Routley. Dialectical logic, semantics and metamathematics. *Erkenntnis*, 14(3):301–331, 1979.
18. Alasdair Urquhart. Distributive lattices with a dual homomorphic operation. *Studia Logica*, 38(2):201–209, 1979.


[16] I think that there may be other negations beyond logic's negation but I leave this aside.

# Default Rules in the Logic of First-Degree Entailments

Katalin Bimbó

**Abstract** Monotonicity is a property of the consequence relation of classical logic, which has been questioned for a range of reasons, some of which we overview. *Default logic*, which is a non-monotonic logic, was introduced to model defeasible conclusions that may be justified by default rules. In this paper, we scrutinize certain features of non-monotonicity of a consequence relation to legitimize it. The *logic of first-degree entailments*, which makes more distinctions than 2-valued logics does is especially suitable as the base logic for a system of default logic, because the applicability of default rules depends on the absence of formulas from a theory. We define two notions of *extensions* with default rules and illustrate their use by some standard as well as new examples.

**Keywords** Consequence relation • Default logic • Entailment • Monotonicity • Relevance logic

## 1 Introduction

Default logic augments a base logic with *default rules*, which allow conclusions to be drawn, when a piece of information is known and a certain other piece of information is not known. The latter means that default rules are not like many other inference rules, which are applicable whenever something is known. Clearly, the accumulation of information (or the expansion of the set of premises) may lead to a loss of certain conclusions that followed before by default rules. Due to this phenomenon, default logic is called a *non-monotonic logic*. There are other logics, for example, relevance logics that support a non-monotonic consequence relation.

*Relevance logics* typically do not allow arbitrary conclusions to be drawn from a pair of sentences that are each other's negations. In classical logic, the presence of a contradiction is tantamount to a catastrophe: the tiniest contradiction licenses any other statement as its conclusion. While it may be preposterous to include both a statement and its negation into our collection of beliefs *on purpose*, it would be rather unfortunate, if our reasoner would go into a hysterical rage every time a contradiction is discovered. Contradictions could pop up in the process of taking stock of the consequences of earlier beliefs, or they could result from new information being received. Guaranteeing consistency at all times may be cumbersome or costly.

As Kautz and Selman note about the added complications brought upon by default logic, "it has been tacitly assumed that this additional complexity arises from the use of consistency tests." ([25, p. 244]) They claim that "the most efficient use of default information is to 'flesh out' the missing detail in a knowledge base in a 'brave' manner, a process

Katalin Bimbó
Department of Philosophy, University of Alberta, Edmonton, AB, T6G 2E7, Canada
e-mail: bimbo@ualberta.ca

 
H. Omori, H. Wansing (eds.), *New Essays on Belnap-Dunn Logic*,
Synthese Library 418, https://doi.org/10.1007/978-3-030-31136-0_9

that corresponds to finding an extension." (ibid.) The problem of dealing with inconsistencies motivates [26], where it is said that an unsatisfactory aspect of Reiter's default logic is that "a default theory has only a trivial default extension that includes every formula once the axioms in the default theory have contradictions" [26, p. 253]. Then they proceed to build a default logic by modifying the resolution calculus, and obtain a generalization of Reiter's default logic. A similar motivation leads in [29] to changing the underlying logic from 2-valued logic to a 3-valued one, namely, to Łukasiewicz's 3-valued logic ($Ł_3$). As [29, p. 150] states, "[t]he problem is that non-monotonic formalisms based on classical logic lead sometimes to too strong conclusions." $Ł_3$ was originally invented to capture the modal concept of contingency, and the logic allows one to define "possibility" and "necessity" as modal operators. In this approach, triviality is avoided by modalizing default rules as well as by modifying the applicability of a default rule to include an additional consistency requirement. The latter means that if $C$ is the presumption in a default rule, then $C$ is not impossible relative to the set of sentences to which the default rule is applied. Finally, we mention [27] that introduces the system IDL. "The main idea behind the conception of IDL is the realization that the disposition to perform inferences under partial evidences lead to contradictions." (ibid.) In IDL, the default rules include the operator ?, which may be interpreted as an **S5**-type possibility operator. Each of these approaches changes the logic used in a default system; however, none of them uses the logic we do.

This paper replaces the background logic for default rules, namely, it uses the logic of first-degree entailments (**fde**) in place of the 2-valued propositional logic. This is a step toward basing default logic on a non-monotonic logic such as one of the main relevance logics (**T**, **R** and **E**). In this paper, we use a fragment shared by the aforementioned logics, namely, **fde**. The latter is decidable, which is an advantage if we want to define extensions of belief bases constructively. It also allows us to retain absolute consistency — even if we have to put up with a few local inconsistencies in the form of some sentences together with their negations.

In Section 2, we overview versions of monotonicity and non-monotonicity for consequence relations. Our aim is to convince the reader that there are good reasons to consider non-monotonic consequence relations. We provide a series of informal examples and we point out that relevance logic (with formally developed systems) has a well-defined consequence relation that is non-monotonic. Section 3 introduces **fde** in several ways in order to make it as familiar as the 2-valued propositional logic is. Finally, in Section 4, we define extensions in two ways, and show that absolute inconsistency of an extension is the exception rather than the typical situation. We end the paper with some remarks in Section 5.

## 2 Monotonicity and non-monotonicity of consequence relations

Consequence relations have been investigated both in concrete and abstract contexts. In [33], Tarski investigated properties of the *consequence operator* ($Cn$) for classical logic, which may be seen to emerge from the consequence relation of an axiomatic system for classical logic. He distinguished between postulates that are specific to the connectives of two-valued logic (in particular, to negation and implication), and postulates that characterize consequences of arbitrary sets of sentences.

The type of $Cn$, the consequence operator, is $Cn\colon \mathscr{P}(\text{wff}) \longrightarrow \mathscr{P}(\text{wff})$, where wff is the set of well-formed formulas. That is, $Cn$ is a total function that assigns a set of formulas to a set of formulas. $Cn$ is a *closure operator* in the usual sense, that is, the following clauses hold. ($X$ and $Y$ are sets of formulas.)

(1) $X \subseteq Cn(X)$;
(2) $X \subseteq Y$ implies $Cn(X) \subseteq Cn(Y)$;
(3) $Cn(Cn(X)) \subseteq Cn(X)$.

Additionally, Tarski stipulated that every set $Cn(X)$ is generated by a finite set, and that there is an "absolutely contradictory" sentence, the consequences of which coincide with the set of all sentences. Absolutely contradictory sentences are easily definable in classical logic as $\mathcal{A} \wedge \neg\mathcal{A}$ (a formula and its negation as the conjuncts of a formula) or as $\boldsymbol{F}$ (a zero-ary constant formula). In the context of this paper, we view as a classical feature of a logic, when $x \wedge -x$ is an absolutely contradictory formula (where $\wedge$ and $-$ are connectives of a particular logic).

Returning to (1)–(3), we note that (3) may be "strengthened" to identity, because of (1). The natural reading of $Cn(Cn(X)) = Cn(X)$ is that for any $X$, $Cn(X)$ is the *fixed point* of the $Cn$ operator. Another reading of the equation is that the consequence relation is *transitive*. This is a property that is normally accepted in various logics. In different proof systems, this may correspond to a cut rule, or an explicit rule of transitivity for implication such as the rule form of prefixing or suffixing. In an axiomatic system — assuming the usual definitions of proof and derivation — (3) corresponds to the absence of a bound on the length of finite sequences of formulas, which constitute derivations and proofs.

The first property is also widely accepted in logic; (1) expresses the *reflexivity* of the consequence relation. Some people find it counterintuitive that every sentence implies itself, because they feel that this is like to "beg the question." Of course, the problem with concluding $\mathcal{A}$ from $\mathcal{A}$ is not that a sentence does not imply itself; rather in a disputation, a different proof is hoped for, notably, one without the conclusion being an explicit assumption.

Property (2) is usually called *monotonicity*, and this is the most frequently questioned property of a consequence operator. In proof systems (for classical logic), (2) can be seen to correspond to some of the following. An axiomatic calculus, with the usual definition of a proof permits the insertion of provable formulas into proofs — whether they are used or not in obtaining the final formula in the proof. In tableau proofs, similar additions are allowed, and the splitting rule of synthetic tableaux is a rule that amounts to the law of excluded middle (for arbitrary formulas). Many natural deduction systems license the vacuous discharge of premises, and some sequent calculi include a rule that is called *weakening* or *thinning* on the left. Sequent calculi, tableaux and natural deduction systems are the easiest to modify so that monotonicity is omitted; indeed, scores of substructural logics are defined this way. The notion of a proof in an axiomatic calculus does not have the resources to track the use of formulas, however, the addition of a bookkeeping device such as dependency indices on the formulas may lead to a tighter definition of a proof. (The notion of traversing proofs in [7, §3.1] is an example of proofs in an axiomatic system with no slack.)

The last two assumptions postulated by Tarski are incorporated into the consequence operators induced by some logics, but not by all. For instance, an $\omega$-rule violates the finiteness assumption. The inclusion of a sentential constant $\boldsymbol{F}$ may be possible but undesirable; for example, the logic of relevant implication **R** can be conservatively extended with $\boldsymbol{F}$ and its axiom $\boldsymbol{F} \to \mathcal{A}$ (where $\to$ is implication). However, $\boldsymbol{F}$ is not definable as $\mathcal{A} \wedge {\sim}\mathcal{A}$ (where $\sim$ is the original negation of **R**), and $\boldsymbol{F}$ is (informally) called "absurdity," which indicates the disagreeable character of $\boldsymbol{F}$.

Some more abstract investigations of consequence relations, which may place less emphasis on the consequence relation emerging from an axiomatic formulation of classical logic, often characterize a consequence relation itself (see e.g., [32] and [15, Ch. 6]). We use $\vdash$ as general notation for a consequence relation. If $\vdash\ \subseteq \mathscr{P}(\mathrm{wff}) \times \mathscr{P}(\mathrm{wff})$, that is, $\vdash$ is a *symmetric consequence relation* (cf. [15, Ch. 6]), then (4)–(6) for $\vdash$ parallel (1)–(3) for $Cn$.

(4) $Y \subseteq X$ implies $X \vdash Y$;
(5) $X \vdash Y$ implies $X \cup Z \vdash Y$;
(6) if for all $y \in Y$, $X \vdash y$ and $Y \vdash Z$, then $X \vdash Z$.

Of course, this is a rather impoverished relation — though it is a pre-order. An analogy with sequent calculi would suggest that the sets of formulas appearing on the left- and on the right-hand sides of the $\vdash$ should be considered differently with respect to how the elements of the sets are thought to combine.

*Asymmetric consequence relations* only allow formulas on the right-hand side of the $\vdash$ ; that is, $\vdash\ \subseteq \mathscr{P}(\mathrm{wff}) \times \mathrm{wff}$. The problem of combining conclusions is avoided, but it remains a question as to how the elements of the set of formulas on the left-hand side of the $\vdash$ are combined. For classical logic, there is no choice really; however, in logics with richer languages (e.g., in **R**) there are at least two possibilities which can lead to remarkably different consequence relations. We will return to this question later in this section. Let us now consider some properties for $\vdash$ that are similar to the properties of $Cn$.

(7) $X \vdash x$ for any $x \in X$;
(8) $X \vdash y$ implies $X \cup Z \vdash y$;
(9) if for all $y \in Y$, $X \vdash y$, and $Y \vdash z$, then $X \vdash z$.

*Monotonicity*, in the form of (5) or (8), is the most problematic feature of $\vdash$ from among (4)–(9). Sometimes, monotonicity is abandoned or it is replaced by weakened variants such as (10) or (11) (below).

(10) $X \vdash x$ and $X \vdash y$ imply $X \cup \{\, x \,\} \vdash y$.

The property expressed by (10) is called *cautious monotonicity*. The expansion of a set of premises is not impossible, however, it is (severely) restricted, because only formulas that are already derivable from the premise set may become new premises. Of course, (10) is a special case of monotonicity, and it is unproblematic for the consequence relation of classical logic as an axiomatic calculus. However, even cautious monotonicity is "too much monotonicity" from the point of view of certain relevance logics.

In terms of proofs in an axiomatic calculus, the insertion of $x$ is perfectly OK. Indeed, it does not matter that $x$ may be derived from $X$. Monotonicity — in all its forms — is an organic feature of axiomatic calculi with the standard notion of proofs, and it reflects the paucity of control over proofs in those proof systems.

As a contrast, in the Fitch-style natural deduction system $\mathbf{FT}_{\rightarrow}$ (see [1] and [2, pp. 42-50]) or in the sequent calculus $\mathbf{LT}^{t}_{\rightarrow}$ (see [7, §3.2]), $X \vdash x$ and $X \vdash y$ do not need to imply $X, x \vdash y$, where $\vdash$ is a consequence relation emerging from one of those calculi. Of course, in these calculi, treating $X$ as a set is not reasonable either, in other words, "$X, x$" is not simply $X \cup \{\, x \,\}$.

Another weakening of monotonicity assumes the presence of a negation connective (that we denote by $-$ here). The symbol $\nvdash$ indicates that the consequence relation does not hold.

(11) $X \nvdash -x$ and $X \vdash y$ imply $X \cup \{x\} \vdash y$.

This principle is called *rational monotonicity*. Rationality amounts to the prohibition of expanding a consistent premise set to an inconsistent one. (Inconsistency means negation inconsistency here, that is, a set of formulas is inconsistent when a formula and its negation are both in the set.)

Rational monotonicity may be seen as a component of the core idea behind the *Lindenbaum lemma* in the metatheory of classical logic. Given a consistent set of formulas that set may be extended to a consistent set which is maximal in the sense that the addition of any further formulas would make the set inconsistent. The lemma is often proved via first defining a Lindenbaum construction that contains as a step the addition of $x$ when $-x$ is not a consequence of the set at that stage. The Lindenbaum lemma does not stress that no conclusions of the starting set are lost, because sooner or later all of them are included into the expanding set of formulas — until nothing else can be added, hence the set is maximal. Beyond the connection between (11) and the Lindenbaum construction, the latter is interesting for our purposes, because it is the inspiration for Definitions 9 and 10.

Speaking of classical logic, (10) and (11), the two principles we mentioned above seem to be the only possible refinements of the monotonicity principle. However, monotonicity can simply *fail*.

The latter can happen, because given some new information, an old conclusion no longer follows. This kind of non-monotonicity is intrinsic to the original consequence relation of *relevance logics*: the addition of a new premise that is irrelevant to the conclusion prevents the proof of the same conclusion, where the premises are thought to be fused. Similarly, in *default logic*, the inclusion of the negation of the justification of a default rule into a belief set prevents an application of the rule.

For example, "Tweety flies" may be obtained from "Tweety is a bird" and "Birds fly." Then, "Tweety does not fly" either might be added directly, or it might be concluded from "Tweety is a penguin" and "Penguins do not fly."[1] Some users of default logics argue for consistency checks, but maintaining consistency will simply disallow the use of some default rules. This amounts to *ignoring* some of the available information *instead of using* it.

## 2.1 Information tolerance

*Example #1.* One might think that in an area such as mathematics all reasoning must be monotonic, that is, if something has been proved, then it never can (or should) be revoked. Constructive mathematics is sometimes viewed as a subsystem of "classical" mathematics, because it omits certain axioms that are constructively not acceptable. (Primary examples of the latter are the double negation elimination rule and the axiom of choice (AC).) Everything provable in Peano arithmetic (PA) is provable (via a translation) in Heyting arithmetic (HA). However, HA distinguishes between formulas that are provably equivalent in PA; that is, the expansion of HA to PA results in a loss of distinctions. (Cf. [21].) In analysis, certain constructively true theorems no longer hold, because AC introduces

[1] We remain silent on whether penguins are birds, but we definitely tackle Tweety. Cf. [25, p. 246].

objects that do not exist — constructively speaking. For instance, not all functions from reals into reals are continuous, if AC is an axiom.[2]

*Example #2.* Newtonian physics has been superseded by the theory of relativity. Nonetheless, it remains in use for a restricted domain of problems. As long as a moving object's speed ($v$) is small compared to $c$ (the speed of light in vacuum), the simplicity of the calculations within Newtonian physics compensates for the negligible imprecision (that results from the point of view of relativity theory). Depending on how a concrete calculation is described, we may see that a new added piece of information, namely, that $v \approx c$ either prevents the same results to be deduced or leads to a contradiction. If $v \lll c$ is a proviso of our Newtonian physics, then it is no longer applicable; otherwise, we have contradictory results derived from the two theories.

*Example #3.* In artificial intelligence (AI), many examples derive from everyday reasoning problems with a little extra abstraction. A robot that has to move from point $a$ to point $b$, first, might (try to) proceed in a straight line, and this may just do as a rule in a lab with smooth carpet flooring. However, on a rugged terrain (like the surface of the Mars), the robot might have to check that there is no boulder or crater filled with water in front of it. The problem the robot is facing is not completely unlike the problem a human driver may encounter when she tries to drive to the airport in a big city during rush hour. The robot could solve the problem of getting from point $a$ to $b$ by applying a general rule (if the situation lacks certain features), together with a series of more specific rules.

*Example #4.* Our last example illustrates the dynamic character of reasoning, which is less important for grand scientific theories such as mechanics or set theory, which take many years or decades to develop. Default logic does not address the process of changing an agent's beliefs. As McDermott and Doyle stated in 1980: "Classical symbolic logic lacks tools for describing how to revise a formal theory to deal with inconsistencies caused by new information. This lack is due to a recognition that the general problem of finding and selecting among alternate revisions is very hard." (See [28, pp. 42-43].) The predicted difficulties notwithstanding, some researchers took up the idea of modeling the effect that a new piece of information has on an agent. This area became known as *belief revision*, and originally, it focused on regaining consistency in the presence of new information. The *revision operation* actually happens to prefer the new piece of information over the existing collection of beliefs. (See [19] and [31] for detailed expositions.) Default logic — either based on classical logic or on **fde** — does not aim at describing the dynamics of drawing defeasible conclusions and withdrawing some of them when necessary. But an upshot of both approaches is that new information is not ignored.

Despite typical examples that may suggest otherwise, non-monotonicity in default logic amounts to the non-monotonicity of extensions. (Cf. [30, p. 91].) *Extensions* in default logic are the analogs of *theories* in relevance logics. Given a set of formulas, extensions and theories comprise the consequences of those formulas. A distinctive feature of extensions is that some of the consequences — indeed, the fascinating consequences — result by applications of default rules. The latter are not like any other inference rule, because their applicability depends on a formula *not being derivable*. This immediately explains the non-monotonicity of extensions. If $X \subseteq Y$, then it is possible that the same set of default rules produces a smaller extension from $Y$ than from $X$, because some default rules are no longer applicable given what is in $Y$. The unusual requirement of the non-provability of a formula might compel a user to constantly check for consistency. If default logic is based on 2-valued logic, then even a "small" contradiction is a double whammy: it prevents ap-

[2] See also Gödel's [20].

plications of default rules but it trivializes the extension by equating it with the set of all formulas.[3]

The interaction of negation and monotonicity differs in relevance and classical logics. Let us assume that *fusion* (denoted by $\circ$) is in the language either as a primitive or as defined. In **R**, $\sim(\mathcal{A} \to \sim\mathcal{B})$ can be taken to be $\mathcal{A} \circ \mathcal{B}$. Fusion is similar to conjunction, sometimes, even called intensional conjunction. But they are not the same connective; for instance, $(\mathcal{A} \circ \mathcal{B}) \to \mathcal{A}$ is not a theorem of **R**. Premises in **R**, viewed as a multiset of formulas, may be thought to be combined by $\wedge$ or $\circ$, and the emerging consequence relations will differ. (We indicate that the premises are fused by listing them on the left-hand side of the $\vdash$ separated by ; — replicating the notation that was introduced in [11] and is used in sequent calculi for positive fragments of **R** and **T**.)

*Example* 2.1. Let us assume that $p$ and $q$ are a pair of distinct propositional variables. $p; p; p \vdash_{\mathbf{R}} (p \circ p) \circ p$, whereas $p; p; p; q \nvdash_{\mathbf{R}} (p \circ p) \circ p$ and $p; p; p; p \nvdash_{\mathbf{R}} (p \circ p) \circ p$. Hence, in general, $\mathcal{A}_1; \ldots; \mathcal{A}_n \vdash_{\mathbf{R}} \mathcal{B}$ does not imply that $\mathcal{A}_1; \ldots; \mathcal{A}_n; C \vdash_{\mathbf{R}} \mathcal{B}$.

The formula $C$ may not contain a negation at all as the example illustrates. However, even if it does, moreover, if $C$ is $\sim\mathcal{B}$, it still does not follow that $C$ may be added to the premises.

*Example* 2.2. Let us assume that $p$ is a propositional variable. $p \vdash_{\mathbf{R}} p \circ p$, whereas $p \circ \sim(p \circ p) \nvdash_{\mathbf{R}} p \circ p$. Both the provability and the non-provability claims can be shown using a sequent calculus for $\mathbf{R}_{\underset{\to}{\sim}}$, which is decidable. (See [14] or [18].) The non-provability also follows from metatheorems about **R** — together with assigning +0 to the propositional variable $p$ and calculating the value of the following formula in $M_0$ using R. K. Meyer's implication matrix.

$$\underset{+0}{(p} \underset{-0}{\to} \underset{-0}{\sim} \underset{+0}{p)} \underset{-3}{\to} \underset{+0}{((p} \underset{-0}{\to} \underset{-0}{\sim} \underset{+0}{p)} \underset{+0}{\to} \underset{-0}{\sim} \underset{+0}{p)}$$

As a contrast between how fused premises in **R** and conjoined premises in classical logic behave with the negation of a conclusion placed among the premises, we have (13) and (12).

(12) $\mathcal{A}_1, \ldots, \mathcal{A}_n \vdash_{\text{FOL}} \mathcal{B}$ implies $\mathcal{A}_1, \ldots, \mathcal{A}_n, \neg\mathcal{B} \vdash_{\text{FOL}} \mathcal{B}$.
(13) $\mathcal{A}_1; \ldots; \mathcal{A}_n \vdash_{\mathbf{R}} \mathcal{B}$ *does not* imply $\mathcal{A}_1; \ldots; \mathcal{A}_n; \sim\mathcal{B} \vdash_{\mathbf{R}} \mathcal{B}$.

This discrepancy survives generalization to a new conclusion. (14) is obvious in the case of **R**, because $C$ may be completely irrelevant to the $\mathcal{A}$'s or $\mathcal{B}$. (15) is equally obvious in the case of FOL, because $\mathcal{B} \wedge \neg\mathcal{B}$ implies an arbitrary $C$.

(14) $\mathcal{A}_1; \ldots; \mathcal{A}_n \vdash_{\mathbf{R}} \mathcal{B}$ *does not* imply $\mathcal{A}_1; \ldots; \mathcal{A}_n; \sim\mathcal{B} \vdash_{\mathbf{R}} C$.
(15) $\mathcal{A}_1, \ldots, \mathcal{A}_n \vdash_{\text{FOL}} \mathcal{B}$ implies $\mathcal{A}_1, \ldots, \mathcal{A}_n, \neg\mathcal{B} \vdash_{\text{FOL}} C$.

FOL cannot distinguish between a set of formulas that is inconsistent in the sense of containing both $\mathcal{A}$ and $\neg\mathcal{A}$ and a set of formulas that contains all formulas. In relevance logics, however, if $\mathcal{A}$ and $\sim\mathcal{A}$ are in a set of formulas, another formula $\mathcal{B}$ may very well be left out. We think that it is useful to take advantage of this feature of relevance logic when default reasoning is concerned. The distinction between *absolute inconsistency* and *negation inconsistency* is already present in **fde**. We think that it is plausible that the occurrence

[3] To the objection that "There is no need to apply any default rules, when the belief set is already trivial," we reply that we agree. However, we think that the way forward is through bypassing the absolute inconsistency of an extension when a couple of sentences negate each other.

of $\mathcal{A}$ and $\sim\mathcal{A}$ at the same time neither eliminates nor inflates the information content of the set of formulas. To provide an informal argument, we modify the example of "the two firefighters" from [16].

*Example #5.* (Two firefighters) It is easy to imagine that having checked in at a hotel, a person neglects to study the fire escape route. (We can contemplate a variety of reasons how and why this happens: Fires happen rarely nowadays, so knowing the exits seems unimportant; There is an automatic sprinkle system visible on the ceilings, so there is more time to exit if there is a fire; The floor map with the route is missing from the back of door; Our traveler is dead-tired after a long flight; and so on.) Furthermore, the hotel may have an elaborate system of corridors, possibly, with some signage missing.

To put it briefly, when a fire starts late in the night, our traveler has no idea how to leave the building using a staircase and there are no signs in the corridor or leaflets in the room to help. Then, in quick succession, two firefighters show up at the door. The first one claims that the staircase to the left is the safe way to exit, and the staircase to the right is filled with smoke, hence, that's not safe. The second one claims the opposite, namely, that the staircase to the right is the safe route, while the staircase to the left is filled with smoke and not safe.

We can formalize the claims in various ways, but the two firefighters clearly contradict each other. All things being equal — both firefighters appear to be serious fellows doing their job — there is no reason to assume that either tries to mislead or endanger a hotel guest. By FOL, anything follows now, including that the best option is to go back to sleep and that the best option is to fashion all the bed linen into a makeshift rope and climb onto the balcony below, etc. More interestingly, FOL also licenses the claim that "*There are no staircases to the left or to the right*," which contradicts the useful information that can be extracted from *both* firefighters' claims. Importantly, by relevance logic, it does not follow that there are no staircases to the left or to the right. One might think that once our traveler knows that there are (at least) two staircases, and where they are, the next reasonable step could be to examine their condition.

We hope that we have provided sufficient motivation that a non-monotonic logic might be related to actual reasoning. As a first step toward basing a default system on a relevance logic (such as **T**, **E** and **R**), we start with a smaller fragment shared by these logics (as well as by some others like **B**). That is, we will use the logic of first-degree entailments (**fde**) in place of 2-valued logic.

## 3 First-degree entailments and consequence

Three important distinguishing features of relevance logics are already present in the fragment called *first-degree entailments*: (i) $\mathcal{A} \wedge \sim\mathcal{A}$ does not imply an arbitrary $\mathcal{B}$; (ii) if $\mathcal{A}$ implies $\mathcal{B}$, then $\mathcal{A}$ and $\mathcal{B}$ share a variable; (iii) the so-called "disjunctive syllogism" fails, that is, $\mathcal{A} \wedge (\sim\mathcal{A} \vee \mathcal{B})$ does not imply $\mathcal{B}$. At the same time, **fde** is simpler than relevance logics of higher (finite) degrees.

*Definition* 1. The *language* of **fde** includes four connectives: $\sim$ (negation), $\wedge$ (conjunction), $\vee$ (disjunction) and $\rightarrow$ (entailment). The arity of the connectives is as usual (negation is unary, the others are binary), and we assume that there is a set of propositional variables (of cardinality $\leq \aleph_0$). *Zero-degree formulas* are inductively generated from propositional variables by $\sim$, $\wedge$ and $\vee$.[4] *First-degree formulas* are those that are formed from a pair of zero-degree formulas by $\rightarrow$.

In other words, first-degree formulas have degree one with respect to entailments, moreover, $\rightarrow$ is the main connective of those formulas. We call first-degree entailments those first-degree formulas that are theorems of the proof systems or valid in the semantics given below.

The logic of first-degree entailments has been characterized in various ways. Syntactic characterizations include a Hilbert-style calculus (that we recall below), normal forms, as well as a tableaux characterization by pairs of trees. (See e.g., [2], [12] and [14].) Algebraic semantics for **fde** was provided by Belnap, using the matrix $M_0$ with truth sets, and by Dunn, using the lattice **4**. (See [2, §18], [6] and [14].) Other semantics were given by Białynicki–Birula and Rasiowa (for pseudo-Boolean algebras), and by Dunn (for De Morgan lattices). (See [14, §§4.3-4.4].) Both of the latter use sets to represent the algebra of **fde**. First-degree entailments turn out to be the finer 2-valued tautologies when those are sifted (cf. [13]).

We will briefly mention a syntactic system, an algebra and a set-theoretic semantics for **fde** to introduce various aspects of **fde**.

*Definition* 2. The axioms and rules of **fde** are the following.

(A1) $(\mathcal{A} \wedge \mathcal{B}) \rightarrow \mathcal{A}$
(A2) $(\mathcal{A} \wedge \mathcal{B}) \rightarrow \mathcal{B}$
(A3) $\mathcal{A} \rightarrow (\mathcal{A} \vee \mathcal{B})$
(A4) $\mathcal{A} \rightarrow (\mathcal{B} \vee \mathcal{A})$
(A5) $(\mathcal{A} \wedge (\mathcal{B} \vee C)) \rightarrow ((\mathcal{A} \wedge \mathcal{B}) \vee C)$
(A6) $\mathcal{A} \rightarrow \sim\sim\mathcal{A}$
(A7) $\sim\sim\mathcal{A} \rightarrow \mathcal{A}$
(R1) $\mathcal{A} \rightarrow \mathcal{B}$ and $\mathcal{B} \rightarrow C$ imply $\mathcal{A} \rightarrow C$
(R2) $\mathcal{A} \rightarrow \mathcal{B}$ and $\mathcal{A} \rightarrow C$ imply $\mathcal{A} \rightarrow (\mathcal{B} \wedge C)$
(R3) $\mathcal{A} \rightarrow \mathcal{B}$ and $C \rightarrow \mathcal{B}$ imply $(\mathcal{A} \vee C) \rightarrow \mathcal{B}$
(R4) $\mathcal{A} \rightarrow \mathcal{B}$ implies $\sim\mathcal{B} \rightarrow \sim\mathcal{A}$

Given the axiomatization, we can define the notion of a theorem. All the formulas that occur in the axioms and in the rules are first-degree entailments. This restriction means that proofs may be defined as usual, since, obviously, no analogue of the deduction theorem can be sought. Instead, we think of a theorem of **fde** as expressing that the consequent is a consequence of the antecedent. That is, a *theorem* is a formula that has a proof, which is a finite sequence of first-degree entailments that are either axioms or obtained from previous formulas in the sequence by a rule.

The theorems of **fde** have the *perfect interpolation* property, which nicely captures, moreover strengthens variable sharing. That is, if $\vdash_{\mathbf{fde}} \mathcal{A} \rightarrow \mathcal{B}$, then there is a zero-degree formula containing no other propositional variables than those common to $\mathcal{A}$ and $\mathcal{B}$ such that $\vdash_{\mathbf{fde}} \mathcal{A} \rightarrow C$ and $\vdash_{\mathbf{fde}} C \rightarrow \mathcal{B}$. (See [2, §15].)

The logic of first-degree entailments is *decidable* (see [2] or [12]), which is clearly advantageous for the base logic of a default system. An axiomatic formulation is not necessarily the best calculus to use when one intends to decide if a zero-degree formula entails another one. However, we assume that one of the equivalent tableaux can be used instead. (See [12] or [7] for tableaux.)

[4] "Zero-degree" is meant to indicate that the degree of these formulas, when only entailments count, is null.

**Fde** can be algebraized without difficulty. The main connective of each theorem is $\rightarrow$, and equivalence classes may be formed by collecting together formulas that are mutually provable. **Fde** yields an especially attractive structure, namely, the class of De Morgan lattices. A short definition of De Morgan lattices is that they are distributive lattices with a De Morgan negation. The following finite set of equations defines De Morgan lattices, which puts them into a well-understood class of algebras.

*Definition* 3. A *De Morgan lattice* is $\mathfrak{A} = \langle A; \wedge, \vee, \sim\rangle$ (of type $\langle 2, 2, 1\rangle$) in which equations $(a1)$–$(a11)$ hold.

| | | | |
|---|---|---|---|
| $(a1)$ | $a \wedge b = b \wedge a$ | $(a2)$ | $a \vee b = b \vee a$ |
| $(a3)$ | $a \wedge (b \wedge c) = (a \wedge b) \wedge c$ | $(a4)$ | $a \vee (b \vee c) = (a \vee b) \vee c$ |
| $(a5)$ | $a \wedge (b \vee a) = a$ | $(a6)$ | $a \vee (b \wedge a) = a$ |
| $(a7)$ | $a \wedge (b \vee c) = (a \wedge b) \vee (a \wedge c)$ | $(a8)$ | $a \vee (b \wedge c) = (a \vee b) \wedge (a \vee c)$ |
| $(a9)$ | $\sim(a \vee b) = \sim a \wedge \sim b$ | $(a10)$ | $\sim(a \wedge b) = \sim a \vee \sim b$ |
| $(a11)$ | $\sim\sim a = a$ | | |

De Morgan lattices are self-dual, and $\sim$ is a dual isomorphism.

A semantics for **fde** may be defined on $V$, a ring of subsets of $U$, that is, a subset of the power set of $U$ that is closed under intersection ($\cap$) and union ($\cup$). An interpretation $I$ (comprising a pair of functions $\langle I_1, I_2\rangle$) maps $p$ into a pair of elements of $V$. Zero-degree formulas are interpreted straightforwardly according to (I1)–(I3), using $\cap$ and $\cup$, and the permutation of $I_1$ and $I_2$ as operations.

(I1) $I(\mathcal{A} \wedge \mathcal{B}) = \langle I_1(\mathcal{A}) \cap I_1(\mathcal{B}), I_2(\mathcal{A}) \cup I_2(\mathcal{B})\rangle$
(I2) $I(\mathcal{A} \vee \mathcal{B}) = \langle I_1(\mathcal{A}) \cup I_1(\mathcal{B}), I_2(\mathcal{A}) \cap I_2(\mathcal{B})\rangle$
(I3) $I(\sim\mathcal{A}) = \langle I_2(\mathcal{A}), I_1(\mathcal{A})\rangle$

An **fde** formula $\mathcal{A} \rightarrow \mathcal{B}$ is valid iff in any $\langle U; \cap, \cup\rangle$, $I(\mathcal{A}) \leq I(\mathcal{B})$, where the latter is spelled out as $I_1(\mathcal{A}) \subseteq I_1(\mathcal{B})$ and $I_2(\mathcal{B}) \subseteq I_2(\mathcal{A})$. This semantics is adequate for **fde**. (See [10] and [14].)

*Example* 3.1. As an illustration, we construct a falsifying interpretation for a notoriously irrelevant implication. $(p \wedge \sim p) \rightarrow (q \vee \sim q)$ is, of course, not a theorem of **fde**, when $p$ is not $q$, and it is easy to see that this is how it should be. Let $A \cup B \in V$ the largest set in $V$, with $A, B \in V$, but $A \not\subseteq B$ and $B \not\subseteq A$. Letting $I(p) = \langle A, A\rangle$ and $I(q) = \langle B, B\rangle$, $I(p \wedge \sim p) = \langle A, A\rangle$, whereas $I(q \vee \sim q) = \langle B, B\rangle$. However, neither $I_1(p \wedge \sim p) \leq I_1(q \vee \sim q)$, nor $I_2(q \vee \sim q) \leq I_2(p \wedge \sim p)$.

First-degree entailments allow us to define consequence from provable formulas. This yields a notion of a binary consequence: zero-degree formulas are, by definition, finite. The notions of conjunctive and disjunctive normal forms can be defined in **fde**, with the result that every theorem may be thought to have a normal form in which the antecedent is in conjunctive normal form and the consequent is in disjunctive normal form. Then, we can define a symmetric consequence relation from **fde** theorems that would involve finitely many zero-degree formulas.

*Definition* 4 (Finite **fde** consequence). Let $\nabla$ and $\Delta$ be finite sets of zero-degree formulas. $\nabla \vdash^{\text{fin}}_{\mathbf{fde}} \Delta$ if and only if $\bigwedge \nabla \rightarrow \bigvee \Delta$ is an **fde** theorem, where $\bigwedge \nabla$ is a conjunction of all the formulas in $\nabla$, and $\bigvee \Delta$ is a disjunction of all the formulas in $\Delta$.

Although **fde** theorems are finite, we can define a consequence relation that is a relation between an arbitrary set of formulas and a formula, by relying on $\vdash^{\text{fin}}_{\mathbf{fde}}$ to provide the witness for the relation.

*Definition* 5 (**Fde** consequence). Let $\nabla$ be a set of zero degree formulas and $\mathcal{A}$ be a zero-degree formula. $\nabla \vdash_{\mathbf{fde}} \mathcal{A}$ iff there are formulas $\mathcal{B}_1, \ldots, \mathcal{B}_n$ (where $n \in \omega$) in $\nabla$ such that $(\mathcal{B}_1 \wedge \cdots \wedge \mathcal{B}_n) \rightarrow \mathcal{A}$ is an **fde** theorem.

The latter notion of consequence parallels the usual consequence relation of FOL, because it allows $\nabla$ to be infinite. Just as proofs are finite objects in an axiomatization of FOL, the antecedent of the **fde** theorem that sanctions the consequence is finite.

## 4 Default logic based on fde

The choice of **fde** as the base logic places certain constraints on the sort of defaults we might consider. Obviously, there are no quantifiers in **fde** formulas, which means that, if we were to handle defaults that involve universally quantified statements (or generics), then we would have to use suitable instances of those formulas instead.

**Fde** allows more variations than classical logic does on how a default rule could be applied. In order to show this, we quickly recall the way default rules are formulated and applied in 2-valued contexts. A default rule is an ordered triple of formulas $\langle \mathcal{A}, \mathcal{B}, C \rangle$ (or $\mathcal{A}\colon \mathcal{B}/C$), where $\mathcal{A}$ is *the elicitor* (or trigger) formula, $\mathcal{B}$ is *the justification* (or prerequisite) formula and $C$ is *the presumption* (or default conclusion) formula.[5] Informally, the idea is that if $\mathcal{A} \in \Gamma$ (where $\Gamma$ is a set of classical first-order formulas), and $\neg\mathcal{B} \notin \Gamma$, then $\Gamma$ may be extended by $C$.

A few variations on default rules have been also considered. First, $\mathcal{A}$ may be absent, that is, the default rule may be of the shape $\langle\ , \mathcal{B}, C \rangle$. Rules like this are always applicable as long as $\neg\mathcal{B} \notin \Gamma$. If the prerequisite is omitted, that is, the rule is of the form $\langle \mathcal{A},\ , C \rangle$, then the default rule behaves like a one-premise inference rule; whenever, $\mathcal{A} \in \Gamma$, $C$ may be added to $\Gamma$. A yet another version of a default rule is when $\mathcal{B}$ and $C$ are the same formula. Notice that this is not simply reiterating something, because $\langle \mathcal{A}, \mathcal{B}, \mathcal{B} \rangle$ has the effect of permitting the addition of $\mathcal{B}$ to $\Gamma$, when $\mathcal{A} \in \Gamma$ and $\neg\mathcal{B} \notin \Gamma$. Default rules of this latter kind are called *normal*, and they capture the idea that if $\mathcal{A}$ is known, but $\mathcal{B}$ is not known to be false, then $\mathcal{B}$ may be presumed.

When we think of actual beliefs of an agent or of implementations of beliefs of an agent, it seems more feasible to work with a finite set of formulas than with theories comprising beliefs, which may not have a finite base set. If $\Gamma$ is a theory, then the application of $\langle \mathcal{A}, \mathcal{B}, C \rangle$ may be equivalently formulated as if $\Gamma \vdash_{\text{FOL}} \mathcal{A}$ and $\Gamma \nvdash_{\text{FOL}} \neg\mathcal{B}$, then we may replace $\Gamma$ by $\Gamma \cup \{C\}$. Then, the application of a normal default rule with no elicitor, $\langle\ , \mathcal{B}, \mathcal{B} \rangle$ is highly reminiscent of a step in the Lindenbaum construction, because if $\Gamma \nvdash_{\text{FOL}} \neg\mathcal{B}$, then $\Gamma$ may be expanded with $\mathcal{B}$ while consistency is preserved. It speaks to the generality of the approach that some special default rules are usual inference rules or steps in maximizing consistent theories. However, the default rules that appear in examples in the literature and typically lead to non-monotonicity are not these special ones.

Considerations of consistency, which is encoded in normal default rules, lead to further complications when more than one default rule is applied. If there are multiple default rules (of arbitrary shape) that are potentially applicable, then it is easy to end up with an inconsistent set of formulas.

[5] Calling $\mathcal{B}$ the justification is a bit misleading, because it is not that $\mathcal{B} \in \Gamma$ is required. Rather, we need that $\neg\mathcal{B} \notin \Gamma$. However, "justification" is standard terminology.

*Example* 4.1. Let $\Gamma = \{\mathcal{A}, \mathcal{D}\}$, and let us have two default rules, $\langle \mathcal{A}, \mathcal{B}, C\rangle$ and $\langle \mathcal{D}, \neg\mathcal{B}, \neg C\rangle$. Assuming that there are no inferential connections between the formulas (below the surface, so to speak), $\Gamma \vdash_{\mathrm{FOL}} \mathcal{A}$, and $\Gamma \vdash_{\mathrm{FOL}} \mathcal{D}$, but neither $\Gamma \vdash_{\mathrm{FOL}} \mathcal{B}$ nor $\Gamma \vdash_{\mathrm{FOL}} \neg\mathcal{B}$. The application of the two default rules allow us to expand $\Gamma$ to $\{\mathcal{A}, C, \neg C, \mathcal{D}\}$.

One might think that default rules and sets of default rules should be examined before any application. It seems reasonable — and easy to check — that no default rules should be considered seriously, if they are of the form $\langle \mathcal{A}, \mathcal{B}, \neg\mathcal{A}\rangle$, because this in effect says that "if $\mathcal{A}$ is known and something else is the case ($\neg\mathcal{B}$ is not known), then we may include $\neg\mathcal{A}$ into our set." (We could call these rules *ab*normal.) Perhaps, even a pair of default rules as above is suspicious, because of the presence of $C$ and $\neg C$. However, this is not as a clear cut situation as the *ab*normal rules are. It is very well possible that we would never expect to know both $\mathcal{A}$ and $\mathcal{D}$ (using the formulas from the above example), and so $\Gamma$ represents an anomaly. If we have more default rules with a more varied relationship between the formulas in the rules, then scrutinizing the rules in advance as a measure to avoid producing an inconsistent expansion from a $\Gamma$ is unfeasible. If $\Gamma$ is much larger than in Example 4.1, let us say, it contains a couple of thousands of formulas with tens of thousands of propositional variables, then it may become practically impossible to check that $C$ or $\neg C$ is not implied by $\Gamma$.

We will disregard the problem of constantly re-checking negation inconsistency of extensions, but we will draw on the anatomy of the Lindenbaum construction. Returning to the question of non-monotonicity for a moment, we should recall, that in a default logic, it amounts to starting with a larger set of formulas (plus a set of default rules) and not having as large an extension as in the case of a smaller set of formulas. It is interesting to note that monotonicity *does not need to obtain* in a similar sense in the Lindenbaum construction either. Maximally consistent sets of formulas do not support an attractive (or useful) order relation, because no two maximally consistent sets of formulas (in the same language) have $\subseteq$ holding between them. (If $\Gamma_1 \subseteq \Gamma_2$ where the $\Gamma$'s are maximal, then $\Gamma_1 = \Gamma_2$. That is, the proper subset relation ($\subset$) turns the set of all maximally consistent sets of formulas into an anti-chain.) Now let us fix an enumeration of all formulas, and let us consider two sets $X_1$ and $X_2$ such that $X_1 \subseteq X_2$ (and $X_2$ is consistent). If $\Gamma_1$ and $\Gamma_2$, respectively, are their maximal extensions (using the fixed enumeration of all formulas), then it is very well possible that $\Gamma_1 \nsubseteq \Gamma_2$. Consider a simple example: if $\neg\mathcal{A} \notin X_1$ but $\neg\mathcal{A} \in X_2$, then, if $\mathcal{A}$ precedes $\neg\mathcal{A}$ in the enumeration, then $\mathcal{A} \in \Gamma_1$, but $\mathcal{A} \notin \Gamma_2$ (if no other formulas are interfering), because $\Gamma_2$ is consistent. Thus, there is no principal problem with incorporating the gist of the Lindenbaum construction into the definition of extensions by default rules.

In order to build a default logic on **fde**, we have to clarify how we intend to capture the informal understanding of the application of a default rule using **fde**, which is not a 2-valued logic. The algebraization of **fde** results in a De Morgan lattice. Although the free De Morgan lattice, generated by a finite or infinite set of elements does not have a fixed point for negation, there are De Morgan lattices that do. (Cf. [14] and [22].) We used such a lattice, formed from sets, with two fixed points in Example 3.1. This suggests that an interpretation has to cater for formulas that are *both* true and false, or dually, *neither* true nor false.[6] **Fde** has four-valued interpretations where the four truth values may be called "true," "false," "neither" and "both." (See the interpretation given by Dunn in [12], as well as [14], [5] and [17].) To reiterate, a collection of data records or even a data base may easily contain pieces of information that — explicitly, or somewhat implicitly

[6] Truth and falsity do not need be thought of in some "deep metaphysical or ontological sense." In our view, logic has more to do with information and reasoning than with picturing the deep structure of reality.

— contradict each other. The primary example is the World Wide Web, which is teaming with inconsistencies of the most trivial type.

The first step in defining a default system based on **fde** is to specify default rules using the language of **fde**.

*Definition* 6. Let $\mathcal{A}, \mathcal{B}$ and $C$ be zero-degree formulas. Then $\langle \mathcal{A}, \mathcal{B}, C \rangle$ is a *default rule* in the language of **fde**.

The application of a default rule given a set of formulas is defined as follows. We preserve the role of $\sim$ as an indication that the negation is known.

*Definition* 7. Let $\nabla$ be a set of zero-degree formulas, and let $\langle \mathcal{A}, \mathcal{B}, C \rangle$ be a default rule, which we denote by $\delta$. The result of an *application of the default rule* $\delta$ to $\nabla$, denoted by $\nabla \lessdot \delta$, is $\nabla$ if $\nabla \nvdash_{\mathbf{fde}} \mathcal{A}$ or $\nabla \vdash_{\mathbf{fde}} {\sim}\mathcal{B}$, and $\nabla \cup \{ C \}$ if $\nabla \vdash_{\mathbf{fde}} \mathcal{A}$ and $\nabla \nvdash_{\mathbf{fde}} {\sim}\mathcal{B}$.

If $C$ is added to a set of formulas as a result of an application of the default rule $\delta$, then we might say that the application of $\delta$ has been *successful*.

*Definition* 8. Let $\nabla$ be a set of zero-degree formulas, and let $\delta$ be the default rule $\langle \mathcal{A}, \mathcal{B}, C \rangle$. $\delta$ is *apposite* to $\nabla$, if all propositional variables occurring in $C$ occur in some element of $\nabla$.

One of the advantages of using **fde** instead of classical logic is reflected in the next two lemmas. The statement of the lemma is quite obvious once **fde** and its features that we tried to convey in Section 3 are well understood.

*Lemma* 1 (Non-triviality, 1). Given $\nabla$ such that some propositional variable $p$ does not occur in any element of $\nabla$ (hence, $\{ \mathcal{D} \colon \nabla \vdash_{\mathbf{fde}} \mathcal{D} \} \neq$ wff), and given a default rule $\delta$ which is apposite to $\nabla$, $\nabla \lessdot \delta$ is not trivial, nor is $\{ \mathcal{D} \colon \nabla \lessdot \delta \vdash_{\mathbf{fde}} \mathcal{D} \}$ trivial.

*Proof.* Since $p$ does not occur in $\nabla$, we have that $\nabla \nvdash p$ and $\nabla \nvdash {\sim}p$. The same can be said about $\nabla \lessdot \delta$, because $p$ does not occur in $\delta$ either. In other words, $p$ is irrelevant to $\nabla$ before and after the application of $\delta$ (with respect to which $p$ is also irrelevant).

There is no limitation on the cardinality of the set of formulas in Definition 7, that is, $\nabla$ may not be a finite set. Allowing infinite sets is reasonable, in general, because we might want to consider sets of formulas for $\nabla$ that are already closed under consequence; or we might consider some other infinite sets, that adhere to some pattern. On the other hand, if we do not restrict $\nabla$ in any way, then the lemma is obviously false. If we require that $\nabla$ has a negation consistent set of consequences (i.e., for no $\mathcal{A}$, $\mathcal{A}, {\sim}\mathcal{A} \in \{ \mathcal{D} \colon \nabla \vdash_{\mathbf{fde}} \mathcal{D} \}$), then we have the following.

*Lemma* 2 (Non-triviality, 2). If $\{ \mathcal{D} \colon \nabla \vdash_{\mathbf{fde}} \mathcal{D} \}$ is a negation consistent set of zero-degree formulas, then $\nabla \lessdot \delta$ is not trivial, that is, there is a formula $\mathcal{E}$ such that $\mathcal{E} \notin \{ \mathcal{D} \colon \nabla \lessdot \delta \vdash_{\mathbf{fde}} \mathcal{D} \}$.

*Proof.* Let us assume that $\delta$ is $\langle \mathcal{A}, \mathcal{B}, C \rangle$. $\{ \mathcal{D} \colon \nabla \lessdot \delta \vdash_{\mathbf{fde}} \mathcal{D} \}$ may be negation inconsistent, for instance, if $\nabla \lessdot \delta$ is successful, and ${\sim}C$ is a consequence of $\nabla$. However, there is a formula $\mathcal{G}$ such that either $\mathcal{G} \notin \{ \mathcal{D} \colon \nabla \vdash_{\mathbf{fde}} \mathcal{D} \}$ or ${\sim}\mathcal{G} \notin \{ \mathcal{D} \colon \nabla \vdash_{\mathbf{fde}} \mathcal{D} \}$, and $C$ and $\mathcal{G}$ do not have any propositional variables in common. Then unchangeably, $\nabla \lessdot \delta$ does not imply $\mathcal{G}$ or ${\sim}\mathcal{G}$.

Of course, we might want to consider a set of defaults with more than a single element. We borrow from the Lindenbaum procedure the idea of an *enumeration*. We assume that the default rules, that is, certain triplets of formulas are enumerated, and that each default rule $\delta_i$ has the form $\langle \mathcal{A}_i, \mathcal{B}_i, C_i \rangle$.

*Remark* 1. We introduce the enumeration to gain more control over $\Delta$. This move is not motivated by the use of *entrenchments* in belief revision or by ordering default rules to deal with problems in default logic based on classical logic (cf. [24]).

*Definition* 9. Let $\nabla$ be a set of zero-degree formulas, and let $\Delta_\epsilon$ be a set of defaults with a fixed enumeration $\epsilon$. Then the *extension* of $\nabla$ according to $\Delta_\epsilon$, denoted by $\nabla \lessdot \Delta_\epsilon$, is defined as follows.

1. $\nabla_0 = \nabla \quad \text{and} \quad I_0 = \emptyset$
2. $\nabla_{n+1} = \begin{cases} \nabla_n \text{ and } I_{n+1} = I_n & \text{if } \nabla_n \nvdash_{\mathbf{fde}} \mathcal{A}_{n+1} \text{ or } \nabla_n, C_{n+1} \vdash_{\mathbf{fde}} \bigvee\{\,{\sim}\mathcal{B}_{i \in I \cup \{n+1\}}\,\}, \\ \nabla_n \cup \{\, C_{n+1} \,\} \text{ and} & \\ I_{n+1} = I_n \cup \{\, n+1 \,\} & \text{otherwise.} \end{cases}$
3. $\nabla \lessdot \Delta_\epsilon = \bigcup_{n \in \omega} \nabla_n$

A nice feature of this definition is that it is *constructive* in the sense that the above stepwise construction is an effective procedure in virtue of $\vdash_{\mathbf{fde}}$ being decidable. Also, step 2 excludes the possibility of a default rule of the form $\langle \mathcal{A}, \mathcal{B}, {\sim}\mathcal{B} \rangle$ to be ever successfully applied, which seems to be a good idea. Just because ${\sim}\mathcal{B}$ is not provable, that does not mean that it should be added into a set of formulas. If at the same time $\mathcal{B}$ is provable, then the successful application of such a rule would immediately lead to negation inconsistency.

*Remark* 2. Default rules are not formulas, rather formulas with a pair of conditions attached to them. Definition 9 closely parallels a Lindenbaum construction, and does not take into account that a condition may not be satisfied at step $l$ but become true at a later step.[7] We will introduce an improved $\lessdot$ in Definition 10, but first, we note some properties of the extension that results by $\lessdot$.

*Lemma* 3. Given $\nabla$ and $\Delta_\epsilon$, the *extension* $\nabla \lessdot \Delta_\epsilon$ always exists. Furthermore, $\nabla \lessdot \Delta_\epsilon$ has the following properties.

(1) $\nabla \lessdot \Delta_\epsilon$ is unique.
(2) No prerequisite of any default rule that has been successfully applied from $\Delta_\epsilon$ is negated in $\nabla \lessdot \Delta_\epsilon$.

*Proof.* For the first claim, we note that $\vdash_{\mathbf{fde}}$ is decidable and step 2 in Definition 9 is deterministic.

The claim in (2) is true in virtue of step 2 in Definition 9: if the negation of the prerequisite of any already applied rule would become provable once $C_{n+1}$ is added, then that default rule is not applied.

Perhaps, it would be also desirable for us to be able to say that $\nabla \lessdot \Delta_\epsilon$ is "maximal" in some sense. Maximality, clearly, cannot mean that the set of formulas $\nabla \lessdot \Delta_\epsilon$ is maximal (or maximally consistent) in an absolute sense — $\nabla$ may be finite and $\Delta$ may be a singleton set of default rules. Moreover, $\nabla \lessdot \Delta_\epsilon$ is not maximal in the sense that the enumeration of the default rules may prevent the (successful) application of a default rule, because a prerequisite of a default rule is added at a later stage than the default was considered. This latter observation motivates an enhancement of the definition of $\nabla \lessdot \Delta_\epsilon$.

[7] This property of Definition 9 may lead to some puzzling results, when a few defaults are applied to a small set $\nabla$.

*Definition* 10. Let $\nabla$ be a set of zero-degree formulas, and let $\Delta_\epsilon$ be a set of default rules ($\delta_i$'s) with an enumeration $\epsilon$. Then $\nabla \overset{*}{\lessdot} \Delta_\epsilon$ is defined as follows. (The enumeration is inherited by $\Theta$ from $\Delta_\epsilon$.)

1. $\nabla_0 = \nabla$ and $\Theta_0 = \{ \delta_1 \}$
2. $\nabla_{n+1} = \nabla_n \lessdot \Theta_n$ and $\Theta_{n+1} = (\Theta_n \cup \{ \delta_{n+2} \}) - \{ \delta_{i \leq n+1} : \nabla_{n+1} \vdash_{\mathbf{fde}} {\sim}\mathcal{B}_i \}$
3. $\nabla \overset{*}{\lessdot} \Delta_\epsilon = \bigcup_{n \in \omega} \nabla_n$

The idea behind this definition is that a formula $C_{i+1}$ (from a default rule) $\delta_{i+1}$ might be left out (of $\nabla \lessdot \Delta_\epsilon$) for two reasons. First, $\mathcal{A}_{i+1}$ may not be implied by $\nabla_i$, second, ${\sim}\mathcal{B}_{i+1}$ may be implied by $\nabla_i$. If the latter, then $\delta_i$ does not need to be reconsidered, because $\cup$ is a monotone operation on sets. However, if the former, then we retain $\delta_{i+1}$ as an element of $\Theta_{i+1}$ to be considered again at the next stage. The construction remains effective, because at each finite stage $\Theta_i$ is finite.

*Theorem* 4.1. Given $\nabla$ and $\Delta_\epsilon$, the *extension* $\nabla \overset{*}{\lessdot} \Delta_\epsilon$ has the properties of $\nabla \lessdot \Delta_\epsilon$ in Lemma 3. Furthermore, $\nabla \overset{*}{\lessdot} \Delta_\epsilon$ is maximal with respect to $\Delta_\epsilon$, that is, all the default rules in $\Delta_\epsilon$ that could have been applied have been applied.

*Proof.* Let us assume that a default rule $\delta_i$ could have been applied, meaning that $\nabla \overset{*}{\lessdot} \Delta_\epsilon \vdash_{\mathbf{fde}} \mathcal{A}_i$, $\nabla \overset{*}{\lessdot} \Delta_\epsilon \nvdash_{\mathbf{fde}} {\sim}\mathcal{B}_i$ but $C_i \notin \nabla \overset{*}{\lessdot} \Delta_\epsilon$. Then $\nabla_i \nvdash_{\mathbf{fde}} \mathcal{A}_i$ must be the case, for otherwise, $C_i \in \nabla_{i+1}$ would be the case. We know that for all $n \in \omega$, $\nabla_n \subseteq \nabla \overset{*}{\lessdot} \Delta_\epsilon$, and so if $\nabla \overset{*}{\lessdot} \Delta_\epsilon \nvdash_{\mathbf{fde}} {\sim}\mathcal{B}_i$, then $\nabla_n \nvdash_{\mathbf{fde}} {\sim}\mathcal{B}_i$. Thus, $\delta_i \in \Theta_{i+1}$. Since proofs are finite, there is a $j$ such that $\nabla_j \vdash_{\mathbf{fde}} \mathcal{A}_i$ (where $i < j$ due to the reasoning so far). We know that $\nabla_j \nvdash_{\mathbf{fde}} {\sim}\mathcal{B}_i$, hence, $C_i \in \nabla_{j+1}$ which contradicts our assumption.

We should note that one could exploit the four-valued character of **fde** explicitly within the definition of an application of a default given a set of formulas. Let us assume again that $\nabla$ is a set of zero-degree formulas, and $\delta$ is the default $\langle \mathcal{A}, \mathcal{B}, C \rangle$. Since a pair of formulas such as ${\sim}\mathcal{B}$ and $\mathcal{B}$ do not lead to triviality, $\nabla \nvdash_{\mathbf{fde}} {\sim}\mathcal{B}$ could seem not to be crucial; hence, something that could be ignored. This would certainly change the resulting extension, because, in general, more defaults would be applicable. However, it seems that this modification would not be in the intended spirit of the default rules. The prerequisite is thought to capture the idea that $\mathcal{B}$ is consistent or possible with respect to $\nabla$ in some sense. Although, in **fde**, if ${\sim}\mathcal{B}$ is derivable, then inconsistency in the absolute sense does not result, in general, by considering $\mathcal{B}$ too, it would seem that an argument would be required why we can ignore ${\sim}\mathcal{B}$ in an application of $\delta$. The inclusion of $C$ by this default rule relies on the possibility of believing in $\mathcal{B}$, which may be undermined by the derivability of ${\sim}\mathcal{B}$. In any case, a modification that would allow $\delta$ to be applied independently of whether ${\sim}\mathcal{B}$ is or is not derivable would have the same effect as deleting the prerequisites from the default rules. This would make defaults much less interesting, hence, we do not pursue this potential modification here.

Now, let us consider what happens with some concrete examples. We will use instances of generic (or universally quantified) statements.

*Example* 4.2 (Tweety). Let propositional variables and literals stand for the following sentences.

1. $p$: Tweety is a bird.
2. $q$: Tweety flies.
3. $r$: Tweety is a penguin.
4. ${\sim}q$: Tweety does not fly.

We may have two normal defaults $\delta_1$ and $\delta_2$, $\langle p, q, q \rangle$ and $\langle r, {\sim}q, {\sim}q \rangle$, respectively, which are instances of "Birds are presumed to fly unless they are known not to fly," and "Penguins are presumed not to fly unless they are known to fly." As before, we do not assume anything about the relationship between penguins and birds.

We might have a set of formulas $\nabla$ which does not contain or imply one or the other of $p$ and $r$, in which case the corresponding default is not considered at all. If both $p$ and $r$ are elements of $\nabla$, then the issue is whether $\nabla \vdash_{\mathbf{fde}} q$ or $\nabla \vdash_{\mathbf{fde}} {\sim}q$. If $\varDelta = \{ \delta_1, \delta_2 \}$ and $\nabla = \{ p, r \}$, then $\epsilon$ will determine which default rule is considered first. Since both default rules are normal, a successful application of $\delta_1$ prevents a successful application of $\delta_2$, and vice versa. In other words, $\{ p, r \} \overset{*}{\lessdot} \langle \delta_1, \delta_2 \rangle = \{ p, q, r \}$, because $\{ p, q, r \} \lessdot \langle \delta_2 \rangle = \{ p, q, r \}$, whereas $\{ p, r \} \overset{*}{\lessdot} \langle \delta_2, \delta_1 \rangle = \{ p, {\sim}q, r \}$, because $\{ p, {\sim}q, r \} \lessdot \langle \delta_1 \rangle = \{ p, {\sim}q, r \}$. That is, if neither $\nabla \vdash_{\mathbf{fde}} q$, nor $\nabla \vdash_{\mathbf{fde}} {\sim}q$, then the enumeration $\epsilon$ is what determines whether $q$ or ${\sim}q$ end up being included into the expansion. If $\nabla \nvdash_{\mathbf{fde}} p$ and $\nabla \nvdash_{\mathbf{fde}} r$, then neither $\delta_1$ nor $\delta_2$ is productive: $\{ p \vee r \} \overset{*}{\lessdot} \langle \delta_1, \delta_2 \rangle = \{ p \vee r \}$, but also $\{ p \vee r, q, {\sim}q \} \overset{*}{\lessdot} \langle \delta_1, \delta_2 \rangle = \{ p \vee r, q, {\sim}q \}$.

We have not linked $\epsilon$ to preferences an agent may have over the set of default rules that are available to him. However, one could suppose that the defaults that are more specific should precede defaults that are more general (and which might have a greater number of exceptions). A fleeting conclusion from this observation is that some general statements such as "Penguins are birds," which we have completely avoided in our examples, are perhaps better treated as affecting the ordering (or the enumeration) of the default rules rather than merely being default rules themselves. A software implementation, for instance, could include a simple ontology borrowing from biological taxonomy. Then, penguins would form a subset of the set of the birds, and $\langle \delta_2, \delta_1 \rangle$ would be either the only permitted ordering or it would be the preferred one.

As another example, we consider a Nixon example, which is a further variation on the already modified original Nixon example from the literature (cf. [24, p. 177].)

*Example* 4.3 (Nixon). We take propositional variables to stand for the following sentences.

1. $q$: Nixon is a quaker.
2. $r$: Nixon is a republican.
3. $p$: Nixon is a pacifist.
4. $g$: Nixon is a gun enthusiast.
5. $o$: Nixon is a politician.

Examples involving Nixon, have been used to illustrate floating conclusions. Let $\delta_1$ be $\langle q, \ , {\sim}r \rangle$ and $\delta_2$ be $\langle r, \ , {\sim}q \rangle$, that is, "Nixon is not a republican, if he is a quaker," and conversely, "Nixon is not a quaker, if he is a republican." It is commonly agreed that $\delta_1$ and $\delta_2$ are "hard" defaults, in the sense that they do not require a prerequisite and establish the incompatibility of a pair of statements.

The next pair of defaults characterizes quakers and republicans by normal defaults, $\delta_3$ and $\delta_4$. These are $\langle q, p, p \rangle$ and $\langle r, g, g \rangle$. These can be rendered in English as "Nixon is a pacifist, if he is a quaker who is not known not to be a pacifist" and "Nixon is a gun enthusiast, if he is a republican who is not known not to be a gun enthusiast."

The last two default rules we add are $\delta_5$ and $\delta_6$, which lead to the float, namely, that "Nixon is a politician." $\delta_5$ is $\langle p, o, o \rangle$, "Nixon is a politician, save we have information to the contrary, provided he is a pacifist." $\delta_6$ is $\langle g, o, o \rangle$, "Nixon is a politician, save we have information to the contrary, provided he is a gun enthusiast."

Given our defaults, which we gather into $\varDelta$, it is not difficult to see that the floating conclusion that "Nixon is a politician" will be added to a set of sentences $\nabla$, if either $p$

or $g$ is an element of $\nabla$ (or follows from the elements of $\nabla$). Since we have not defined any operations on extensions or the set of extensions that emerge when extensions are generated with different enumerations of a set of default rules, we do not need to worry about $o$ being excluded.

*Remark* 3. There is no agreement between people about floating conclusions. We think that it is a plus that we can add floating conclusions *directly* — together with traces of their sources, and all this can be done without trivializing the starting set of our beliefs. (Cf. [3, §1.5].)

A, perhaps, traditional picture of Nixon would start with $\nabla = \{ r \}$. Let us assume that $\varDelta_\epsilon$ is $\varDelta$, where the indices of the default rules coincide with their ordering. Then, $\nabla \overset{*}{\lessdot} \varDelta_\epsilon = \{ r, {\sim}q, g, o \}$, where we listed the elements of the set in the order of their addition. On the other hand, if Nixon would be known to be a republican (let us say, a member of the GOP), who is not a politician, then we could have $\nabla = \{ r, {\sim}o \}$. Now $\nabla \overset{*}{\lessdot} \varDelta_\epsilon = \{ r, {\sim}q, g \}$, which would illustrate both the non-monotonicity between extensions, and also that **fde**'s tolerance toward negation inconsistency does not mean indulgence in arbitrary inconsistencies.

If Nixon would be known to be both a quaker and a republican, then we would end up with the largest set of sentences (within the limits of this small example). Indeed, every default rule will be successfully applied, and despite the unlikely (and contradictory) characteristics of Nixon, the floating conclusion will not be lost. Let $\nabla = \{ q, r \}$. The extension $\nabla \overset{*}{\lessdot} \varDelta_\epsilon$ is $\{ q, r, {\sim}r, {\sim}q, p, g, o \}$. We could consider further initial sets of sentences about Nixon, but the last extension illustrates an advantage **fde** provides — even without dealing with a range of various extensions — in modeling floating conclusions. Since negation inconsistency does not destroy an extension, the floating conclusion is retained when we have inconsistent information concerning Nixon's leanings.

*Remark* 4. We note that our Nixon example differs from the "dove–hawk–extremist" version not as much in the labels, but in allowing negation inconsistent sets. To exclude the floating conclusion $o$, the default rules $\delta_5$ and $\delta_6$ would have to be changed, for example, to $\delta'_5 = \langle p, {\sim}g, o \rangle$ and $\delta'_6 = \langle g, {\sim}p, o \rangle$. These defaults are not very plausible. The first means, that "Nixon is a politician, if he is a pacifist and not a gun enthusiast." The second switches "pacifist" and "gun enthusiast." These are very different defaults than $\langle p, \ , {\sim}g \rangle$ and $\langle g, \ , {\sim}p \rangle$, which would not exclude $o$, if the rules are $\delta_5$ and $\delta_6$.

As a final example, we consider the case of Plat.[8]

*Example* 4.4 (Plat). The following propositional variables mnemonically abbreviate sentences.

1. $w$: Plat has webbed feet and a beak.
2. $f$: Plat has fur.
3. $v$: Plat is venomous and lays eggs.
4. $b$: Plat is a bird.
5. $m$: Plat is a mammal.
6. $s$: Plat is a snake.
7. $a$: Plat is an animal.

It is reasonable to stipulate the following default rules: $\delta_1 = \langle w, a, b \rangle$, $\delta_2 = \langle f, a, m \rangle$ and $\delta_3 = \langle v, a, s \rangle$. These default rules are not normal rules, but they are reasonable. For instance, Plat could have fur and be a toy rather than an animal.

[8] Plat is not a nickname for Plato, even if he considered featherless bipedal creatures.

If $\nabla = \{ w, f, v \}$, then $\nabla \overset{*}{\lessdot} \langle \delta_1, \delta_2, \delta_3 \rangle = \{ w, f, v, b, m, s \}$. Informally, the extension is inconsistent, however, formally, the sentence letters are unrelated. To capture that "No birds are mammals," "No snakes are mammals" and "No birds are snakes," we could add instances of these sentences with Plat as hard default rules. (The next six rules are similar to $\delta_1$ and $\delta_2$ in Example 4.3.) $\delta_4 = \langle b, \ , {\sim}m \rangle$, $\delta_5 = \langle b, \ , {\sim}s \rangle$, $\delta_6 = \langle m, \ , {\sim}b \rangle$, $\delta_7 = \langle m, \ , {\sim}s \rangle$, $\delta_8 = \langle s, \ , {\sim}b \rangle$ and $\delta_9 = \langle s, \ , {\sim}m \rangle$. Let the set of the default rules be $\Delta = \{ \delta_i \colon 1 \leq i \leq 9 \}$ and let $\epsilon$ be $<$ on the indices. Then $\{ w, f, v \} \overset{*}{\lessdot} \Delta_\epsilon$ is no longer formally consistent, but it does not include (does not even imply) all literals over 1–7. The extension $\{ w, f, v, b, m, s, {\sim}b, {\sim}m, {\sim}s \}$ may represent the state of the mind of some biologists after their first encounter with *platypuses*. From a logical point of view, the interesting facts are that $\{ w, f, v \} \overset{*}{\lessdot} \Delta_\epsilon \nvdash_{\mathbf{fde}} {\sim}w$, $\{ w, f, v \} \overset{*}{\lessdot} \Delta_\epsilon \nvdash_{\mathbf{fde}} {\sim}f$ and $\{ w, f, v \} \overset{*}{\lessdot} \Delta_\epsilon \nvdash_{\mathbf{fde}} {\sim}v$. The non-provability of the latter three formulas shows that useful information is retained, similarly, as in the informal Example #5.

## 5 Conclusions

We considered some ideas and expectations with respect to non-monotonicity. Relevance logics are non-monotonic with respect to some of their consequence relations. Default logic (based on classical logic) yields non-monotonicity with respect to extensions.

It seems to us that it is a straightforward idea to take default rules and use a relevance logic to regulate the application of default rules. We considered **fde**, which is a well-investigated and decidable fragment of some well-known relevance logics. At the same time, **fde** has a vocabulary that is straightforwardly comparable to that of classical propositional logic.

We defined default rules and their applications using **fde**. Then, we introduced constructive extensions, and showed that they often are non-trivial. The Tweety example may be treated in the usual fashion in the present framework, whereas floating conclusions can be retained in a negation inconsistent set (unlike in classical default logic). The distinction between negation inconsistency and triviality also allows us to retain useful information in extension that become negation inconsistent after the application of default rules.

We emphasized the usefulness of $\mathcal{A} \land {\sim}\mathcal{A} \nvdash_{\mathbf{fde}} \mathcal{B}$, where $\mathcal{A}$ and $\mathcal{B}$ are irrelevant to each other. However, $\mathcal{A} \nvdash_{\mathbf{fde}} \mathcal{B} \lor {\sim}\mathcal{B}$ is also the case. Then, a particular instance of $\mathcal{A} \lor {\sim}\mathcal{A}$ could be used in a default rule to select the appropriate context. We leave this idea to be worked out in detail in a future paper, but we give a quick illustration of what we mean. Let us say that $F_{27}$ is a field. It may be a parcel of land or a mathematical object. If we have default rules that deal with agriculture and default rules that deal with algebra, then the choice between the areas may be achieved by adding an innocuous piece of information about $F_{27}$. "$F_{27}$ is or isn't plowed" will guide the reasoning toward farming, whereas "$F_{27}$ is or isn't Abelian" will point to generalizations about algebraic structures.

**Acknowledgements** I am grateful to the editors of this volume for inviting me to contribute a paper. I would like to thank two anonymous referees for their comments on this paper. I would also like to thank M. Carlson who commented on a very short and very early version of this paper at the Central Division Meeting of the American Philosophical Association in Minneapolis, MN, in 2011.

## References


1. Alan R. Anderson. Entailment shorn of modality, (abstract). *Journal of Symbolic Logic*, 25:388, 1960.
2. Alan R. Anderson and Nuel D. Belnap. *Entailment: The Logic of Relevance and Necessity*, volume I. Princeton University Press, Princeton, NJ, 1975.
3. Aldo G. Antonelli. *Grounded Consequence for Defeasible Logic.* Cambridge University Press, Cambridge, UK, 2005.
4. Jc Beall. Multiple-conclusion LP and default classicality. *Review of Symbolic Logic*, 4:326–336, 2011.
5. Nuel D. Belnap. A useful four-valued logic. In J. M. Dunn and G. Epstein, editors, *Modern Uses of Multiple-valued Logic*, pages 8–37. Reidel Publishing Company, Dordrecht, 1977.
6. Nuel D. Belnap and Joel H. Spencer. Intensionally complemented distributive lattices. *Portugalie Mathematica*, 25:99–104, 1966.
7. Katalin Bimbó. Relevance logics. In D. Jacquette, editor, *Philosophy of Logic*, volume 5 of *Handbook of the Philosophy of Science* (D. Gabbay, P. Thagard and J. Woods, eds.), pages 723–789. Elsevier (North-Holland), Amsterdam, 2007.
8. Katalin Bimbó. *Proof Theory: Sequent Calculi and Related Formalisms.* CRC Press, Boca Raton, FL, 2015.
9. Gerhard Brewka. Cumulative default logic: In defense of nonmonotonic inference rules. *Artificial Intelligence*, 50:183–205, 1991.
10. J. Michael Dunn. *The Algebra of Intensional Logics.* PhD thesis, University of Pittsburgh, Ann Arbor (UMI), 1966.
11. J. Michael Dunn. A 'Gentzen system' for positive relevant implication, (abstract). *Journal of Symbolic Logic*, 38:356–357, 1973. doi: 10.2307/2272113.
12. J. Michael Dunn. Intuitive semantics for first-degree entailments and 'coupled trees'. *Philosophical Studies*, 29:149–168, 1976.
13. J. Michael Dunn. A sieve for entailments. *Journal of Philosophical Logic*, 9:41–57, 1980.
14. J. Michael Dunn. Relevance logic and entailment. In D. Gabbay and F. Guenthner, editors, *Handbook of Philosophical Logic*, volume 3, pages 117–224. D. Reidel, Dordrecht, 1st edition, 1986.
15. J. Michael Dunn and Gary M. Hardegree. *Algebraic Methods in Philosophical Logic*, volume 41 of *Oxford Logic Guides*. Oxford University Press, Oxford, UK, 2001.
16. J. Michael Dunn and Nicholas M. Kiefer. The Paradox of the Two Firefighters. Contradictory Information: Better than Nothing?, in C. Baskent and T. Ferguson, eds., *Graham Priest on Dialetheism and Paraconsistency*, Outstanding Contributions to Logic, Springer Nature, (16 pages, forthcoming).
17. Ronald Fagin, Joseph Y. Halpern, Yoram Moses, and Moshe Y. Vardi. *Reasoning about Knowledge.* MIT Press, 2003.
18. Alejandro J. García and Guillermo R. Simari. Defeasible logic programming: An argumentative approach. *Theory and Practice of Logic Programming*, 4:95–138, 2004.
19. Peter Gärdenfors and Hans Rott. Belief revision. In *Handbook of Logic in Artificial Intelligence and Logic Programming*, pages 35–132. Oxford University Press, Oxford, UK, 1995.
20. Kurt Gödel. Some basic theorems on the foundations of mathematics and their implications. In S. Feferman, editor, *Collected Works*, volume III, pages 304–323. Oxford University Press and Clarendon Press, New York, NY and Oxford, UK, 1995 1951.
21. Kurt Gödel. Zur intuitionistischen Arithmetik und Zahlentheorie. In S. Feferman, editor, *Collected Works*, volume I, pages 286–295. Oxford University Press and Clarendon Press, New York, NY and Oxford, UK, 1986.
22. Paul Halmos and Steven Givant. *Logic as Algebra.* Number 21 in Dolciani Mathematical Expositions. Mathematical Association of America, 1998.
23. John F. Horty. Skepticism and floating conclusions. *Artificial Intelligence*, 135:55–72, 2002.
24. John F. Horty. *Reasons as Defaults.* Oxford University Press, Oxford, (UK), 2012.
25. Henry A. Kautz and Bart Selman. Hard problems for simple default logics. *Artificial Intelligence*, 49:243–279, 1991.
26. Zhangang Lin, Yue Ma, and Zuoquan Lin. A fault-tolerant default logic. In M. Fisher, W. van der Hoek, B. Konev, and A. Lisitsa, editors, *Logics in Artificial Intelligence. JELIA 2006*, number 4160 in Lecture Notes in Artificial Intelligence, pages 253–265, Berlin, 2006. Springer.
27. Ana Teresa Martins and Tarcísio Pequeno. A sequent calculus for a paraconsistent nonmonotonic logic. citeseerx.ist.psu.edu/viewdoc/summary?doi=10.1.1.52.2641, June 2007.
28. Drew McDermott and Jon Doyle. Non-monotonic logic I. *Artificial Intelligence*, 13:41–72, 1980.

29. Anna Radzikowska. A three-valued approach to default logic. *Journal of Applied NonClassical Logics*, 6:149–190, 1996.
30. Ray Reiter. A logic for default reasoning. *Artificial Intelligence*, 13:81–132, 1980.
31. Hans Rott. *Change, Choice and Inference. A Study of Belief Revision and Nonmonotonic Reasoning*, volume 42 of *Oxford Logic Guides*. Oxford University Press, Oxford, UK, 2001.
32. D. J. Shoesmith and T. J. Smiley. *Multiple-conclusion Logic*. Cambridge University Press, Cambridge, UK, 1978.
33. Alfred Tarski. On some fundamental concepts of metamathematics. In *Logic, Semantics, Metamathematics*, pages 30–37. Clarendon, Oxford, UK, 1956.
34. Wyane Wobcke. Belief revision, conditional logic and nonmonotonic reasoning. *Notre Dame Journal of Formal Logic*, 36:55–102, 1995.

# Belnap and Nāgārjuna on How Computers and Sentient Beings Should Think: Truth, Trust and the Catuṣkoṭi

Jay L. Garfield

**Abstract** The Meyer-Dunn semantics for First Degree Entailment and the Belnap four-valued data base logic are strikingly similar to the Buddhist catuṣkoṭi, or four-cornered logic deployed by Nāgārjuna. I show that we can exploit this similarity to better understand the nature of truth.

**Keywords** Catuṣkoṭi • First-degree entailment • Nāgārjuna • Belnap • Buddhist logic • Truth

## 1 Introduction

There are obvious superficial similarities between the classical Indian Buddhist *catuṣkoṭi* (four corner) partition of logical space and the approximation lattice (or Hasse diagram) used in Belnap (1977a,b) and Dunn (1976) to provide a semantics for First Degree Entailment (FDE) and the four-valued logic (B4) for computer database reasoning presented in Belnap (1977a). Most obviously, both have four corners! Each draws on the same basic set of truth values {**T**,**F**}, and we find the same valuations at the four corners, *viz.*, **T**, **F**, **Both (B)**, **Neither (N)**.[1] There is, that is, an obvious similarity in the ways in which early Buddhist philosophers – including the historical Buddha himself (c. 5th c BCE) and the influential Madhyamaka philosopher Nāgārjuna (c. 2nd c CE) – partition logical space.[2]

Priest and Routley (1989) note Nāgārjuna's willingness to countenance contradictions (p. 15) and cite him as an ancestor of paraconsistent thinking. In a series of articles (Deguchi, Garfield and Priest 2008, 2013a,b,c; Garfield 2008; Garfield and Priest 2003, 2009) we have defended a version of that initial insight. That is, we have argued

Jay L. Garfield
Department of Philosophy, Smith College, Northampton, MA 01063, USA
e-mail: jgarfield@smith.edu

[1] I use *truth value* to refer to the elements of the basic set of truth values and *valuation* to refer to the truth set that is assigned to a sentence (one of the four subsets of the set of truth values). This reflects the intuition common to Belnap's and the Buddhist framework that there are only two truth values, but that a sentence may be assigned more or less than one of them. So, when Belnap calls B4 "a useful four valued logic," I think that this is a mis-statement. It does no harm so long as one is not worrying about the nature of *truth* itself; the logic can be represented as a four-valued or as a four-valuational logic as one wishes. But when one starts to talk about truth more systematically – as Belnap himself does, and as we will do below – it is useful to keep truth values and truth evalaluations separate.

[2] It is worth noting that this particular use of the catuṣkoṭi in assorting truth valuations is an instance of a broader commitment to the 4-way partition as a schema for thinking about alternatives. For instance, in Chapter 1 of *Mūlamadhyamakakārikā (Fundamental Verses on the Middle Way)*, Nāgārjuna asks whether things are self-caused, caused by something else, both, or neither; in the 22nd chapter, he asks whether the Buddha exists, does not exist, both or neither after death. (Garfield 1995) There are many other such examples. So, the application of the catuṣkoṭi to the valuation of sentences is a special case of a broader analytical strategy.

 
H. Omori, H. Wansing (eds.), *New Essays on Belnap-Dunn Logic*,
Synthese Library 418, https://doi.org/10.1007/978-3-030-31136-0_10

that Nāgārjuna (among others in the Buddhist tradition) is committed not only to an in-principle four-way partition of logical space, but also to the truth of certain contradictions.[3] Priest (2010) explores in detail how to model the catuṣkoṭi using FDE.

Nonetheless, as Priest (2010) also notes, there are also certain superficial differences between Nāgārjuna's and other Buddhist philosophers' deployment of this rubric and the Belnap-Dunn semantics for FDE and B4. The most obvious is that in the Belnap-Dunn framework, the four valuations are both exclusive and *exhaustive*. That is, every sentence is assigned one and only one valuation. Nāgārjuna, however, like other Buddhist philosophers, deploys both positive and *negative* instances of the *catuṣkoṭi*. In the latter form of argument, all four corners are denied, and the truth is declared to be ineffable. Garfield and Priest (2009) argue that this amounts to the addition of a fifth valuation (none of the above, including the empty set!) corresponding to ineffability. Nothing like this is present in FDE or B4. It might then appear to be overhasty to identify early Buddhist logic with these Belnap-Dunn logics. Nonetheless, Priest shows (*Ibid.*) it is easy to extend FDE to accommodate the negative *catuṣkoṭi* as well. We can take Priest's framework for granted as providing a good model for this aspect of early Buddhist logic, and there is no need to revisit these issues here. Another difference that has been noted between early Buddhist logic and the Belnap-Dunn logics is that Nāgārjuna offers no theory of inference. In the Buddhist world the earliest discussion of inference is that of Dignāga.[4] (Garfield 2015)

There is one last domain in which comparison of classical Buddhist logic and the Belnap-Dunn logics is appropriate, and that is the theory of truth they each reflect. Belnap (1977a, pp. 47-48) emphasizes that B4 reflects an *epistemic* account of truth, which he contrasts with an *ontological* account of truth. That is, he argues, when a computer (and this is his intended inferential agent) evaluates a sentence, it is effectively regarding that sentence as *told true, told false, told both* or *told neither*. These are, therefore, effectively *information states* (and not necessarily belief states, although they may be that as well). The computer has no access to the world independent of those who input information, and so it can only draw inferences based upon what it is told. Ontological truth, on this view, transcends what one has been merely told; epistemic truth then is merely transcendental truth according to one's sources. Belnap suggests that B4 is a logic only for the latter; the implicature is that ontological truth is merely two-valuational. This is a real difference between the Belnap-Dunn approach to truth and valuation and the Buddhist approach, one to which we will return below.

[3] As we acknowledge, this enlightened dialetheism in classical India was short-lived (if 900 years or so is short). For when Dignāga (4th c CE) introduces the Nyāya logic to Buddhism, it pretty much takes over, carrying with it a commitment to only two possible valuations, a categorial model of inference and a commitment to consistency. Here I am not concerned with this subsequent decline in logical sophistication in the Buddhist tradition. I should also note that this reading of Nāgārjuna is controversial, and has been challenged by Siderits (2013), Tillemans (2013), Tanaka (2013), and Williams (2018), among others. We need not enter into the complex historical and exegetical terrain here; suffice it to say that we think that we have provided a highly plausible account of Nāgārjuna's own approach to loigical space, and at the very least a compelling rational reconstruction of his view.

[4] We should not infer from this lack of an explicit account of inference or entailment that Nāgārjuna had no views about the matter. It is not clear, however, what his views were, or should have been. This is an issue that no commentator to date has addressed.

## 2 FDE Meyer-Dunn lattice and the Catuṣkoṭi

The Meyer-Dunn lattice for FDE that Belnap takes over for the semantics of B4 has *Truth* at the top, *Falsehood* at the bottom and *None* and *Both* on the wings. Negation takes truth to falsity and vice versa, and is fixed at none and both. Conjunction and disjunction have intuitive definitions: a conjunction has truth in its valuation iff each conjunct does, and falsity iff at least one conjunct does; a disjunction has truth in its valuation iff at least one disjunct does, and falsity iff each disjunct does. (Belnap 1977a, p.42) This is equivalent to saying that the valuation of a disjunction is the least upper bound of the valuations of the disjuncts, and that the valuation of a conjunction is the greatest lower bound of the valuations of the conjuncts, which is also very intuitive.[5]

The *catuṣkoṭi* is not described with as much structure. But when it is deployed to evaluate sentences, we see the same quartet of valuations, built from the same set of truth values. And it is plain that when we consider the full quartet, each of them is taken seriously as a position in logical space, even when all are rejected. So, when Nāgārjuna says at *Mūlamadhyamakakārikā (MMK)* XVIII:8

> Everything is real; and is not real;
> Both real and unreal;
> Neither real nor unreal.
> That is the Lord Buddha's teaching (Tsongkhapa 2006, p. 383)

it is plain that he takes each of these four positions seriously. And when he says at XX:11 that

> We do not assert "Empty."
> Nor do we assert "Nonempty."
> We neither assert both nor neither.
> They are used only for the purpose of designation (*Ibid.*, p. 447)

It is obvious that he is not only denying all four corners, but denying things that someone might assert. The logical space he sketches, then, is very much the space that Belnap and Dunn limn about 1800 years later.

As I have noted, there are two distinct forms of the *catuṣkoṭi* to be found in Madhyamaka texts. In some circumstances, Nāgārjuna suggests that each of the four corners is cogent, or even true. (e.g. XVIII:8 above), and in others (e.g. XX:11 above) he argues that none are assertable. Priest (*op. cit.*) has shown that variants of the FDE semantics can be used to understand the cogency of each of these. The relational semantics for FDE is adequate to the positive *catuṣkoṭi*, and the same semantics, shorn of the requirement that the assignment function is exhaustive validates the FDE inferences in the context of the negative *catuṣkoṭi*. So, Priest has shown, there is a cogent interpretation of the *catuṣkoṭi*, and it is FDE, and that the inferential logic of FDE is valid in the context of both kinds of *catuṣkoṭi* deployed by Nāgārjuna. In what follows, I will ask, what theory of *truth* underlies Nāgārjuna's project, and what it means to say that a statement may be both true and not true, or neither.

[5] Note that these intuitive – and even classical – definitions lead to two counterintuitive results: The conjunction of a formula with valuations *both* and *none* is false; the disjunction of formulae with those valuations is true.

## 3 Nāgārjuna and Belnap on Truth

The Sanskrit term translated into English as *truth* is *satya*. That term is also translated into English as *reality*. The term itself derives from *sat* – *to be* – with a nominalizing suffice. So, it means *how things are*. It has become common for modern commentators on Buddhist philosophy to say that *satya* is ambiguous between *truth* and *reality*. But this is wrong, and seeing that *satya* is unambiguous can lead us to see better how to understand *truth* in English as well. Nāgārjuna, like many other Buddhist philosophers, glosses *true* as *nondeceptive*, and *false* as *deceptive*. At MMK XIII.2, Nāgārjuna asks, "if whatever is deceptive is false, what deceives?" and replies, "Saying that, the transcendent lord perfectly presents emptiness." (Tsongkhapa 2006, p. 291) Emptiness is true because it is non-deceptive; compounded phenomena, although real – as Candrakīrti and Tsongkhapa emphasize – deceive, and so are false. Truth and falsity are often explained in Indian literature through metaphors of illusion. (Westerhoff 2010) Mirages are false because they are deceptive; actual water is true water, because it is non-deceptive. A rope is truly a rope, but is a false snake.

This is often cashed out in Buddhist literature in terms of the concordance or discordance between the mode of existence and the mode of appearance of thing. When something exists as it appears, it is true; when it appears in one way and exists in another, it is false. (Cowherds 2010; Garfield 2015) And this account of the meaning of *satya* runs through the entire Buddhist tradition. It is because compounded, or conventional, phenomena *exist* only conventionally but *appear* to exist ultimately that they are deceptive, and hence false; it is because emptiness, when apprehended directly, both *is* and *appears* to be empty that it is nondeceptive, and hence true.

To take sentences or beliefs to be true or to be false is then simply a special case of a more general phenomenon. Sentences or beliefs that deceive us – that do not lead to the fulfillment of our epistemic purposes – are false; those that conduce to epistemic success are true.[6] Nāgārjuna's view of truth is therefore explicitly epistemic.[7]

Is this a strange Indian notion? I do not think so, and I think that the fascination of recent Western philosophy with language and epistemology has obscured our thinking about truth. *Truth* in English, as the Oxford English dictionary confirms, is cognate with *trust*. To be true is to be trustworthy, or to be real, and this is the primary meaning of the term in Middle English. And traces of these much older uses of *true* remain in our language. We can talk about *true coin of the realm, a true Scotsman*, about *being true to one's principles*, or *to one's spouse*. When we marry we *plight our troth, etc...*

While etymology is not an argument, it can direct our gaze in the right place. Just as in the case of *satya* in Sanskrit, in English, the use of *truth* as a value of sentences or beliefs is not the *primary* use of the term, but a special case. True sentences are those that we can

---

[6] Dharmakīrti, for instance, understands this explicitly in terms of the use of sentences in the pursuit of human aims (the *puruṣārthas*) and argues that in the absence of such a teleological grounding no sense could be made of the distinction between truth and falsity. Since he rejects the reality of the external world, he argues that correspondence cannot capture the meaning of truth. (Guerrero 2014) Whether this account of the semantic range of *satya* and *sat* applies to *all* Indian philosophers – or even *all* Buddhist philosophers – writing in Sanskrit is an open question.

[7] This also suggests that in this framework truth may be a contextual notion. That is, what we can rely on may vary from context to context. If we think of contexts as partially ordered, with larger contexts taking precedence over smaller contexts, and so warrant in larger contexts trumping warrant in more restricted contexts, we might end up with something like the Gupta-Belnap revision theory of truth (Gupta and Belnap, 1993), but that is a matter for another day.

trust, like the true water in a lake, as opposed to the false water in a mirage, like a true friend instead of a false friend, a true banknote as opposed to a fake. Truth, that is, is an epistemic notion.

Belnap also endorses an epistemic understanding of truth. He writes:

> My four values are unabashedly epistemic. According to my instructions, sentence are to be marked with either a ***T*** or an ***F***, a ***None*** or a ***Both***, according as to what the computer has been told; or, with only a slight metaphor, according to what it believes or knows. Does this make the enterprise wrong-headed? Or not logic? No. Of course these sentences have truth-values independently of what the computer has been told; but who can gainsay that the computer cannot *use* the actual truth-value of the sentences in which it is interested? All it can possibly *use* as a basis for inference is what it knows or believes, i.e., what it has been told. (1977b, p. 47, emphasis in original)

That is, as Belnap conceives the truth values as they are deployed in B4 – a logic for a database inference engine – they are to be understood as *told true, told false, told both* and *told neither* (where the computer *trusts* what it has been told).[8] Now, Belnap also countenances what he calls an "ontological" sense of truth (Ibid.), something more than "told true," but, wisely, he never specifies what that is. For the purposes of inference, this doesn't matter. We, like computers, draw inferences not on the basis of any relation between sentences and reality, but on the basis of what we are told, what we know, on the basis of premises in which we can *trust*. In the context of any inference, that is – whether that of a computer or of a person – truth is an epistemic notion.

## 4 Truth, Valuations, and Logical Space

Nāgārjuna's (or, indeed, any Buddhist's) and Belnap's analyses of truth hence constitute another affinity of the *catuṣkoṭi* to B4. This is a point in favor of Nāgārjuna's implicit theory of how a sentient being ought to reason, and so an argument that B4 is not just for computers, but for all of us. The catuṣkoṭi is at bottom a canon for reasoning, for analysis. While it recognizes only two truth *values*, it partitions logical space into four regions by drawing on all of the subsets of the set of those values, and directs us to consider all four regions of logical space.

As we have seen, those two truth values are in turn interpreted as assessments of the trustworthiness of sentences; that is, they evaluate sentences not much directly with reference to a world to which they are meant to correspond; instead, they evaluate sentences with respect to the degree to which we can rely on them in inference and in action. That is, they are essentially epistemic. And our only gauge of the trustworthiness of a sentence can be the epistemic instrument – or pramāṇa – that delivers it to us.

Most Buddhist philosophers recognize only two *pramāṇas*, viz., perception and inference. Mādhyamikas in the tradition of Nāgārjuna and Candrakīrti include in that set the other two standard Indian *pramāṇas*, testimony and analogy. We need not quarrel over whether these two are reducible to the first two as later Buddhist epistemologists argue; the more general point is that all that we know, or have a right to trust, is what comes to us through a very specific set of channels. Those channels might deliver univocal information, conflicting information, or no information at all; hence the partition of logical space.

---

[8] When Priest (2010, pp. 30-31) talks about *true, false, both* and *neither* as "status predicates," he has something similar in mind.

But this means that we are in exactly the same situation as Belnap's computer. When Belnap says "that the computer cannot *use* the actual truth-value of the sentences in which it is interested. All it can possibly *use* as a basis for inference is what it knows or believes, i.e., what it has been told," he could be talking about any human epistemic agent as well. So, since we can only use as the basis of our inferences those things we believe, or are "told," in an extended sense of that term, by our epistemic instruments, the logic appropriate to Belnap's computer is the logic appropriate to us, viz., FDE or an appropriate variant, with the valuations available for sentences being those represented by the *catuṣkoṭi* or the Belnap-Dunn approximation lattice.

This also tells us something about what we can mean by *truth*. Belnap, unlike Nāgārjuna, holds out hope for what he calls an "ontological" sense of truth, which he distinguishes from being *told true*. What could this be? One plausible answer might be *correspondence* with *reality*. But it is far from easy to specify the nature of that correspondence, or how we could have access to it. We would have to have independent access to our beliefs and to the reality to which they are supposed to correspond, and we would have to be in possession of a correspondence relation. Moreover, language itself is surely part of reality, not something standing over against it which might or might not correspond to it.

Deflationists respond to this consideration by arguing that truth is nothing more than disquotation, or satisfaction of the Tarski biconditional schema. That may be to go so far as to be uninformative, and in any case fails to motivate the four valuational approach that makes so much sense. A Buddhist theory of truth as nondeceptiveness or trustworthiness has a bit more content, and explains the normative force of *truth*. It also not only makes sense of the complex semantic ranges of the Sanskrit word *satya* and the English word *truth*, but helps to explain why two truth values engender four valuations, why the *catuṣkoṭi* is such a rich analytical framework, and why FDE is such a good model of inference.

## References


1. Belnap, N.D. (1977a). "How a Computer Should Think," in: G. Ryle (ed.), *Contemporary Aspects of Philosophy*. Stocksfield: Oriel Press, pp. 30-56.
2. Belnap, N.D. (1977b). "A Useful Four-Valued Logic," in: J. M. Dunn and G. Epstein (eds.), *Modern Uses of Multiple-Valued Logic*. Dordrecht: Reidel, pp. 5-37
3. Cowherds. (2010). *Moonshadows: Conventional Truth in Buddhist Philosophy*. New York: Oxford University Press.
4. Deguchi, Y., J. Garfield and G. Priest. (2008). "The Way of the Dialetheist: Contradictions in Buddhist Philosophy", *Philosophy East and West* 58: 3, pp. 395-402.
5. Deguchi, Y., J. Garfield and G. Priest. (2013a). "How We Think Mādhyamikas Think: Reply to Tillemans," *Philosophy East and West* 63:3, pp. 427-436.
6. Deguchi, Y., J. Garfield and G. Priest. (2013b). "Does a Table Have Buddha-Nature? A Moment of Yes and No. Answer! But Not in Words or Signs: Reply to Siderits," *Philosophy East and West* 63:3, pp. 387-398.
7. Deguchi, Y., J. Garfield and G. Priest. (2013c). "A Mountain By Any Other Name: Reply to Tanaka," *Philosophy East and West* 63:3, pp. 335-343.
8. Dunn, J.M. (1976). "Intuitive Semantics for First-Degree Entailments and 'Coupled Trees'," *Philosophical Studies* 29, pp.149-168.
9. Garfield, J. (1995). *Fundamental Wisdom of the Middle Way: Nāgārjuna's Mūlamadhyamakakārikā*. New York: Oxford University Press.
10. Garfield, J. (2008). "Turning a Madhyamaka Trick: Reply to Huntington," *Journal of Indian Philosophy* 36:4, pp. 428-449.
11. Garfield, J. (2015). *Engaging Buddhism: Why it Matters to Philosophy.* New York: Oxford University Press.

12. Garfield, J. and G. Priest. (2003). "Nāgārjuna and the Limits of Thought," *Philosophy East and West* 53:1, pp. 1-21.
13. Garfield, J. and G. Priest. (2009). "Mountains are Just Mountains", in M D'Amato, J Garfield and T Tillemans, (eds.), "Pointing at the Moon: Buddhism, Logic, Analysis", New York: Oxford University Press, pp 71-82.
14. Guerrero, L. (2014). *Truth for the Rest of Us: Dharmakīrti's Theory of Truth.* PhD dissertation, University of New Mexico.
15. Gupta, N. and N. Belnap. (1993). *The Revision Theory of Truth.* Cambridge: MIT Press.
16. Jones, R. (2018). "Dialetheism, Paradox, and Nāgārjuna's Way of Thinking," *Comparative Philosophy* 9:2, pp. 41-68.
17. Priest, G. (2010). "The Logic of the Catuṣkoṭi," *Comparative Philosophy* 1:2, pp. 24-54.
18. Priest, G. and R. Routley. (1989). "First Historical Introduction: A Preliminary History of Paraconsistent and Dialethic Approaches," in G. Priest, R. Routley, and J. Norman (eds), *Paraconsistent Logic: Essays on the Inconsistent.* Düsseldorf: Philosophia Verlag, pp. 3- 75.
19. Siderits, M. (2013). "Does a Table Have Buddha Nature," *Philosophy East and West* 63:3, pp. 373-386.
20. Tillemans, T. (2013). "How Do Mādhyamikas Thin? Revisited," *Philosophy East and West* 63:3, pp. 417-425.
21. Tsongkhapa. (2006). *Ocean of Reasoning: A Great Commentary on Nāgārjuna's Mūlamadhyamakakārikā.* (N. Samten and J. Garfield, trans.). New York: Oxford University Press.
22. Westerhoff, J. (2010). *Twelve Examples of Illusion.* New York: Oxford University Press.

# K3, Ł3, LP, RM3, A3, FDE, M: How to Make Many-Valued Logics Work for You*

Allen P. Hazen and Francis Jeffry Pelletier

**Abstract** We investigate some well-known (and a few not-so-well-known) many-valued logics that have a small number (3 or 4) of truth values. For some of them we complain that they do not have any *logical* use (despite their perhaps having some intuitive semantic interest) and we look at ways to add features so as to make them useful, while retaining their intuitive appeal. At the end, we show some surprising results in the system FDE, and its relationships with features of other logics. We close with some new examples of "synonymous logics." An Appendix contains a Fitch-style natural deduction system for our augmented FDE, and proofs of soundness and completeness. These rules and proofs are easily modifiable for all the logics we discuss.

**Keywords** Three-valid logics • Four-valued logics • Material implication • Propositional quantification • Synonymous logics

## 1 Introduction

This article is about some many-valued logics that have a relatively small number of truth values – either three or four. This topic has its modern (and Western) start with the work of Post (1921), Łukasiewicz (1920) and Łukasiewicz and Tarski (1930), which slowly gained interest of other logicians and saw an increase in writings during the 1930s (Wajsberg, 1931; Bochvar, 1939,1943; Słupecki, 1936, 1939a,b; Kleene, 1938). But it remained somewhat of a niche enterprise until the early 1950s, when two works brought the topic somewhat more into the "mainstream" of formal logic (Rosser and Turquette, 1952; Kleene, 1952). The general survey book, Rescher (1969), brought the topic of many-valued logics to the attention of a wider group within philosophy. Also in the 1960s there was an interest in free logics, where singular terms were not presumed to designate existing entities (Lambert, 1967; van Fraassen, 1966, 1969). The works of van Fraassen in particular promoted the notion of "supervaluations", which were seen (by some) as a way to retain two-valued logic while allowing for a third value of a "gap" – no truth value at all. More recently there was a spate of work . . . still continuing . . . that seems to have begun in the late 1970s (da

* The first author was fortunate to have both Michael Dunn and Nuel Belnap as teachers, and though painfully aware that the current paper is hardly significant enough for the purpose, wishes to thank and honour them.

Allen P. Hazen
Department of Philosophy, University of Alberta, Edmonton, Canada
e-mail: aphazen@ualberta.ca

Francis Jeffry Pelletier
Simon Fraser University, Burnaby, B.C., Canada
e-mail: jeffpell@sfu.ca



H. Omori, H. Wansing (eds.), *New Essays on Belnap-Dunn Logic*,
Synthese Library 418, https://doi.org/10.1007/978-3-030-31136-0_11

Costa, 1974, 1982; Urquhart, 1986; Rozoner, 1989; Avron, 1986, 1991; Arieli and Avron, 1998; Bolc and Borowick, 1992).

Our purpose is not to recount the development of this type of many-valued logic in its generality, but instead to focus on one specific group of these logics and show some surprising connections amongst them. Of course, in the totality of the writings on the various logics, many – perhaps even most (although we think not all) – of these connections have been noted. We will remark on some of these notices as we go. But we think these connections have not been put together in the same way as we do. Most of the work on these logics with small numbers of truth values have focused on the three-valued case, adding or comparing different connectives that different authors have proposed. In contrast, we will start with a four-valued logic and add one connective to it. And using that we will show how it might be applied to various three-valued logics. A further difference between our approach and many of the others in the area is that we mostly concentrate on the truth-valued semantic descriptions of the logics. Other approaches tend to emphasize boolean or de Morgan lattice-theoretic approaches for their semantics. Related to this last, these other approaches tend to develop a syntactic theory that supports their lattice-theoretic semantics by favouring sequent calculus formulations or sometimes (especially in older works) a Hilbert-style axiomatic syntax. Even ones that mention natural deduction are pretty much uniformly working within a Gentzen-style formulation, whereas we provide pedagogically more attractive Fitch-style natural deduction systems for all the logics under consideration (in an Appendix).

In particular, we try to show how one interesting four-valued logic can absorb one new connective and that this can lead us to see many of the previously-noted connections in a new light, namely, as the natural application of this connective to differing background logics. In this way we think of our development as "bottom up", starting with a particular logic and adding a specific connective. And that this can be extended in various ways to other specific logics. This is in contrast to works that proceed in a "top down" manner by considering whole classes of logics and considering all the possible connectives that might have some properties in common, and then remarking that some of these combinations have been employed by previous authors. It may well be that the conclusions reached by the two ways are identical, but we think ours might be found to be more intuitive and memorable, and easier for students (and non-logicians) to grasp, even if not so wide-reaching as what the top-down method might provide.

## 2 Truth-Values: Many, versus Gaps and Gluts

Everyone knows and loves the two "classical" truth-values, **T**rue and **F**alse. In "classical" logic[1], every sentence is either **T** or **F** but not both. Before getting to the detailed issues of this paper, we pause to make four comments of a "philosophical" nature with the intent of setting the issues aside. First, there is the long-standing issue of whether a logic contains sentences, statements, or propositions. We take no position on this issue: what we think of as the *logical* point is neutral on this issue. Related to this, perhaps, is the

[1] There are those (e.g., Priest, 2006) who claim that this common terminology of 'classical' betrays a bias and is in fact incorrect as an account of the history of thought about logic. We won't attempt to deal with this, and will just use the common terminology. For a more comprehensive review of topics and issues relevant to truth values see Wannsing (2017).

issue of whether a logic should be considered a set of its theorems or defined instead by the valid inferences it accepts. This topic *does* mark a distinction in the three-valued cases we consider, and we will remark on this as the occasion arises. Second, there are logics – perhaps not "classical" ones – that would prefer to claim the objects of interest are (e.g.) obligations, or commands, or questions, or ethical/epistemic items, rather than **T** and **F**. (E.g., Rescher, 1966; Adler, 1980; Searle and Vanderveken, 1985; Vanderveken, 1990; Hale, 1993; Wiśniewski, 1995, and Vranas, 2008. See Schroeder (2008) for a discussion of the logical issues concerning non-truth-valued arguments.) Perhaps the items that have such values are not to be considered 'sentences' but rather exhortations or promises or…? If so, then we say that we are restricting our attention to (declarative) sentences. Third, while 'sentence' seems appropriate for propositional logic, some further accommodation is required for formulas of predicate logic. The present paper is mostly concerned with propositional logic, and so we will simply use 'sentence'. And finally, there is the philosophical issue of whether **T** and **F** (and any other truth-values) should be considered to be *objects*– perhaps existing in a "third realm" – or as *properties* of sentences. Our view is that this distinction – which is maybe of considerable philosophical interest – does not affect any of the claims we will be making, and so we just will pass over it in silence.

So, given that **T** and **F** are so revered, where does this leave many-valued logic? One intuition that has been held by very many theorists is that, despite the fact that **T** and **F** are clearly excellent truth-values, there simply are more than just the **T** and **F** truth values. Some of these theorists hold that various phenomena show that there is another category beyond **T** and **F**: perhaps we should call it **I** for **I**ndefinite. Presupposition failures in a formula might determine that it is **I**; vagueness might determine that a formula is **I**; the unactualized (as yet) future might determine that some formulas are **I** (now); fictional discourse seems neither straightforwardly true or false, so perhaps sentences about fictional objects also should designate a third truth-value, **I**. And there no doubt are other types of formulas that could intuitively be thought to be "indeterminate".

(1) a. The present Queen of Germany is happy.
b. That 45-year-old man is old.
c. There will be a Presidential impeachment in the next 30 years.
d. James Tiberius Kirk is captain of the starship *Enterprise*.

Some theorists, agreeing with the just-expressed intuition that the sentences in (1) are neither **T** nor **F**, nonetheless wish to deny that they designate a third value. Rather, they say, such sentences *lack* a truth-value (i.e., are *neither* **T** nor **F**; nor are they **I**, since they deny such a value). Such sentences *have no truth-value*: they express a truth-value *gap*. And logics that have this understanding are usually called "gap logics."

Formally speaking, there is little difference between the two attitudes of expressing **I** or being a gap (in a simple three-valued logic where there is just one non-classical truth value). In both cases one wishes to know how to evaluate complex sentences that contain one of these types of sentences as a part. That is, we need a way to express the truth-table-like properties of this third option. In the remainder of this paper we will use **N** for this attitude, whether it is viewed as expressing **I** or as being a gap. (The **N** can be viewed as expressing the notion "neither **T** nor **F**", which is neutral between expressing **I** or expressing a gap.) And we will call the logics generated with either one of these understandings of **N**, "gap logics."

Gap logics – whether taken to allow for truth values other than **T** and **F**, or instead to allow for some sentences not to have a truth value – are only one way to look at some recalcitrant natural language cases. Other theorists point to some different phenomena in their justification of a third truth value. The semantic paradoxes, such as the Liar Paradox (where the sentence "This sentence is false" is false if it is true, and is true if it is false) suggest that some statements should be treated as *both* true and false. It is difficult to deal with this paradox in a gap logic, because saying that "This sentence is false" is neither true nor false leads immediately to the strengthened Liar Paradox: "Either this sentence is false or else it is neither true nor false". Here it seems that claiming that the sentence expresses a gap leads to truth, while saying that it is true leads to falsity and saying that it is false leads to truth. In contrast, to say that the sentence "This sentence is false" is *both* true and false does not lead to this regress: The sentence "This sentence is either false or else is both true and false" leads only to the conclusion that it is both true and false. This *dialetheic* theory – that some sentences are both true and false – is called a "glut theory" because of the presence of both truth values characterizing some sentences. The intermediate value, on this conception, can be called **B** (for "both"), and we will use this. Again, there is perhaps a distinction between some sentence's having a third value **B**, or of partaking of both truth and falsity. Again we remain neutral and employ **B** for this intermediate possibility.

There are also examples from naïve set theory such as the Paradox of Well-Founded sets, the Paradox of a Universal Set, Russell's Paradox, and Richard's Paradox (see Cantini, 2014, for a historical overview of these and others). A desire on the part of some theorists in the philosophy of mathematics is to reinstate naïve set theory and use it in the development of mathematics, instead of (for example) Zermelo-Frankel set theory. Such theorists perhaps will feel encouraged in their belief that the glut-style of resolution of the Liar Paradox might be usable in the case of these other set theoretic paradoxes, and that naïve set theory can once again be used.

Various theorists have pointed out that even vagueness needn't be viewed as *missing* a truth-value; it is equally plausible to think of vagueness as manifesting *both* the positive character and its absence. A middle-height person can be seen as *both* short and tall, equally as plausibly as being *neither* short nor tall. (See Hyde, 1997; Beall and Colyvan, 2001; Hyde and Colyvan 2008.) As well, certain psychological studies seem to indicate that the "dialetheic answer" is more common in general conversation than the "gap" answer, at least in a wide variety of cases (see Alxatib and Pelletier, 2011; Ripley, 2011, for studies that examine people's answers to such cases).

There are also other categories of examples that have been discussed in philosophy over the ages: Can God make a stone so heavy that He can't lift it? Perhaps the answer is, "Well, yes and no." There are cases of vague predicates such as *green* and *religion*, e.g., a necklace's turquoise stone can be both green and not green, and a belief system such as Theravada Buddhism or Marxism can be both a religion and not a religion. There are legal systems that are inconsistent, in which an action is both legal and illegal. In the physical world, there is a point when a person is walking through a doorway at which the person is both in and not in the room. And so on. As a result, the glut theory has its share of advocates.

A way to bring the gap and glut views of the truth values under the same conceptual roof is to think of the method that assigns a truth value to formulas as a *relation* rather than a *function*. A function, by definition, assigns exactly one value to a formula – either **T** or **F** or, in the context of our 3-valued logics, it may also assign **B** or **N**. A relation, however, can assign more than one value: we may now think that there are exactly two truth values, **T** and **F**, but the assignment *relation* can assign subsets of $\{\mathbf{T}, \mathbf{F}\}$. A gap logic allows assignment of $\{\mathbf{T}\}$, $\{\mathbf{F}\}$, and $\emptyset$. A glut logic allows $\{\mathbf{T}\}$, $\{\mathbf{F}\}$, and $\{\mathbf{T}, \mathbf{F}\}$.

But both the gap and the glut logics seem to have difficulties of a logical nature that we will examine in §5.

## 3 K3 and LP: The Basic 3-Valued Logics

In all of the 3-valued logics we consider, the 3-valued matrices for $\wedge, \vee, \neg$ are the same, except for what the "third value" is called: in K3 (Kleene, 1938; Kleene, 1952, §64) we call it **N** but in LP (Asenjo, 1966; Asenjo and Tamburino, 1975; Priest, 1979) we call it **B**.

One way to understand why the K3 truth values for these connectives are correctly given in the following tables is to think of **N** as meaning "has one of the values **T** or **F**, but I don't know which one". In this understanding, a negation of **N** would also have to have the value **N**; a disjunction with one disjunct valued **N** and the other **T** would as a whole be **T**... but if one were valued **N** and the other **F**, then as a whole it would have to be **N**. Similar considerations will show that the $\wedge$ truth table is also in accord with this understanding of the **N** value. (This interpretation of the truth values presumes that one pays no attention to the *interaction* of the connectives: $(p \wedge \neg p)$ might seem to be false and $(p \vee \neg p)$ seem true, but that's because of the interaction of two connectives and their arguments.)

On the other hand, if the middle value is understood as being *both* true and false as in the Logic of Paradox, LP, we will still get those same truth tables, with only a change of letter from **N** to **B**: negating something that is both true and false will yield something that is both false and true. If a disjunction had one disjunct valued **B** and the other **T**, then the entire disjunction would be valued **T**. But if that second disjunct were valued **F** instead, then the entire disjunction would be valued **B**. Similar considerations will show that the $\wedge$ truth table is also in accord with this understanding of the **B** value.

| $\wedge$ | **T** | **N/B** | **F** |
|---|---|---|---|
| **T** | **T** | **N/B** | **F** |
| **N/B** | **N/B** | **N/B** | **F** |
| **F** | **F** | **F** | **F** |

| $\vee$ | **T** | **N/B** | **F** |
|---|---|---|---|
| **T** | **T** | **T** | **T** |
| **N/B** | **T** | **N/B** | **N/B** |
| **F** | **T** | **N/B** | **F** |

| $\neg$ | |
|---|---|
| **T** | **F** |
| **N/B** | **N/B** |
| **F** | **T** |

Despite the apparent identity of K3 and LP (other than the mere change of names of **N** and **B**), the two logics are in fact different by virtue of their differing accounts of what semantic values are to be considered as privileged. That is, which values are to be the *designated values* for the logic – those values that are semantically said to be the ones that the logic is concerned to manifest in a positive light. In both logics this positivity results from the value(s) that exhibit truth, or at least some degree of it: but in K3 only **T** has that property whereas in LP both **T** and **B** manifest some degree of truth. And therefore LP treats the semantically privileged values – the *designated values* – to be both **T** and **B**, while in K3 the only designated value is **T**.

Although both K3 and LP have no formula whose value is always **T**, LP (unlike K3) does have formulas whose semantic value is always designated. In fact, the class of propositional LP formulas that are designated is identical to that of propositional classical 2-valued logic, a fact to which we return in §5.

Note that these truth tables give the classical values for compound formulas with only classically-valued components, and give the intermediate value for formulas all of whose components have that value. Thus neither logic is functionally complete, that is, neither logic can describe all the possible relations that their semantics allows to hold among their truth-values. For, no truth function giving a non-classical (classical) value for uniformly classical (non-classical) arguments can be represented by them.

## 4 FDE: A Four-Valued Logic

The logic FDE was described in Belnap (1992) and Dunn (1976). It is a four-valued system: the values are **T**, **F**, **B**, and **N**. . . the four values we have already encountered (although the intuitive semantics behind these values, as given by Belnap, are perhaps somewhat different than we encountered above). K3 and LP agree with each other (and with classical logic) as regards the values **T** and **F**, so in combining them as a way to form FDE, it is natural to identify FDE's **T** and **F** with those of K3 and LP. FDE then agrees with K3 about **N**: agrees that it is not a designated value, and agrees on the values of combinations of **N** with **T** and **F**. FDE also agrees with LP about **B**: agrees that it is a designated value, and agrees on the values of combinations of **B** with **T** and **F**. Neither of the three-valued logics allow combinations of **B** with **N**, of course, so these combinations need a bit more thought. The choice made is perhaps most easily described by using the "relational" account of truth assignment, where the four values are *sets* of the two classical values: **T**={T}, **B**={T,F}, **N**={ }, **F**={F} (where of course the bold-face values stand for FDE values and the normal text values stand for classical ones!). A given classical value then goes into the set of values of a compound formula just in case the component formulas have values containing classical values which, by the classical rules, would yield the given value for the compound. (This way of thinking also makes the designation status of **B** and **N** seem natural: an FDE value is designated just in case it, thought of as a set, contains the classical value T.)

This treatment of the four values and their ordering is perhaps best visualized by the diagram:

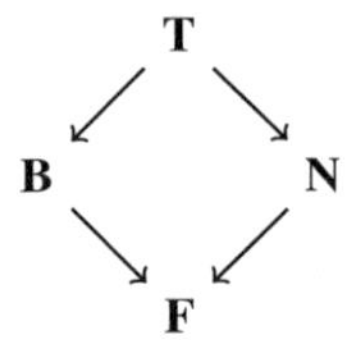


The four values form a lattice, with the value of a conjunction (disjunction) being the meet (join) of the values of its conjuncts (disjuncts), and with negation interpreted as inverting the lattice order. Considered only as a lattice (so, thinking only about meet and join) this is isomorphic to the four-element Boolean algebra, but the negation operator is not the Boolean complement: **B** and **N** are fixed points for the negation of FDE, but the "intermediate" elements of the four-valued Boolean algebra are each other's Boolean complements. In working with this, the only potentially surprising points concern conjunctions and disjunctions of **B** with **N**: the conjunction is **F** even though neither conjunct is **F**, and the disjunct is **T** without either disjunct being **T**.

More formally, we take the basic truth-tables for $\wedge, \vee$ and $\neg$ to be straightforwardly taken from K3 and LP, for the values that do not involve only **B** and **N** together. For those values we interpolate, using meet and join, as given in the truth tables in Table 1 for the FDE connectives.

| $\varphi$ | $\psi$ | $\neg\varphi$ | $(\varphi \wedge \psi)$ | $(\varphi \vee \psi)$ |
|---|---|---|---|---|
| **T** | **T** | **F** | **T** | **T** |
| **T** | **B** | | **B** | **T** |
| **T** | **N** | | **N** | **T** |
| **T** | **F** | | **F** | **T** |
| **B** | **T** | **B** | **B** | **T** |
| **B** | **B** | | **B** | **B** |
| **B** | **N** | | **F** | **T** |
| **B** | **F** | | **F** | **B** |
| **N** | **T** | **N** | **N** | **T** |
| **N** | **B** | | **F** | **T** |
| **N** | **N** | | **N** | **N** |
| **N** | **F** | | **F** | **N** |
| **F** | **T** | **T** | **F** | **T** |
| **F** | **B** | | **F** | **B** |
| **F** | **N** | | **F** | **N** |
| **F** | **F** | | **F** | **F** |

**Table 1** Four-valued truth-tables for the basic connectives of FDE

Most, if not all, of the concerns mentioned in §3 as motivating interest in LP or K3 can be elaborated in ways to motivate the use of FDE. A number of authors have argued specifically for the appropriateness of FDE for some applications: for example, Belnap, 1992, which first appeared as two articles in 1977, has argued that FDE is appropriately used in connection with some kinds of data base (or better: knowledge base) systems, where different sources of data entry might contain contradictory information, and where some information is missing but the knowledge base is required nevertheless to be able to reason about such cases; and Camp (2002) has argued that it might be useful in argumentation where the references of our interlocutors' (or our own!) terms are indeterminate, and for that reason can generate statements that are **N** and others that are **B**. Arieli and Avron (2017) refer to a number of other areas, such as knowledge-base integration, fuzzy logic, relevance logics, self-reference, and preferential modelling.

## 5 What's Wrong with K3, LP, and FDE

We start with K3 and LP. Note that the $\wedge, \vee, \neg$ truth-functions always yield a **N** (or a **B**) whenever all the input values are **N** (or **B**, respectively). This means that there are no formulas that always take the value **T** in these logics. Since these three truth functions are the only ones available from the primitive connectives, the same fact also implies that not every truth function is definable in K3 (or LP). And if $\varphi$'s being semantically valid means that $\varphi$ always takes the value **T** for all input values, as it does in K3, then there are no semantically valid formulas in K3.

And given that these are all the primitive truth functions, then the only plausible candidate for being a conditional in these logics comes by way of the classical definition, $(\varphi \supset \psi) =_{df} (\neg\varphi \vee \psi)$, whose truth tables in K3 and LP are given in Table 2:

| $\varphi$ | $\psi$ | $(\varphi \supset_{K3} \psi)$ | $(\varphi \supset_{LP} \psi)$ |
|---|---|---|---|
| **T** | **T** | **T** | **T** |
| **T** | **N/B** | **N** | **B** |
| **T** | **F** | **F** | **F** |
| **N/B** | **T** | **T** | **T** |
| **N/B** | **N/B** | **N** | **B** |
| **N/B** | **F** | **N** | **B** |
| **F** | **T** | **T** | **T** |
| **F** | **N/B** | **T** | **T** |
| **F** | **F** | **T** | **T** |

**Table 2** Conditionals available in K3 and LP, using the classical definition

The only difference between $\supset_{K3}$ and $\supset_{LP}$ is that in K3 the "third value" is understood as indicating a truth-value gap, i.e., as *neither* **T** nor **F**, whereas whereas in LP it is understood as indicating a truth-value glut, i.e., as *both* **T** and **F**.

One notes that the rule of inference, Modus Ponens (MP), $(\varphi \supset_{K3} \psi), \varphi \vDash \psi$, is a valid rule in K3: if both premises are **T**, then the conclusion will also be, as can be seen from the truth table for $\supset_{K3}$ in Table 2. Despite this, the *statement* of MP, $((\varphi \supset_{K3} \psi) \wedge \varphi) \supset_{K3} \psi$, does *not* always exhibit the value **T** (since no formula of K3 always takes the value **T**). Thus the deduction theorem [if $\Gamma, \varphi \vDash \psi$ then $\Gamma \vDash (\varphi \supset_{K3} \psi)$] does not hold for K3. Matters are different for the logic LP: the rule of MP is *invalid*, as can be seen by making $\varphi$ be **B** and $\psi$ be **F**. In that case both $(\varphi \supset_{LP} \psi)$ and $\varphi$ have designated values (both are **B**), but yet the conclusion, $\psi$, is **F**. On the other hand, the *statement* of MP, $((\varphi \supset_{LP} \psi) \wedge \varphi) \supset_{LP} \psi$, is always designated (either **T** or **B**). (One can see that the counterexample to the *rule* MP makes both conjuncts of the antecedent be **B** and hence that antecedent is **B**. But a **B** antecedent with a **F** consequent is evaluated **B**. Hence the statement is designated.)

Not having a conditional that enforces MP (not having what is often called "a detachable conditional") is a serious drawback. It means that the language cannot state what inferences it takes to be designated, or at least, if it does claim some conditional to be designated then it cannot apply MP to that conditional, even when the antecedent is designated. This has the consequence that one cannot state what inferences can be chained together, that is, the formula $(((\varphi \supset \psi) \wedge (\psi \supset \theta)) \supset (\varphi \supset \theta))$ is not designated in K3 (suppose all its parts are **N**); and the argument $((\varphi \supset \psi) \wedge (\psi \supset \theta)) \vDash (\varphi \supset \theta)$ is invalid in LP (let $[\![\varphi]\!] = \mathbf{T}, [\![\psi]\!] = \mathbf{B}, [\![\theta]\!] = \mathbf{F}$). In essence, these sorts of shortcomings mean that no reasoning can be carried out *within* such logics – although perhaps some reasoning *about* the logics can be carried out.[2]

As we remarked above, the theorems of LP are identical to those of classical logic – LP does nothing more than divide up the classical notion of truth into two parts: the "true-only" and the "true-and-also-false" (but has no way of distinguishing or separating these two subparts of LP's theorems). As both types of truth are designated values in the logic, the theorems of the two are the same, and hence LP is not different from classical logic so far as logical truth goes. Where it does differ is in the class of valid rules of inference,

[2] For example, perhaps the pair of inferences in one or another of these logics is that $\varphi \vDash \psi$ is semantically valid, and so is $\psi \vDash \theta$. And therefore the inference $\varphi \vDash \theta$ must also be valid. But note that this would be a judgement made *about* the logic and not *within* the logic. The logic itself can't be used in this or any equivalent way.

as we have noted. A different description applies to K3: whereas the deduction theorem holds and modus ponens fails for LP, in K3 the deduction theorem fails but modus ponens holds.

So, it seems that both K3 and LP are not very useful as guides to reasoning, despite their apparent (to some) virtues in accounting for intuitive semantic values of sentences that exhibit some troublesome features (such as vagueness or semantic paradox). After discussing the logical state of K3 and LP, and the disappointing properties of the conditional operators available in them, Soloman Feferman put it:

> Multiplying such examples, I conclude that *nothing like sustained ordinary reasoning can be carried on in either logic.* (Feferman, 1984, p. 264, italics in original.)

On the topic of non-classical logics more generally, van Fraassen (1969) expresses the worry in more picturesque terminology. New logics, he worries, along with "the appearance of wonderful new 'logical' connectives, and of rules of 'deduction' resembling the prescriptions to be read in *The Key of Solomon*" will make "standard logic texts read like witches' grimoires".

Turning now to FDE, it is clearly a merger of K3 and LP, so shares all their shortcomings. Since FDE includes the feature of K3 that there are no truth functions from all-atomics-having-the-value-**N** to any other truth value, it follows that there are no formulas that always take one or the other of the designated values **T** or **B**. And similarly, FDE is functionally incomplete since there is no formula that will yield one of **T**, **B**, or **F** if all its atomic letters are assigned **N**. (A similar situation holds holds for LP and FDE: Although LP does have formulas that always take designated values, there is no truth function in FDE that can generate **T**, **N**, or **F**, when all the atomic letters are assigned **B**.) Since these logics fail to be functionally complete, there are features of the semantic model that cannot be expressed in the language – such as that there are possible functions that take sets of **N**-valued (or **B**-valued) atomics to classically-valued formulas. Surely a logic has a weakness if it can't express what its semantic metatheory says is a feature of its domain. And so far, all the logics we have discussed contain this weakness. Like K3 and LP, FDE does not have a usable conditional and hence there is no sense in which it is a logic that one can reason with. Although Feferman (1984) didn't include FDE in his disparaging remarks we just quoted, since FDE is simply a "gluing together " of K3 and LP – the two logics Feferman did complain about – it is clear that he would have held the same opinion about FDE.

We wish to show that matters are not so dire as Feferman seems to think for the FDE-related logics we are investigating. And so we now turn to possible ways to fix these shortcomings. We start with the topic of adding an appropriate conditional; later we consider the issue of functional completeness.

## 6 FDE Fixes

Various researchers have added conditionals to K3 and LP, with the thought of being able to do "sustained reasoning" in these logics. We might mention here Łukaseiewicz's conditional added to K3[3], with the intent of accommodating "future contingents." This yields

[3] Although of course this is not the historical reason he came up with his conditional, since he didn't have K3 before him at the time.

a logic that is usually called Ł3. And conditionals can be added to LP with the intent of allowing reasoning to take place in the context of dialetheism. For this we will look here at the more closely at the logics RM3 and a logic we will call $LP^{\rightarrow}$ that are generated by adding different conditionals to LP.[4]

There are other ways to approach the problem of a lack of inferential capacity. One could, for example, change the interpretation of the connectives ¬, ∨, ∧ so that somehow together they could define a connective that allows for inferences. Or, one could add some other types of connectives: De and Omori (2015) add what they call an extensional or exclusive negation, ∼, alongside the ¬ that is part of FDE.[5] A conditional connective, ⥽, can then be defined: $(A \multimap B) =_{df} (\sim A \vee B)$. Their ⥽ has the following truth matrix (p. 835):

| ⥽ | T | B | N | F |
|---|---|---|---|---|
| T | T | B | N | F |
| B | T | T | N | N |
| N | T | B | T | B |
| F | T | T | T | T |

**Table 3** Conditional of De and Omori, 2015

There have been many suggestions for combining one or another intensional conditional with the other connectives (and quantifiers) to form logics tolerating truth value gaps or gluts. Intuitionistic logic famously allows for gaps, and its motivation has occasionally been explained as appealing to a sort of vagueness in mathematical concepts (as for example in Dummett, 1963). Newton da Costa (da Costa, 1974, 1982) has proposed a family of logics which are closely related to what is known as "Dual Intuitionistic Logic" (Wansing, 2008, 2012; Shramko, 2016), and tolerate gluts in the sense that the inference from $\Phi \wedge \neg\Phi$ to arbitrary $\Psi$ is not derivable in them. These systems are not extensions of FDE: various De Morgan equivalences fail in them. However Nelson (1949) proposed a variant of Intuitionistic logic with a different negation operator for which all the classical double negation and De Morgan inferences hold: his system is gap-tolerant in the same way as ordinary Intuitionistic logic (and has been recommended (Almukdad and Nelson, 1984) as a logic of vagueness). Then there are the "Relevant" or "Relevance" logics (an introductory survey is Dunn and Restall, 2002): In system R (**R**elevant implication), though not generally in weaker relevant logics, the relevant conditional is definable as $(\neg A \oplus B)$ or $\neg(A \otimes \neg B)$, where $\oplus$ is the relevant logic's intensional "fission" and $\otimes$ is its intensional "fu-

[4] Tedder (2015) uses the name A3, in honour of Avron (1991), for what we call $LP^{\rightarrow}$. Since we are comparing a number of logics, we prefer our more systematic nomenclature of adding the relevant conditional as a superscript. Our citation of Avron (1991) in an earlier draft did not consider the remark of Avron on p.283fn6, that the conditional being added to form $LP^{\rightarrow}$ was already defined in Monteiro (1967); Wójcicki (1984). Avron himself thanked his referee for that information. Heinrich Wansing kindly sent us a paper (Batens and De Clerq, 2004) which makes the claim in its first paragraph that the logic we are calling $LP^{\rightarrow}$ was first proposed by Schütte (1960, pp. 73-75), referring the reader to Batens (1980a), which in turn refers the reader to Batens (1980b), where a proof is given on pp. 516. Batens gives a slightly different but equivalent axiomatic system for $LP^{\rightarrow}$, and includes the propositional constant ⊥ with obvious axioms. The conclusion is that Schütte's systems $\Phi$, $\Phi_v$, and $\Phi_r$ are in fact Hilbert-style formulations for the logics we will call $FDE^{\rightarrow}$, $LP^{\rightarrow}$ and $K3^{\rightarrow}$, respectively. (We note that the sections of Schütte, 1960, that contain this material are omitted from the English translation, Schütte, 1977.)

[5] We discuss their ∼ below in §7.3.

sion", using the standard negation. (For details, see Anderson et al., 1992.) And after all, Anderson and Belnap (1962) initially isolated FDE as the **F**irst **D**egree fragment of their relevance system E (for **E**ntailment), and it is a common fragment of most of the systems in the Relevant family. Weak systems of this family, lacking the Contraction inference (which we discuss below) and so diverging radically from the classical behaviour of he conditional, have been used in formulating negation-inconsistent but non-trivial versions of "naïve" set theory, with unrestricted axiom schemes of contradiction.[6]

All these have intensional, in the sense of non-truth-functional, conditionals. In contrast, we propose a system with a truth-functional conditional, one as much like the 2-valued material conditional of classical logic as possible. We take the connectives of FDE as given, making no changes to their logic: the extensions of FDE (and of the related logics LP and K3) with our conditional are thus, in technical terms, *conservative* extensions. They are, however, proper extensions: the conditional is not definable in terms of the connectives of FDE (for a proof that no conditional definable in those terms has the basic properties of classical material implication, see Beall et al., 2013).

There are, of course, drawbacks. The classical nature of our conditional means that our logics cannot be used, as weak relevance logics can, as the basis for systems of naïve set theory: Curry's Paradox would be immediately derivable. We do not see our approach as in competition with the intensional approaches, but rather as adapted for different applications. Tedder's (2015) success in axiomatizing the truths of 3-valued inconsistent models, it seems to us, gives grounds for hope that logics like ours will prove useful in some applications.

Many, possibly most, writers on the FDE family of logics start by considering 3-valued logics such as K3 or LP, and consider the effects of adding conditionals of one sort or another to them. But rather than starting with these 3-valued logics, adding one or more conditionals to it, and then extending one or another of the conditionals to the 4-valued FDE, our strategy here will be to start with the general case of FDE, propose our conditional $\rightarrow$ for it, which generates the logic we call FDE$^{\rightarrow}$, and to look at how that conditional behaves in the K3$^{\rightarrow}$ and LP$^{\rightarrow}$ "components" of FDE$^{\rightarrow}$. We find many interesting and unexpected connections with previous work.[7]

Here is the conditional we propose: we call it the "classical material implication" and we symbolize it $\rightarrow$ in Table 4.[8]

This conditional has been suggested before – indeed, Omori and Wansing (2017, p. 1036) say it is "the most well-known, as well as well-studied, implication". Be that as it may, we propose to further investigate the resulting FDE$^{\rightarrow}$ logic, as well as some further possible additions to it. But prior to that we will look at extending this logic by restricting the possible truth-values to three.

In §4 we remarked that the natural understanding of FDE's semantic values is that the designated ones are **T** and **B**, while the undesignated ones are **N** and **F**. As can be seen

[6] See Brady, 2006, for work in which this approach is used in formulating a paraconsistent "naïve" set theory. Many of the papers in Priest and Sylvan (1989) are on the topic of naïve set theory, and investigations of set theories with naïve comprehension, formulated in weak relevance logics, have continued: e.g. Weber (2010).

[7] For a survey of work done on describing and enhancing FDE, including various possibilities of adding conditionals, see Omori and Wansing (2017). This is also an overview of the other works on FDE that are reported in this special issue of *Studia Logica*, titled "40 Years of FDE".

[8] Omori and Wansing (2017, p. 10) call it an extensional operator, saying it can be "seen as a material implication defined in terms of exclusion negation and disjunction".

| → | T | B | N | F |
|---|---|---|---|---|
| T | T | B | N | F |
| B | T | B | N | F |
| N | T | T | T | T |
| F | T | T | T | T |

$$[\![\varphi \to \psi]\!] = [\![\psi]\!], \text{ if } [\![\varphi]\!] \text{ is designated}$$
$$= \mathbf{T}, \text{ otherwise}$$

**Table 4** Truth matrix for our conditional, →.

in Table 4, any conditional statement that takes a designated value has the feature that if its antecedent is also designated, then its consequent will be. It can also be seen that if the values are restricted to the classical **T** and **F**, then the conditional mirrors the classical ⊃. We also obtain the truth table for the classical ⊃ if we look at Table 4 with blurred vision, so that the two designated values blur into one and the two undesignated values similarly coalesce, showing that the valid formulas (i.e., those taking a designated value on every assignment of truth values to their atoms) and inferences (i.e., those preserving designation on every assignment) are precisely those of the pure ⊃ fragment of classical logic. Combined with a similarly blurred view of the truth tables for ∧ and ∨ we can extend this observation to formulas having these connectives as well as →, and we can extend it further to quantifiers if we think of them as generalized conjunctions and disjunctions. Thus:

*Proposition* 1. For *positive* formulas (i.e., those not containing ¬), the valid formulas and inferences of $FDE^{\to}$ are exactly those of classical logic.

The 4-valued logic $FDE^{\to}$ has two obvious 3-valued extensions, defined semantically by restricting the set of truth values allowed: $K3^{\to}$ defined by reference to {**T**,**N**,**F**} and $LP^{\to}$ defined by reference to {**T**,**B**,**F**}. These logics result from the addition of a conditional connective, →, defined by the relevant rows/columns of Table 4, to the conditional-free logics K3 and LP.

## *6.1 $FDE^{\to}$*

The classical nature of the conditional in $FDE^{\to}$ makes it a much more pleasant item to work with than many of the previously proposed conditionals for many-valued logics: something like sustained ordinary reasoning *can* be carried out in $FDE^{\to}$and in its 3-valued extensions.[9] However, although the positive logic is completely classical, there are some surprises in the interaction of → with negation. One unpleasant one is that the principle of Contraposition fails:

$$\nvDash_{FDE^{\to}} (\varphi \to \psi) \to (\neg\psi \to \neg\varphi)$$

if $[\![\varphi]\!] = \mathbf{T}$ and $[\![\psi]\!] = \mathbf{B}$, then $[\![\varphi \to \psi]\!]$ takes the designated value **B**, but $[\![\neg\psi \to \neg\varphi]\!]$ is **F**. And also, if $[\![\varphi]\!] = \mathbf{N}$ and $[\![\psi]\!] = \mathbf{F}$, then $[\![\varphi \to \psi]\!]$=**T**, but $[\![\neg\psi \to \neg\varphi]\!]$= **F**. Since a

[9] Omori and Wansing (2017, p. 1036) mention three other conditionals that each have their own interesting properties, although these properties detract from the classical nature of these conditionals. Sutcliffe et al. (2017) give a different four-valued conditional using some "intuitive" considerations of "meaning", and employ an automated theorem proving system to contrast the different conclusions that can be drawn from each.

counterexample can be obtained using either of the nonclassical values, Contraposition is invalid not only in $FDE^{\rightarrow}$ but also in both of its 3-valued extensions $K3^{\rightarrow}$ and $LP^{\rightarrow}$.

We can, of course, define a new conditional connective for which Contraposition holds:

$$(\varphi \Rightarrow \psi) =_{df} ((\varphi \rightarrow \psi) \wedge (\neg\psi \rightarrow \neg\varphi))$$

This is a useful connective! It has the truth matrix of Table 5.[10]

| ⇒ | T | B | N | F |
|---|---|---|---|---|
| T | T | F | N | F |
| B | T | B | N | F |
| N | T | N | T | N |
| F | T | T | T | T |

**Table 5** Truth matrix for a defined conditional, ⇒.

The corresponding biconditional

$$(\varphi \Leftrightarrow \psi) =_{df} ((\varphi \Rightarrow \psi) \wedge (\psi \Rightarrow \varphi))$$

has the truth matrix of Table 6 and takes a designated value if and only if $[\![\varphi]\!] = [\![\psi]\!]$. (Note that $[\![(\varphi \Leftrightarrow \psi)]\!]$=**B** if $[\![\varphi]\!] = [\![\psi]\!] =$ **B**, and =**T** if $[\![\varphi]\!]$ and $[\![\psi]\!]$ have the same one of the other three values.)

| ⇔ | T | B | N | F |
|---|---|---|---|---|
| T | T | F | N | F |
| B | F | B | N | F |
| N | N | N | T | N |
| F | F | F | N | T |

**Table 6** Truth matrix for a defined biconditional, ⇔.

As a result, it supports a principle of Substitution: formulas of the form

$$(\varphi \Leftrightarrow \psi) \rightarrow ((\cdots\varphi\cdots) \Leftrightarrow (\cdots\psi\cdots))$$

are valid in $FDE^{\rightarrow}$, and hence in any of its 3-valued extensions. In contrast, the biconditional similarly defined in terms of our basic conditional yields a designated value just in case either both of its terms have designated values or both have undesignated values. As a result, it does not license substitution: If $\varphi$ and $\psi$ have *different* designated values, and also if they have different undesignated values, $(\varphi \leftrightarrow \psi)$ will have a designated value, but $(\neg\varphi \leftrightarrow \neg\psi)$ will not. However, $\leftrightarrow$ can be added to the list of "positive" connectives, $\wedge, \vee, \rightarrow$, whose logic is exactly classical. We will return to the usefulness of $\Rightarrow$ for FDE in §7.

On the other hand, the $\Rightarrow$ connective has certain undesirable features which militate against its adoption as the basic conditional operator of a logic designed for use. Principles analogous to some of the *structural* rules of Gentzen (1934), easily derivable by the conventional natural deduction rules for the conditional, fail for it. Some fail for $\Rightarrow$ in $K3^{\rightarrow}$, and others fail for $\Rightarrow$ in $LP^{\rightarrow}$, as we shall see.

[10] Omori and Wansing (2017, p. 1036) cite this conditional as one of their four "interesting" conditionals that could be added to FDE. We will make use of the corresponding biconditional below, in §7.

## *6.2 On the K3$^{\to}$ Side*

The truth tables for $\to$ and $\Rightarrow$ differ on only one of the nine lines for K3$^{\to}$, as indicated in Table 7: if $[\![\varphi]\!]$=**N** and $[\![\psi]\!]$=**F**, then $[\![\varphi \to \psi]\!]$=**T** but $[\![\varphi \Rightarrow \psi]\!]$=**N**.

| $\varphi$ | $\psi$ | $(\varphi \to \psi)$ | $(\varphi \Rightarrow \psi)$ | |
|---|---|---|---|---|
| **T** | **T** | **T** | **T** | |
| **T** | **N** | **N** | **N** | |
| **T** | **F** | **F** | **F** | |
| **N** | **T** | **T** | **T** | |
| **N** | **N** | **T** | **T** | |
| **N** | **F** | **T** | **N** | $\Leftarrow$ |
| **F** | **T** | **T** | **T** | |
| **F** | **N** | **T** | **T** | |
| **F** | **F** | **T** | **T** | |

**Table 7** Comparison of two conditionals in K3$^{\to}$

This is enough, however, to invalidate the principle of Contraction for $\Rightarrow$:

$$\nvDash_{K3^{\to}} (\varphi \Rightarrow (\varphi \Rightarrow \psi)) \Rightarrow (\varphi \Rightarrow \psi)$$

This should be old news! The derived truth table for the $\Rightarrow$ of K3$^{\to}$ in Table 7 is exactly that of the conditional of Łukasiewicz's 3-valued logic, Ł3, for which Contraction failure is familiar. Since the interpretations of the other connectives are the same for K3$^{\to}$and Ł3, we have

*Proposition* 2. Łukasiewicz's 3-valued logic Ł3 is faithfully interpretable in K3$^{\to}$.

In fact, we have the converse as well[11]: our $\to$ can be defined in Ł3 by

$$(\varphi \to \psi) =_{df} (\varphi \Rightarrow (\varphi \Rightarrow \psi)),$$

and hence

*Proposition* 3. K3$^{\to}$ is faithfully interpretable in Ł3.

K3$^{\to}$ and Ł3 can thus be seen as, in effect, alternative formulations of a single logic (see further discussion in Section 8).[12] We think the classical nature of $\to$ (other than some of its interaction with $\neg$, such as the failure of contraposition) makes it easier to use than $\Rightarrow$. It also has a straightforward natural deduction formulation (see Appendix I), and we recommend translation into K3$^{\to}$ to anyone interested in actually proving theorems in Ł3.

[11] Nelson (1959) should be credited with the observations of the failure of contraposition, its brute force restoration, and contraction failure in his system of constructible falsity. . . which is essentially the addition of intuitionistic implication to K3. What our observation adds to this is that these phenomena do not depend on the intuitionistic nature of Nelson's implication, but arise already in the 3-valued and 4-valued logics. In the latter part of his 1959 paper, Nelson defines another variant logic that can be seen as a 3-valued paraconsistent logic. This is discussed in Kamide (2017).

[12] Mutual faithful interpretability is not in general sufficient to establish identity of logics, but the particulars of our cases will add on what further is required, as we discuss in §8.

### 6.3 On the LP$^{\rightarrow}$ Side

Now looking at the logic with truth values {**T**,**B**,**F**}, we see in Table 8 that $\Rightarrow$ again differs from $\rightarrow$ on only one of the nine lines: but this time it is the line where $[\![\varphi]\!]$=**T** and $[\![\psi]\!]$=**B**. Here $[\![\varphi \rightarrow \psi]\!]$=**B** but $[\![\varphi \Rightarrow \psi]\!]$=**F**.

| $\varphi$ | $\psi$ | $(\varphi \rightarrow \psi)$ | $(\varphi \Rightarrow \psi)$ | |
|---|---|---|---|---|
| **T** | **T** | **T** | **T** | |
| **T** | **B** | **B** | **F** | $\Leftarrow$ |
| **T** | **F** | **F** | **F** | |
| **B** | **T** | **T** | **T** | |
| **B** | **B** | **B** | **B** | |
| **B** | **F** | **F** | **F** | |
| **F** | **T** | **T** | **T** | |
| **F** | **B** | **T** | **T** | |
| **F** | **F** | **T** | **T** | |

**Table 8** Comparison of two conditionals in LP

This difference in truth table is enough to invalidate the principle of Thinning:

$\nvDash_{LP^{\rightarrow}} \varphi \Rightarrow (\psi \Rightarrow \varphi)$

For, if $[\![\varphi]\!]$ = **B** and $[\![\psi]\!]$ = **T**, then $[\![\psi \Rightarrow \varphi]\!]$ and hence $[\![\varphi \Rightarrow (\psi \Rightarrow \varphi)]\!]$ will be **F**. As Tedder (2015) notes[13], $\Rightarrow$ in LP$^{\rightarrow}$ has exactly the truth table of the logic RM3 (and, since RM3 is a cousin of the relevance family of logics, failure of Thinning in it is just what one would expect). RM3 and LP agree on $\{\wedge, \vee, \neg\}$, and so – as in the analogous case on the K3 side – we have

*Proposition* 4. RM3 can be faithfully interpreted in LP$^{\rightarrow}$.

Again, the converse is also true: $\rightarrow$ can be defined in terms of $\Rightarrow$ in RM3 by

$$(\varphi \rightarrow \psi) =_{df} ((\varphi \Rightarrow \psi) \vee \psi)$$

Hence

*Proposition* 5. LP$^{\rightarrow}$ can be faithfully interpreted in RM3.

So in the same sense as with Ł3 and K3$^{\rightarrow}$, RM3 and LP$^{\rightarrow}$ can be thought of as alternative formulations of a single logic (again, see discussion in Section 8), and again, we think adoption of $\rightarrow$ as primitive is likely to be more convenient. For, although the metamathematical study of RM3 has been fruitful in the study of relevance logics, it seems to us that LP$^{\rightarrow}$ is likely to be a more convenient to *use* in the task of actually trying to prove theorems of RM3 – particularly in light of Tedder's (2015) employment of it in formulating mathematically interesting axiomatic theories.

[13] Citing Avron (1991).

### 6.4 Putting the Two Sides Together

#### 6.4.1 About $M^{\rightarrow}$

Given that K3 and LP are obtained semantically from FDE by adding different conditions on valuations, and syntactically by adding different rules, one might incautiously conjecture that the FDE consequence relation is simply the intersection of the K3 and LP relations. Not so: $(\varphi \wedge \neg\varphi)$ implies $(\psi \vee \neg\psi)$ in both 3-valued logics, but not in FDE.[14] There is thus a fifth logic in the neighbourhood, which is sometimes[15] called *Mingle* and which we will unimaginatively call M.

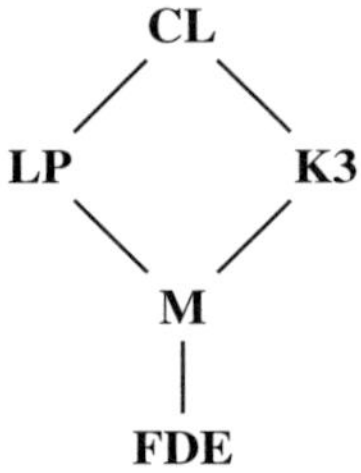


It is characterized semantically as the set of inferences preserving designation on every FDE valuation which does not assign **B** to one formula and **N** to another, and syntactically by adding the above-mentioned inference as a general rule to a formulation of FDE.

Consider now the addition of $\rightarrow$ to M: Since $M^{\rightarrow}$ contains $\rightarrow$, it of course has a deduction theorem: if $\Gamma, \varphi \vdash \psi$ then $\Gamma \vdash (\varphi \rightarrow \psi)$. The structural rules of thinning and contraction fail for $\Rightarrow$ : counterexamples to thinning need only gluts (it fails in RM3), and counterexamples to contraction require only gaps (it fails in Ł3). So both fail in $M^{\rightarrow}$ – but not in the same model! We note in passing that this means that the following formula is a theorem:

$$\vDash_{M^{\rightarrow}} (C \Rightarrow (D \Rightarrow C)) \vee ((A \Rightarrow (A \Rightarrow B)) \Rightarrow (A \Rightarrow B))$$

Note that this is a provable disjunction with no variable in common between the two disjuncts, yet neither disjunct by itself is provable![16] Some might think this shows that $M^{\rightarrow}$ is silly and unreasonable.

We too think M and $M^{\rightarrow}$ seem like a rather silly logics (why should the presence of a single truth-value glut rule out the existence of any truth-value gaps, or vice versa?[17]), but record here that $M^{\rightarrow}$ can be formulated in the obvious way, and that our completeness

[14] In K3, $(\varphi \wedge \neg\varphi)$ is never designated, so $(\varphi \wedge \neg\varphi) \vDash_{K3} (\psi \vee \neg\psi)$ will (vacuously) hold; in LP, $(\psi \vee \neg\psi)$ is always designated, so $(\varphi \wedge \neg\varphi) \vDash_{LP} (\psi \vee \neg\psi)$ always holds. However, in FDE, if $[\![\varphi]\!] = \mathbf{B}$ then $(\varphi \wedge \neg\varphi) = \mathbf{B}$ and hence is designated; but if $[\![\psi]\!] = \mathbf{N}$, then $[\![(\psi \vee \neg\psi)]\!] = \mathbf{N}$ and is undesignated. So $(\varphi \wedge \neg\varphi) \nvDash_{FDE} (\psi \vee \neg\psi)$.

[15] See Humberstone (2011, p. 334) for a discussion of the Mingle axiom in the context of relevant logic and why the present logic could also be named Mingle. Beall et al. (2013, §5) generate a 3-valued version of this logic by considering a "dual requirement" on validity: that not only should all valuations that make every premise true also make the conclusion true, but also that any valuation that falsifies the conclusion must also falsify one of the premises.

[16] The (very lengthy) proof in our Fitch-system of the Appendix, uses our "classicalizing" principle (in Appendix 1) that a conclusion follows *tout court* if it follows from $\phi$ and also from $\phi \rightarrow \psi$, for any instances of $\phi$ and $\psi$. The actual proof requires two further embeddings of different instances of the classicalizing rule. It is quite a tedious proof!

[17] Well, perhaps there is someone who is convinced that a three-valued logic is the right response to the paradoxes and the puzzles about indeterminacy, but is not sure whether it should go the glut route or the

proof in Appendix II for $\text{FDE}^{\rightarrow}$, using the natural deduction system of Appendix I, extends without difficulty to cover it.

*Proposition* 6. $\text{M}^{\rightarrow}$ and $\text{M}^{\Rightarrow}$ are mutually faithfully interpretable in each other

### 6.4.2 About $\text{FDE}^{\rightarrow}$

We have not explored the behaviour of $\Rightarrow$ in $\text{FDE}^{\rightarrow}$, and, in contrast to the situation with the 3-valued logics, do not know of a historically proposed equivalent for it. Omori and Wansing (2017, p. 1036) list it as one of four possible conditionals to add to FDE, alongside $\rightarrow$, but our result shows that this does not yield a different logic. Contraction and Thinning will, of course, both fail for $\Rightarrow$ in the 4-valued logic. With no relevant insight, we have resorted to construction of truth tables to verify that $\rightarrow$ can be recovered from $\Rightarrow$ in the 4-valued environment by the double-barrelled definition.

$$(\varphi \rightarrow \psi) =_{df} ((\varphi \Rightarrow (\varphi \Rightarrow \psi)) \vee \psi)$$

*Proposition* 7. $\text{FDE}^{\rightarrow}$ and $\text{FDE}^{\Rightarrow}$ are mutually faithfully interpretable in each other

Thus showing, with the addition of the considerations we give below in §8, that as in the 3-valued extensions of FDE, adding either of the 4-valued conditionals $\rightarrow$ or $\Rightarrow$ to basic FDE generates the identical logic as adding the other.

## 7 More and Less Drastic Expansions

### *7.1 Functional Completeness by Adding Constants and Parametric Operators*

In classical two-valued logic, a standard way to prove functional completeness (i.e., that every two-valued truth function can be described by a given set of connectives... we will restrict our attention to the functions $\wedge, \vee, \neg$) is to show that any arbitrary truth table can be described by a formula using those connectives. Using these connectives, we can generate a formula in disjunctive normal form. In textbooks this is normally "proved" by providing an example that seems clearly to be generalizable, rather like this: Our example will have three atomic letters, *p, q, r*. There are two cases. Either the resulting truth table has at least one **T** value or it doesn't. In the latter case the formula is a contradiction and that truth table can be described by the formula $(p \wedge \neg p \wedge q \wedge r)$. Otherwise the example truth table will look something like Table 9. From this table we construct a formula as follows. For each row where the value is **T** – as for example in the first row – write a formula that "describes" the values of the three atomic arguments. If an atomic argument is **T** in that row, simply use the letter itself; if it is **F** in that row, use its negation; and conjoin the results. So, the formula for the first row would be $(p \wedge q \wedge r)$. The formula for the fifth row would be $(\neg p \wedge q \wedge r)$, etc. After constructing these "descriptive" conjunctions, the final formula is the disjunction of them all. In this example case, it is

$$(p \wedge q \wedge r) \vee (\neg p \wedge q \wedge r) \vee (\neg p \wedge q \wedge \neg r) \vee (\neg p \wedge \neg q \wedge \neg r)$$

---

gap route. But this person is sure that only one of these ways will be correct and that having both would be philosophically distasteful.

This formula clearly has the same truth table as that of Table 9. And the student is now expected to generalize this example.

| p | q | r | |
|---|---|---|---|
| **T** | **T** | **T** | **T** $(p \wedge q \wedge r)$ |
| **T** | **T** | **F** | **F** |
| **T** | **F** | **T** | **F** |
| **T** | **F** | **F** | **F** |
| **F** | **T** | **T** | **T** $(\neg p \wedge q \wedge r)$ |
| **F** | **T** | **F** | **T** $(\neg p \wedge q \wedge \neg r)$ |
| **F** | **F** | **T** | **F** |
| **F** | **F** | **F** | **T** $(\neg p \wedge \neg q \wedge \neg r)$ |

**Table 9** Example of 2-valued Functional Completeness

A generalization of this method can be employed for many-valued logics, as described in Rosser and Turquette (1952), works in the following way. We might extend the language of $\text{FDE}^{\rightarrow}$ by adding constants (0-ary operators that always take a fixed value) for each of the four truth values. We use the names **t**, **b**, **n**, and **f** for these four operators whose fixed values are **T**, **B**, **N**, and **F** respectively. We can also add four unary operators, $\text{J}_\mathbf{t}(\phi)$, $\text{J}_\mathbf{b}(\phi)$, $\text{J}_\mathbf{n}(\phi)$, $\text{J}_\mathbf{f}(\phi)$, whose values are **T** just in case the argument $\phi$ takes the value **t**, **b**, **n**, **f**, respectively, and otherwise takes the value **F**. In two-valued logic, $\neg$ takes the place of $\text{J}_\mathbf{f}$, while an unadorned formula takes the place of $\text{J}_\mathbf{t}$. But now that we have further values, we need a more generalized set of parametric operators. Similarly, in the two-valued case, having the "descriptive formula" for a row indicates that the desired value is **T** for that row, and not having it indicates that we want **F**. With more values to choose from, this isn't enough, and we need a slightly more complex construction using the four constants **t, b, n, f**.

Even a simple example in FDE will be difficult to display: with three atomic letters we have $4^3 = 64$ rows of a truth table. And each row needs to be "described", not just the **T** rows. As in the two-valued case, each row is a conjunction, and if we have just three atomic letters, there are just three conjuncts. Each row of the truth table in the FDE case will have the form

$v_a \quad v_b \quad v_c \quad | \quad v_k$

where the $v_a, v_b, v_c$ are the values of the atomics $p, q, r$ respectively and the $v_k$ is the value assigned to the to-be-described formula in that row. The "descriptive formula" for that row is

$(J_{v_a}(p) \wedge J_{v_b}(q) \wedge J_{v_c}(r) \wedge \mathbf{k})$

where **k** is the constant that names the value $v_k$. The resultant formula that has the same truth table as the arbitrary example given, is the disjunction of all the descriptive formulas for every row (not just the **t** rows, as in the two-valued case). Again, the example is supposed to be generalized and cover any possible FDE truth table. Since the method uses only the $\wedge, \vee, \neg$ of basic FDE plus the four constants and four parametric operators, this shows that such a logic is functionally complete for the given four-valued semantics.

Note that this method works for any of the logics under discussion, assuming that the appropriate J-operators and constants for the relevant truth values are present. (And also note that $\neg$ is not required in any of them.)

We turn our attention now to other ways to generate functional completeness in our logics of interest. That was a topic of interest in the 1930s, especially for the 3-valued logics, and especially for the particular case of Łukasiewicz Ł3. (See in particular Wajsberg, 1931, who added to Ł3 a one-place operator that yielded the value **N** no matter what the truth value of the formula that was its argument.) However we will follow a different path, and show how to alter the Rosser-Turquette method just employed, by using features of our four-valued $\mathrm{FDE}^{\rightarrow}$; we do this by showing how to define or otherwise simplify the J-operators and truth-value constants in some different ways.

## *7.2 Some Definitional Possibilities*

As we remarked above, adding $\rightarrow$ to FDE, yielding $\mathrm{FDE}^{\rightarrow}$, does not yield a functionally complete logic. But we can use this modified FDE that contains our $\rightarrow$ to lessen the requirement of needing all eight of the new primitives we added to basic FDE in the previous subsection.

We start by retaining the four truth-constants, **t**, **b**, **n**, **f** as a primitive. Now consider the formula

$(\varphi \rightarrow \mathbf{f})$

If $[\![\varphi]\!]$ is designated, i.e., is either **T** or **B**, then this formula is will be **F**. And if $[\![\varphi]\!]$ is undesignated, i.e., is either **N** or **F**, then this formula is will be **T**, as can be checked by truth tables.

Recall from §6.1 that we used $\rightarrow$ and $\neg$ to define $\Rightarrow$, and then used $\Rightarrow$ and $\wedge$ to define $\Leftrightarrow$. Finally, recall also that $[\![(\varphi \Leftrightarrow \psi)]\!]$ is designated if and only if $[\![\varphi]\!] = [\![\psi]\!]$. Putting this together, the biconditional $(\varphi \Leftrightarrow \mathbf{t})$ defines $J_{\mathbf{t}}(\varphi)$: it takes the value **T** if $[\![\varphi]\!] = \mathbf{T}$, and takes the value **F** otherwise. Similarly, $(\varphi \Leftrightarrow \mathbf{b})$, $(\varphi \Leftrightarrow \mathbf{n})$ and $(\varphi \Leftrightarrow \mathbf{f})$ define $J_{\mathbf{b}}(\varphi)$, $J_{\mathbf{n}}(\varphi)$, $J_{\mathbf{f}}(\varphi)$ respectively.

So, by adding the $\rightarrow$ conditional to FDE, we have thus eliminated four of the eight primitives we added to FDE in the previous subsection by using the Rosser-Turquette method. We could also eliminate one or the other of **t** and **f**, since $[\![\mathbf{t}]\!] = [\![\neg\mathbf{f}]\!]$ and $[\![\mathbf{f}]\!] = [\![\neg\mathbf{t}]\!]$. (However, this would make $\neg$ be a required primitive, whereas the Rosser-Turquette method did not employ $\neg$.)

## *7.3 Employing Second-Order Logic*[18]

In Hazen and Pelletier (2018a) it is shown that a Second Order logic based on LP was surprisingly weak. This was due to the limited expressive power of the language with no conditional operator. In contrast, $\mathrm{FDE}^{\rightarrow}$, with its classical conditional, is expressively strong enough to provide a full base for classical Second Order Logic. Using propositional quantification, we can define a *falsum* propositional constant:

$\mathbf{f} = \forall p(p)$

[18] For more information on second order quantification in FDE and related arguments see (Hazen and Pelletier 2018b)

(Obviously set/property quantification would do as well: $\mathbf{f} = \forall X \forall x(X(x))$.) This constant and the conditional give us in effect an ersatz "classical" negation: the conditional $(\varphi \to \mathbf{f})$ takes the value **F** when $[\![\varphi]\!]$ is designated and the value **T** when $[\![\varphi]\!]$ is undesignated, as we remarked in the previous subsection. Call a predicate (monadic or relational) *classical* just in case no atomic formula in it, on any assignment to the individual variables, takes either of the intermediate truth values **N** or **B**. **N** can be ruled out by the First Order formula $\forall x(\psi(x) \vee \neg\psi(x))$, which will not take a designated value if the predicate yields the value **N** for any individual in the domain. The possibility that $\psi$ somewhere yields the value **B**, however, cannot be excluded by a purely First Order formula: if $\psi$ yields **B** for every individual, then every First Order formula in which $\psi$ is the only bit of non-logical vocabulary occurring will have the designated value **B**. Using our ersatz classical negation, however, we can say $((\exists x(\psi(x) \wedge \neg\psi(x)) \to \mathbf{f})$. Finally, then, the classicality of a predicate, can by expressed by the conjunction of these two formulas. And so we may then interpret classical Second Order Logic in Second Order FDE$^{\to}$ by simply restricting all Second Order quantifiers to classical predicates.

Coming back down to the fragment of Second Order FDE$^{\to}$ with only propositional quantification, we can define the other three propositional constants. *Verum* is easy:

$$\mathbf{t} = \exists p(p)$$

Defining a constant for **B** is perhaps less obvious. $\exists p(p \wedge \neg p)$ would work in the 3-valued logics LP and LP$^{\to}$, but when **N** is also available it fails: $[\![p]\!] = \mathbf{B}$ gives $[\![p \wedge \neg p]\!] = \mathbf{B}$, $[\![p]\!] = \mathbf{N}$ gives $[\![p \wedge \neg p]\!] = \mathbf{N}$, and the existential quantification will have as value the join of these in the truth-value lattice: **T**![19] However, $[\![p \to p]\!]=\mathbf{B}$ when $[\![p]\!]=\mathbf{B}$, and is **T** otherwise, so we may define

$$\mathbf{b} = \forall p(p \to p)$$

Defining a constant for **N** in FDE$^{\to}$is altogether harder (though $\exists p(p \wedge \neg p)$ would work in the 3-valued logic K3$^{\to}$). Indeed, it can be shown that

*Theorem* 1. No *definiens* with, in prenex form, a single block of propositional quantifiers (all universal or all existential) is able to define a constant for **N**.

*Proof.* Consider a purely propositional formula of FDE$^{\to}$. An assignment giving the value **B** to all of its atomic variables will give the formula the value **B**. Thus the set of values assumed by the formula on different assignments to its propositional variables will include **B**. The value of the sentence formed by binding its variables by existential quantifiers will be the lattice join of the values in this set, and so must be either **B** or **T**. The value of the sentence formed by binding its variables by universal quantifiers will be the lattice meet of the values in this set, and so must be either **B** or **F**. In neither case will the quantified formula serve as a definiens for the constant **n**.

The propositional constant **n** can, however, be defined by a sentence of more complicated quantificational structure. Recall that $[\![(\varphi \Leftrightarrow \psi)]\!]$ is designated if and only if $[\![\varphi]\!] = [\![\psi]\!]$. It can thus be thought of as expressing identity of truth value. The conditional $((\varphi \Leftrightarrow \mathbf{t}) \to \mathbf{f})$, then, says that the value of $\varphi$ is *not* **T**, and similarly for $((\varphi \Leftrightarrow \mathbf{b}) \to \mathbf{f})$ and $((\varphi \Leftrightarrow \mathbf{f}) \to \mathbf{f})$. So we may define **n** by

[19] The existentially quantified formula takes the join of the values **N**, **B**, and **F**... which, perhaps surprisingly, is **T**. One doesn't expect a disjunction to take a higher value than any of its disjuncts, but in this case, because the four values are not linearly ordered, it does.

$$\mathbf{n} =_{df} \exists q(((q \Leftrightarrow \mathbf{t}) \to \mathbf{f}) \land ((q \Leftrightarrow \mathbf{b}) \to \mathbf{f}) \land ((q \Leftrightarrow \mathbf{f}) \to \mathbf{f}) \land q)$$

For, if $q$ in the matrix is assigned one of the values **T**, **B**, or **F**, one of the first three conjuncts, and so the whole, will have the value **F**. If $[\![q]\!]$=**N**, however, the first three conjuncts will all have the value **T**, but the fourth, $q$, and so the whole conjunction, will have the value **N**. Since the join of **F** and **N** is **N**, the value of the full, quantified, sentence is **N**. By extracting the quantifiers concealed in the propositional constants in the contained biconditionals in the right order, we can put the definiens into prenex form with only one alternation of quantifiers (so, it will be $\Sigma_2$).

One immediate application of the propositional constants is to allow the definition or a "truly classical negation" operator, for which we will use $\backsim$, and define as:

$$\backsim \phi =_{df} (\phi \to \mathbf{f})$$

using *this* negation, our classical material implication is definable in terms of disjunction by the familiar definition:

$$(\phi \to \psi) =_{df} (\backsim \phi \lor \psi)$$

Note that this is not $\sim$, the classical negation of De and Omori (2015). Table 10 compares their $\sim$ and our $\backsim$. That same familiar definition, but using their negation $\sim$ yields their $\multimap$ conditional instead.

| $\phi$ | $\sim \phi$ | $\backsim \phi$ |
|---|---|---|
| **T** | **F** | **F** |
| **B** | **N** | **F** |
| **N** | **B** | **T** |
| **F** | **T** | **T** |

**Table 10** Comparison of two "classical negation operators" in FDE

## 8 Synonymous Logics

In Pelletier (1984) and Pelletier and Urquhart (2003), the notion of "translational equivalence" was introduced and defined, and used to describe a concept of *synonymous logics*. This concept was intended to describe cases where two logic systems were "really the same system" despite having different formulations, different vocabulary, and possibly having such different formulation that it would not be at all obvious that the logics were "really the same." This notion was shown to be different from various other conceptions in the literature, such as *mutual interpretability* and having *exact translations between logics*, which were shown to be weaker; other notions, such as having identical definitional extensions, were shown to be sufficient conditions for synonymy.[20] Two logics are *mutually interpretable* iff there are definable mappings taking valid formulas and arguments of each to valid formulas and arguments of the other. They are *faithfully interpretable* iff the same

[20] Further aspects of the notion, as well as formal details, are in Pelletier and Urquhart, 2003.

mappings take invalid formulas and arguments to invalid ones. Note that the definitions of mutual interpretability and even mutual faithful interpretability do not in any obvious way imply synonymy of the logics, although the counterexamples generally seem very complex and "artificial". (See French, 2010, Chapter 3, for a categorization of the different types of interpretability.)

Two logics, $\mathcal{L}_1$ and $\mathcal{L}_2$, are translationally equivalent if and only if there are translation schemes $t_1$ from $\mathcal{L}_1$ into $\mathcal{L}_2$ and $t_2$ from $\mathcal{L}_2$ into $\mathcal{L}_1$ such that

1. if $\Gamma \vdash_{\mathcal{L}_1} \varphi$ then $\Gamma \vdash_{\mathcal{L}_2} \varphi^{t_1}$
2. if $\Gamma \vdash_{\mathcal{L}_2} \varphi$ then $\Gamma \vdash_{\mathcal{L}_1} \varphi^{t_2}$
3. for any formula $\varphi$ in $\mathcal{L}_1$, $(\varphi^{t_1})^{t_2}$ is equivalent to $\varphi$ (in $\mathcal{L}_1$)
4. for any formula $\varphi$ in $\mathcal{L}_2$, $(\varphi^{t_2})^{t_1}$ is equivalent to $\varphi$ (in $\mathcal{L}_2$)[21]

We are in a position to show some new results of synonymity of logics. Since the languages of our different logics are identical except for their conditionals (and biconditionals), we will employ the following translations for all the portions of the logics involved except the differing (bi)conditionals in the logics:

- For $\varphi$ an atomic sentence, $(\varphi)^t$ is $\varphi$
- For negated formulas, $(\neg\varphi)^t$ is $\neg(\varphi)^t$
- If $\circ$ is any binary operator other than a (bi)conditional, $(\varphi \circ \psi)^t$ is $((\varphi)^t \circ (\psi)^t)$

In §6.2 we showed that systems K3$^{\rightarrow}$ and Ł3 could be faithfully interpreted in each other by means of the following "translations". It should be understood that these translation functions apply recursively, to any embedded formulas that also have the conditionals in question.[22] (As for example in Case 2 of the proof of Theorem 2.)

- $t_1 : (\varphi \rightarrow \psi)^{t_1} =_{df} ((\varphi)^{t_1} \Rightarrow ((\varphi)^{t_1} \Rightarrow (\psi)^{t_1}))$
- $t_2 : (\varphi \Rightarrow \psi)^{t_2} =_{df} (((\varphi)^{t_2} \rightarrow (\psi)^{t_2}) \wedge ((\neg\psi)^{t_2} \rightarrow (\neg\varphi)^{t_2}))$

We can now prove a stronger result. The earlier proof of mutual interpretability showed that each of K3$^{\rightarrow}$ (whose conditional is $\rightarrow$) and Ł3 (whose conditional is $\Rightarrow$) could define a formula that had the same truth table as the conditional of the other (and since all other connectives were identical, that is all that is required to establish mutual interpretability). What is further needed to show synonymy is that applying the definition in one logic to generate the conditional of the other logic yields a result such that we can apply the other logic's definition to it and generate a formula semantically equivalent to the conditional in the first logic. (And vice versa, of course). That is to say: where $t_1$ and $t_2$ are the translations we used in §6.2, we need to be able to prove

[21] In Pelletier and Urquhart (2003) it was assumed that the logics in question had a "biconditional equivalence connective" and the third and fourth conditions were expressed in terms of the biconditional being a theorem in the appropriate logics. In the context of that paper, the logics were classical except for modal operators, and so there were always such equivalence operators in each logic. In the present context, we cannot assume that the biconditionals of the various logics will operate in the same way, and so we envisage checking the "equivalent to" conditions semantically, by simply looking at the relevant truth tables. Although defining translational equivalence in full generality for non-truth-functional logics is quite difficult (as Nelson, 1959, p.216, remarks, since the only logics we are discussing have a finite number of truth values, the relevant kind of equivalence of formulas is that of having (on each assignment) the same truth value. And in fact, our logics allow us to define a connective, $\Leftrightarrow$, which can reasonably be taken to express the relation of having the same truth value.

[22] This follows from what Kuhn and Weatherson (2018) call the "compositionality" requirement of Pelletier and Urquhart's (2003) definition of translational equivalence.

- $((\varphi \Rightarrow \psi)^{t_2})^{t_1}$ is semantically equivalent to $(\varphi \Rightarrow \psi)$ in Ł3
- $((\varphi \rightarrow \psi)^{t_1})^{t_2}$ is semantically equivalent to $(\varphi \rightarrow \psi)$ in $K3^{\rightarrow}$

*Theorem* 2. **$K3^{\rightarrow}$ and Ł3 are synonymous logics**

*Proof.*
Case 1:

$$((\varphi \Rightarrow \psi)^{t_2})^{t_1} = [(\varphi \rightarrow \psi) \wedge (\neg\psi \rightarrow \neg\varphi)]^{t_1} \tag{1}$$

$$= ((\varphi \Rightarrow (\varphi \Rightarrow \psi)) \wedge (\neg\psi \Rightarrow (\neg\psi \Rightarrow \neg\varphi)) \tag{2}$$

$$= (\varphi \Rightarrow \psi) \tag{3}$$

Case 2:

$$((\varphi \rightarrow \psi)^{t_1})^{t_2} = (\varphi \Rightarrow (\varphi \Rightarrow \psi))^{t_2} \tag{4}$$

$$= (\varphi \rightarrow (\varphi \Rightarrow \psi)^{t_2}) \wedge (\neg(\varphi \Rightarrow \psi)^{t_2} \rightarrow \neg\varphi) \tag{5}$$

$$= [[\varphi \rightarrow (((\varphi \rightarrow \psi) \wedge (\neg\psi \rightarrow \varphi))] \wedge [\neg((\varphi \rightarrow \psi) \wedge (\neg\psi \rightarrow \neg\varphi)) \rightarrow \neg\varphi]] \tag{6}$$

$$= (\varphi \rightarrow \psi) \tag{7}$$

It can be seen from the truth table in Table 11 that formulas (2) [the left-most $\Rightarrow$] and (3) [the $\wedge$ formula] are semantically the same, and from the truth table in Table 12 that formulas (6) [the $\wedge$ column] and (7) [the left-most $\rightarrow$ column] are semantically the same.

| $\varphi$ $\psi$ | $\varphi \Rightarrow \psi$ | $\neg\psi \Rightarrow \neg\varphi$ | $[(\varphi \Rightarrow (\varphi \Rightarrow \psi)))$ | $\wedge$ | $(\neg\psi \Rightarrow (\neg\psi \Rightarrow \neg\varphi))]$ |
|---|---|---|---|---|---|
| **T T** | **T** | **T** | **T** | **T** | **T** |
| **T N** | **N** | **T** | **N** | **N** | **T** |
| **T F** | **F** | **T** | **F** | **F** | **T** |
| **N T** | **T** | **F** | **T** | **T** | **T** |
| **N N** | **T** | **T** | **T** | **T** | **T** |
| **N F** | **N** | **N** | **T** | **N** | **N** |
| **F T** | **T** | **T** | **T** | **T** | **T** |
| **F N** | **T** | **T** | **T** | **T** | **T** |
| **F F** | **T** | **T** | **T** | **T** | **T** |

**Table 11** $\varphi \Rightarrow \psi$ of Ł3 (left column) is equivalent to its "double translation" (the $\wedge$ column)

| $\varphi$ $\psi$ | $\varphi \rightarrow \psi$ | $\neg\psi \rightarrow \neg\varphi$ | $[[\varphi \rightarrow (((\varphi \rightarrow \psi) \wedge (\neg\psi \rightarrow \varphi))]$ | $\wedge$ | $[\neg((\varphi \rightarrow \psi) \wedge (\neg\psi \rightarrow \neg\varphi)) \rightarrow \neg\varphi]]$ |
|---|---|---|---|---|---|
| **T T** | **T** | **T** | **T** | **T** | **T** |
| **T N** | **N** | **N** | **N** | **N** | **T** |
| **T F** | **F** | **F** | **F** | **F** | **F** |
| **N T** | **T** | **T** | **T** | **T** | **T** |
| **N N** | **T** | **T** | **T** | **T** | **T** |
| **N F** | **T** | **N** | **T** | **T** | **T** |
| **F T** | **T** | **T** | **T** | **T** | **T** |
| **F N** | **T** | **T** | **T** | **T** | **T** |
| **F F** | **T** | **T** | **T** | **T** | **T** |

**Table 12** $\varphi \rightarrow \psi$ of $K3^{\rightarrow}$ (left column) is equivalent to its "double translation" (the $\wedge$ column)

In §6.3 we showed that systems $LP^{\rightarrow}$ and RM3 could be faithfully interpreted in each other by means of the following "translations".

- $t_1 : (\varphi \rightarrow \psi)^{t_1} =_{df} (((\varphi)^{t_1} \Rightarrow (\psi)^{t_1}) \vee (\psi)^{t_1})$
- $t_2 : (\varphi \Rightarrow \psi)^{t_2} =_{df} (((\varphi)^{t_2} \rightarrow (\psi)^{t_2}) \wedge ((\neg\psi)^{t_2} \rightarrow (\neg\varphi)^{t_2}))$

Again we can now prove a stronger result. The earlier proof of mutual interpretability showed that each of $LP^{\rightarrow}$ (whose conditional is $\rightarrow$) and RM3 (whose conditional is $\Rightarrow$) could define a formula that had the same truth table as the conditional of the other (and since all other connectives were identical, that is all that is required to establish mutual interpretability). What is further needed to show synonymy is that applying the definition in one logic to generate the conditional of the other logic yields a result such that we can apply the other logic's definition to it and generate a formula semantically equivalent to the conditional in the first logic. (And vice versa, of course). That is to say: where $t_1$ and $t_2$ are the translations we used in §6.3, we need to be able to prove

- $((\varphi \Rightarrow \psi)^{t_2})^{t_1}$ is semantically equivalent to $(\varphi \Rightarrow \psi)$ in RM3
- $((\varphi \rightarrow \psi)^{t_1})^{t_2}$ is semantically equivalent to $(\varphi \rightarrow \psi)$ in $LP^{\rightarrow}$

*Theorem* 3. **$LP^{\rightarrow}$ and RM3 are synonymous logics**

*Proof.*
Case 1:

$$((\varphi \Rightarrow \psi)^{t_2})^{t_1} = [(\varphi \rightarrow \psi) \wedge (\neg\psi \rightarrow \neg\varphi)]^{t_1} \tag{8}$$

$$= (((\varphi \Rightarrow \psi) \vee \psi) \wedge ((\neg\psi \Rightarrow \neg\varphi) \vee \neg\varphi)) \tag{9}$$

$$= (\varphi \Rightarrow \psi) \tag{10}$$

Case 2:

$$((\varphi \rightarrow \psi)^{t_1})^{t_2} = ((\varphi \Rightarrow \psi) \vee \psi)^{t_2} \tag{11}$$

$$= (((\varphi \rightarrow \psi) \wedge (\neg\psi \rightarrow \neg\varphi)) \vee \psi) \tag{12}$$

$$= (\varphi \rightarrow \psi) \tag{13}$$

It can be seen from the truth table in Table 13 that formulas (9) [the $\wedge$ column] and (10) [the left-most $\Rightarrow$] are semantically the same, and from the truth table in Table 14 that formulas (12) [the $\vee$ column] and (13) [the left-most $\rightarrow$] are semantically the same.

These instances of synonymous logics strike us as both somewhat unexpected and also as "cleaner" versions of synonymy of logics than the one(s) displayed in Pelletier and Urquhart (2003), which employed a propositional constant in one of the logics. As remarked in that article (following De Bouvère, 1965), two logics are translationally equivalent in this way if they have a common definitional extension. Note then that for each of our pairs of synonymous logics, the appropriate 3-valued logic with two conditional operators, $\rightarrow$ and $\Rightarrow$, is a definitional extension of both members of the pair. And thus the two logics are translationally equivalent to each other.

Lastly, and as mentioned in §§6.4.1 and 6.4.2, we can state two final theorems concerning the addition of $\rightarrow$ or $\Rightarrow$ to M and to FDE itself:

| $\varphi$ | $\psi$ | $\varphi \rightarrow \psi$ | $\neg\psi \rightarrow \neg\varphi$ | $\varphi \Rightarrow \psi$ | $[(\varphi \rightarrow \psi) \wedge (\neg\psi \rightarrow \neg\varphi)]$ | $[((\varphi \Rightarrow \psi) \vee \psi)$ | $\wedge$ | $((\neg\psi \Rightarrow \neg\varphi) \vee \neg\varphi)]$ |
|---|---|---|---|---|---|---|---|---|
| **T** | **T** | **T** | **T** | **T** | **T** | **T** | **T** | **T** |
| **T** | **B** | **B** | **F** | **F** | **F** | **B** | **F** | **F** |
| **T** | **F** | **F** | **F** | **F** | **F** | **F** | **F** | **F** |
| **B** | **T** | **T** | **T** | **T** | **T** | **T** | **T** | **T** |
| **B** | **B** | **B** | **B** | **B** | **B** | **B** | **B** | **B** |
| **B** | **F** | **F** | **B** | **F** | **F** | **F** | **F** | **B** |
| **F** | **T** | **T** | **T** | **T** | **T** | **T** | **T** | **T** |
| **F** | **B** | **T** | **T** | **T** | **T** | **T** | **T** | **T** |
| **F** | **F** | **T** | **T** | **T** | **T** | **T** | **T** | **T** |

**Table 13** $\varphi \Rightarrow \psi$ of RM3 (third column) is equivalent to its "double translation" (the $\wedge$ column)

| $\varphi$ | $\psi$ | $\varphi \rightarrow \psi$ | $\neg\psi \rightarrow \neg\varphi$ | $\varphi \Rightarrow \psi$ | $[(\varphi \rightarrow \psi) \wedge (\neg\psi \rightarrow \neg\varphi)]$ | $[((\varphi \rightarrow \psi) \wedge (\neg\psi \rightarrow \neg\varphi)) \vee \psi]$ |
|---|---|---|---|---|---|---|
| **T** | **T** | **T** | **T** | **T** | **T** | **T** |
| **T** | **B** | **B** | **F** | **F** | **F** | **B** |
| **T** | **F** | **F** | **F** | **F** | **F** | **F** |
| **B** | **T** | **T** | **T** | **T** | **T** | **T** |
| **B** | **B** | **B** | **B** | **B** | **B** | **B** |
| **B** | **F** | **F** | **B** | **F** | **F** | **F** |
| **F** | **T** | **T** | **T** | **T** | **T** | **T** |
| **F** | **B** | **T** | **T** | **T** | **T** | **T** |
| **F** | **F** | **T** | **T** | **T** | **T** | **T** |

**Table 14** $\varphi \rightarrow \psi$ of $LP^{\rightarrow}$ (left column) is equivalent to its "double translation" (the rightmost column)

*Theorem* 4. $M^{\rightarrow}$ and $M^{\Rightarrow}$ are synonymous logics.

*Theorem* 5. $FDE^{\rightarrow}$ and $FDE^{\Rightarrow}$ are synonymous logics.

## 9 Concluding Remarks

We have discussed a family of logics that are related to FDE, showing how they are related to each other and also to some other logics that have populated the literature (such as Ł3 and RM3). We also identified a 4-valued logic $M^{\rightarrow}$ that is stronger than $FDE^{\rightarrow}$ but weaker than each of the two 3-valued extensions of $FDE^{\rightarrow}$. (That is, weaker than both $K3^{\rightarrow}$ and $LP^{\rightarrow}$.) Surprisingly, perhaps, we are also able to show some new and "cleaner" examples of synonymous logics (in the sense of Pelletier and Urquhart, 2003). Natural deduction systems for the various logics, as well as soundness and completeness proofs are in the Appendices.

## References


1. Adler, M. (1980). *Pragmatic Logic for Commands.* Amsterdam: John Benjamin's.
2. Almukdad, A. and D. Nelson (1984). Constructible falsity and inexact predicates. *Journal of Symbolic Logic*, 49, 231–233.
3. Alxatib, S. and F. J. Pelletier (2011). The psychology of vagueness: Borderline cases and contradictions. *Mind and Language*, 26, 287–326.

4. Anderson, A., N. Belnap, and M. Dunn (1992). *Entailment: The Logic of Relevance and Necessity*, Volume II. Princeton: Princeton UP.
5. Anderson, A. and N. Belnap (1962). Tautological entailments. *Philosophical Studies*, 13, 9–24.
6. Arieli, O. and A. Avron (1998). The value of the four values. *Artificial Intelligence*, 102, 97–141.
7. Arieli, O. and A. Avron (2017). Four-valued paradefinite logics. *Studia Logica*, 105, 1087–1122.
8. Asenjo, F. (1966). A calculus of antinomies. *Notre Dame Journal of Formal Logic*, 7, 103–105.
9. Asenjo, F. G. and J. Tamburino (1975). Logic of antinomies. *Notre Dame Journal of Formal Logic*, 16, 17–44.
10. Avron, A. (1986). On an implication connective of RM. *Notre Dame Journal of Formal Logic*, 7, 201–209.
11. Avron, A. (1991). Natural 3-valued logics: Characterization and proof theory. *Journal of Symbolic Logic*, 56, 276–294.
12. Batens, D. (1980a). A completeness-proof method for extensions of the implicational fragment of the propositional calculus. *Notre Dame Journal of Formal Logic*, 21, 509–517.
13. Batens, D. (1980b). Paraconsistent extensional propositional logics. *Logique et Analyse*, 90-91, 195–234.
14. Batens, D. and K. De Clerq (2004). A rich paraconsistent extension of full positive logic. *Logique et Analyse*, 47, 227–257.
15. Beall, J. and M. Colyvan (2001). From heaps of gluts to Hyde-ing the sorites. *Mind*, 110, 401–408.
16. Beall, J., T. Forster, and J. Seligman (2013). A note on freedom from detachment in the logic of paradox. *Notre Dame Journal of Formal Logic*, 54, 15–20.
17. Belnap, N. (1992). A useful four-valued logic: How a computer should think. In A. Anderson, N. Belnap, and J. Dunn (Eds.), *Entailment: The Logic of Relevance and Necessity, Volume II*, pp. 506–541. Princeton: Princeton UP. First appeared as "A Useful Four-valued Logic" Modern Use of Multiple-valued Logic J.M. Dunn and G. Epstein (eds.), pp. 3-37; Dordrecht: D. Reidel, 1977; and "How a Computer Should Think" Contemporary Aspects of Philosophy G. Ryle (ed.), pp. 30-56; Oriel Press, 1977.
18. Bochvar, D. (1939). On a 3-valued logic calculus and its application to the analysis of contradictions. *Matématicéskij sbornik*, 4, 287–308. English translation by Merrie Bergmann, 1981, *History and Philosophy of Logic*, 2, pp. 87–112.
19. Bochvar, D. (1943). On the consistency of a 3-valued calculus. *Matématicéskij sbornik*, 12, 353–369.
20. Bolc, L. and P. Borowick (1992). *Many-Valued Logics: Vol. 1, Theoretical Foundations*. Berlin: Springer-Verlag.
21. Brady, R. (2006). *Universal Logic*. Stanford: CSLI Press.
22. Camp, J. (2002). *Confusion*. Cambridge, MA: Harvard UP.
23. Cantini, A. (2014). Paradoxes and Contemporary Logic. In E. Zalta (Ed.), *The Stanford Encyclopedia of Philosophy*. http://plato.stanford.edu/archives/fall2014/entries/ paradoxes-contemporary-logic/.
24. da Costa, N. (1974). On the theory of inconsistent formal systems. *Notre Dame Journal of Formal Logic*, 15, 497–510.
25. da Costa, N. (1982). The philosophical import of paraconsistent logic. *Journal of Non-Classical Logic*, 1, 1–19.
26. De, M. and H. Omori (2015). Classical negation and expansions of Belnap-Dunn logic. *Studia Logica*, 103, 825–851.
27. de Bouvère, K. (1965). Synonymous theories. In J. Addison, L. Henkin, and A. Tarski (Eds.), *Symposium on the Theory of Models*, pp. 402– 406. Amsterdam: North-Holland.
28. Dummett, M. (1963). The philosophical significance of Gödel's theorem. *Ratio*, 5, 140–155. Reprinted in M. Dummett, *Truth and Other Enigmas*, Chap. 12., 1978. Cambridge: Harvard University Press.
29. Dunn, J. (1976). Intuitive semantics for first degree entailment and coupled trees. *Philosophical Studies*, 29, 149–168.
30. Dunn, M. and G. Restall (2002). Relevance logic. In D. Gabbay and F. Guenthner (Eds.), *Handbook of Philosophical Logic*, 2nd Edition, Volume 6, pp. 1–136. Kluwer.
31. Feferman, S. (1984). Toward useful type-free theories, I. *Journal of Symbolic Logic*, 40, 75–111. Reprinted in Robert L. Martin (ed.), *Recent Essays on Truth and the Liar Paradox*, pp. 237–288, Oxford: Clarendon Press. Page references to this reprinting.
32. Fitch, F. (1952). *Symbolic Logic: An Introduction*. NY: Ronald Press.
33. French, R. (2010). *Translational Embeddings in Modal Logic*. Ph.D. thesis, Monash University, Department of Philosophy.

34. Gentzen, G. (1934). Untersuchungen über das logische Schließen, I and II. *Mathematische Zeitschrift*, 39, 176–210, 405–431. English translation "Investigations into Logical Deduction" published in *American Philosophical Quarterly*, 1, 288–306 (1964), and 2, 204–218 (1965). Reprinted in M. E. Szabo (ed.) (1969) *The Collected Papers of Gerhard Gentzen*, North-Holland, Amsterdam, pp. 68–131. Page references to the APQ version.
35. Hale, B. (1993). Could there be a logic of attitudes? In J. Haldane and C. Wright (Eds.), *Reality, Representation, and Projection*, pp. 337–364. New York: Oxford UP.
36. Hazen, A. P. and F. J. Pelletier (2014). Gentzen and Jaskowski natural deduction: Fundamentally similar but importantly different. *Studia Logica*, 102, 1103–1142.
37. Hazen, A. P. and F. J. Pelletier (2018a). Second-order logic of paradox. *Notre Dame Journal of Formal Logic*, 59, 547-558.
38. Hazen, A. P. and F. J. Pelletier (2018b). Pecularities of some three- and four-valued second order logics. *Logica Universalis*, 12, 493–509.
39. Humberstone, L. (2011). *The Connectives*. Cambridge, MA: MIT Press.
40. Hyde, D. (1997). From heaps and gaps to heaps of gluts. *Mind*, 106, 641–660.
41. Hyde, D. and M. Colyvan (2008). Paraconsistent vagueness: Why not? *Australasian Journal of Logic*, 6, 107–121.
42. Jaskowski, S. (1934). On the rules of suppositions in formal logic. *Studia Logica*, 1, 5–32. Reprinted in S. McCall (1967), Polish Logic 1920–1939 Oxford UP, pp. 232–258.
43. Kamide, N. (2017). Paraconsistent double negations as classical and intuitionistic negations. *Studia Logica*, 105, 1167–1191.
44. Kleene, S. (1938). On a notation for ordinal numbers. *Journal of Symbolic Logic*, 3, 150–155.
45. Kleene, S. (1952). *Introduction to Metamathematics*. Amsterdam: North-Holland.
46. Kuhn, S. and B. Weatherson (2018). Notes on some ideas in Lloyd Humberstone's Philosophical Applications of Modal Logic. *Australasian Journal of Logic*, 15, 1–18.
47. Lambert, K. (1967). Free logics and the concept of existence. *Notre Dame Journal of Formal Logic*, 8, 133–144.
48. Łukasiewicz, J. (1920). On three-valued logic. *Ruch Filozoficny*, 5, 170–171. English translation in *J. Łukasiewicz Selected Works* (ed. L. Borkowski), Amsterdam: North-Holland, 1970, pp. 87–88.
49. Łukasiewicz, J. and A. Tarski (1930). Investigations into the sentential calculus. *Compètes rendus des séances de la Société et des Lettres de Varsovie*, 23, 30–50. English translation in A. Tarski *Logic, Semantics Metamathematics*, 2nd edition, Hackett: Indianapolis, 1983, pp. 38–59.
50. Monteiro, A. (1967). Construction des algebres de Łukasiewicz trivalentes dans les algèbres de Boole monadiques. *Mathematica Japonica*, 12, 1–23.
51. Nelson, D. (1949). Constructible falsity. *Journal of Symbolic Logic*, 14, 16–26.
52. Nelson, D. (1959). Negation and separation of concepts in constructive systems. In A. Heyting (Ed.), *Constructivity in Mathematics: Proceedings of the Colloquium Held at Amsterdam*, 1957, pp. 208–225. Amsterdam: North Holland.
53. Omori, H. and H. Wansing (2017). 40 years of FDE: An introductory overview. *Studia Logica*, 105, 1021–1049.
54. Pelletier, F. J. (1984). Six problems in translational equivalence. *Logique et Analyse*, 108, 423–434.
55. Pelletier, F. J. and A. Urquhart (2003). Synonymous logics. *Journal of Philosophical Logic*, 32, 259–285. See also the authors' "Synonymous Logics: A Correction", *JPL*, 37, 95–100, 2008.
56. Post, E. (1921). Introduction to a general theory of elementary propositions. *American Journal of Mathematics*, 43, 163–185. Reprinted in J. von Heijenoort (ed.), *From Frege to Gödel: A Source Book in Mathematical Logic*, 1979–1931, pp.265–293.
57. Přenosil, A. (2017). Cut elimination, identity elimination, and interpolation in super-Belnap logics. *Studia Logica*, 105, 1255–1289.
58. Priest, G. (1979). The logic of paradox. *Journal of Philosophical Logic*, 8, 219–241.
59. Priest, G. (2006). *In Contradiction: A Study of the Transconsistent*, 2nd Ed. Oxford: Oxford University Press.
60. Priest, G. (2018). Natural deduction systems for logics in the FDE family. This volume.
61. Priest, G. and R. Sylvan (Eds.) (1989). *Paraconsistent Logic: Essays on the Inconsistent*. Munich: Philosophia Verlag.
62. Rescher, N. (1966). *The Logic of Commands*. New York: Routledge and Kegan Paul.
63. Rescher, N. (1969). *Many-valued Logic*. New York: McGraw-Hill.
64. Ripley, D. (2011). Contradiction at the borders. In R. Nouwen, R. van Rooij, U. Sauerland, and H.-C. Schmitz (Eds.), *Vagueness in Communication*, pp. 169–188. Dordrecht: Springer.

65. Rivieccio, U. (2012). An infinity of super-Belnap logics. *Journal of Applied Non-Classical Logics*, 22, 319–335.
66. Rosser, J. B. and A. Turquette (1952). *Many-Valued Logics*. Amsterdam: North Holland.
67. Rozoner, L. (1989). On the interpretation of inconsistent theories. *Information Sciences*, 47, 243–266.
68. Schroeder, M. (2008). What is the Frege-Geach problem? *Philosophy Compass*, 3, 703–720.
69. Schütte, K. (1960). *Beweistheorie*. Berlin: Springer.
70. Schütte, K. (1977). *Proof Theory*. New York: Springer-Verlag. Extensively revised Second Edition of Schütte 1960, with an English translation by J. N. Crossley.
71. Searle, J. and D. Vanderveken (1985). *Foundations of Illocutionary Logic*. Cambridge, UK: Cambridge UP.
72. Shramko, Y. (2016). A modal translation for dual-intuitionistic logic. *Review of Symbolic Logic*, 9, 251–364.
73. Shramko, Y. and H. Wansing (2017). Truth values. In E. N. Zalta (Ed.), *The Stanford Encyclopedia of Philosophy* (Summer 2017 ed.). Metaphysics Research Lab, Stanford University. `http://plato.stanford.edu/entries/truth-values`
74. Słupecki, J. (1936). The full 3-valued propositional calculus. *Compètes rendus des séances de la Société et des Lettres de Varsovie, Classe III*, 29, 9–11. Translated in S. McCall (ed.) Polish Logic: 1920–1939, (Oxford) 1967, pp. 335–337.
75. Słupecki, J. (1939a). A criterion of completeness of many-valued systems of propositional calculus. *Compètes rendus des séances de la Société et des Lettres de Varsovie, Classe III*, 32, 102–110. English translation in Studia Logica, v. 30, pp. 153–157, 1972.
76. Słupecki, J. (1939b). Proof of the axiomatizability of full many-valued systems of propositional calculus. *Compètes rendus des séances de la Société et des Lettres de Varsovie, Classe III*, 32, 110–128.
77. Sutcliffe, G., F. J. Pelletier, and A. P. Hazen (2018, May). Making Belnap's "useful four-valued logic" useful. In K. Brawner and V. Rus (Eds.), *Proceedings of the 31st International FLAIRS Conference, Palo Alto*, pp. 116–121. AAAI Press.
78. Tedder, A. (2014). *Paraconsistent logic for dialethic arithmetics*. Master's thesis, University of Alberta, Philosophy Department, Edmonton, Alberta, Canada. Available at https://www.library.ualberta.ca/catalog/6796277
79. Tedder, A. (2015). Axioms for finite collapse models of arithmetic. *Review of Symbolic Logic*, 8, 529–539.
80. Urquhart, A. (1986). Basic many-valued logic. In D. Gabbay and F. Guenthner (Eds.), *Handbook of Philosophical Logic*, Vol. 2, pp. 249–295. Dordrecht: Reidel.
81. van Fraassen, B. (1966). Singular terms, truth-value gaps, and free logic. *Journal of Philosophy*, 63, 481–495.
82. van Fraassen, B. (1969). Presuppositions, supervaluations and free logic. In K. Lambert (Ed.), *The Logical Way of Doing Things*, pp. 67–92. New Haven: Yale UP.
83. Vanderveken, D. (1990). *Meaning and Speech Acts*. Cambridge, UK: Cambridge UP.
84. Vranas, P. (2008). New foundations for imperative logic, I: Logical connectives, consistency, and quantifiers. *Noûs*, 42, 529–572.
85. Wajsberg, M. (1931). Axiomatization of the 3-valued propositional calculus. *Compètes rendus des séances de la Société et des Lettres de Varsovie, Classe III*, 24, 126–148. Translated in S. McCall (ed.), *Polish Logic: 1920–1939*, pp. 264–284, 1967. Oxford: Oxford University Press.
86. Wansing, H. (2008). Constructive negation, implication, and co-implication. *Journal of Applied Non-Classical Logics*, 18, 341–364.
87. Wansing, H. (2012). A non-inferentialist, anti-realistic conception of logical truth and falsity. *Topoi*, 31, 93–100.
88. Weber, Z. (2010). Extensionality and restriction in naïve set theory. *Studia Logica*, 94, 87–104.
89. Wisniewski, A. (1995). *The Posing of Questions: Logical Foundations of Erotetic Inferences*. Dordrecht: Kluwer.
90. Wojcicki, R. (1984). *Lectures on Propositional Calculi*. Warsaw: Ossolineum.

## Appendix I: Deductive Systems

One direction that has been taken in the literature is the development of "super-Belnap" logics, that is, extensions of FDE. See Rivieccio (2012) and Přenosil (2017) for examples and elaboration. One development in this tradition is the generation of proof systems for such logics, which have pretty much been in the tradition of Gentzen-style sequent calculi (Gentzen, 1934), usually with multiple conclusion sequents although some single conclusion systems have also been given. Sometimes this is paired with a comparison of a Hilbert-style axiomatic development of the same logics, and sometimes a natural deduction system of the Gentzen type is presented. We think that various natural deduction systems of the Jaśkowski style (Jaśkowski, 1934) are preferable for a number of "philosophical" reasons (many recounted in Hazen and Pelletier, 2014) as well as for pedagogical clarity, and in particular we will present a Fitch-style version (Fitch, 1952) of the Jaśkowski method.

Tedder (2014) contains a Hilbert-style axiomatic system, and (more interestingly) a (multiple succedent) sequent calculus for $LP^{\rightarrow}$. (Tedder, 2015, presents only the Hilbert system.) The sequent calculus closely follows Gentzen's (1934) system LP for classical logic, with the following changes:

(i) Gentzen's rules for negation ("change the side and change the sign") are dropped,
(ii) double negation rules are added, allowing a sequent to be followed by one like it except that one of its formulas–on either side–is doubly negated,
(iii) negative rules for conjunction and disjunction (and, analogously, for quantifiers) are added, allowing the insertion of a negated conjunction or disjunction under the same conditions as allow the insertion of its De Morgan equivalent disjunction or conjunction,
(iv) negative conditional rules are added, allowing $\neg(\varphi \rightarrow \psi)$ to be inserted under the same conditions as $(\varphi \wedge \neg\psi)$, and
(v) Gentzen's identity axioms, $\varphi \vdash \varphi$, are supplemented with "Gap excluding" axioms of the form $\vdash \varphi, \neg\varphi$.

Cut elimination is proven by a straightforward adaptation of Gentzen's method.

Dropping the Gap exclusion axioms from this system yields a sequent calculus for $FDE^{\rightarrow}$. Replacing them with "Glut excluding axioms" of the form $\varphi, \neg\varphi \vdash$ gives one for $K3^{\rightarrow}$, and replacing them instead with Mingle axioms of the form $\varphi, \neg\varphi \vdash \psi, \neg\psi$ gives one for $M^{\rightarrow}$.

A natural deduction system, either in the style of Gentzen's NK or in the Fitch-style presentation of many American textbooks, for any of these logics will include

(i) the standard Introduction and Elimination rules for $\wedge, \vee, \rightarrow$ (and for the quantifiers in a First Order system),
(ii) Double Negation Introduction and Elimination rules, by which a formula and its double negation may each be inferred from the other,
(iii) Negative Introduction and Elimination rules for $\wedge$ and $\vee$ (and, analogously for the quantifiers), as in Fitch (1952) enforcing the interdeducibility of negated conjunctions and disjunctions with their De Morgan equivalent disjunctions and conjunctions,
(iv) Negative Introduction and Elimination rules for $\rightarrow$, enforcing the interdeducibility of $\neg(\varphi \rightarrow \psi)$ with $(\varphi \wedge \neg\psi)$.

As is well-known, the standard Introduction and Elimination rules for $\rightarrow$ give only the Intuitionistic logic of the conditional, so these have to be supplemented with some classicizing postulate to give the full classical logic of the positive connectives. Addition of Peirce's Law, $(((\varphi \rightarrow \psi) \rightarrow \varphi) \rightarrow \varphi)$ as an axiom scheme is one conventional way of doing this, but an alternative axiom scheme that we call Dilemma, $(\varphi \vee (\varphi \rightarrow \psi))$, seems a good deal easier to work with, and can be converted into a moderately elegant natural deduction rule:

(v) A formula, $\chi$, may be asserted if it is derivable both from the hypothesis $\varphi$ and from the hypothesis $(\varphi \rightarrow \psi)$.

These rules suffice for the propositional logic $\mathrm{FDE}^{\rightarrow}$. A First Order system would have to supplement the standard and negative Introduction and Elimination Rules for the quantifiers with something to guarantee the "constant domain" inference

$$\forall x(\varphi \vee \Psi(x)) \vdash (\varphi \vee \forall x(\Psi(x))$$

(see Fitch, 1952, §21.31.) Systems for the other logics are obtained by adding rules to this basic system:

1) Ex falso quodlibet ("explosion"), for $\mathrm{K3}^{\rightarrow}$;
2) Excluded middle: $\chi$ may be asserted if it is derivable both from the hypothesis $\varphi$ and from the hypothesis $\neg\varphi$, for $\mathrm{LP}^{\rightarrow}$;
3) Mingle: $(\psi \vee \neg\psi)$ may be inferred from $(\varphi \wedge \neg\varphi)$, for $\mathrm{M}^{\rightarrow}$.

More visually: Natural deduction rules for $\mathrm{FDE}^{\rightarrow}$will be double negation and both the positive and negative IntElim rules for $\wedge$ and $\vee$. Additionally, there is a series of rules for our $\rightarrow$ operator.

Double Negation:

| | | |
|---|---|---|
| 1 | $A$ | |
| 2 | $\neg\neg A$ | $\neg\neg$ Int |

| | | |
|---|---|---|
| 1 | $\neg\neg A$ | |
| 2 | $A$ | $\neg\neg$ Elim |

Standard rules for lattice connectives:

| | | |
|---|---|---|
| 1 | $A \wedge B$ | |
| 2 | $A$ | $\wedge$ Elim |

| | | |
|---|---|---|
| 1 | $A \wedge B$ | |
| 2 | $B$ | $\wedge$ Elim |

| | | |
|---|---|---|
| 1 | $A$ | |
| 2 | $B$ | |
| 3 | $A \wedge B$ | $\wedge$ Int |

$$
\begin{array}{r|l l}
1 & A \vee B & \\
\hline
2 & \quad | \; A & \\
3 & \quad | \; \ldots & \\
4 & \quad | \; C & \\
5 & \quad | \; B & \\
6 & \quad | \; \ldots & \\
7 & \quad | \; C & \\
8 & C & \vee \text{ Elim}
\end{array}
\qquad
\begin{array}{r|l l}
1 & A & \\
\hline
2 & A \vee B & \vee \text{ Int}
\end{array}
\qquad
\begin{array}{r|l l}
1 & B & \\
\hline
2 & A \vee B & \vee \text{ Int}
\end{array}
$$

Negative rules for the lattice connectives[23]:

$$
\begin{array}{r|l l}
1 & \neg(A \vee B) & \\
\hline
2 & \neg A & \neg\vee \text{ Elim}
\end{array}
\qquad
\begin{array}{r|l l}
1 & \neg(A \vee B) & \\
\hline
2 & \neg B & \neg\vee \text{ Elim}
\end{array}
$$

$$
\begin{array}{r|l l}
1 & \neg A & \\
2 & \neg B & \\
\hline
3 & \neg(A \vee B) & \neg\vee \text{ Int}
\end{array}
$$

$$
\begin{array}{r|l l}
1 & \neg A & \\
\hline
2 & \neg(A \wedge B) & \neg\wedge \text{ Int}
\end{array}
\qquad
\begin{array}{r|l l}
1 & \neg B & \\
\hline
2 & \neg(A \wedge B) & \neg\wedge \text{ Int}
\end{array}
$$

$$
\begin{array}{r|l l}
1 & \neg(A \wedge B) & \\
\hline
2 & \quad | \; \neg A & \\
3 & \quad | \; \ldots & \\
4 & \quad | \; C & \\
5 & \quad | \; \neg B & \\
6 & \quad | \; \ldots & \\
7 & \quad | \; C & \\
8 & C & \neg\wedge \text{ Elim}
\end{array}
$$

Rules for Conditional[24]

[23] In the context of the positive rules, these negative rules are equivalent to Fitch's (1952) original negation IntElim rules.

[24] The Dilemma rule, if added to, say, Intuitionistic Logic, would yield a formulation of full classical logic. It does not collapse $\text{FDE}^{\rightarrow}$ into classical logic because the usual ¬-Introduction rule (*reductio*) of Intuitionistic Logic is absent.

| | | |
|---|---|---|
| 1 | $(A \to B)$ | |
| 2 | $A$ | |
| 3 | $B$ | $\to$ Elim |

| | | |
|---|---|---|
| 1 | $A$ (hypothesis) | |
| 2 | $\ldots$ | |
| 3 | $B$ | |
| 4 | $A \to B$ | $\to$ Int |

| | | |
|---|---|---|
| 1 | $\neg(A \to B)$ | |
| 2 | $A$ | $\neg \to$ Elim |

| | | |
|---|---|---|
| 1 | $\neg(A \to B)$ | |
| 2 | $\neg B$ | $\neg \to$ Elim |

| | | |
|---|---|---|
| 1 | $A$ (hypothesis) | |
| 2 | $\ldots$ | |
| 3 | $B$ | |
| 4 | $A \to C$ (hypothesis) | |
| 5 | $\ldots$ | |
| 6 | $B$ | |
| 7 | $B$ | Dilemma |

| | | |
|---|---|---|
| 1 | $A$ | |
| 2 | $\neg B$ | |
| 3 | $\neg(A \to B)$ | $\neg \to$ Int |

## Appendix II: Soundness and Completeness

The natural deduction system is provably sound and complete, where soundness is taken to mean that if all the premisses of a derivation have designated values on an assignment, the conclusion will as well.

Soundness can be verified in the usual way, arguing by induction on the size of the derivation after establishing that each rule is sound. For rules in which a conclusion is inferred directly from one or two premisses, this is immediate, by inspection of the truth tables. A rule in which a hypothesis is discharged (that is, in the terminology of Fitch-style natural deduction, a rule involving one or two *subordinate proofs*) is considered sound just in case, if all the undischarged hypotheses above the conclusion of the rule (in Fitch-style: the formulas *reiterated* into subordinate proofs) have designated values, the conclusion also has a designated value. It is easy to see that the subproof rules of the system will be sound *provided that* the reasoning within the subordinate proofs is sound. The full soundness proof, then, will take the form of a double induction, on the length, and on the depth of nesting of subordinate proofs within, a proof. The overall strategy is perfectly standard for soundness proofs of natural deduction systems.

Completeness can be proven by a variant of Henkin's method, similar to that used in Priest (2018). We desire to show that, if a formula A is not derivable from a set of premisses $\Gamma$, then there is an assignment on which A takes an undesignated value but every member of $\Gamma$ is designated. In a Henkin-style proof this is done in two stages. In the first, it is shown that $\Gamma$ can be extended to an eligible set $\Gamma^*$ which still does not (syntactically) imply A, where a set of formulas is said to be eligible if it has some of the formal characteristics of the set of formulas taking designated values on some assignment. In the second it is shown that the eligible set is actually elected: an assignment is defined on which all and only the members of the set take designated values.

In applying Henkin's method to a classical system, eligible sets are simply *complete theories*, a.k.a. *maximal consistent sets* of formulas. In logics tolerating contradictions, consistency is obviously not a requirement, and in logics tolerating truth value "gaps" maximality is also not to be hoped for! An appropriate notion of eligibility for our purposes counts a set of formulas as eligible if and only if (i) it is deductively closed (and so, in particular, contains a conjunction if and only if it contains both conjuncts, and contains a disjunction if it contains either disjunct), and (ii) contains at least one of the disjuncts of each disjunction it contains. Given A not derivable from $\Gamma$, it is readily seen that a set maximal with respect to the property of containing $\Gamma$ but not implying A will be eligible in this sense, and the existence of such a maximal set follows from standard set-theoretic considerations (Teichmüller-Tukey lemma). (In general these maximal sets will not be the only eligible supersets of $\Gamma$ not implying A, and an alternative proof adding formulas to $\Gamma$ only if they are required by clause (ii) may yield a smaller eligible set.)

Given an eligible set $\Gamma^*$, we define an assignment to propositional variables by setting

- $v(p) = \mathbf{T}$ iff $p$ is a member but $\neg p$ is not a member of $\Gamma^*$,
- $v(p) = \mathbf{B}$ iff $p$ and $\neg p$ both belong to $\Gamma^*$,
- $v(p) = \mathbf{N}$ iff neither $p$ nor $\neg p$ belongs to $\Gamma^*$, and
- $v(p) = \mathbf{F}$ iff $\neg p$ but not $p$ is a member of $\Gamma^*$.

(Note that a variable takes a designated value if and only if it belongs to $\Gamma^*$.) It remains to verify (by induction on formula complexity) that arbitrary formulas take values on this assignment under the same conditions of their, and their negations', membership in $\Gamma^*$. None of the cases are hard; those not immediately obvious usually become obvious when one remembers that formulas of the form $A \vee (A \rightarrow B)$ are provable in the system.

In a bit more detail:

**Soundness:**

The notion of validity we want is:

*Definition* 1. **Validity:** A system of rules is sound for one of our logics if and only if: If all the assumptions[25] of a (possibly subordinate) proof have designated values, then every formula occurring diretly as an item of that derivation (as opposed to occurring in a subordinated derivation subordinate to it) will also have a designated value.

One could easily check that all the rules are classically valid, and conclude that *of course* the FDE$^{\rightarrow}$system is sound. But perhaps some of the *negative* rules are worth checking.

A quick check of the $\neg\wedge$-Int rules against this truth table makes it clear that these rules are sound. The presence of **B** means that $\neg\wedge$-Elim takes a bit longer, but is clearly correct also.

In full formal detail, the soundness proof (as is usual for a soundness proof of Fitch-style natural deduction systems) is a double induction on

1. Length: the number of non-assumption formula items in the derivation.

[25] Where formulas reiterated into a subproof are counted among that subproof's assumptions.

| ∧ | T | B | N | F |
|---|---|---|---|---|
| T | T | B | N | F |
| B | B | B | F | F |
| N | N | F | N | F |
| F | F | F | F | F |

**Table 15** ∧ truth-table

2. Depth: the depth of nesting of subproofs.

The induction step of the Depth induction goes:

> Suppose A comes by a subproof-using rule. (Now, a number of cases.) By hypothesis of induction on Length, anything reiterated into the subproof is ok (since it would occur in the same proof (or one to which it is subordinate) above this step). By hypothesis of induction on Depth [we call this the "key step"], the subproof is sound. So, if the hypothesis of the subproof is ok, so is its "active" last item. So – now verifying each case – A is ok.
>
> This gives us the cases for subproof-involving rules in the inductive step of the Length induction for derivations of depth $\leq n$. So derivations at this depth are sound. In a derivation of depth $n + 1$, every subproof is of depth $\leq n$, so the "key step" is guaranteed.

**Completeness Construction:**

Define a set of formulas to be *saturated* if and only if

1. It is *deductively closed*, and so automatically contains
   a. A conjunction if and only if it contains both conjuncts, and
   b. A disjunction if it contains at least one disjunct
2. It contains at least one disjunct of each of its disjunctions

Note that for *classical* logic, in which by deductive closure $(A \vee \neg A)$ belongs to every saturated set, the consistent saturated sets are just the maximal consistent ones: saturated sets are the generalization for non-classical logics of the "maximal consistent sets" familiar from Henkin proofs for classic(-ally based) logic(s).

What we have to prove is:

*Lemma* 1. For any set of formulas, $\Gamma$, and for any formula, $A$, not derivable from $\Gamma$, there is a *saturated* superset of $\Gamma$, $\Gamma^*$, not containing $A$. (The term *saturated set* is often defined to require consistency, but this has been in the context of discussing logics which, though allowing truth-value gaps, don't allow for gluts.)

*Proof.* (By a version of the standard Lindenbaum construction.)

For classical logic, where we want *maximal* consistent sets anyway, it is normal to consider *all* the formulas of the language (in some order), tossing each one in if its addition doesn't permit the derivation of THE BAD THING. But this can tend to *undesirably stuffed* sets of formulas! For example, the *maximal* consistent sets of intuitionistic propositional logic are *classically* maximal! So we prefer a more cautious addition of formulas.

*Definition* 2. An ordinally indexed series of sets of formulas:

> Let $\Gamma_0 = \Gamma$
>
> Assume some fixed well-ordering of the formulas of the language:
>
> For *odd successors*, $\alpha$,

- If $\Gamma_\alpha$ is saturated, stop
- if not, pick the *first* (in the assumed ordering) disjunction in $\Gamma_\alpha$ for which neither disjunct is in $\Gamma_\alpha$. By the $\vee$-Elim rule, if $\Gamma_\alpha$ does not imply $A$, at least one of these disjuncts can be added to $\Gamma_\alpha$ without permitting the derivation of A. Let $\Gamma_{\alpha+1}$ be the result of adding the first such disjunct (or the only one, if only one can be added) to $\Gamma_\alpha$.

For *even successors* (and for 0), let $\Gamma_{\alpha+1}$ be the deductive closure of $\Gamma_\alpha$.

For *limit ordinals* $\lambda$, let $\Gamma_\lambda = \bigcup_{\alpha<\lambda}(\Gamma_\alpha)$.

By cardinality considerations (there are more ordinals than formulas, so eventually we will run out of formulas to add), this sequence must reach a fixed point, $\Gamma_\Omega$. By the usual Henkin arguments, $\Gamma_\Omega$ is saturated and does not imply $A$.

Now, for *atomic* $p$, define

- $p$ has the value **T** iff $p$ does and $\neg p$ does not belong to $\Gamma_\Omega$
- $p$ has the value **B** iff both $p$ and $\neg p$ belong to $\Gamma_\Omega$
- $p$ has the value **N** iff neither $p$ nor $\neg p$ belong to $\Gamma_\Omega$
- $p$ has value **F** iff $\neg p$ but not $p$ belongs to $\Gamma_\Omega$

We now in a position to *verify* that for arbitrary formulas $A$, the same correlation holds between value and status with respect to membership in $\Gamma_\Omega$

## *Completeness Verification:*

*Theorem* 6. Given a *saturated* set S, if we define, for atomic $p$:
$p \rightsquigarrow$**T** iff $p \in S$ and $\neg p \notin S$
$p \rightsquigarrow$**B** iff $p \in S$ and $\neg p \in S$
$p \rightsquigarrow$**N** iff $p \notin S$ and $\neg p \notin S$
$p \rightsquigarrow$**F** iff $p \notin S$ and $\neg p \in S$
we will have the same coincidence of values and S-status for all formulas

*Proof.* By induction on formula structures. We omit proofs for the $\neg, \vee, \wedge$ connectives, which are obvious. So we consider $\rightarrow$:
**The way up** Suppose equivalence holds for $A$ and $B$, we show it holds for $A \rightarrow B$.
There are 16 combinations of truth values for $A, B$. For each, by hypothesis of induction, assume the right S-membership status.

- **Cases**: Left column. $B$ is **T**, so is in S. So, $A \rightarrow B$ is in S. $\neg B$ is *not* in S, so by simple $\neg\rightarrow$-Elim, $\neg(A \rightarrow B)$ can't be either.
- **Cases**: Bottom two rows. A is either **F** or **N**, so by hypothesis of induction, A is *not* in S. By Dilemma, $A \vee (A \rightarrow B)$ is in S, so by saturation $(A \rightarrow B) \in S$. In the other direction, simple $\neg\rightarrow$-Elim rule would get from $\neg(A \rightarrow B)$ to $A$, so $\neg(A \rightarrow B)$ can't be in S.
- **Cases**: Top row: $A$ has value **T**, so $A \in S, \neg A \notin S$
  - **subcase**: $B$ has value **B**, so both $B \in S$ and $\neg B \in S$. Since $B \in S$, $\rightarrow$-Int gives $(A \rightarrow B) \in S$. Since $A \in S$ and $\neg B \in S$, the simple $\neg\rightarrow$-Int rule gives $\neg(A \rightarrow B) \in S$.

- **subcase**: $B$ has value **N**, so neither $B \in S$ nor $\neg B \in S$. If $(A \to B)$ were in S, $\to$-Elim would put $B$ in, so $(A \to B) \notin S$. If $\neg(A \to B)$ were in S, $\neg \to$-Elim rule would put $\neg B$ in S, so $\neg(A \to B) \notin S$.
- **subcase**: $B$ has value **F**, so $\neg B \in S$ and $B \notin S$. If $(A \to B)$ were in S, $\to$-Elim would put $B$ in, so $(A \to B) \notin S$. On the other hand, since $A$ and $\neg B$ are both available, $\neg \to$-Int gives $\neg(A \to B) \in S$.

- **Cases**: Second row: $A$ has value **B**, so both $A \in S$ and $\neg A \in S$
  - **subcase**: $B$ has value **B**, so by hypothesis of induction, $B$ and $\neg B$ are both in S. Since $B$ is available, $\to$-Int gets $(A \to B) \in S$. Since $A$ and $\neg B$ are both available, $\neg \to$-Int puts $\neg(A \to B) \in S$.
  - **subcase**: $B$ has value **N**, so neither $B$ nor $\neg B$ is in S. Since $A$ is available, if $(A \to B)$ were in S, $\to$-Elim would put $B$ in S. If $\neg(A \to B)$ were in S, simple $\neg \to$-Elim would put $\neg B$ in S. So neither $(A \to B)$ nor $\neg(A \to B$ is in S.
  - **subcase**: $B$ has value **F**, so $\neg B \in S$ and $B \notin S$. Since $A$ and $\neg B$ are available, $\neg \to$-Int puts $\neg(A \to B) \in S$. Since $A$ is available, if $A \to B$ were in S, $B$ would be also.

Normally, in a Henkin-style completeness proof, one defines truth for atoms as membership in the saturated set, and the *verification* stage checks that complex formulas are true if and only if they are members. This requires *two* arguments: one that *if* a formula is true it is a member, and one that *if* a formula belongs to the set, then it is true.

Because the semantic clauses for **T**, **B**, **N**, **F** are more complex, these two parts are now intermingled. If a formula has a certain value, then both the membership *and non-membership* of the formula and its negation (one for each) in the set have to be checked. Thus **the way up** (verifying that formulas with certain values have the right membership statuses) already in effect includes **the way back down** (verifying that formulas with certain membership statuses have the right values).

# FDE as a Base for Constructive Logic

Andreas Kapsner

**Abstract** In this paper I explore whether First Degree Entailment (FDE) might serve as a suitable base for a constructive logic and a constructive account of language roughly along the lines of Michael Dummett's project. I present three interpretation of FDE's truth values that hold some promise in this setting. The paper also includes an extensive critical discussion of work by Shramko, Dunn and Takenaka on constructive readings of FDE-like values and the possible need to further extend the space of truth values.

**Keywords** FDE • First Degree Entailment • Constructive Logic • Intuitionistic Logic • Michael Dummett • Truth Value Gluts • Tri-lattice SIXTEEN

## 1 Introduction

In this paper, I want to explore the question whether First Degree Entailment (FDE) might serve as a suitable base for a constructive logic.[1] There are several ways to understand what a "constructive logic" might be. What I am concerned with here is a logic that can handle the inferential aspects of a broader constructive account of language, roughly in the ballpark of what Michael Dummett had in mind. We will see that we will have to deviate quite a bit from the specifics of Dummett's account, but I will try to make the best case for each of these deviations.

In the course of the paper, I will see whether the four-valued account of FDE can find some intuitive justification in a constructive account of language. I will begin by giving a very short overview of Dummett's philosophy of language and how it relates to logic. I will then introduce FDE and go quickly over its original interpretation. I will then look closer at how to make sense of gaps and gluts (in other words, of values $N$ and $B$) in a constructive setting, even though Dummett himself was firmly against both. There are three ways in which I can see a general constructive strategy succeeding at this, and I will try to make the best case for each one of them.

In the second part of this paper, I will look in quite some detail at a paper by Yaroslav Shramko, J. Michael Dunn and Tatsutoshi Takenaka, *The Trilattice of Constructive Truth Values* ([9]). They supply the first and most detailed discussion of a constructive reading of FDE-like values, and therefore their paper is most important for my topic. However, the main thrust of their argument is that the truth value space should be even further expanded, going beyond the already relatively high number of four truth values that FDE has. I will

Andreas Kapsner
Faculty of Philosophy, Munich Center for Mathematical Philosophy, Munich, Germany
e-mail: Andreas.Kapsner@lrz.uni-muenchen.de

[1] I would like to thank the editors and two referees for helpful comments. Moreover, Yaroslav Shramko and J. Michael Dunn have been very helpful and kind in discussing my critical comments about their work that make up much of he second part of this paper.



H. Omori, H. Wansing (eds.), *New Essays on Belnap-Dunn Logic*,
Synthese Library 418, https://doi.org/10.1007/978-3-030-31136-0_12

critically discuss their ideas and then see how they relate to my own ideas expressed in the first section of the paper.

It turned out that a full answer to the question whether FDE can serve as a *logic* for constructivism was too much to cover in one paper. I will only get to the very first stage of this project here, namely an account of how to interpret the values. The full argument will have to be finished in future work, which I am currently drafting. That future work will have two main themes: First, the question of which of these values should be designated, for there might be reasons to make a different choice from FDE's original choice, $T$ and $B$. Second, I plan to discuss how negations, conjunctions and disjunctions receive their values in FDE, and how well that fits with the intended constructive interpretation.

Many of the original ideas in this paper overlap with material from my book *Logics and Falsifications*, [7], as well. Where they do, I will strive to make the idea clear enough to be fully understood here; the reader who wants to furthermore be compelled (or even convinced) might have to turn to the book in these cases. Also, my summary of Dummett's views in the next section and thereafter will draw on more extensive discussions in the book that I have cut short as far as possible.[2]

## 2 Constructive Semantics

A widely accepted way of specifying the meaning of statements is to equate them with their truth conditions. To understand a statement is to apprehend under which circumstances it would be true. Michael Dummett was, for various reasons, unconvinced by the traditional account of meaning in terms of truth conditions, at least in so far as these might be beyond our epistemic grasp. What does it mean to say that "to understand a statement is to know under which circumstances it would be true," if these circumstances are such that we couldn't, even in principle, know that they obtain? He thought that this was implausible for a host of connected reasons that have to do with how we acquire language and how we can show that we understand our own language. He thought that a constructive account of language was much more promising.

Dummett's prime example of a constructive semantic theory is the account of mathematical statements in Brouwer's intuitionism: Such a statement is correctly assertible/true only if it is provable, and we cannot assume the correct assertibility/truth of a theorem if we cannot give a proof, or at least a method of constructing such a proof.

In general, the constructivist holds that a statement cannot be true unless we are, at least in principle, able to come to know of its truth. In mathematics, "to come to know of its truth" means to find a proof of a statement; in empirical contexts, it means to verify the statement. Put another way, a statement is true iff it is verifiable.

There are many subtleties in this account and in its exegesis, most of which I cannot touch on (such as Dummett's going back and forth between spelling out truth in terms of verifiability and his trying to completely avoid the concept of truth in favor of the notion of correct assertibility; the exact type of modality in "is verifi*able*"; etc, much of which I discuss in [7]). For this piece, the following points are central (and they, too, receive much more detailed discussion in [7]):

---

[2] I will only quote directly from one of his pieces, one which I had not read at the time of writing the book, but which presents some pertinent ideas very clearly.

1. In some of his most prominent papers, Dummett argued that the notion of verification (and verifiability) cannot be the only basic building block of a semantic theory. Instead, it would be necessary to consider the notions of *falsification* and *falsifiability*, as well. Verifications and falsifications make up the most central concepts, though more than once he changed his mindwhich should ultimately be the one that decides which assertions are to be considered correct (i.e., the verifiable ones or the non-falsifiables).
2. For Dummett, verifications and falsifications are *certain* affairs, in quite the way proofs in mathematics are usually thought to be beyond any possible doubt. A faulty verification cannot exist, only something we maybe mistook for a verification, and the same goes for falsifications.

## 3 First Degree Entailment

In this section I will introduce FDE and its main interpretation in terms of pieces of information. I will move swiftly, because the story is well known. In the context of this collection in particular, it would be especially tedious to repeat it in too much detail. I will then turn to the question how that story can be modified to serve a constructive cause, as laid out in the last section.

Nuel Belnap introduced the four-valued semantics of FDE in two influential papers in the 1970s.[3] One is entitled "A useful four-valued logic", the other "How a computer should think", both reprinted in this issue, and its interpretation was further developed in the papers by J. Michael Dunn that are likewise reprinted in this issue. The logic introduces two new truth values in addition to the classical $T$ (True) and $F$ (False). These values are called $B$ (Both) and $N$ (Neither).

Algebraically, these four values can be arranged in the following bi-lattice, with conjunction as the meet, disjunction as the join, and negation as an operator that flips $T$ and $F$ but is a fixed-point operator for $B$ and $N$ (the second lattice structure is induced by the information order, displayed horizontally).

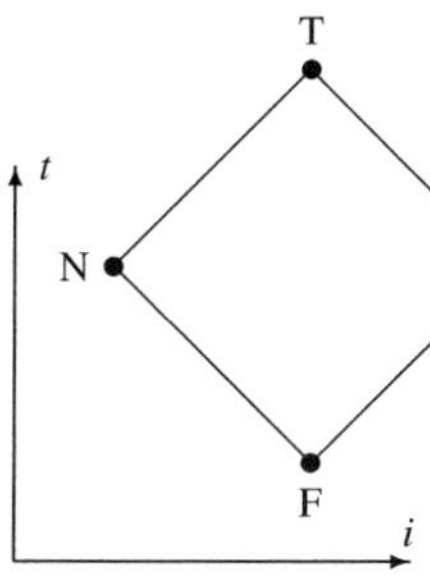


**Fig. 1** Bi-lattice *FOUR*

[3] As one of the reviewers rightly points out, this is not the first occurrence of FDE as a logic, which was first described as the first degree fragment of the relevant logic $R$ (hence the name).

What we are concerned with in this paper is an intuitive interpretation of the semantics of FDE that makes it suitable for the constructivist enterprise. To this end, the best starting point is in the interpretation introduced in the second paper mentioned above, the computer / database interpretation.

In this interpretation, Belnap thinks of the valuations as recording what information a computer has received about different statements. The computer is given input in the form of statements that are labeled as true or false. As many people are supposed to be building up the database of the computer, it is not impossible that one person might enter a statement as true, while another enters it as false. We then end up with four possibilities for each statement:

N : The computer has received no information pertaining to the statement.
F : The computer has received the information that the statement is false.
T : The computer has received the information that the statement is true.
B : The computer has received the information that the statement is true and the information that it is false.

The job of the computer now is to compute the values of complex statements and draw suitable inferences.

In [7], I proposed that a closely related example is especially useful to think about, although it is not Belnap's example. Instead of a computer, think about a criminal trial. Let us suppose that only testimonial evidence is available, then the transition is rather straightforward; just substitute "court" for "computer".

In either case, it is not clear that these four values need to have much to do with any substantial notion of truth at all. The computer might have been fed false data, and the court might have been lied to. Belnap sometimes calls his semantic values "epistemic" values and marks the distinction between them and what he calls "ontological" truth.

In the upcoming discussion of these matters, it is sometimes convenient to relabel the four values in a manner suggested by Dunn. The four corners will be furnished by members of the power set of the two basic semantic values $t$ and $f$. These are known as"told-truth" values, "told-true", and "told-false." $N$ corresponds to the empty set $\{\}$, $F$ to $\{f\}$, $T$ to $\{t\}$, and $B$ to $\{t, f\}$.

The idea is not so much that there are again four semantic values that merely have different names. Rather, this way of looking at the logic suggests that there are only two basic semantic values, just like in classical logic (modulo the "told-" prefix). However, valuations are not, as in classical semantics, total functions, but rather relations. That is, statements will not be assigned one and only one value, but may receive none, one or two of the values on offer (told-true and told-false).

Whether the semantics is a four-valued functional one or a two-valued relational one does not make too much of a difference when it comes to formal properties. The consequence relation that will be induced will be the same one, provided the corresponding values are chosen to be designated.

## 4 Three Constructive Readings of the Values

Now, what could a constructive interpretation of the four values be? In this section, I will explore three distinct answers to this question, starting right where we left off in the last section.

## *4.1 Told Truth as Constructive Truth*

Well, then, is there something in the told-truth / told-falsity interpretation that might interest the constructivist? It seems that it is a far cry from the ideas of *verifications* and *falsifications* as outlined above. Talk is cheap, verifications, at least some of them, are costly.

However, Dummett observed that we have a natural tendency to take what others tell us to indeed be true, and, more to the point, that we are perfectly justified in this tendency. He writes:

> We need no particular reason to take things to be as others inform us that they are, save when we have some weaker contrary ground for not so taking them; but we always need a particular ground for declining to take them to be as we are told that they are. ([2] p.265)

What he is saying is that we can, and indeed that we should, take others at their words, unless we have a good reason to doubt them. He goes on:

> To view the matter otherwise is to subvert the whole institution of language: to subvert it in something of the way that the liar does, but more far-reachingly. If we subvert the institution of language, we dissolve our own being. ([2] p. 266)

This strong conclusion is grounded in a reflection on how children learn their first language and thereby come to occupy their place in an environment of social animals. Such thoughts on language acquisition play a central part in his philosophy of language.

> We acquired our language from others, and could only have done so; for a language is a social institution, and cannot exist save as a social institution. It is not a rule of etiquette, or a device for saving time, that we should accept what others tell us: it is fundamental to the entire institution of language. ([2] p. 266)

> In this way, a disposition to accept what others tell one is central and fundamental to acquiring language at all: unless one does so, one cannot be said to understand language. ([2] p.267)

Given all that, it might indeed make sense to start our semantics with told-truth values, and thereby to build the tendency to take others by their words right into the semantics. No matter whether that tendency is innate or something that has to be acquired along the way, we cannot be said to understand our first language unless we exhibit it.

Fitting this idea into the larger picture of Dummett's account, it seems plausible to take the truth values to be, not quite "Told-Truth"-values, but rather "Told-Constructive-Truth"-values. To accept something one hears would be to accept that it is verifiable, not epistemically unconstrainedly true.

What we would end up with would plausibly lead us to something like FDE, where the value $B$ stands for situations that might lead to a later development in the acquisition of language:

> As one's mastery of language deepens, one learns to curb this fundamental disposition in certain cases. One learns the various possibilities of error on the part of others; one learns also how language may be employed dishonestly. One recognises then that one may sometimes be justified in not accepting what one is told: one recognises also that, in some cases, a special reason is required for accepting it. But the foundation stone cannot actually be removed without causing the entire edifice to collapse. The institution of language, and, with it, the existence of human society our membership in which goes to constitute us as human beings and thus as what each of us is, rest upon certain fundamental assumptions: and one of these is that knowledge is transmitted by means of language. To make this assumption is to treat accepting what we are told, without the need for any further special reason for accepting it, as the normal case. About this, there is a strong impulse to say: we cannot but do so. ([2] p. 267)

It is when we are confronted with sources that contradict each other that we learn that we cannot always follow our natural tendency to believe what is said. But this does not mean that we lose this tendency, nor that we *should* lose it.

There might be a way, then, to tell the constructive story from the point of view of told-constructive-truth. However, it would certainly take a longer breath than I have time to take here to explore this fully in order to fit it to what Dummett wrote in other publications in a plausible way, and I am by no means convinced that this can be achieved at all.

The proposal also does not fit too easily with the account that I have given in [7], where much of the emphasis is on analysing in which situations it can be correct or incorrect to assert statements that take the value N. In the understanding of the book, these are statements that are neither presently verifiable nor presently falsifiable, such as future tense statements. On the other hand, in the told-truth scenario, there seem to be only *propositions* that can take value N, but no asserted statements that do so. Just by asserting it, one elevates the proposition to value $T$ (or to $B$, in case it took value $F$ before).

I don't know what a fully formed constructive view along told-truth lines would look like, but it would have to deviate considerably from the view I have a much clearer picture of, the one I developed in the book. Therefore, let me leave this path to be further explored at another time, and get to an interpretation of FDE's values that is closer to what Dummett was aiming at as well as closer to what I have laid out before.

## 4.2 Verifiable, Falsifiable, Both and Neither

So, let us now move on to an interpretation that is as close as possible to Dummett's own way of putting things, as well as my analysis of it in [7]. Remembering that the constructivist wants to rephrase semantics in terms of verifications and falsifications, a natural move would be to directly hook up the semantic values $T$ and $F$ to "verifiable" and "falsifiable", respectively.

But what about $N$ and $B$?

The analogy to the informational interpretation of FDE would suggest that $N$ is a value taken by statements that are neither verifiable nor falsifiable, and that $B$ is a value that attaches to statements that are both verified and falsified. (In terms of the two-valued relational semantics, what we are doing is to hook up value $t$ to "verifiable" and value $f$ to "falsifiable"). Is there conceptual space for these options?

Dummett thought not. He explicitly rejected both verification/falsification gaps and gluts alike. In this section, I will look at his arguments and suggest possibilities to avoid their conclusions in turn, starting with verification/falsification gaps.

### 4.2.1 Constructive Gaps

On the face of it, it would seem that a gap between verifiability and falsifiablity should be quite congenial to the constructivists. Isn't their whole point one of epistemic humility, of recognizing the limitations of our cognitive capacities? What better way to make this admission than by saying that there are (or even that there just might be) statements that are neither verifiable nor falsifiable?

Indeed, I have argued that there is nothing wrong with allowing gaps into a constructive semantics in [7]. However, as I just noted, this goes against what Dummett himself thought. He objected to gaps for two tenuously related reasons: First, his preferred logic, intuitionistic logic, can't easily be interpreted to be a gappy logic. Though the law of excluded middle, $A \vee \neg A$, is not provable in intuitionistic logic, one can prove the law of excluded third, $\neg\neg(A \vee \neg A)$. This would seem, under a natural reading, to be incompatible with statements that are recognizable verification/falsification gaps.

I have argued that taking this as a reason for rejecting gaps is putting the cart in front of the horse. We should first get clear about our semantic primitives and only then inquire what logic their interplay gives us, not prejudge this issue by claiming that there can only be one appropriate logic for the constructive enterprise, namely intuitionistic logic. That intuitionistic logic is not the only, and, indeed, not even the most attractive logic for this endeavor was one of the main points of [7]. The upshot of that line of thought should be that we move to a different logic, not to a different semantic theory only in order to save our logic. (Of course, as a matter of methodology we can start with a particular logic and work our way backwards in order to see how well it can be motivated by certain semantic considerations. In fact, that is just what I am doing in this paper. However, if the result of that investigation is that the fit is bad, it should be the logic that has to budge. This is not just my personal view, it is also the direction in which Dummett was taking things in attacking classical logic.) So, the first reason Dummett had for rejecting gaps, based as it was on the particular features of intuitionistic logic, seems moot to me.

The second reason Dummett was not enthusiastic about gaps is that they seemed to stand in the way of certain deep metaphysical conclusions about realism and anti-realism that he wanted to draw from his study of language and logic. This argument linking the philosophy of language to metaphysics is one of the most distinctive features of Dummett's philosophy, but I have argued (and still believe) that if our best semantic theory does not allow us far-reaching metaphysical conclusions, then this is not a good reason to dismiss it but rather to go on looking for other ways to tackle the metaphysical questions.

To sum up, I have discussed and set aside both of these objections against gaps in [7] and will not go into them further here. Verification/falsification gluts, on the other hand, are another story.

### 4.2.2 Constructive Gluts

To repeat what I have said above, these gluts were also rejected by Dummett, and this time I agreed with him in [7].

Actually, Dummett does not really give much of a reason for his rejection of gluts (at least not in the texts I am aware of): He just states that no statement could ever be both verified and falsified as an obvious fact and moves on.

This seems understandable for two reasons: First, when he was writing his most influential works, discussions of gluts were conducted far removed from the logical mainstream. Today, it is not at all unusual to ponder the application of glutty logics to many philosophical problems, but that is very much a recent phenomenon. That there is a way to make such logics work and behave might not at all have been obvious at the time.

Second and more importantly, it seems that allowing verification/falsification gluts would go against the most fundamental understanding of the terms themselves. Verifications are things that establish a statement beyond any further doubt, and falsifications things that rule out a statement definitively and forever. What possible sense could be made of a statement that is both verified and falsified?

There are two potential answers to this question. First, one might take one or more leafs out of the book of the dialetheists (also hardly a force to be reckoned with at the publication time of Dummett's classic papers). That is, one could take the arguments G. Priest, R. Sylvan and others have been giving for true contradictions in realms such as semantic paradoxes, strange physical phenomena or Buddhist conceptions of ultimate reality, and rephrase them in constructivist terms.

Dialetheism does not carry the stigma of absurdity it once did, though it remains a fringe position that few are willing to fully endorse. Moreover, it seems to me that on the whole, the constructive version of dialetheism will look not too dissimilar to the original version, though it might turn out to have some idiosyncratic twists and turns here or there. Therefore, I will just note here that a constructivist who is attracted to dialetheism (or a dialetheist who is attracted to constructivism) will probably find something she likes in FDE under the interpretation at hand; I will, however, not delve into that topic further here.

Instead, I will concentrate on the second reason for voting for verification/falsification gluts, a reason that is both more distinctive of the constructive enterprise and, at least to my mind, more compelling.

This case for gluts is based on the fact that, in the empirical realm, we must always be prepared to finding our best corroborated hypotheses to fail, and to being forced to accept something we had formerly thought was ruled out by the evidence. Verifications and falsifications are based on evidence, and evidence is inherently defeasible, no matter how secure it might seem. Pigs *might* turn out to be camouflaged unicorns; or, if you like more fanciful examples, Trump *might* turn out to be a political mastermind who is covertly advancing the good of humanity in general and of his voter base in particular. Given this, are we really well advised to model our semantics on "conclusive" verifications and falsifications? It seems that that is to mistake the nature of empirical statements and to draw too close a parallel to mathematics, where proof indeed might be taken to be conclusive and definitive.

The line of argument that leads from this observation to constructive gluts is pursued by Jon Cogburn in [1], which is the most explicit defense of such gluts I know of. (In fact, he argues that verificationists of Dummett's ilk should embrace not only gluts in the semantics, but also an outright dialetheism; I find it more natural to mean only the topics I discussed above by the term "dialetheism", but nothing much hinges on that choice of terminology.)

The central claim of the paper can be summarized thus: Once we move out of the pristine realm of mathematics and step into the rough and dirty empirical world, there are no absolutely certain and conclusive proofs, which he calls *strong verifications*, any more. The most we may hope for is very good but still defeasible pieces of evidence, which he calls *weak verifications*.

But then, it is hard to deny that there might be situations where we are confronted with very good evidence for a statement A and equally good evidence against it. That is, as weak verifiability is spelled out in terms of very good evidence for a statement, and weak falsifiability in terms of very good evidence against a statement, we must face the possibility of weak verification/weak falsification gluts. Though Cogburn makes no attempt at pointing to any particular logic that might be motivated this way, it is clear that it will have to be paraconsistent. Beyond that, it seems that FDE would be at least a natural candidate for him to consider. If we are convinced by Cogburn's argument, we may rest our case and move on to the next section.

However, there seems to be a rather obvious response to Cogburn's line of argument, one he is well aware of: If the evidence we are dealing with here is supposed to be defeasible, why not say that, if we have good evidence *for* as well as very good evidence *against* the same statement A, these bodies of evidence cancel each other out, so that A is neither verified nor falsified?

There are several answers Cogburn gives to this, the most interesting of which I take to be the last one:

> [T]he claim that a warrant for *P* automatically undermines warrants for $\neg P$ involves an ugly equivocation between not having enough evidence for a claim and having too much evidence for the claim and its negation. ([1], p. 234)

Indeed, I believe that the distinction is an important one to keep. However, I have my doubts that Cogburn is taking things in the right direction by modifying our understanding of what a verification is (namely, for the empirical realm, what he calls a weak verification, as strong verifications do not exist there). The right upshot of Cogburn's argument seems to me to be that verifications and falsifications are actually not the right primitives to use for our semantics, rather to say that we have misconstrued what a verification or a falsification amounts to.

First of all, I think that this adjustment of the meaning of "verification" is so far removed from common usage that it would be better to use another word altogether. Locutions like

> "Until recently, scientists had verified that ulcers were caused by stress, but then they verified that stress had not much to do with it."

seem close to incoherent.[4]

One might be tempted to see this as a merely verbal issue, albeit one on which most who use the terms seem to land on my side. That is, the side that takes verifications and falsifications to be definitive, and that thinks that when a supposedly verified(falsified) statement turns out to be in fact not verified(falsified), it never actually was in the first place. As I mentioned before, Dummett is with me on this side. But even a philosopher like Popper, who famously thought that verified scientific theories do not even exist (for reasons very similar to Cogburn's) believed that truly *falsified* theories never come back.

However, it goes beyond that merely verbal level: I also believe that we still have use of the notion of a strong verification, as Cogburn would call it (I will continue from here on to simply call these *verifications*), even if we might be pushed to reach for a different set of semantic primitives. How that might play out is the topic of the next section.

## 4.3 Information with a Capital I

Both an interpretation of FDE in terms of told-truth/falsity or one in terms of verifications and falsifications seems interesting to me. But, as will have become clear, I am not completely sure that a Dummettian philosopher will be on board with either of them. Hence, I

[4] In [7] (p. 152) I toyed with the idea of taking verifications to be "good evidence plus the absence of good counterevidence". That removes the need for allowing verification/falsification gluts, but still leaves the unwelcome possibility of a verified statement becoming unverified later on. Instead of this, I ultimately went with a view that Cogburn would probably categorize as an naive strong-verifiability (and -falsifiability) view.

will now try to find something useful somewhere in the middle between the two interpretations, so to speak.

To that end, let us think about the values in Dunn's relational semantics, and let them stand for the following now:

Statement $A$ takes value $t$ if we have Information supporting it.

Statement $A$ takes value $f$ if we have Information against it.

This, admittedly, looks pretty close to Belnap and Dunn's original interpretation. The crucial difference lies in what I intend *Information* with a capital I to mean. Basically, it is simply very good and trustworthy information.

I don't mean this to be a incredibly original new concept, and one might justifiably wonder if a new piece of terminology is really needed here. For example, Information is surely very similar to is the notion of *warrant* that often appears in these kinds of discussion. However, in general I am not quite sure about what people exactly mean by warrant, and that is one reason why I stay clear of it.

In addition, most authors would seem to be averse to the idea that we could have both warrant for and warrant against a statement, which would be the situation that would be denoted by $\{t, f\}$ if Information were warrant; this, of course, is a combination I want to allow.

I also believe that Information is reasonably close to what Cogburn called *weak verification*, but I have given my reasons for staying away from that label in the last section. Also, I am not perfectly sure that he would consider exquisite quality of the evidence to be a necessary condition for weak verifiability, though I think he does.

To be a bit more precise now about what I want to call Information: I intend to be very broad in the question what kinds of things can be Information: Thus I include, besides testimonial information: Direct perceptual evidence, information retrieved from memory, inferential information and mixed forms, such as the information that a stotting gazelle relays to a predator.[5]

However, the point of the concept is that not all grades of quality of information are fit to be Information. What I want Information to be understood as is information that is so trustworthy that we won't give up belief in it unless the counterevidence we receive is also Information. And even in that case, the case modeled by the value $\{t, f\}$ in our semantics, we might not give up belief in it, at least not easily. In such a case, the proper response is an at least mild epistemic crisis, not a mere shrug of the shoulders, as would be appropriate when we find two otherwise identical cans of soup with different price stickers on it.

One reviewer remarked that she or he found the preceding passage rather unconvincing. At almost any other point in this paper, I would have tried to meekly reiterate that I am merely trying to make the best case for a philosophical position that I am by no means convinced of myself. However, here is one of the few places that I would like to dig in my heels. That a small or big epistemic crisis is a more suitable reaction to an Information glut than almost any other I can think of seems correct to me. Take as an example contradicting statements from the two persons you trust most in your life, about a matter which they both clearly know the truth about and which is of great practical and moral importance to you all. Should you drop belief in both sources? In the one that entails the least changes in

[5] "Stotting" is when a gazelle jumps up and down in a seemingly pointless fashion, in order to inform a potential predator that it is strong enough to outrun her in case she should choose to start the hunt. Normally this information is meant for lions, but it will also be of interest to you when you're out on spear-safari, and you will have to integrate perceptual and inferential information to make use of it.

the network of your other beliefs? Or should you pause and feel the need to sit down and have a drink? The latter strikes me as the rational thing to do. Conversely, experiencing such an epistemic crisis is a good indicator that the pieces of information making up the glut really deserve to be called Information, at least in subjects that are not excessively prone to epistemic crises.[6]

Of course, that still does not give a clear decision rule to apply in order to find out whether a piece of information is also Information, and there cannot plausibly be such a rule.[7] However, here is another heuristic: The quality of Information is in the ballpark of information from direct perception in good conditions. It takes a lot to get us to distrust such evidence, but it can happen, as we know from optical illusions and other phenomena.

Let us consider testimony in more detail: Just any old piece of testimony we come across, such as it appears in the told-truth account of FDE we explored above, will surely not suffice to amount to Information. Not even in the example of things asserted under oath at a court would this be the case (or maybe better: *especially not* in this case).

That is *not* to say that testimony could never add up to Information. Dummett, heavily concerned with language acquisition as we saw him being, could certainly not exclude testimony as a possible source of Informtion with a capital I. Much of what a child learns about the world, about language and about the relation between the two rests on her taking the validity of testimony, especially by her parents and other caregivers, for granted. The reason they take this information to be Information is that they trust the source so much, and even after they come to learn that even their parents may have to be distrusted in some circumstances, trust in the integrity of a particular source will be the main reason that information is elevated to Information.

However, it might also be the case that a number of lesser trusted sources, all independently corroborating a statement, might collectively amount to Information. That I allow this possibility is to say that I am willing to commit the fabulously badly named "faggot fallacy"[8]: Badly named not just for its offensive sounding first half (the intended picture is actually that many weak pieces of evidence are like twigs (faggots) that together don't make up a strong piece of wood capable of bearing any weight), but also for its second half: It is, in general, precisely *not* a fallacy, though it can be in special circumstances (where, for example, all sources ultimately trace back to one primary source that is not particularly trustworthy).

But, again, even though Information can surely come to us via testimony, just because someone said something does not mean that you need to think that you have Information backing it up (though, given what we have read about the natural tendency to take others by their words, you maybe should defeasably believe it to be true/verifiable. It is just that it won't necessarily take much to defeat that belief.) This leaves room for assertions of statements that have (and continue to have after the assertion is made) value $N$. I find that plausible, and it will allow me to deploy my thoughts about such statements, as developed in [7], in my projected future work on this topic, where I plan to think about the right choice of designated values.

---

[6] After submitting this paper, I realized that my notion of Information has a lot in common with the certainties Wittgenstein discussed in *On Certainty*.

[7] We might try go for a picture where the quality of information is weighted continuously, and where there is a cut-off point marking where Information starts, ending up with a picture that will look much like what Dunn proposes in [3]. This would, at least in theory, tell us where the line between information and Information lies; in practice, however, the hard cases will for the most part remain hard cases.

[8] See [5] for critical discussion.

There is something inherently subjective in the notion of Information that is lacking in the notion of plain information. It is quite an objective question whether or not you have been told something. On the other hand, whether something deserves the status of Information involves questions about how trustworthy the source is, and that call is going to be quite subjective, indeed. That such subjectivity should creep into semantics might be a worrying or absurd thought to some. On the other hand, it is a common observation that successful communication gets more and more unlikely the more such subjective standards are diverging between the parties of the conversation. It can get to the point where the feeling ensues that one is not speaking the same language as the interlocutor. The picture emerging form a semantic theory that is Information-based would support that feeling: Indeed, in most cases two different speakers will speak subtly different languages, but these differences are normally negligible enough to make communication possible.

A possible way to at least dilute (if not to fully dissolve) the inherent subjectivity in question would be to relativize the values in FDE to a relevant group of speakers (or reasoners, or epistemic peers, or whatever else might seem plausible): One might want to say that a statement takes value $T$ iff all speakers of the group take the information supporting it to be Information (and take none of the information against it, should there be any, to be Information). This might, if the group of discussants is large, be a too high bar to pass, so alternatively, one might ask only that some of the speakers do so, or, to let an element of democracy creep into semantics, that the majority of the speakers does so.

In discussing Cogburn's argument, I suggested that there might be still a role to be played for verifications and falsifications (or, in his terminology, strong verifications and strong falsifications), even if one accepts his point about the unavoidable fallibility of empirical knowledge.

Here is how I take Information values hook up to "verifiable" and "falsifiable": Value $\{t\}$ means that you are given defeasible warrant to conclude that a statement is verifiable. That kind of warrant can increase in strength (through things like corroborating evidence) to the point where it becomes almost unthinkable that the statement was not verifiable. But knowing beyond any possible doubt that a statement is verifiable stays an idealized limit point that is never actually achieved.

One might think that that point needs to be actually achievable to give the constructive argument about recognition its bite, but I think that isn't necessarily the case. What needs to be recognizable in this picture is not anymore that a given construction verifies a statement (*that* cannot be recognized), but of each verification that it constitutes Information about the statement (and the same, mutatis mutandis, for falsifications).

Radically skeptic arguments aside, we know that we have verified many statements, even though we might not know of any individual one about which we have supporting Information whether it is verified. It is exactly this feature that would, it seems to me, guide us towards Information instead of verifications/falsifications as the primitives of our semantics, just along the lines of Cogburn's argument. However, to give up verifications and falsifications will not allow a story to be told in which our beliefs in that which we have Information about are justified. (In this picture, the nature of verification becomes vaguely similar to Crispin Wright's "superassertability". They are not quite the same, but I don't have the space to discuss the differences here.)

Whether any of the three proposals in this section present a fruitful way of conceiving of language is an interesting question, but surely one that would take much more space to fully answer than I have here. Let me just say that, if FDE is indeed to be seen as a useful logic for the constructive enterprise, then these are the most promising interpretations of its values that I have been able to come up with, and I believe the last one to hold most promise. If these interpretations are inherently unappealing to the Dummettian constructivist, I don't know how to whet her appetite for FDE further. I, for one, have become much more open to the idea that a logic like FDE might supply a suitable base for the constructive endeavor than I was at the time of writing my book.

I have done, then, the best I can to motivate the uncommonly high number of *four* constructive truth values. Next, I will explore the question whether maybe we could use even *more* values than that.

## 5 Non-constructive Values and More Lattices

We are now at a point where we can discuss the extended lattices in Dunn, Shramko and Takenaka's paper ([9]). Their work represents, I believe, the closest in sprit to what I am after in this paper that is to be found in the literature.

Their main attention is actually not on the bilattice FOUR, but that there is a very close relation between their paper and the ideas above will become obvious presently. I will, in the end of this paper, discuss this relation in detail; before that I will introduce the content of [9] and supply some critical remarks. I am very grateful to the authors (whom I will allow myself to refer to as "SDT" in what follows) for giving me detailed feedback on my criticism. At this point, I think we still don't see eye to eye on all matters, but many things have been cleared up, and I will draw on those clarifications in what follows. Up front, it is worth mentioning that they told me that their main interest was viewing the values from an algebraical perspective, where mine is, as I said in the introduction, a stoutly depraved interest. This might well be the source of some of our different views on the matter.

### *5.1 Overview of the Paper*

The main item SDT discuss is a sixteen-valued logic that can be represented by not only a bi-, but indeed a tri-lattice.

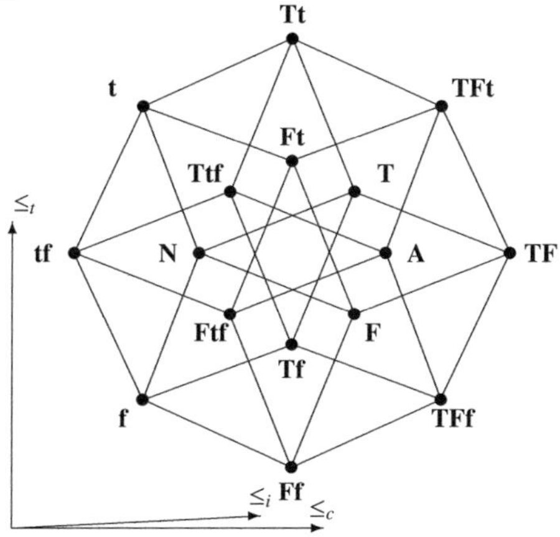


**Fig. 2** Trilattice $SIXTEEN_3$

By now, this structure has become the object of intense study under different interpretations, most of them disconnected from the constructive enterprise (see [10]).

However, in its original guise as it appears in SDT's paper, the values can be thought of as the powerset of the values $\{T, F, t, f\}$ ($N$ stands for the empty set, $A$ for $\{T, F, t, f\}$). $T$ and $F$ are constructive truth and falsity, evidently close to what I am calling verifiability and falsifiability here. $t$ and $f$ are more mysterious.[9] They are what SDT call "non-constructive" values. What they mean by this is not, as one might suppose, classical truth values. It is something more intricate, and, as will become clear below, it is my view that not everything they say about non-constructive values can easily be reconciled in one single interpretation of the term.

The inspiration for the notion of a non-constructive value comes from a detailed consideration of intuitionistic logic and the interpretation of its Kripke semantics.[10] More precisely, SDT start with a consideration of the difference between intuitionistic non-truth ($A$ receives value 0 in a Kripke model) and the truth of a negated statement in intuitionistic logic ($\neg A$ receives value 1 in a Kripke model).[11] SDT argue that it is absence of proof that is the right way to conceive of falsity in intuitionistic logic.

$f$ is the value that a statement receives if it is not provable; it corresponds to value 0 in the standard Kripke semantics. Evidently inspired by this, SDT note that a non-constructive value, such as $f$, will be projected backwards in this style of semantics. In the case of intuitionistic logic, this follows straightforwardly from the usual forwards projection of value 1, and the fact that there are no other values in this setup.

SDT go on to bring in the idea that falsity should also be a constructive achievement, introducing the value $F$. This value records "a presence of some "negative" construction that directly refutes A." ([9, p. 771]) Though they don't really talk about why the truth of intuitionistic negations should not qualify for that job, I am on board with that move for the reasons explained extensively in [7], especially where empirical statements are concerned. Instead of their *refutations*, I usually talk about *falsifications* in such contexts, but I don't think that that terminology reveals a great conceptual difference between SDT and me.

However, at this point they do not dispense with the non-constructive notion of the absence of proof that is signalled by $f$, and they wish to keep calling it a form of falsity ([9, p.773]).[12]

Then ([9, p.773]) they bring in, for balance, the value $t$, which, for all we have read so far, must mean that a statement that receives this value is not falsified (yet). Interest-

---

[9] Besides being mysterious, they might engender confusion with the relational semantics due also to Dunn which was discussed in the beginning of this paper. I considered deviating from SDT's notation to avoid that confusion and write $\bcancel{F}$ instead of $t$ and $\bcancel{T}$ instead of $f$. It will become clear soon why I think that that would be a suitable modification, but in the end I thought it slightly dubious to do so. SDT argue that $f$ is a form of falsity and $t$ a form of truth, and I will argue against that. Choice of formal terminology seems to guide the reader strongly here, and I hope my point is made strongly enough not to require that guidance.

[10] I assume familiarity with this semantic framework; see either [9] or [7] for details.

[11] They call the first semantic falsehood and the second syntactic falsehood ([9, p.769]), but the terminology is slightly uneven. Later, the first is called "intuitionistic falsehood", while something like truth of Nelson's strong negation is called "constructive falsehood" ([9, p.770]); at this point, the truth of the intuitionistic negation receives no more mention.

[12] At this point, at the latest, I would opt for different terminology. Where there is a explicit constructive notion of falsity in play, I fail to see any rationale for calling the absence of proof "falsity". Indeed, even before that point, back in the discussion of intuitionistic logic, I would have preferred to call the truth of the negation the "real" notion of intuitionistic falsity, but I concede that things are more messy there. In a way, this is a verbal dispute, but, as far as verbal disputes go, a pretty serious one. Not to disturb the expository nature of this section, I will leave that dispute in this footnote for now.

ingly, though, they never explicitly introduce it as that. Instead, they write that $t$ denotes "acceptability", and $f$ is from here on talked about as "rejectability". They explain: ([9, p.773])

> [T]o make the picture complete, we have to introduce among the basic truth-values the notion of a non-constructive truth. We can understand such non-constructive truth as a specific epistemic attitude, when one considers a sentence acceptable, although it may well be that no proof of the sentence is so far present. This is a notion of truth as (maybe temporal) acceptability. If we (for some reason) need a sentence, we might accept it, although later we can change our mind and decide that the sentence is not acceptable any more (e.g. when we obtain a refutation of this sentence)[13] This motivates introduction of a new truth-value that can be labelled as $t$.
>
> The expression $\alpha \vdash_t A$ can be interpreted as '$A$ is acceptable within a' (or maybe ' A is tolerated within a').

They go on to state that $t$, being a non-constructive truth value just as $f$ is, should be preserved backwards as well. And indeed, in as far as "acceptability" means "absence of refutation", this seems intuitive; we shall, however, come back to this point.

SDT then make the move that lets the impressive structure depicted above come to life: They consider the whole range of combinations of these truth values and place them in their tri-lattice.

It is a *tri*-lattice because there are three lattice orderings on the sixteen values. In addition to the truth order and the information order, there is a "constructivity"-order. The diagram above shows a projection in which truth and constructivity orders are relatively easy to see. The information order is harder to make out, but also the least mysterious one; I shall have little to say about it here. The truth order is also pretty straightforward. It treats the constructive and non-constructive parts of a value separately and puts the result back together again (so to speak): The meet of $Tf$ and $Tt$, for example, is $Tf$.

The idea behind the constructivity order is that a value is higher up, the more constructive values and the less non-constructive values it contains. $TF$ is the top element, $tf$ is the bottom ( cf. [9, p. 775]).

Now, the space of sixteen truth values seems quite big. It includes items like $Tf$ and $Ft$, for which an intuitive understanding might be hard to come by. SDT motivate their inclusion as follows ([9, p. 774]):

> A multivaluational approach implies 'independent' treatment of all four truth values. Thus, 'abnormal' situations can occur, when a sentence is refuted, but still marked as acceptable, as well as not acceptable although proven. Such situations may well happen when we have an overflow of information, inconsistent data, or when an agent behaves not in a rational way.

Even though they thus argue for the intelligibility of values that combine $T$ and $f$ (or $F$ and $t$), they later go on to discuss sensible restrictions of the value space (Section 6). Of most interest, in my view, is the structure resulting from disallowing these very combinations. We will get back to that below, after some critical remarks about the cohesiveness of all the things I have just summarized.

---

[13] This parenthetical remark is the closest SDT get to suggest that $t$ is to be thought of as absence of refutation (but of course that $A$ rules out $B$ does not generally mean that $B$ is equivalent to absence of $A$).

## 5.2 Problems with the Interpretation, and Two Alternatives

As I mentioned before, from reading what SDT have to say, I don't quite see why one would call the absence of proof a form of falsity, and likewise the absence of falsification a form of truth. As it happens, I myself have argued in [7] that in many contexts, the absence of a falsification can suffice to make an statement correctly assertible. If now one wanted to equate correct assertibility with truth, one would arrive at a picture that would underwrite half of SDT's terminology: $t$, i.e. absence of falsification, could indeed be read as a form of truth. On the other hand, that $f$, i.e. absence of proof, should be read as a form of falsity does not fit comfortably into this line of thought. In any case, if something broadly along those lines was what SDT had in mind, it came out in the paper in a rather inchoate form. I personally would advise to avoid talking about $t$ and $f$ as forms of truth and falsity.

But even if we free ourselves from the stricture of having to call them kinds of truth and falsity, it seems hard to me to come up with an interpretation of the non-constructive truth values that chimes with all the formal properties SDT list. In this section, I will suggest three different readings of the new notions that all have virtues and vices in that respect. Moreover, all three interpretations can inform the larger project of my present paper in different ways, as will be explained in the next section.

Let us start (as we should) with the original understanding, the one intended by the authors.[14]: $f$ stands for "not verified/proven yet" and $t$ for "not falsified/refuted yet". Call it the *Absence Interpretation*.

At first blush, backwards projection presents no problem for this interpretation (though I will find a problem with it a couple of paragraphs down the line). However, the truth order might strike us as somewhat strange: Why would we put $T$ and $t$ at the same level of truth? If truth is to be understood as "having a proof/verification", then $T$ is all the way there, while $t$, "not yet falsified", has a long way to go.

Also, the constructivity order seems rather ad hoc under the *Absence Interpretation*. We see that $N$ is more constructive than either $f$ and $t$. But what is $N$ if not the absence of constructive achievement? Why should it reside anywhere but at the very bottom of the constructivity order?

Related to this, why is there even a difference between $N$ and $tf$? SDT at no point consider conflating them, in contrast to them considering the equivalence of $T$ and $Tt$. In personal communication, SDT have suggested that $N$ should be read as "not considered yet", while $t$ and $f$ should be read as "attempted to be proven, but failed so far" and "attempted to be refuted, but failed so far".

Those are interesting glosses that make a distinction plausible, but now the backwards projection property seems to become questionable. The natural progression seems, for example, to be "A has not been considered yet", "A was attempted to be proven, but failed so far" and then "A was proven". In that progression, the non-constructive value $f$ is *not* projected backwards. In response to this observation, the authors have suggested that $N$ might best be thought of a pre-theoretic value, while $t$ and $f$ only make sense in terms of a given theory (where they, then, will be projected backwards). Though I am not quite sure, I believe that that either amounts to considerably restricting what they say about backwards projection in the paper, or to introduce a distinction between lattices of meta-theoretic and

[14] The two alternative proposals below were also suggested to me by their paper, but SDT have made it clear in personal communication that they were not what they had in mind.

lattices of intra-theoretic values, only the latter of which the remarks about backwards projection apply to.

Let me introduce the second interpretation, one that certain passages of the paper suggested to me, but which SDT tell me was *not* their intended target. Call this the *Told-(Non)Constructive Interpretation*. The idea is simply to apply the "told" interpretation of FDE to the *Absence Interpretation*:$T$ means that we have been told that a proof of the statement exists, $F$ means that we have been told that a refutation of the statement exists, $f$ means that we have been told that no proof of the statement exists, and so on.

The nice thing about this interpretation is that the question which combinations of truth values make sense is easily answered: They all do, we can revel in the beauty of the full sixteen element lattice.

Also, this is the only interpretation I can think of under which the constructivity order makes some sense. $N$ means that we've been told nothing about a statement, while $f$ and $t$ mean that we've been told that certain constructive achievements are missing. If we choose to believe that information, we have reason to believe that certain constructions have not been achieved, while in the case of $N$ it might well be the case that proofs or refutations of the statement exist (we just haven't been told about it). The sum total, so to speak, of constructive achievements we have reason to believe exist is lower, the more $t$s and $f$s as opposed to $N$s we are presented with.

However, the backwards projection property makes little sense on this interpretation. If any projection behavior is to be expected at all, then it is forward projection for all values except $N$. This is only if we assume that information is never lost, of course, but giving up that proviso would mean to lose all projections.

There is a third interpretation I want to discuss, one that, once again, is suggested by some passages of [9]. However, again, it does not seem to be the intended interpretation, and neither does it fit to *everything* in the paper. It seems that this interpretation fits pretty well to what they say from around the time they introduce the "acceptability" and "rejectability" terminology, so let us call this accordingly the *Acceptability/Rejectability Interpretation*.

As I noted above, at this point talk of absences of constructive achievements quiets down in the paper, and almost never resurfaces. It does not strike me as *that* obvious that acceptability and the absence of refutation should come to the same thing,[15] so it is worthwhile to explore whether there is some other understanding of acceptability and rejectability that could make sense here that has nothing to do with the absence of proofs or refutations.

Well, then: $t$ is not an absence, but a presence of *something* that makes the statement acceptable, even though it has (maybe) not been proven. What might that be? There are surely many options here, and I will lay one out in the last section of this paper that seems especially relevant in the larger context of this piece. Let me mention another one here, just as an illustration (I will not get back to it): We could go for a combination of the two earlier interpretations. Let $T$ mean that we have a proof/verification, let $t$ mean that we have merely been told there is one. If we only have the latter, we may indeed be justified in accepting the statement (maybe on grounds such as those discussed in section 4.1, "Told Truth as Constructive Truth").

Whatever the details might be, the *Acceptability/Rejectability Interpretation* interpretation has the virtue of being interesting in its own right, and making it easier to understand

[15] Though I have to admit that this idea resonates with much of what I wrote in [7].

some of the combinations of truth values (see below). Also, it is the only interpretation under which it strikes me as obvious why $t$ should be a species of truth, and $f$ a kind of falsehood (see above). On the other hand, it does poorly with respect of fulfilling SDT's views about backward projection and the constructivity order; I see no way to make either plausible under this interpretation.

To sum up, I have my doubts that the following three aspects can be reconciled in any interpretation, at least in quite the way SDT discuss them:

1. The backwards transmission of non-constructive truth values in Kripke models
2. The position of values in the constructivity order
3. The explication of certain combinations of truth values

What I propose to do in view of this is simply the following: I will, at least in this paper, ignore the first two aspects. Backwards projection seems something that goes naturally with the Absence Interpretation, but there seems to be no reason to complain if it fails to hold in another sensible interpretation of non-constructive values. In fact, it is not the time to worry about projection properties at all, as they make no difference to the negation, conjunction and disjunction fragment that FDE constitutes.[16] The topic will need to be picked up when the question of how to add a constructive conditional comes on the table, but that point lies, at least for me, well ahead in the future.

Furthermore, while I won't mind a constructivity order where it can be found, I won't be concerned if it cannot. For my project, it does not seem to me to play a role of great importance, just as the information order. What is of utmost interest is the truth order. And to get the right truth order, we first have to decide which of the sixteen values really make sense on our interpretations.

## 5.3 The Right Size of the Value Space

The three interpretations of the last section, I believe, plausibly all lead to different ways in which the truth value space should be restricted. To get the easy case out of the way: All sixteen values make, as I have already noted, perfect sense under the *Told-(Non)Constructive Interpretation*. For the other two interpretations, the case is not that clear.

The questions we should ask here come in two flavors: First, do certain combinations make sense at all, given the intended interpretation? Items like $Tf$ and $Ft$ are on top of the list here. And second, shouldn't certain values, even if they make sense, be identified with each other, given their interpretation? Here, we are looking at pairs such as $Tt$ and $T$, or at $tf$ and $N$.

### 5.3.1 *TF*

I have spent many pages in the earlier parts of this paper trying to make the combination of both constructive values plausible. In contrast, SDT seem to see no problem with this combination; it is not a feature they spend much time motivating. There is, of course, no

[16] This, of course, is in contrast with intuitionistic logic, where negation is modelled as an intensional notion.

point in repeating here what I wrote above in much detail, so let us move on (note that as there are no non-constructive values in $TF$, there is no pertinent new idea that wasn't on the table before this section).

### 5.3.2 $tf$

There is nothing mysterious about $tf$ under the *Absence Interpretation*: It is just the state in which we have achieved neither proof nor refutation for a statement (as I noted, the question might arise why we should make a difference between this value and $N$, but I leave that for later).

Under the *Acceptability/Rejectability Interpretation*, on the other hand, we might well ask whether it makes sense to call a statement acceptable and rejectable at the same time. It will depend on the exact account we give of acceptability and rejectability, but to give one example in which the idea seems to make sense: In chapter 6 of [7], I discussed areas of discourse which allow for "faultless disagreements", that is, context in which it may well be correct for me to say "A", and equally correct for you to say "Not-A". Aesthetic disputes are often discussed as examples of this, but in the book I extended the idea to range over a broad range of cases. If saying "Not-A" is tantamount to rejecting "A", then this seems to hook up nicely with value $tf$ under the *Acceptability/Rejectability Interpretation*, but I can easily imagine how the *Acceptability/Rejectability Interpretation* might be developed in a different way that make $tf$ dubious.

### 5.3.3 $Tf$ and $Ft$

Even if we are willing to take statements that take both constructive values seriously, surely a combination like $Tf$ is even harder to swallow. What does it mean that there is and there is not a proof for a statement? And indeed, as we saw, SDT see the need to motivate the intelligibility of these kinds of combinations. We saw them suggest that such values describe situations "when we have an overflow of information, inconsistent data, or when an agent behaves not in a rational way."

The latter idea about irrationality moves by rather quickly. Maybe there's something there, but it is rather unconnected to the rest of the paper, and, moreover, it is hard to see which logic *couldn't* be motivated by such a remark if the kind of irrationality at play is not further specified.[17]

The idea of overflows of data and inconsistent data (I take them to be a single idea, and not two different issues) seems more connected to the rest of the paper. However, to have too much data seems to lead to a situation in which $TF$ is appropriate, not $Tf$. Say we have a very reliable test for the presence of a virus in a body, and a very reliable test for the absence of that very virus. Now imagine both are positive. What this means (if anything involving the notion of "is verified" is appropriate at all) is that we have verified and falsified the claim that there is this particular virus in the body, not that we have verified it and not verified it.

I am not sure that these remarks can underwrite the intelligibility of these values, at least not on the *Absence Interpretation*. As it happens, the remarks of overflowing / inconsistent

[17] To give an example from quite a different area of research, contrast this with *prospect theory* ([6]), where it is fairly clear what patterns of (arguably) irrational behaviour the theory tries to explain.

data were the point at which I first thought of the *Told-(Non)Constructive Interpretation*, under which they strike me as completely natural; but again, this is not the interpretation the authors were actually having in mind.

As I have discussed in section 4.2, one might think that some sort of dialetheism about verifications and falsifications, or proofs and refutations might be defensible. If so, then this defense might also turn out to be expansible to an argument for the (non)-existence of proofs and refutations. Alternatively, one could argue that there is some kind of vagueness that needs accounting for (where something might be a proof if squinted at from over here, but not from over there). As indicated above, I am not sure whether that it is a very promising route that leads that way, but in private communication SDT have shown some sympathy for these ideas. They also mention that having a proof in an inconsistent theory (such as naive set theory) might qualify for this value.

All this deserves further reflection, but it seems fair to say that $Tf$, $Ft$ and all larger values containing them remain problematic under the *Absence Interpretation*, a fact the authors were evidently conscious of when they discussed sublattices *not* containing these values, which we will get to presently.

Under the *Acceptability/Rejectability Interpretation*, the values look even more problematic to me. It would be a strange interpretation of "is acceptable" if it were allowed to accept falsified statements, or one that seems off-topic (for example, "acceptable for the sake of a counterfactual argument", an interpretation that seems to have little to do with what SDT are after). It looks like they will have to go if we favor this interpretation.

### 5.3.4 $T$ and $Tt$, $F$ and $Ff$

The second restriction explicitly considered by SDT is to disallow combinations of constructive and non-constructive versions of the same value.

The idea is that "constructive values, being stronger, "swallow up" their non-constructive counterparts" ([9, p. 784]). The idea seems attractive on the *Absence Interpretation*: It is not so much that it is obvious that a proven statement cannot be refuted (we are explicitly allowing this possibility by allowing $TF$). It is that it makes no sense to distinguish between $T$ and $Tt$. Consider a proven statement. Either we have an additional refutation, or we don't (constructivist worries about the law of excluded middle seem misplaced here; proofs and refutations are things that are by definition recognizable, so we know whether we have one or not). In the first case, $TF$ is the appropriate value, in the second $T$ (or $Tt$, if you wish to choose this one and let $T$ go).

Under the *Acceptability/Rejectability Interpretation*, the redundancy seems likewise obvious. As a verified statement is surely acceptable, there is no point of distinguishing $T$ and $Tt$.

### 5.3.5 $N$, $tf$, $t$ and $f$

As I mentioned, the idea that there might be a question why $N$ should be anything different from $tf$ receives no discussion in [9], though it seems natural to ask it under the *Absence Interpretation*. As I mentioned, the authors suggested in private communication the following way to understand the difference between them: $N$ should be understood as "not considered yet", while $tf$ should be read as "both proof and refutation has been attempted, but both failed".

I have no other suggestion to make of how to motivate the difference. Whether this is a difference that we would value to be able to make in our semantics will depend on the application, but note the following: If we decide that we would not want to make it, and to identify $N$ with $tf$, it would seem that also the need for $t$ and $f$ would become questionable: $t$ means that a refutation is missing. What about a proof? Either it is missing (value $tf$) or not (value $Tt$) (once again, qualms about excluded middle seem inappropriate).

On the other hand, even if we decide to get rid of $tf$ on the *Acceptability/Rejectability Interpretation*, $t$ (and $f$) will still be useful: It says that something exists that makes the statement acceptable, though it is not yet proven, and that makes perfect sense.

## 5.4 More Lattices

So, we have a range of options when we consider how the value space should be cut down to size. Let us begin at the point SDT got to in the paper. They considered, as mentioned, disallowing $Tf$ and $Ft$ and all values that contain these combinations. This restriction leads to what they call a Bi-and-a-half-lattice named $NINE_{2.5}$ (unlike for the other lattices in this paper, the following displays the truth order horizontally, as in [9]):

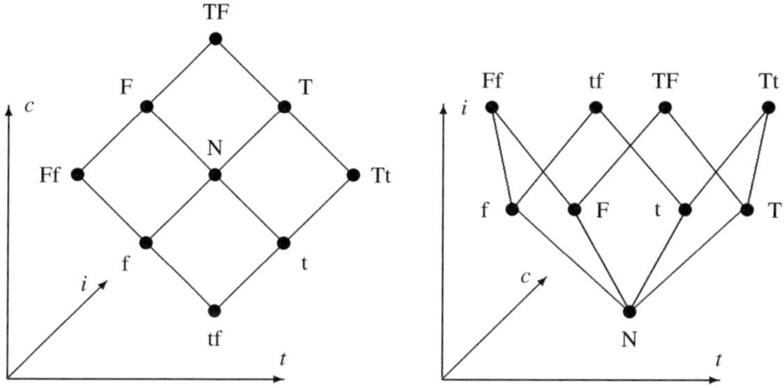


**Fig. 3** Bi-and-a-half-lattice $NINE_{2.5}$

It is a bi-and-a-half-lattice because it has a full lattice order for both truth and constructivity, but only a semi-lattice order for information (with $N$ as bottom but no top).

Next up they consider the restriction of the value space that disallows simultaneous occurrences of $T$ and $t$ ($F$ and $f$) in one and the same truth value. The idea is that the constructive value "swallows" the non-constructive one. The restriction (made independently of the earlier one), they claim, leads to "another version of $NINE_{2.5}$" ([9, p.784]).

This actually doesn't seem to be the case, at least not quite. If we simply swap the now disallowed $Tt$ and $Ff$ for the now rehabilitated $Tf$ and $Ft$, we find that the truth order is messed up:

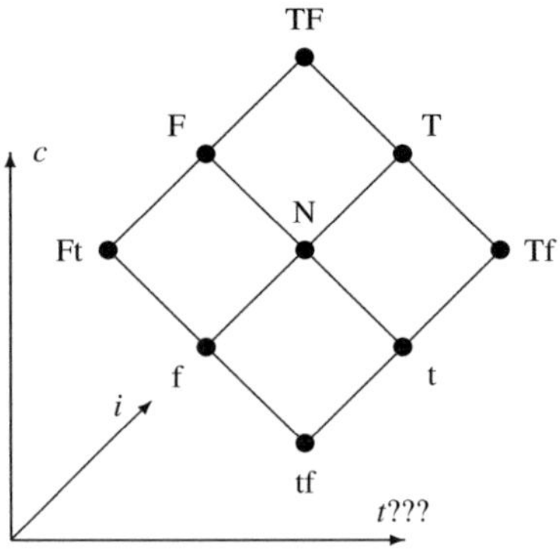


**Fig. 4** A broken version of $NINE_{2.5}$

There is no easy way to rectify this, simply because now we are missing our top and bottom elements for the truth order.

It is, however, possible to rearrange the values in the following way (by having $t$ and $f$ swap places):

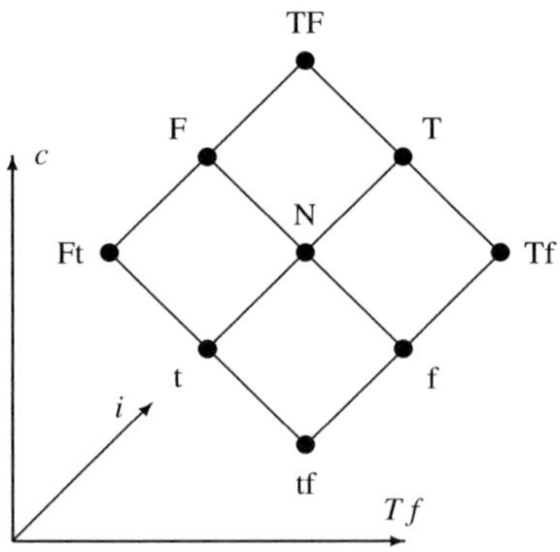


**Fig. 5** Bi-and-a-half-lattice $NINE_{2.5}$ with a $Tf$ order

This, however, is clearly not a truth order any more. It is, if anything, an order that is a constructive-truth-and-non-constructive-falsity-order. In a (not closely, but still somewhat) related context, SDT mention a grue-like feeling that certain properties give rise to; this order surely does. It is clearly of less use and interest than the truth order, and therefore the lattice in total is less attractive, even though the restriction indeed seems attractive.

We can do a little better than that, though. Conceptually, the idea is that there is no difference between $T$ and $Tt$. It then does not matter much which of the two we use. We can, instead of $T$ "swallowing up" $Tt$, have $Tt$ devour $T$. That is, purge the space of all values that contain $T$ but not $t$ (and, of course, that contain $F$ but not $f$). We then get our top and bottom elements for a truth order back. However, it does not seem to me to be possible to fit the values easily into the structure $NINE_{2.5}$ without doing some damage to the intuitions of the underlying interpretation.

However, we may well ask whether that is great problem. We have seen above that there is something to say in favor of both restrictions, and we might just as well try to apply them simultaneously. We would be left with a space of seven values. Here is a plausible way to arrange them, displayed in two naming schemes corresponding to the above ideas:

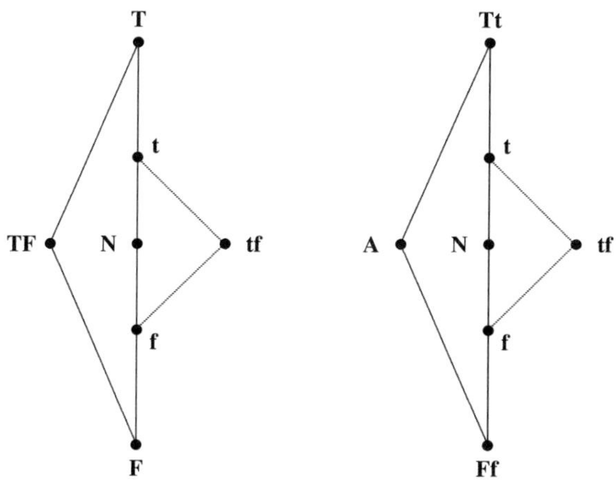


**Fig. 6** Lattice *SEVEN*

Let us simply call this *lattice SEVEN* (NB: The names for the lattices in this paper are chosen in accordance with SDT's naming scheme, but they might need further superscripts to differentiate them from other structures in the literature that I am unaware of.).

The right version has two advantages: It can be seen as a lattice and a half, with half an information order in addition to the truth order. And, moreover, the truth order can be defined just as in SDT's paper.

In the lattice on the left, there seems to be a problem in finding a truth order again, as we would have two candidates for the top element, $T$ and $t$. However, we can see the order as embodying the thought that absence of refutation is not as sturdy a form of truth as the availability of a proof is.

Now, we also saw that certain ways of reading our interpretations would suggest to identify $N$ and $tf$ (in the *Acceptability/Rejectability Interpretation*, $N$ might be the natural one to keep, while in the *Absence Interpretation*, we might want to go with $tf$). This would lead us to a version of bilattice FOUR, with additional occurrences of $t$ and $f$ between $N$ and $T$, and between $N$ and $F$, respectively (alternative names in brackets, the nature of the truth order to be fixed in harmony with your choice).

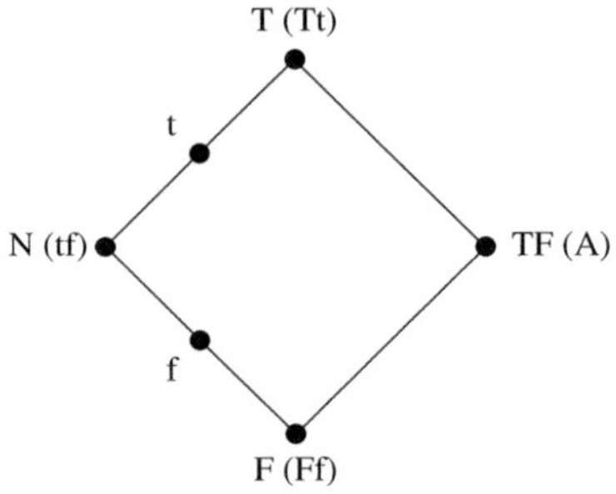


**Fig. 7** Bi-lattice $SIX$

These six values would make sense on the *Acceptability/Rejectability Interpretation*, but we saw that under the *Absence Interpretation*, even the simple $t$ and $f$ were getting dubious unless we brought in the "not yet considered" gloss to save the difference between $N$ and $tf$. Of course, when we take them out, we are finally back at our original bilattice $FOUR$.

One last item needs to be considered: What would happen if we chose to keep combinations like $Tf$, but otherwise follow the considerations in the last paragraph that seemed to lead back to bilattice $FOUR$ under the *Absence Interpretation*? We would be back up to six values, albeit with different names and in a different plausible arrangement, this time a simple lattice we will call lattice $SIX$:[18]

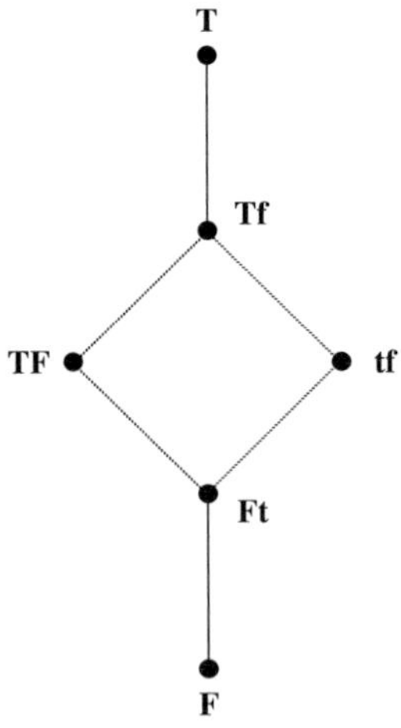


**Fig. 8** Lattice $SIX$

[18] One of the referees spotted that this lattice, albeit under a very different interpretation, appears in [4].

To sum up:

- The *Absence Interpretation*, for all we have said up to now, seems to be most in harmony with lattice $SEVEN$, at least if we employ the "not yet considered"-gloss for $N$. If we don't, we seem to tumble back into bilattice $FOUR$. However, lattice $SIX$ will be seen to supply a possible solution to a problem that will be treated in Part 2.
- The *Acceptability/Rejectability Interpretation* seems to go well either with lattice $SEVEN$ as well, or with the six-valued bilattice $SIX$ left by cutting out $tf$.
- The *Told-(Non)Constructive Interpretation*, finally, is the only natural interpretation that seems to me to lead to the full tri-lattice $SIXTEEN$.

## 6 Bringing it all together, and looking ahead

We started out with three different ways of understanding constructive values to make the four valued space palatable to a Dummettian semanticist. We then explored three ways to understand "non-constructive" values in the last section. How do these interpretations fare when we bring them together?

Though there are surely ways to go for hybrid variations, there is a pleasingly simple way to pair them up one-to-one. The "told-truth as constructive truth" (4.1) idea obviously needs to go with the *Told-(Non)Constructive Interpretation*, and thus will lead us to look at tri-lattice $SIXTEEN$.

The interpretation in section "Verified, Falsified, Both and Neither" (4.2), likewise rather obviously fits to the *Absence Interpretation* for non-constructive values. We have seen a range of lattices that might be useful for this interpretation, starting at nine values and, with a decreasing sense of excitement maybe, trickling down to seven and finally back to four values. Again, $SEVEN$ and $FOUR$ seem to me to hold the most promise.

Lastly, the I/information interpretation of 4.3 of constructive values. This interpretation could actually very reasonably be treated along the lines of the *Told-(Non)Constructive Interpretation* as well, with $T$ meaning that we have Information supporting a statement, and $t$ meaning that we have $t$ information supporting it.

However, to get to the nice one-to-one pairing, let us make an effort to bring this interpretation together with the *Acceptability/Rejectability Interpretation*. It's not much of a stretch, either: A statement for which we have only supporting information (no information against it, and no Information either way) should be judged "acceptable", and thus receive value $t$. We are right to accept what receives value $t$ until further notice, but it will always be trumped by $F$. Therefore, having $Tf$ and $Ft$ in our value space seems problematic, which is in harmony with what we already have seen in discussing the *Acceptability/Rejectability Interpretation*. The lattice to look at here would, again, seem to be lattice $SEVEN$.

As will have become obvious, I find the conceptual space SDT have opened up fascinating and exciting to explore. Is the notion of an non-constructive truth value really essential for the Dummettian project, though? I think, for the most part, probably not. The ability to talk about absences of verifications and falsifications is surely important to have, but I see no dire need to bring these notions into a semantics at the most basic level. A possible exception here might concern the combination of the I/information interpretation and the Acceptability/Rejectability Interpretation. I think that here there might be a use for the non-constructive notions, but I will have to leave the development of that thought to future work.

To end this piece, let me further indicate what I hope to do in the future to bring the project to a fuller end. We have, at this point, rather thoroughly discussed how the truth values we find in FDE might be understood. To get to an actual logic from here, we need to decide which of those values should be designated, that is, preserved from premises to conclusions in a valid inference. I think that there are reasons to make a different choice from FDE's original choice of designated values, $T$ and $B$, for reasons similar to the argument given in [8].

After this question is decided, we have to see how well the constructive interpretation of complex statements involving negations, conjunctions and disjunctions can be squared with what $FOUR$ (and the other plausible lattices discussed here) presents us with. In thinking about this question, I found that this is more intricate than I had assumed.

Something similar can be said about the whole project of judging the merits of FDE as the base of a constructive logic: On the one hand, I found it to be more challenging to think through the issue than I expected. On the other hand, it now strikes me more clearly than before as a project worthwhile pursuing.

## References


1. Jon Cogburn. The philosophical basis of what? The anti-realist case for dialethism. *The Law of Non-Contradiction*, pages 503–24, 2004.
2. Michael Dummett. Testimony and memory. In *Knowing from words*, pages 251–272. Springer, 1994.
3. J Michael Dunn. Contradictory information: Too much of a good thing. *Journal of Philosophical Logic*, 39(4):425–452, 2010.
4. Josep Maria Font. Belnap's four-valued logic and de morgan lattices. *Logic Journal of IGPL*, 5(3):1–29, 1997.
5. Susan Haack. Trial and error: the supreme court's philosophy of science. *American Journal of Public Health*, 95(S1):S66–S73, 2005.
6. Daniel Kahneman and Amos Tversky. Prospect theory: An analysis of decisions under risk. *Econometrica*, 47:278, 1979.
7. Andreas Kapsner. *Logics and Falsifications*, volume 40 of *Trends in Logic*. Springer, 2014.
8. Andreas Pietz and Umberto Rivieccio. Nothing but the truth. *Journal of Philosophical Logic*, 42(1): 125-135, 2013.
9. Yaroslav Shramko, J Michael Dunn, and Tatsutoshi Takenaka. The trilattice of constructive truth values. *Journal of Logic and Computation*, 11(6):761–788, 2001.
10. Yaroslav Shramko and Heinrich Wansing. *Truth and falsehood: An inquiry into generalized logical values*, volume 36. Springer Science & Business Media, 2011.

# Bridging the Two Plans in the Semantics for Relevant Logic

Takuro Onishi

**Abstract** This paper considers how the two plans in the semantics for relevant logic are related to each other. The so-called American plan, classical-style four-valued semantics, is intuitive, but weak. The so-called Australian plan, two-valued frame semantics, is very powerful, but the semantic devices employed need some explanation. Examining R. Routley's 1984 paper 'American plan completed,' this paper argues that the American plan provides an explanatory and ontological basis for the Australian plan, and that the latter is just a developed form of the former.

**Keywords** Relevant logic • American plan • Australian plan • Routley star • Negation as modality • First Degree Entailment

## 1 Introduction

There have been two "plans" in the semantics for relevant logic. One is the tradition of Belnap-Dunn style four-valued semantics [8, 1, 2],[1] which is called the *American plan*. The other is the *Australian plan* that is two-valued relational (frame) semantics, also known as Routley-Meyer semantics [24, 25, 23]. The present paper considers how these two plans are related to each other and concludes that the Australian plan can be regarded as a developed form of the American plan.

I start with R. Routley's 1984 paper "American plan completed," in which he examined how to extend the American plan to accommodate full relevant implication [22]. His attitude in the paper is indeed interesting. While he condemns the extended American plan for being hopelessly complicated, he also suggests that it might perhaps provide an explanatory basis for the star function employed in the Australian plan, which has been often criticized as unintuitive. In this paper I try to make clear the import of this suggestion referring to the recent discussion on negative modality.

## 2 Routley's American plan

This section quickly reviews the two plans in the semantics of relevant logic, and introduces Routley's extended American plan.

Takuro Onishi
Unit of Kyoto Initiatives for the Humanities and Social Sciences, Kyoto University, Kyoto, Japan
e-mail: takuro.onishi@gmail.com

[1] In addition to these original works, I refer to [10, 19, 11, 14] for equivalent formulations of FDE and helpful references.

 
H. Omori, H. Wansing (eds.), *New Essays on Belnap-Dunn Logic*,
Synthese Library 418, https://doi.org/10.1007/978-3-030-31136-0_13

## 2.1 The two plans

The American plan is a four-valued semantics for FDE (First Degree Entailment) that assigns to each formula one of the four values, *true, false, neither true nor false*, and *both true and false*. It is equivalently formulated as a semantics in which a valuation assigns truth and falsity separately to each formula, and hence there are four cases (true but not false, false but not true, neither, and both) corresponding to the above four values.

Given a valuation over atomic formulas, truth and falsity of complex formulas are determined by *classical* semantic clauses, especially for negation, we have:

$$\sim A \text{ is true} \iff A \text{ is false}$$
$$\sim A \text{ is false} \iff A \text{ is true}$$

So there is a sense in which the meaning of negation (and the other connectives, conjunction and disjunction) is the same as that in classical logic. The only difference is that while in classical logic truth and falsity are mutually exclusive and jointly exhaustive, they are not in FDE.

The Australian plan is a two-valued frame semantics in which truth and falsity (i.e. non-truth) are relativized to states (worlds or situations). The main feature is the semantic clauses for implication and negation:

$$x \models A \to B \iff \forall y, z : Rxyz \,\&\, y \models A \Rightarrow z \models B;$$
$$x \models \sim B \iff x^* \not\models A,$$

where $R$ is a ternary relation, $*$ a function on a frame, and $x \models A$ means that $A$ is true at a state $x$.

The ternary relation is a very powerful tool to represent various principles concerning relevant implication as corresponding to properties of it, which is impossible in the American plan. On the other hand, the intuitiveness of the classical semantic clauses is lost in the Australian plan. In particular, the so-called *star function* or *Routley star* $*$ that models relevant negation is often criticized for its lack of intuitive meaning [5, 26].

One may think that it would be great if we could put together the two plans to obtain a semantics that has both ternary relation for implication and the intuitive classical clauses for negation, avoiding the use of Routley star. This is what Routley attempted in the 1984 paper. Let us look at where it lead him.

## 2.2 Four-valued relational semantics

We call it the *extended American plan*. It is a four-valued frame semantics in which each model is equipped with two ternary relations, $R$ and $S$, and a twin valuation $\langle \models^+, \models^- \rangle$ that determines truth and falsity of formulas separately. The additional ternary relation $S$ is for the falsity condition of implication. Truth and falsity are relativized to states, but the semantic clauses for conjunction, disjunction, and importantly, negation preserve classicality, which means we use no star function.

Below are some definitions and basic facts concerning the semantics based on Routley's presentation, some details are changed though.

*Definition* 1 (Language). We work with the set $L$ of formulas built from the connectives $\wedge, \vee, \sim, \rightarrow$ over the set $PV$ of propositional variables.

*Definition* 2 (Frames and models). A structure $\langle 0, 0^*, U, V, K, R, S \rangle$, where $K$ is a non-empty set of states, $0 \in U \subseteq K$, $0^* \in V \subseteq K$, and $R, S \subseteq K^3$, is a **B-frame** if

- $\exists u \in U : Ruxy \iff \exists v \in V : Sxvy$
  - We write $x \leq y$ if either one of these holds.
- $x \leq x$; [2]
- $Rxyz \,\&\, x' \leq x \Rightarrow Rx'yz$; $Syxz \,\&\, x \leq x' \Rightarrow Syx'z$.

Let $\models^+, \models^- \subseteq K \times PV$ be *hereditary* valuations on a **B**-frame $\mathcal{F} = \langle 0, 0^*, U, V, K, R, S \rangle$ such that for any $x, y \in K$ and any propositional variable $p$,

$$x \models^+ p \,\&\, x \leq y \Rightarrow y \models^+ p; \quad x \models^- p \,\&\, x \leq y \Rightarrow y \models^- p.$$

We call $\langle \mathcal{F}, \langle \models^+, \models^- \rangle \rangle$ a **B-model**.

Given a **B**-model $\mathcal{M} = \langle 0, 0^*, U, V, K, R, S, \langle \models^+, \models^- \rangle \rangle$, the twin valuation $\langle \models^+, \models^- \rangle$ is extended to arbitrary formulas as follows:

$$\begin{aligned}
x \models^+ A \wedge B &\iff x \models^+ A \,\&\, x \models^+ B \\
x \models^- A \wedge B &\iff x \models^- A \text{ or } x \models^- B \\
x \models^+ A \vee B &\iff x \models^+ A \text{ or } x \models^+ B \\
x \models^- A \vee B &\iff x \models^- A \,\&\, x \models^- B \\
x \models^+ \sim A &\iff x \models^- A \\
x \models^- \sim A &\iff x \models^+ A \\
x \models^+ A \rightarrow B &\iff \forall y, z : Rxyz \,\&\, y \models^+ A \Rightarrow z \models^+ B \\
x \models^- A \rightarrow B &\iff \exists y, z : Syxz \,\&\, y \models^- B \,\&\, z \not\models^- A.
\end{aligned}$$

We write like $\mathcal{M}, x \models A$ to make explicit which model is in question.

The (twin) heredity property of valuations is extended to all formulas.

*Proposition* 3. For any formula $A \in L$ and any states $x, y$ in a **B**-model,

$$x \models^+ A \,\&\, x \leq y \Rightarrow y \models^+ A; \quad x \models^- A \,\&\, x \leq y \Rightarrow y \models^- A.$$

*Definition* 4 (Entailments and validities). We define different notions of entailment and validity at different level. Let $\mathcal{F}$ denote an arbitrary **B**-frame $\langle 0, 0^*, U, V, K, R, S \rangle$, and $\mathcal{M}$ an arbitrary **B**-model on $\mathcal{F}$.

$$\begin{aligned}
A \models^{\mathcal{M}^+} B &\iff \forall x \in K : \mathcal{M}, x \models^+ A \Rightarrow \mathcal{M}, x \models^+ B \\
A \models^{\mathcal{M}^-} B &\iff \forall x \in K : \mathcal{M}, x \models^- B \Rightarrow \mathcal{M}, x \models^- A \\
A \models^{\mathcal{M}} B &\iff A \models^{\mathcal{M}^+} B \,\&\, A \models^{\mathcal{M}^-} B \\
A \models^{\mathcal{F}^+} B &\iff A \models^{\mathcal{M}^+} B \text{ for any } \mathcal{M} \text{ on } \mathcal{F} \\
A \models^{\mathcal{F}^-} B &\iff A \models^{\mathcal{M}^-} B \text{ for any } \mathcal{M} \text{ on } \mathcal{F}
\end{aligned}$$

[2] In what follows we assume that the binary relation $\leq$ so defined is always a partial order, namely that it is not only reflexive but also transitive and anti-symmetrical. As usual, this does not affect the logic.

$$\begin{aligned}
A \models^{\mathcal{F}} B &\iff A \models^{\mathcal{F}^+} B \,\&\, A \models^{\mathcal{F}^-} B\\
A \models^{+} B &\iff A \models^{\mathcal{F}^+} B \text{ for any } \mathcal{F}\\
A \models^{-} B &\iff A \models^{\mathcal{F}^-} B \text{ for any } \mathcal{F}\\
A \models B &\iff A \models^{+} B \,\&\, A \models^{-} B.
\end{aligned}$$

We say that $A$ *t-entails* (*f-entails*, or *2-entails*) $B$ in a model $\mathcal{M}$ when $A \models^{\mathcal{M}^+} B$ ($A \models^{\mathcal{M}^-} B$ or $A \models^{\mathcal{M}} B$ resp.). Similarly for entailment in a frame and entailment *simpliciter* (i.e. $\models^+, \models^-$ and $\models$).

A formula $A$ is said to be *t-valid* (*f-valid*) *in a model* $\mathcal{M}$ if $0 \models^+ A$ ($0^* \not\models^- A$) in $\mathcal{M}$. If $A$ is both $t$-valid and $f$-valid in $\mathcal{M}$, it is *2-valid* in $\mathcal{M}$. Validity in a frame and validity *simpliciter* are defined in an obvious way.

*Proposition* 5. For any **B**-model $\mathcal{M}$,

$$\begin{aligned}
A \models^{\mathcal{M}^+} B &\iff 0 \models^{+} A \to B\\
A \models^{\mathcal{M}^-} B &\iff 0^* \not\models^{-} A \to B.
\end{aligned}$$

**Theorem 1 (Routley 1984).** *A formula A is 2-valid if and only if it is provable in the axiom system of* ***B***.

## 2.3 Extensions

Routley's extended American plan captures the basic relevant logic **B** employing two ternary relations but without the star function. The next step is to represent various extensions of **B** in terms of structural properties of frames.

In the usual two-valued framework, we have the following kind of correspondence. For example, *EAssertion* $(A \to (A \to B)) \to B$ is valid in a frame if and only if the frame satisfies the commutativity of $R$:

$$Rxyz \Longrightarrow Ryxz.$$

It is one of the great features of the Australian plan that it enables us to develop a correspondence theory between axiom schemes and frame properties. It is highly desirable that this should be carried over to the extended American plan which aims to unify the two plans.

There is no problem with *positive* extensions, extensions with axiom schemes not involving negation. All that we have to do is to add the dual version of the original frame properties. For EAssertion and commutativity, we add the commutativity of $S$:

$$Sxyz \Longrightarrow Sxzy,$$

which corresponds to f-validity of EAssertion, while $Rxyz \Rightarrow Ryxz$ is now the counterpart of its t-validity.

Things are not this easy with *negative* extensions of **B**, extensions with axiom schemes involving negation. We focus on the axiom form of Contraposition since Routley takes it as "critical, as well as representative" [22, p. 145]. In the Australian plan, the frame condition that corresponds to Contraposition is concisely formulated appealing to the star function:

$$(A \to B) \to (\sim B \to \sim A) \qquad \text{corresponds} \qquad Rxyz \Longrightarrow Rxz^*y^*.$$

How can we force Contraposition in our extended American plan without the star? Routley considered several options and decided to *double* the semantic clauses for implication. That is, he revised them as follows:

$$x \models^+ A \to B \iff \forall y, z : Rxyz \Rightarrow \begin{cases} y \models^+ A \Rightarrow z \models^+ B & \text{and} \\ z \models^- B \Rightarrow y \models^- A \end{cases}$$

$$x \models^- A \to B \iff \exists y, z : Syxz \Rightarrow \begin{cases} y \models^- B \,\&\, z \not\models^- A & \text{and} \\ z \not\models^+ B \,\&\, y \models^+ A \end{cases}$$

The idea is to make Contraposition built in the meaning of implication, which sounds nicely parallel to the definition of 2-entailment we have seen in the previous section.

Unfortunately this causes awful complication. Especially for nested implication schemes, even when considering only t-validity, we would have to take into account the falsity of (embedded) implication, and hence some interaction between the dual ternary relations $R$ and $S$. Worse, there is a case where a structural property of $R$ and $S$ looks helpless however complicated we make it: Consider proving t-validity of EAssertion under the doubled semantic clauses (and some frame property). By proposition 5, we would show that there is no state $x$ in any model such that $x \models^+ A$ and $x \not\models^+ (A \to B) \to B$. Suppose otherwise. Then for some $y, z$ such that $Rxyz$, either (1) $y \models^+ A \to B$ and $z \not\models^+ A$ or (2) $z \models^- B$ and $y \not\models^- A \to B$. Now, in the latter case, how could we derive a contradiction with the assumption $x \models^+ A$? One will immediately see that any structural property of $R$ and $S$ would not be enough. We need some kind of *cross-over between truth and falsity* ($\models^+$ and $\models^-$).

Confronted with this problem, Routley gave up a star-free semantics. He introduced a *star-like relation* or *star-imitation*, which looks like the star function but is not supposed to be a function. We are going to give a closer look to the relation in the next section, but before that, let us review his attempt and consider how we should proceed hereafter.

## *2.4 Changing the strategy*

The original goal was to build a star-free semantics that is powerful and flexible enough to develop a correspondence theory. Now it has turned out that the semantics is far more complicated than the Australian plan, and that the use of star-like thing is unavoidable anyway. Routley concludes the paper saying:

> The American plan is, in the end, when duly enriched with the Australian mateship component [star-like relation], little more than a very circuitous and cumbersome way of avoiding the much more intuitive Australian plan. [22, p. 157]

This does not necessarily mean the victory of the Australian plan over the American plan. Indeed, though Routley seems to think that the Australian plan including the star function is "much more intuitive," many will disagree. Unless the Routley star is made sense, the Australian plan won't be accepted as an appropriate semantics for relevant logic either.

Here I'd like to take notice of a suggestion Routley himself made in the same paper:

> So perhaps a major virtue of the American approach is that it provides another approach to and explanation of——what it was set up partly in opposition to, and was supposed to be getting away from——the star negation rule and its role in the Australian plan. [22, p. 136]

He is suggesting a slightly different strategy, that is, a strategy of using the American plan as conceptual resource to *explain* the star function, rather than to *avoid* it. Unfortunately, he did not fully make clear in the paper how the strategy will develop. Hence it is the aim of the present paper. In what follows I try to make clear how the American plan works as an explanatory basis for the star function.

## 3 Star-like relation(s)

In this section, I introduce Routley's star-like relation and discuss how it is explained on the basis of the American plan.

### 3.1 *Explanatory task*

The star-like relation Routley employed is a binary relation $M$ on a frame such that for any state $x$ and $y$, if $xMy$ holds then we have

$$x \models^{-} A \iff y \not\models^{+} A \tag{$M$}$$

for any formula $A$.[3] As you may notice, the equivalence is similar to the star-negation rule

$$x \models \sim A \iff x^{*} \not\models A.$$

However, $x$'s $M$-correlate $y$ may not be the function of $x$. $M$ is only assumed to be symmetric, which gives an effect similar to that brought by $x = x^{**}$.

In fact, a relation seems to be sufficient to formulate frame properties that correspond to negative extensions. E.g., Contraposition $(A \to B) \to (\sim B \to\sim A)$ corresponds to the frame property

$$\begin{cases} Rxyz \Rightarrow \exists y', z' : yMy' \,\&\, zMz' \,\&\, Rxz'y' \text{ and} \\ Syxz \Rightarrow \exists y', z' : yMy' \,\&\, zMz' \,\&\, Sz'xy'. \end{cases}$$

Reductio $(A \to\sim A) \to\sim A$ is represented by

$$\begin{cases} \exists x' : xMx' \,\&\, Rxx'x \text{ and} \\ \exists x' : xMx' \,\&\, Sxxx' \end{cases}$$

[3] Strictly speaking, it is first assumed that any valuation satisfies the condition for propositional parameters, and then it is proved that the equivalence extends to arbitrary formulas, appealing to several conditions imposed on $R, S$ and $M$.

These are just a relational version of our familiar conditions $Rxyz \Rightarrow Rxz^*y^*$ and $Rxx^*x$. And it is not likely that there is an axiom schemes that requires a function rather than a relation.

Then, we should say that our explanatory task is twofold. First, of course we are obliged to explain the meaning of the star-like relation, that is, what it is for two state to stand in the relation. In addition to that, we need an explanation of the functionhood of the star-function. Why is it supposed to be a function in the standard Routley-Meyer semantics?

## 3.2 Compatibility and exhaustiveness

The story that I am going to tell about the meaning of the star-like relation is like this. As we will see shortly, thanks to the recent study on *negation as modal operator*, we are in a much better position concerning this explanatory task than Routley was when he wrote the paper [9, 21, 13, 15]. But the explanatory strategy of the negation-as-modality approach faces some fundamental objection. Then we will see that, on the framework of the extended American plan, we can employ the useful concepts from the negation-as-modality approach while avoiding the objection. I will argue that this is what Routley suggests as the virtue of the American plan.

First, let us decompose the equivalence ($M$). If $xMy$, we have the following two implications:

$$x \models^- A \Longrightarrow y \not\models^+ A \tag{1}$$

$$x \not\models^- A \Longrightarrow y \models^+ A. \tag{2}$$

(1) is equivalent to "it is excluded that $A$ is false at $x$ while it is true at $y$." In other words, the information contained $x$ and $y$ never clash each other. In this sense, $x$ and $y$ can be said to be *compatible* with each other. The converse direction, (2), is equivalent to "either $A$ is false at $x$ or $A$ is true at $y$." Thus, $x$ and $y$ jointly cover the whole logical possibilities, truth and falsity. They are jointly *exhaustive*. Therefore, for two states to stand in the star-like relation $M$ is for them to be mutually compatible and jointly exhaustive. The star-like relation is the conjunction of compatibility and exhaustiveness.

Some might say that compatibility and exhaustiveness are, properly speaking, relations between propositions or properties, not between states. Here is an example from logic that will motivate our notion of compatibility and exhaustiveness.

A class of models defined by some properties determines (in)validity of formulas. Each can be seen as a state that contains information about validity and a different class of models will carry different information from another. Take as an example the Kripke-style model theory for modal logic. Let $x$ be the class of all Kripke models (with no constraint on accessibility relation). Then we can write, for example, $x \models^- \Box A \supset A$, in the sense that $x$ contains a counter model to the axiom scheme. On the other hand, if $y$ is the class of models in which a binary accessibility relation is reflexive, then we have $y \not\models^- \Box A \supset A$. Similarly, take Hilbert style axiom systems. Each of them may also be regarded as a state. We write $z \models^+ A$ if the system $z$ proves $A$. By adding or dropping axioms we have different states. Now if a proof system $y$ is sound with respect to a class $x$ of models, then (1) holds. Soundness is compatibility between models and proof systems. Conversely, exhaustiveness (2) means completeness in the world of logic.

Models and proof systems form a system of information states which enable us to make a certain range of inferences. If $A$ has a counter model in a class $x$ of models and a proof system $y$ is compatible with (sound with respect to) $y$, one can infer that $A$ is not provable in $y$, etc. We can find such systems in various areas and fields in which different bodies of information are somehow related so that we can infer what is true (or false) in one of them from what is true (or false) in another. I believe that accessibility relations deployed in frame semantics *in general*, including binary relation for necessity and possibility and ternary relation for relevant implication as well as compatibility and exhaustiveness, can be understood in this way, i.e., as inferential relationships between information states. This is not a place to develop a general theory, so I just refer to the notion of situated inference due to Mares [13] and the conception of ternary relation as information channel by Restall [20] as pointing to the direction I am thinking about. I shall go forward assuming that the meaning of star-like relation has been explained enough.

### 3.3 The explanatory virtue of the American plan

The star-like relation $M$ is the conjunction of compatibility and exhaustiveness. Let us divide it explicitly and work with models that have two binary relations $C$ (compatibility) and $E$ (exhaustiveness) such that

$$xCy\,\&\,x \models^- A \Longrightarrow y \not\models^+ A \qquad \text{(Com)}$$

$$xEy\,\&\,x \not\models^- A \Longrightarrow y \models^+ A. \qquad \text{(Exh)}$$

They are assumed to be symmetric as in the case of $M$.

Given binary relations on a frame, we may introduce two *negative modal operators*, $\rhd$ and $\blacktriangleright$, with the semantic clauses:

$$x \models^+ \rhd A \iff \forall y : xCy \Longrightarrow y \not\models^+ A \qquad (\rhd)$$

$$x \models^+ \blacktriangleright A \iff \exists y : xEy\,\&\,y \not\models^+ A. \qquad (\blacktriangleright)$$

These are justifiably called *negative* modality since they support contraposition (If $A \models B$ then $\rhd B \models \rhd A$ and $\blacktriangleright B \models \blacktriangleright A$) and some inferences characteristic to negation. $\rhd$ is *impossibility* operator and $\blacktriangleright$ is understood as *unnecessity*. There is now much work on negative modality of this kind, which are usually based on two-valued semantics: the truth conditions are

$$x \models \rhd A \iff \forall y : xCy \Longrightarrow y \not\models A \qquad (\rhd_2)$$

$$x \models \blacktriangleright A \iff \exists y : xEy\,\&\,y \not\models A. \qquad (\blacktriangleright_2)$$

Kosta Došen [7] initiated the investigation into negative modality parallel to that into positive modality (necessity and possibility), aiming to elaborate a correspondence theory between axiom schemes and frame properties. His base logic was intuitionistic logic. In the relevant tradition, logicians like Dunn, Restall and Mares have been trying to make sense of star-negation in terms of negative modality [9, 21, 13]. According to the *star postulate* they found, the star state is characterized as "maximally compatible state." It seems to be reasonable to say that the star function is now fully made sense by the star postulate. But here we consider an objection that the negative modality approach *in general* must answer.

The negative modality approach aims to explain their target negations as negative modality, and, in turn, negative modality in terms of binary relation. In other words, binary relation is assumed to be conceptually prior to negative modality and negation. De and Omori argue that this conceptual priority is objectionable:

> If there is any kind of explanation as to why these worlds should be incompatible with ours, we can only see that it must ultimately appeal to negation. [6]

The above ($\triangleright_2$) tells us that the negation of $A$ (or impossibility of $A$) holds at a state $x$ if and only if $A$ fails at any state that is compatible with $x$. But then, what is it for two states to be compatible with each other? The only answer available to us seems to be that two states are compatible if truth of a proposition at one of them excludes the *negation* of the proposition at another. This is just another way of reading ($\triangleright_2$) and we are forced to use the notion of negation to explain the notion of compatibility. Thus, the strategy to explain negation in terms of binary relation is circular, unless the notion of compatibility is given some characterization independent from ($\triangleright_2$), which seems unavailable. As far as I know, we have no principle concerning negation and (in)compatibility other than the above ($\triangleright_2$) on the framework of *two-valued* semantics (similarly for $\blacktriangleright_2$ and exhaustiveness).

It might be possible to find such an independent characterization and resolve the worry of circularity. But it does not matter to us anyway. I mean, in the extended American plan, there is no need to care about the worry at all, for we started with four-valued semantics where negation is already defined by the classical clauses. We don't have to use (Com) and (Exh) as an explanation of negation and falsity. (Com) and (Exh) can be used simply as an explanation of $C$ and $E$ in terms of primitive notions of truth and falsity (or negation).

I claim this is the virtue of the American plan that Routley suggested. If you are a member of the Australian camp trying to make sense of the unintuitive star-negation in terms of binary relation on a two-valued frame, you will have a trouble concerning conceptual priority. In contrast, if you start under the American plan, then negation is simply defined by the intuitive classical clauses and there is no obstacle to introduce compatibility and exhaustiveness relations on a frame, which will be useful to make sense of the Routley star. In this way, the American plan works as an explanatory basis for the star function. [4]

## 4 The star function

We have finished the first task of explaining the meaning of star-like relation based on the American plan: the star-like relation is the conjunction of compatibility and exhaustiveness. In this section I shall make clear why the star-function is supposed to be a function.

As I mentioned above, the star function is now understood through the star postulate, formally written as

[4] There is an approach to semantics of negation and logical constants in general which is based on the notion of incompatibility between *propositions* (sentences, statements, and the like) [3, 4, 17, 18]. A similar objection would be raised concerning the conceptual priority to this approach too, and the proponents might answer to it by invoking an explanation of incompatibility in terms of preclusion relation between commitment and entitlement to propositions: roughly speaking, propositions $A$ and $B$ are incompatible if anyone who is committed to $A$ is thereby precluded from counting as entitled to $B$ (and vice versa) [4, p. 120]. This may satisfy the doubt of circularity, or may not. The notion of preclusion might still appear to involve some notion of negation. I am neutral on this point. The point of the present argument is just that we don't need to care about this if we start with the American plan.

$$\forall x \exists y : xCy \,\&\, \forall z : xCz \Rightarrow z \leq y, \tag{S}$$

That is, each state has a mate state that is maximally compatible with it. On a frame satisfying this constraint, we can define the star function by setting $x^*$ to be the maximally compatible state $y$. But I will discuss the functionhood of the star in light of the dualistic version of this postulate, the *dualist star postulate*,

$$\forall x \exists y : xCy \,\&\, xEy \,\&\, (\forall w : xCw \Rightarrow w \leq y) \,\&\, (\forall w : xEw \Rightarrow y \leq w), \tag{DS}$$

which is essentially the conjunction of the uniqueness and existence conditions for the functionhood of the star. [5]

Before going into the discussion, note that to suppose the star to be a function, or equivalently to assume the (dualist) star postulate, does not change the logic. This is seen from the canonical model construction that Routley presented in the 1984 paper. For each prime theory $a$, which is an element of the canonical model, he defined the *opposite* of $a$ to be

$$a^* := \{B \mid\sim B \notin a\}.$$

as a useful "auxiliary notion." Indeed this behaves just as the star. Thus, the star function is *implicitly* there in the canonical model, which implies the assumption of functionhood of the star does not affect the logic as the set of valid formulas and entailments. In the following, I will examine what kind of change is introduced to the entire structure of models by *explicitly* assuming functionhoood of the star.

## 4.1 Models with star-like relations

First, let me define a model equipped with compatibility and exhaustiveness formally.

*Definition* 6 ($B^{CE}$-frames and models). Let $\mathcal{F} = \langle 0, 0^*, U, V, K, R, S\rangle$ be a **B**-frame, and $C, E \subseteq K^2$. A structure $\langle \mathcal{F}, C, E\rangle$ is called a $\mathbf{B}^{CE}$**-frame** if

1. $xCy \Rightarrow yCx$; $xEy \Rightarrow yEx$;
2. $xCy \,\&\, x' \leq x \Rightarrow x'Cy'$; $xEy \,\&\, x \leq x' \Rightarrow x'Ey$;
3. $\exists x' : Szx'y \,\&\, xCx' \Rightarrow \exists y', z' : yEy' \,\&\, zCz' \,\&\, Rxy'z'$;
4. $\exists x' : Rx'yz \,\&\, xEx' \Rightarrow \exists y', z' : yCy' \,\&\, zEz' \,\&\, Sz'xy'$.

If a hereditary twin valuation $\langle \models^+, \models^- \rangle$ on a $\mathbf{B}^{CE}$-frame satisfies the compatibility and exhaustiveness conditions, i.e.

$$xCy \,\&\, x \models^- p \Longrightarrow y \not\models^+ p \tag{Com}$$
$$xEy \,\&\, x \not\models^- p \Longrightarrow y \models^+ p. \tag{Exh}$$

for any propositional parameter $p$, we call $\langle \mathcal{F}, C, E, \langle \models^+, \models^- \rangle$ a $\mathbf{B}^{CE}$**-model**. The notions of validity and entailments at different levels are defined as before.

We add to the language two unary operators $\rhd$ and $\blacktriangleright$ for which the semantic rules are as follows:

[5] I think the dualist version is better than the original one in several aspects. And it is one of its merits that, as we will see, the dualist version allows us to analyze the uniqueness and existence conditions separately. For other merits of the dualist approach, see [15, 16].

$$\begin{aligned}
x \models^+ \rhd A &\iff \forall y : xCy \Longrightarrow y \not\models^+ A \\
x \models^+ \blacktriangleright A &\iff \exists y : xEy \,\&\, y \not\models^+ A \\
x \models^- \rhd A &\iff \exists y : xEy \,\&\, y \not\models^- A \\
x \models^- \blacktriangleright A &\iff \forall y : xCy \Longrightarrow y \not\models^- A.
\end{aligned}$$

These negative modalities are not ultimately necessary for our purpose, but they serve to make things clear.

The three conditions on valuations, heredity, compatibility and exhaustiveness, can be extended to arbitrary formulas as expected:

*Proposition* 7. For any formula $A$ and any states $x$ and $y$ in a $\mathbf{B}^{CE}$-model,

$$\begin{aligned}
x \leq y \,\&\, x \models^+ A &\Longrightarrow y \models^+ A \\
x \leq y \,\&\, x \models^- A &\Longrightarrow y \models^- A \\
xCy \,\&\, x \models^- A &\Longrightarrow y \not\models^+ A && \text{(Com)} \\
xEy \,\&\, x \not\models^- A &\Longrightarrow y \models^+ A. && \text{(Exh)}
\end{aligned}$$

*Proof.* Routine. The condition 2 is for the heredity of $\rhd$ and $\blacktriangleright$, and 3 and 4 for the case of implication in the inductive proof of (Com) and (Exh) respectively. If other intensional connectives like fusion are added, further conditions are required.

## 4.2 Dualist star postulate

Now we shall consider what happens if we add the *dualist star postulate*:

$$\forall x \exists y : xCy \,\&\, xEy \,\&\, (\forall w : xCw \Rightarrow w \leq y) \,\&\, (\forall w : xEw \Rightarrow y \leq w), \qquad \text{(DS)}$$

which claims the existence of "maximally compatible" and "minimally exhaustive" state $y$ for each state $x$. As opposed to the original star postulate as a constraint on $C$, this is dualistic in the sense that it features $E$ as well as $C$. Note that (DS) is equivalent to the conjunction of the following two conditions (modulo the transitivity of $\leq$):

$$\begin{aligned}
&\forall x, y, z : xCy \,\&\, xEz \Longrightarrow y \leq z, && \text{(DS1)} \\
&\forall x \exists y : xCy \,\&\, xEy. && \text{(DS2)}
\end{aligned}$$

Recall that Routley's star-like relation $M$ is the conjunction of compatibility ($C$) and exhaustiveness ($E$). Then clearly (DS2) is the existence condition for the functionhood of $M$, and it should be also obvious that (DS1) is the uniqueness condition: $xMy$ and $xMz$ imply $y = z$ under (DS1) (modulo the anti-symmetry of $\leq$). Let us look at them separately.

First, it seems rather reasonable to assume (DS1):

*Proposition* 8. In a $\mathbf{B}^{CE}$-model, we can assume (DS1) without changing the logic.

*Proof.* If $xCy$ and $xEz$, then, thanks to (Com) and (Exh), we have

$$\begin{aligned}
y \models^+ A \Rightarrow x \not\models^- A \Rightarrow z \models^+ A; \\
y \models^- A \Rightarrow x \not\models^+ A \Rightarrow z \models^- A
\end{aligned}$$

for any formula $A$. Considering that the ordering relation in the canonical model is defined by subset relation between the sets of true formulas and false formulas in different states, we can assume $y \leq z$ here.

As shown in the proof, (DS1) can be seen as a natural by-product of (Com) and (Exh), so I assume that every $\mathbf{B}^{CE}$-model satisfies (DS1) hereafter. In contrast, it is not immediately clear where (DS2) comes from. I will discuss the origin of the existence condition in the next section. Before that, let us see what happens if we further add (DS2). Indeed the change caused is quite big. The following is an application of the result of [15]:

*Proposition* 9. (Cf. [15]). On a $\mathbf{B}^{CE}$-model with (DS1) and (DS2), we can define a function $*$ such that:

$$x \leq y \Longrightarrow y^* \leq x^*; \quad x^{**} = x;$$
$$x \models^+ \sim A \iff x^* \not\models^+ A \iff x \models^+ \rhd A \iff x \models^+ \blacktriangleright A;$$
$$x \models^- \sim A \iff x^* \not\models^- A \iff x \models^- \rhd A \iff x \models^- \blacktriangleright A;$$
$$Rxyz \iff Sz^*x^*y^*.$$

*Proof.* Immediate. See the condition 3 and 4 for the last item.

And more importantly, under (DS2), the model is essentially two-valued. Once having defined the star-function on a $\mathbf{B}^{CE}$-model as above, we can just forget the falsity valuation $\models^-$ and the structure of $0^*, V$ and $S$. The result is a standard two-valued $\mathbf{B}$-model with Routley star. Conversely, given a two-valued $\mathbf{B}$-model with the star, by defining

$$x \models^+ p \iff x \models p$$
$$x \models^- p \iff x \models \sim p$$
$$Sxyz \iff Ry^*z^*x^*$$
$$xCy \iff xEy \iff y = x^*$$

etc., we get a four-valued $\mathbf{B}^{CE}$-model with (DS2). This construction constitutes an isomorphism between the class of two-valued $\mathbf{B}$-model with Routley star and that of four-valued $\mathbf{B}^{CE}$-models with (DS2). And it is immediate that (the $\{\wedge, \vee, \rightarrow, \sim\}$-fragment of) the logic defined by the class of $\mathbf{B}^{CE}$-model is $\mathbf{B}$ (Note that $\rhd$ and $\blacktriangleright$ collapse to $\sim$).

Explicitly assuming the functionhood of star, the whole framework of four-valued semantics collapses to the two-valued one. In other words, we have seen how the Australian plan generates from the American plan: the Australian plan is the extended American plan with (DS2). However, as I noted, the origin of (DS2) is unclear. How can such an assumption that causes a big change be motivated?

### 4.3 Contraposition

I shall now turn to contraposition. It may sound a bit abrupt, but it will shed a light on the structure of the extended American plan. Here we are concerned with rule-contraposition ($A \models B$ implies $\sim B \models \sim A$) rather than formula-contraposition ($(A \rightarrow B) \rightarrow (\sim B \rightarrow \sim A)$).

Let us first observe that validity (entailment preservation) of rule-contraposition is *built in* the definition of entailment in Routley's American plan since it is defined as 2-

entailment, truth preservation *and* backward falsity preservation: $A \models B$ is trivially equivalent to $\sim B \models \sim A$. But a closer look readily reveals that the bi-directional preservation is not necessary.

*Proposition* 10. With the class of $\mathbf{B}^{CE}$-frames, we have

$$A \models^{+} B \iff A \models^{-} B \quad (\iff \sim B \models^{+} \sim A).$$

Recall that $\models^{+}$ and $\models^{-}$ here denote truth preservation and backward falsity preservation in all frames respectively. Also note that the same holds good for the class of $\mathbf{B}$-frames since a $\mathbf{B}$-frame is just a special case of $\mathbf{B}^{CE}$-frame where $C = E = \emptyset$.

*Proof.* It suffices to show any counter-model against $A \models^{+} B$ can be turned into one against $A \models^{-} B$.

Let $\mathcal{F} = \langle 0, 0^{*}, U, V, K, R, S, C, E, \models^{+}, \models^{-} \rangle$ be a $\mathbf{B}^{CE}$-model such that for some $w \in K$,

$$w \models^{+} A \text{ but } w \not\models^{+} B.$$

We define a dualized model $\mathcal{F}_d = \langle 0_d, 0^{*}_d, U_d, V_d, K_d, R_d, S_d, C_d, E_d, \models^{+}_d, \models^{-}_d \rangle$ to be such that:

$$0_d = 0^{*},\ 0^{*}_d = 0,\ U_d = V,\ V_d = U,\ K_d = K,$$

$$R_d xyz \iff Szxy \qquad S_d xyz \iff Ryzx$$
$$xC_d y \iff xEy \qquad xE_d y \iff xCy$$
$$x \models^{+}_d p \iff x \not\models^{-} p \qquad x \models^{-}_d p \iff x \not\models^{+} p.$$

It is immediately seen that $\mathcal{F}_d$ is a $\mathbf{B}^{CE}$-model. Note that we have

$$x \leq_d y \iff y \leq x$$

for the ordering relation $\leq_d$ defined by $x \leq_d y \iff R_d 0_d xy$. The rest of the conditions for $\mathbf{B}^{CE}$-model follow from this.

Then, as expected, we can easily prove by induction that for any formula $A$ and any $x \in K$,

$$x \models^{+}_d C \iff x \not\models^{-} C \qquad x \models^{-}_d C \iff x \not\models^{+} C.$$

Here are a few examples.

$$\begin{aligned} x \models^{+}_d C \to D &\iff \forall y, z : R_d xyz \,\&\, y \models^{+}_d C \Longrightarrow z \models^{+}_d D \\ &\iff \forall y, z : Szxy \,\&\, y \not\models^{-} C \Longrightarrow z \not\models^{-} D \\ &\iff x \not\models^{-} C \to D. \\ x \models^{+}_d \rhd C &\iff \forall y : xC_d y \Longrightarrow y \not\models^{+}_d C \\ &\iff \forall y : xEy \Longrightarrow y \models^{-} C \\ &\iff x \not\models^{-} \rhd C. \end{aligned}$$

Thus, applying this to the above assumption, we can conclude that

$$w \not\models^{-}_d A \text{ but } w \models^{-}_d B,$$

which implies that $\mathcal{F}_d$ is a counter-model against $A \models^- B$.

So, why do we need 2-entailment if the set of entailment is characterized by truth preservation alone? I don't know how Routley would answer. I only see that we need 2-entailment at the level of single model or frame. Indeed, $t$-entailment *in a model* and $f$-entailment *in a model* do not necessarily coincide:

*Proposition* 11. There is a $\mathbf{B}^{CE}$-model $\mathcal{M}$ in which $A \models^+_{\mathcal{M}} B$ but $A \not\models^-_{\mathcal{M}} B$.

*Proof.* Consider a two-element $\mathbf{B}^{CE}$-frame $\langle 0, 0^*, \{0\}, \{0^*\}, \{0, 0^*\}, R, S, C, E\rangle$, where

$$R = \{\langle 0,0,0\rangle, \langle 0,0^*,0^*\rangle\}, S = \{\langle 0,0^*,0\rangle, \langle 0^*,0^*,0^*\rangle, \langle 0,0,0^*\rangle\}.$$

and $C = E = \emptyset$. Define a valuation on this by

$$0 \models^- p, 0 \not\models^- q, 0^* \not\models^- p,$$

and arbitrarily for the other cases. This satisfies the conditions for valuation trivially since $\leq$ is just the identity relation and $C$ and $E$ are empty in this frame. Now since $0 \models^+ p \to p$ and $0^* \models^+ p \to p$, we have $q \models^+ p \to p$. On the other hand, $0 \not\models^- q$ and $0 \models^- p \to p$ imply that $q \not\models^- p \to p$.

Thus, the rationale of the notion of 2-entailment seems to consist in validating rule-contraposition at the level of single model. I am quite indifferent to whether it should be or not, since it holds at the level of entailment *simpliciter* anyway. I observed this just because it will serve to clarify the role that the star function or the star postulate (DS2) plays.

The above contrast between two levels may be understood like this: while the structure of the entire class of $\mathbf{B}^{CE}$-models is rich enough to provide the dual model for each model, a single model may not be that much. Even when you have a structure that gives a counter example to a $t$-entailment in a model, it may not be possible to find the dual structure within the same model that will refute $f$-entailment, and vice versa.

And what the star function or (DS2) provides is that dualizable structure within a single model or frame. Recall that in Proposition 9 we had $Rxyz \iff Sz^*x^*y^*$ in a $\mathbf{B}^{CE}$-model with (DS2), and also that we defined $R_d xyz \iff Szxy$ in the proof of Proposition 10, where we constructed the dualized model. Also compare the equivalence $x \models^+ {\sim} A \iff x^* \not\models^+ A$ in Proposition 9 and $x \models^-_d C \iff x \not\models^+ C$ in Proposition 10. The star function makes possible the dualizing operation within a single model, which is otherwise available only at the level of the entire class of models. As a result, in a $\mathbf{B}^{CE}$-model with (DS2), $t$-entailment in a model and $f$-entailment in a model become equivalent, and hence rule-contraposition is validated at the level of single model without appealing to 2-entailment.

I don't have any strong argument that we *should* introduce such a dualizer at the level of single model. But it should be noted that the dualizable structure itself is *already there* at the level of the class of models, so it is not foreign even to the friend of the (extended) American plan. The star function only internalizes the existing structure down to the level of single model. And it should be added that the dualizer does not change the logic. As we saw above, whereas (DS2) collapses the four-valued structure into the essentially two-valued one, the logic remains to be $\mathbf{B}$. Then, why not?

## 5 Conclusion

I have told a story about how the Australian plan is developed from the (extended) American plan. We started with Routley's four-valued frame semantics. Since it has falsity as a primitive notion, we can explain and introduce the star-like relations without falling into the circularity that De and Omori allege to find in the two-valued framework of the negation-as-modality approach. I claimed that this is what Routley suggested as the virtue of the American plan. It indeed provides an explanatory basis to make sense of the star function, which it is originally purported to avoid.

Then I analyzed the final step towards the Australian plan. We saw that the dualist star postulate on the binary relations of compatibility and exhaustiveness, especially its half (DS2), makes them a function, and in the present case, collapses the four-valued structure into the essentially two-valued one. An examination of rule-contraposition at different levels showed that the step is understood as internalization of dualizable structure that can be found in the class of models. We may perhaps say that this is the *ontological virtue* of the American plan. The structure of the star function as a dualizer is not an invention of the Australian plan. It is essentially there from the beginning, and the star just makes it available within a single model.

And moreover, the final step does not change the logic. So as far as its significance is clearly understood as I have explained, we may say that the step is harmless. Thus, the (extended) American plan provides an explanatory and ontological basis for the Australian plan. The latter is just a developed form of the former.[6]

**Acknowledgement** The author would like to thank the audience at Kyoto Philosophical Logic Workshop (June 2017), Kyoto Nonclassical Logic Workshop III (September 2017), and Melbourne Logic Seminar (March 2018) for the helpful discussions on the earlier versions of the present paper. This work was supported by JSPS KAKENHI Grant Number 17K13317.

## References

1. Belnap, N. How a computer should think. In *Contemporary Aspects of Philosophy*, G. Ryle, Ed. Oriel Press, 1977, pp. 30-55.
2. Belnap, N. D. A useful four-valued logic. In *Modern Uses of Multiplevalued Logic*, J. Dunn and G. Epstein, Eds. Springer, 1977, pp. 5-37.
3. Brandom, R. *Making It Explicit: Reasoning, Representing, and Discursive Commitment*. Harvard University Press, 1998.
4. Brandom, R. *Between Saying and Doing: Towards an Analytic Pragmatism*. Oxford University Press, 2008.

[6] Although we have focused on **B**, it is clear that, for any positive extensions of **B** Routley considered in the paper, the step of adding (DS2) and making the star function definable does not change the logic. So the Australian plan for an extension of the kind can also be taken as just a developed form of the (extended) American plan for *that* extension. On the other hand, as we saw, some negative extensions do not have a decent formulation in the extended American plan. Then, there is a sense in which the Australian plan for, e.g. **R**, which has the axiom of Contraposition, is *not* a developed form of the American plan for it, since the latter does not exist (I am not denying the possibility of "four-valued" semantics for **R** in any sense. See [12]). But there is another sense in which it *is*, since **R** is an extension of **B** for which the Australian plan is a developed form of the American plan.

5. Copeland, J. On When a Semantics is not a Semantics: some reasons for disliking the Routley-Meyer semantics for relevance logic. *Journal of Philosophical Logic* 8 (1979), 399-413.
6. De, M., and Omori, H. There is more to negation than modality. *Journal of Philosophical Logic* 47, 2, 281-299.
7. Došen, K. Negation as a modal operator. *Reports on Mathematical Logic* 20 (1986), 15-27.
8. Dunn, J. M. Intuitive semantics for first-degree entailments and 'coupled trees'. *Philosophical Studies* 29, 3 (1976), 149-168.
9. Dunn, J. M. A comparative study of various model-theoretic treatments of negation: a history of formal negation. In *What is Negation?*, D. M. Gabbay and H. Wansing, Eds. Springer, 1999, pp. 23-51. .
10. Dunn, J. M., and Restall, G. Relevance logic. In *Handbook of Philosophical Logic*, Vol.6, D. Gabbay and F. Guenthner, Eds. Kluwer Academic Publishers, 2002, pp. 1-128.
11. Humberstone, L. Power matrices and Dunn-Belnap semantics: Reflections on a remark of Graham Priest. *The Australasian Journal of Logic* 11, 1 (2014), 14-45.
12. Mares, E. D. "Four-valued semantics" for the relevant logic R. *Journal of Philosophical Logic* 33, 3 (2004), 327-341.
13. Mares, E. D. *Relevant Logic: A Philosophical Interpretation*. Cambridge University Press, 2004.
14. Omori, H., and Wansing, H. 40 years of FDE: An introductory overview. *Studia Logica* 105, 6 (2017), 1021-1049.
15. Onishi, T. Substructural negations. *The Australasian Journal of Logic* 12, 4 (2015), 177-203.
16. Onishi, T. Understanding negation implicationally in the relevant logic R. *Studia Logica* 104, 6 (2016), 1267-1285.
17. Peregrin, J. Logic as based on incompatibility. *The Logica Yearbook (2010)*, 157-168.
18. Peregrin, J. Logic reduced to bare (proof-theoretical) bones. *Journal of Logic, Language and Information* 24, 2 (2015), 193-209.
19. Priest, G. *An Introduction to Non-classical Logic: From If to Is*, 2nd ed. Cambridge University Press, 2008.
20. Restall, G. Information flow and relevant logics. *Logic, language and computation* (1996), 1-14.
21. Restall, G. Negation in relevant logics (how I stopped worrying and learned to love the Routley star). In *What is Negation?*, D. M. Gabbay and H. Wansing, Eds. Springer, 1999, pp. 53-76.
22. Routley, R. The American plan completed: Alternative classical-style semantics, without stars, for relevant and paraconsistent logics. *Studia Logica* 43, 1-2 (1984), 131-158.
23. Routley, R., and Meyer, R. K. The semantics of entailment. In *Truth, Syntax and Modality*, H. Leblanc, Ed. North Holland, 1972, pp. 199-243.
24. Routley, R., and Meyer, R. K. The semantics of entailment II. *Journal of Philosophical Logic* 1, 1 (1972), 53-73.
25. Routley, R., and Routley, V. The semantics of first degree entailment. *Noûs* 6, 4 (1972), 335-359.
26. van Benthem, J. What is dialectical logic? *Erkenntnis* 14, 3 (1979), 333-347.

# Bilattice Logics and Demi-Negation

Francesco Paoli

**Abstract** The logic of distributive bilattices, an offshoot of Belnap and Dunn's four-valued logic, is presented as an appropriate logic for deductive reasoning in situations where *a)* sentences can be accepted and rejected at the same time, and *b)* the semantic value of a sentence $\alpha$ is given, depending on the dominant connective in $\alpha$, in terms of mutually independent acceptance and rejection conditions. Taking our cue from some writings by Lloyd Humberstone, we expand this logic by a *demi-negation* connective, whose behaviour is intermediate between affirmation and negation. The resulting logic is introduced semantically and axiomatised using methods from abstract algebraic logic.

**Keywords** Abstract algebraic logic • Acceptance/Rejection • Assertion/Denial • Belnap-Dunn logic • Bilattice • Demi-negation • Many-valued logic • Negation by iteration

## 1 Introduction

### *1.1 Acceptance, Rejection, and Belnap–Dunn Logic*

Suppose we have reason to believe that, under some perhaps very uncommon circumstances, it may not be incoherent to assert (or to accept[1]) both a sentence and its negation, at the same time and under the same respect. We will keep this supposition (hereafter labelled C0) in the background throughout this section, brushing aside any concern about the motives that may have led us to gain confidence in its plausibility — perhaps we have been persuaded by the dialetheist into buying her story on the paradoxes, or we are after a not-too-idealised model of the structure of rational belief. Be it as it may, C0 entails[2] that we should relinquish at least one of the following two claims:

C1 Denying (rejecting) a sentence is tantamount to asserting (accepting) its negation.

C2 Assertion (acceptance) and denial (rejection) are mutually exclusive speech acts (propositional attitudes): it is always incoherent to assert (accept) and simultaneously deny (reject) any given sentence.

Francesco Paoli
Dept. of Pedagogy, Psychology, Philosophy, University of Cagliari, Cagliari, Italy
e-mail: paoli@unica.it

[1] In this paper, we will assume that whatever relationships hold between the speech acts of assertion and denial also hold between the attendant propositional attitudes of acceptance and rejection. Although it can be argued that important differences set these two contexts apart, for our present purposes we will consider these differences as irrelevant.

[2] Here, we are presupposing some moderate consistency requirements in our metatheoretical discourse, something that most discussants will be prepared to go along with.



H. Omori, H. Wansing (eds.), *New Essays on Belnap-Dunn Logic*,
Synthese Library 418, https://doi.org/10.1007/978-3-030-31136-0_14

Both C1 and C2 have much to recommend them. Partisans of C1 showcase no less than Frege, and indeed much of the inferentialist tradition stemming from Dummett and Prawitz, in their roster. On the other hand, C2 (variously called *pragmatic consistency assumption* [31, p. 173] or *rejection-consistency principle* [7, p. 172] is a shred of classical orthodoxy that has, by and large, survived even in the most heretic circles that practise paraconsistency and dialetheism. A typical representative of this position is Graham Priest, who claims:

> Acceptance and rejection are not exhaustive, but they are exclusive. [...] One can certainly believe something and believe its negation. One might even argue that one can believe something and not believe it, though this is much more dubious. But it seems difficult to argue that one might both believe something and *refuse* to believe it. [36, pp. 98-99]

Given our background supposition C0, however, either C1 or C2 has to go. As a matter of fact, both claims have received their fair share of criticism. The consensus in favour of C1 has tended to fade over the last decades. Beside Priest, philosophers in the *bilateralist* current have mounted a significant case to the effect that acceptance and rejection should be considered as equally primitive states, on a par with each other [35, 43, 40, 37, 38], whence rejection of a sentence cannot be reduced to the acceptance of its negation. On the other hand, the seminal papers where Nuel Belnap [6] and Michael Dunn [20] introduced their four-valued logic B4 have had the merit to propagate the view that it may not be incoherent for a (human or artificial) reasoner to accept and reject the very same sentence $\alpha$, because $\alpha$ can be 'told to be true' and 'told to be false' by different, but equally reliable, sources.[3] A rejection of C2 has been later buttressed from different, though related, angles:

- experiments can directly falsify observational sentences without any inferential detour through their reduction to absurdity, whence sentences can in principle be provided with independent positive and negative evidential supports [23, 44, 45];
- a dialogical (question-answer) semantics has been proposed to formalise an antirealist approach to burden of proofs, in which the questions "Is there evidence for $\alpha$?" and "Is there evidence against $\alpha$?" can receive independent positive or negative answers [17];
- upon construing acceptability (rejectability) of a sentence as the absence of a refutation (of a proof) thereof, one has it that a given sentence may be at least simultaneously accept*able* and reject*able* [41; 42, ch. 3.6];
- a more *ad hominem* strategy aimed at dialetheists is that of arguing that they are rationally committed to accepting and rejecting at least some instances of the noncontradiction law [7].

Belnap and Dunn's idea is not exceedingly distorted if it is presented in the following terms. The semantic value of a sentence $\alpha$ is an ordered pair $(a, b)$ of classical truth values in $\{0, 1\}$. The value $a$ is 1 or 0 according as a certain rational agent $A$ *accepts* $\alpha$ or not; likewise, $b$ is 1 or 0 according as $A$ *rejects* $\alpha$ or not. Since acceptance and rejection are not mutually exclusive (nor, for that matter, exhaustive), all the combinations $(1,1), (1,0), (0,1), (0,0)$ are legitimate. These semantic values are relabelled as $\top, t, f$

[3] It can be objected that Belnap, although explicit in endorsing C0, is silent about C1 and C2. His position, therefore, could be (indeed, it often has been) given a different interpretation than the one we are offering: whenever it is told that a sentence $\alpha$ is true and it is told that $\alpha$ is false, Belnap's computer "accepts" both $\alpha$ and its negation, but it fails to both "accept" and "reject" $\alpha$. On a more careful reading of [6], though, this interpretation proves to be untenable; see e.g. the reprint of the paper in [1, vol. 2], at the beginning of Section 81.2.1.

and $\bot$, respectively, and lattice-ordered in two different ways: under the truth ordering $\leq_t$, we have that $f <_t \top, \bot <_t t$, while under the information ordering $\leq_i$, we have that $\bot <_i f, t <_i \top$. The Hasse diagrams of $\leq_t$ and $\leq_i$ are depicted in the next figure, respectively from left to right and from bottom to top:

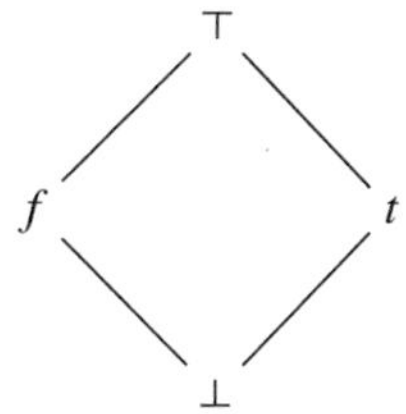


The meet and join of $\leq_t$ are respectively denoted as $\wedge$ and $\vee$, while the meet and join of $\leq_i$ (not considered in either [6] or [20]) are respectively denoted as $\otimes$ and $\oplus$. Thus, in particular:

- Agent $A$ accepts $\alpha \wedge \beta$ if she accepts both $\alpha$ and $\beta$, and rejects $\alpha \wedge \beta$ if she rejects either $\alpha$ or $\beta$.
- Agent $A$ accepts $\alpha \vee \beta$ if she accepts either $\alpha$ or $\beta$, and rejects $\alpha \vee \beta$ if she rejects both $\alpha$ and $\beta$.
- Agent $A$ accepts $\alpha \otimes \beta$ if she accepts both $\alpha$ and $\beta$, and rejects $\alpha \otimes \beta$ if she rejects both $\alpha$ and $\beta$.
- Agent $A$ accepts $\alpha \oplus \beta$ if she accepts either $\alpha$ or $\beta$, and rejects $\alpha \oplus \beta$ if she rejects either $\alpha$ or $\beta$.

Given this picture, there are two natural alternatives for the semantics of *negation*. The connective favoured by Belnap and Dunn, usually called *De Morgan negation* (here noted $-$), is a paraconsistent negation which swaps $t$ and $f$ and has $\top$ and $\bot$ as fixpoints. Alternatively, we can define a Boolean negation (here noted $\neg$) which swaps $t$ and $f$, as well as $\top$ and $\bot$. In terms of acceptance and rejection conditions, we get, respectively:

- Agent $A$ accepts $-\alpha$ whenever she rejects $\alpha$, and rejects $-\alpha$ whenever she accepts $\alpha$.
- Agent $A$ accepts $\neg\alpha$ whenever she fails to accept $\alpha$, and rejects $\neg\alpha$ whenever she fails to reject $\alpha$.

Of course, what we have given so far is just a semantics for propositional connectives, which falls short of defining a logic because no definition of consequence has been provided. Belnap and Dunn essentially defined[4] the logic B4 as a logic of language $\mathcal{L}_0 = (2, 2, 1)$, containing the connectives $\wedge, \vee, -$. Let **B4** be the lattice with universe $\{\top, t, f, \bot\}$ and order $\leq_t$, expanded by the De Morgan negation $-$. The consequence relation of B4 can be defined as follows for all subsets $\Gamma \cup \{\alpha\}$ of the set of formulas $For(\mathcal{L}_0)$ of the language $\mathcal{L}_0$:

$$\Gamma \vdash_{B4} \alpha \text{ iff for all valuations } v : \mathbf{For}(\mathcal{L}_0) \to \mathbf{B4},$$
$$\text{if } v(\gamma) \in \{\top, t\} \text{ for all } \gamma \in \Gamma, \text{ then } v(\alpha) \in \{\top, t\}.$$

[4] This way of putting things is historically very sloppy, if for no other reasons because it completely ignores the relationship between B4 and Anderson and Belnap's *tautological entailments* [1], the first-degree entailment fragment of the relevance logics E and R. We will not go any further into this issue, which would force us to come clean about a number of technicalities.

This choice of designated values is all too natural in light of the above considerations: a value is designated if it attaches to a sentence that is "at least accepted".

Let us take stock. We observed that B4 can be a useful tool for an investigation of deductive reasoning in circumstances where acceptance and rejection can be legitimately viewed as non-exclusive and non-exhaustive, under the additional stipulation that both propositional attitudes behave *classically* — so that, for example, failing to fail to accept $\alpha$ amounts to accepting $\alpha$, and likewise for rejection. Belnap and Dunn formulated B4 in the language $\mathcal{L}_0$, which contains a single conjunction and a single disjunction, both for historical reasons and to ease comparison with classical logic. However, in the framework discussed above, confining ourselves to the conjunction $\wedge$ and the disjunction $\vee$ that stem from the truth ordering, and ignoring their counterparts $\otimes, \oplus$ that originate in the information ordering, can be seen — indeed, it has been seen — as an undue mutilation.

Mending this defect was one of the reasons why Ginsberg [22] introduced the notion of *bilattice* (see Section 2.1 below). As we will see, a bilattice (sometimes called a *pre-bilattice* in the literature) is an algebra $\mathbf{A} = (A, \wedge, \vee, \otimes, \oplus)$ such that $(A, \wedge, \vee)$ and $(A, \otimes, \oplus)$ are both lattices. A bilattice is *distributive* if the twelve distributive laws involving pairs of operations from $\{\wedge, \vee, \otimes, \oplus\}$ all hold. The typical example of distributive bilattice is the expansion $\mathbf{B4}^{\otimes,\oplus}$ of the lattice reduct of Belnap and Dunn's **B4** by operations corresponding to the meet and join of $\leq_i$; actually, this algebra generates the variety of distributive bilattices. There are several studies on the logics of bilattices and distributive bilattices, possibly expanded by a De Morgan negation [2, 3, 13]. The consequence relation of the logic $\mathrm{B4}^{\otimes,\oplus}$ of distributive bilattices (whether it is expanded by a De Morgan negation or not) can be specified along the lines of the definition given above for B4, once we make the necessary adjustments enforced by the specified change of language. Thus, the designated values are still $t$ and $\top$.

## *1.2 Demi-Negation*

Although the logic $\mathrm{B4}^{\otimes,\oplus}$ can be proposed as a suitable model of deductive reasoning under the assumption of non-exclusive acceptance and rejection, one can question its expressive power in so far as there are many connectives, definable in terms of reasonable acceptance and rejection conditions, that its language cannot accommodate. Here is an example. Consider the 4-element distributive bilattice $\mathbf{B4}^{\otimes,\oplus} = (\{\top, \bot, t, f\}, \wedge, \vee, \otimes, \oplus)$, where $\wedge$ and $\otimes$ are the meets (and $\vee, \oplus$ are the joins) of the two lattice orders $\leq_t$ and $\leq_i$. In the sequel, the algebra **4** will denote the expansion of such a bilattice by a constant $\top$ and a unary operation $\sqrt{\neg}$ with the following table:

| $\sqrt{\neg}$ | |
|---|---|
| $\top$ | $t$ |
| $t$ | $\bot$ |
| $\bot$ | $f$ |
| $f$ | $\top$ |

It is clear enough that for all $x$ in $\{\top, \bot, t, f\}$, $\sqrt{\neg}\sqrt{\neg}x$ is the same as the Boolean negation $\neg x$ of $x$, and that the algebras

$$(\{\top, \bot, t, f\}, \wedge, \vee, \neg, t, f), (\{\top, \bot, t, f\}, \otimes, \oplus, \neg, \top, \bot)$$

are both isomorphic to the 4-element Boolean algebra, with $t$ and $\top$ as respective top elements.[5] Moreover, the acceptance and rejection clauses for $\sqrt{\neg}$ that are built in its table assign it a status that is somehow intermediate between an affirmation and a negation of the sentence it is applied to:

- Agent $A$ accepts $\sqrt{\neg}\alpha$ if she rejects $\alpha$, and rejects $\sqrt{\neg}\alpha$ if she fails to accept $\alpha$.

The connective $\sqrt{\neg}$ is importantly related to an intriguing problem discussed by Lloyd Humberstone in a number of papers [26, 27, 28]. Take a logic L whose language includes a unary connective $\neg$ "behaving in some such way as to deserve the name of negation" [26, p. 1]. Is there some conservative expansion L′ of L by a unary connective §, such that §§$\alpha$ turns out to be L′-interderivable with $\neg\alpha$? If so, our original negation would be decomposed in L′ as the iterated application of a connective of *demi-negation*. Apart from the problem in its more abstract incarnation, Humberstone is interested in any demi-negation connectives that can be added to classical logic so as to appropriately decompose the standard truth-functional negation.

Here, of course, we must be careful. In a sense, there is no such connective. In fact, it is not hard to see that there is no unary Boolean function $f$ such that, for all $x \in \{0, 1\}$, we have that $f(f(x)) = 1 - x$ (for a proof, see e.g. [18, Lemma 17.1.11]). In a different sense, though, there is in principle no bar to setting up a sort of modal logic that has the same tautologies, or even the same consequence relation, as classical logic when restricted to the classical language, while housing a demi-negation connective (plus, possibly, additional connectives) in the more comprehensive language of the envisaged conservative expansion. Indeed, one relevant example uses precisely the algebra **4** and the connective $\sqrt{\neg}$ introduced at the beginning of this subsection. Humberstone considers the sentential logic H of language $\mathcal{L} = (2, 2, 2, 2, 1, 0)$ whose consequence relation $\vdash_{\mathrm{H}}$ is the matrix consequence $\vdash_{\mathbb{M}}$ associated with the class of matrices $\mathbb{M} = \{(\mathbf{4}, \{\top, t\}), (\mathbf{4}, \{\top, f\})\}$. In other words, $\Gamma \vdash_{\mathrm{H}} \alpha$ iff: for all valuations $v : \mathbf{For}(\mathcal{L}) \to \mathbf{4}$, if $v(\gamma) \in \{\top, t\}$ for all $\gamma \in \Gamma$, then $v(\alpha) \in \{\top, t\}$, *and* for all valuations $v : \mathbf{For}(\mathcal{L}) \to \mathbf{4}$, if $v(\gamma) \in \{\top, f\}$ for all $\gamma \in \Gamma$, then $v(\alpha) \in \{\top, f\}$. A moment's reflection shows that, in H, $\sqrt{\neg}$ is indeed a demi-negation connective with the requested features. In addition, $\sqrt{\neg}$ is *congruential*: whenever $\alpha \dashv\vdash_{\mathrm{H}} \beta$, we also have that $\sqrt{\neg}\alpha \dashv\vdash_{\mathrm{H}} \sqrt{\neg}\beta$. This last *desideratum* is philosophically crucial for Humberstone, because

> without the latter addition, we could just think of "$\sqrt{\neg}\sqrt{\neg}$" as an idiosyncratically long-winded way of writing "$\neg$". With it, we are insisting that a single occurrence of "$\sqrt{\neg}$" is genuinely responsive to the content of what it applies to. [27, p. 67, notation adapted]
>
> We would like our new connective itself to be congruential [...] because we don't want formulas we have agreed prior to its addition to have the same logical content, to turn out to behave differently in its scope. This is not a requirement of conservative extension [...] though its motivation is a similar distaste for the disruption of prior logical relations. [29, p. 682]

The price to pay for the congruentiality of $\sqrt{\neg}$, however, is rather steep. For a start, H is such that there provably is no single matrix $M$ such that $\vdash_{\mathrm{H}} = \vdash_M$ [26, p. 9]. Furthermore, H conservatively expands classical logic (formulated in the $\{\otimes, \oplus, \neg\}$ language) but not $\mathrm{B4}^{\otimes,\oplus}$ expanded by Boolean negation: for instance, $\alpha \oplus \neg\alpha$ fails to be H-interderivable with $\alpha \vee \neg\alpha$. Finally, since $\top$ and $t$ are the designated values in only one matrix in $\mathbb{M}$, it becomes difficult to justify for H the interpretation of the values in **4** in terms of acceptance and rejection that we had recognised as apt for B4 and $\mathrm{B4}^{\otimes,\oplus}$. Therefore, we cannot read off

[5] The term definable operation on **4** corresponding to the formula $\sqrt{\neg}\sqrt{\neg}\sqrt{\neg}\alpha$ has been investigated in [39].

the table for $\sqrt{\neg}$ the intended meaning conferred upon it by the acceptance and rejection conditions described above.

In light of these reflections, we try to attain an appropriate logic for demi-negation via a different route. From now on, the symbol D will refer to the sentential logic of language $\mathcal{L} = (2,2,2,2,1,0)$ whose consequence relation $\vdash_{\mathrm{D}}$ is the matrix consequence associated with the *single* logical matrix $(\mathbf{4}, \{\top, t\})$. In other words, $\vdash_{\mathrm{D}}$ is so defined for all $\Gamma \cup \{\alpha\} \in For(\mathcal{L})$:

$$\Gamma \vdash_{\mathrm{D}} \alpha \text{ iff for all valuations } v : \mathbf{For}(\mathcal{L}) \to \mathbf{4}, \\ \text{if } v(\gamma) \in \{\top, t\} \text{ for all } \gamma \in \Gamma, \text{ then } v(\alpha) \in \{\top, t\}.$$

Our goal is giving a suitable axiom system for the logic D.

### *1.3 Outline of the Paper*

The rest of the paper is structured as follows. In Section 2, we will review some notions of algebra and abstract algebraic logic that will be used in the sequel. In particular, we include short primers on bilattices and $\tau$-assertional logics. In Section 3 we axiomatise the quasivariety generated by the algebra **4**, combining techniques from the theory of bilattices with tools borrowed from the area of quantum structures; we show that it is, indeed, a (discriminator) variety and that it is term equivalent to a certain variety of bilattices with implications already studied in the literature. This result leads us very close to the desired axiomatisation, since the logic L can be shown to be the $\tau$-assertional logic of this quasivariety for a certain translation $\tau$, and the general theory of translations provides us (under certain conditions) with an algorithm for transforming any known axiomatisation for a quasivariety into an axiomatisation for its $\tau$-assertional logic. In Section 4 we apply this algorithm to our target example; moreover, we streamline the resulting, "ugly" axiom system to a nicer one.

## 2 Preliminaries

### *2.1 Bilattices*

A *bilattice* (sometimes called a *pre-bilattice*) is an algebra $\mathbf{A} = (A, \wedge, \vee, \otimes, \oplus)$ of language $(2,2,2,2)$ such that $(A, \wedge, \vee)$ and $(A, \otimes, \oplus)$ are both lattices. A bilattice is *distributive* if the twelve distributive laws involving pairs of operations from $\{\wedge, \vee, \otimes, \oplus\}$ all hold. On bilattices, see e.g. [2, 4, 22, 21].

There is a standard construction for obtaining bilattices out of pairs of lattices. Let $\mathbf{L}_1 = (L_1, \sqcap_1, \sqcup_1)$ and $\mathbf{L}_2 = (L_2, \sqcap_2, \sqcup_2)$ be two lattices. The *product bilattice* $\mathbf{L}_1 \odot \mathbf{L}_2$ is the algebra $(L_1 \times L_2, \wedge, \vee, \otimes, \oplus)$ of language $(2,2,2,2)$ whose operations are defined as follows:

$$(a_1, a_2) \wedge (b_1, b_2) = (a_1 \sqcap_1 b_1, a_2 \sqcup_2 b_2);$$
$$(a_1, a_2) \vee (b_1, b_2) = (a_1 \sqcup_1 b_1, a_2 \sqcap_2 b_2);$$

$$(a_1, a_2) \otimes (b_1, b_2) = (a_1 \sqcap_1 b_1, a_2 \sqcap_2 b_2);$$
$$(a_1, a_2) \oplus (b_1, b_2) = (a_1 \sqcup_1 b_1, a_2 \sqcup_2 b_2).$$

$\mathbf{L}_1 \odot \mathbf{L}_2$ is indeed a bilattice, and is distributive if and only if both $\mathbf{L}_1$ and $\mathbf{L}_2$ are such. Not every bilattice, however, is representable as a product. This property holds for a class of bilattices that properly includes the distributive bilattices [22, 21, 4, 13]. In particular:

*Theorem* 1. Every distributive bilattice is isomorphic to $\mathbf{L}_1 \odot \mathbf{L}_2$, for some appropriate distributive lattices $\mathbf{L}_1$ and $\mathbf{L}_2$.

As a corollary of the above, we get the next result:

*Corollary* 2. The variety of distributive bilattices is generated by the 4-element algebra $\mathbf{B4}^{\otimes,\oplus}$.

The following arithmetical lemma will be useful in the sequel:

*Lemma* 3. [13, Prop. 3.6] In any distributive bilattice $\mathbf{L}$, for any $a, b \in L$, we have that $a \otimes b = a \vee b$ iff $a \oplus b = a \wedge b$, and $a \otimes b = a \wedge b$ iff $a \oplus b = a \vee b$.

As already mentioned, bilattices can be enriched with a De Morgan negation – and with further connectives, including a binary connective $\supset$ that plays the role of an *implication*. This leads to the next definition. A *Brouwerian bilattice* [14, 12] is an algebra $\mathbf{A} = (A, \wedge, \vee, \otimes, \oplus, \supset, -)$ of language $(2, 2, 2, 2, 2, 1)$ such that the reduct $(A, \wedge, \vee, \otimes, \oplus)$ is a bilattice and the following quasiequations are satisfied:

B1 $x \approx - - x$;
B2 $x \wedge y \approx x \Rightarrow -y \wedge -x \approx -y$;
B3 $x \otimes y \approx x \Rightarrow -y \otimes -x \approx -x$;
B4 $(x \supset x) \supset y \approx y$;
B5 $x \supset (y \supset z) \approx (x \wedge y) \supset z \approx (x \otimes y) \supset z$;
B6 $(x \vee y) \supset z \approx (x \oplus y) \supset z \approx (x \supset z) \wedge (y \supset z)$;
B7 $x \wedge ((x \supset y) \supset (x \otimes y)) \approx x$;
B8 $-(x \supset y) \supset z \approx (x \wedge -y) \supset z$.

Brouwerian bilattices form a variety and are always distributive as bilattices. An algebra

$$\mathbf{A} = (A, \wedge, \vee, \otimes, \oplus, \supset, -, \top, \bot, t, f)$$

is a *bounded Brouwerian bilattice* if $(A, \wedge, \vee, \otimes, \oplus, \supset, -)$ is a Brouwerian bilattice and both $(A, \wedge, \vee, t, f)$ and $(A, \otimes, \oplus, \top, \bot)$ are bounded lattices. Finally, a bounded Brouwerian bilattice is said to be a *bounded classical implicative bilattice* if it satisfies the additional equation

B9 $((x \supset y) \supset x) \supset x \approx x \supset x$.

It can be inferred from the results in [14, 11] that the variety of bounded classical implicative bilattices is generated by the algebra $\mathbf{B4}^{\supset}$, i.e. $\mathbf{B4}^{\otimes,\oplus}$ expanded by the appropriate constants, the De Morgan negation and the binary operation defined by

$$a \supset b = \begin{cases} t, & \text{if } a \in \{f, \bot\}; \\ b, & \text{otherwise.} \end{cases}$$

## 2.2 $\tau$-assertional and algebraisable logics

A (sentential) *logic* is a pair $\mathrm{L} = (\mathbf{For}(\mathcal{L}), \vdash_{\mathrm{L}})$, where $\mathbf{For}(\mathcal{L})$ is the formula algebra of some given language $\mathcal{L}$ and $\vdash_{\mathrm{L}}$ is a substitution-invariant consequence relation over $\mathbf{For}(\mathcal{L})$. Logical languages and algebraic similarity types will be hereafter identified; therefore, "formula" and "term" will be understood as synonymous expressions, while equations will be viewed as pairs of formulas. If $\mathrm{L} = (\mathbf{For}(\mathcal{L}), \vdash_{\mathrm{L}})$ is a logic in the language $\mathcal{L}$, any set $\tau = \{\alpha_i(x) \approx \beta_i(x)\}_{i \in I}$ of equations in a single variable of $\mathcal{L}$ can also be regarded as a function which maps formulas in $For(\mathcal{L})$ to sets of equations of the same language: the behaviour of this function, for any $\gamma \in For(\mathcal{L})$, is given by the condition

$$\tau(\gamma) = \{\alpha_i(x/\gamma) \approx \beta_i(x/\gamma)\}_{i \in I}.$$

This mapping will be called a *formula-equation translation*.

A class $\mathbb{K}$ of similar algebras of language $\mathcal{L}$ is an *algebraic semantics* for L (also of language $\mathcal{L}$) if, for some formula-equation translation $\tau$, the following condition holds for all $\Gamma \cup \{\alpha\} \subseteq For(\mathcal{L})$:

$$\begin{aligned} \Gamma \vdash_{\mathrm{L}} \alpha \quad & \text{iff for every } \mathbf{A} \in \mathbb{K} \text{ and every } \vec{a} \in A^n, \\ & \text{if } \tau(\gamma)^{\mathbf{A}}(\vec{a}) \text{ for all } \gamma \in \Gamma, \text{ then } \tau(\alpha)^{\mathbf{A}}(\vec{a}), \end{aligned}$$

a condition which can be rewritten[6] as

$$\Gamma \vdash_{\mathrm{L}} \alpha \quad \text{iff} \quad \tau(\Gamma) = \bigcup \{\tau(\gamma) : \gamma \in \Gamma\} \vdash_{Eq(\mathbb{K})} \tau(\alpha).$$

If $\mathbb{K}$ is given, we also say that the logic L having the same language as $\mathbb{K}$, whose consequence relation $\vdash_{\mathrm{L}}$ is defined as in the previous display, is the *$\tau$-assertional logic* of $\mathbb{K}$.

Given an equation $\alpha \approx \beta$ and a set of formulas in two variables $\rho = \left\{\gamma_j(x, y)\right\}_{j \in J}$, we use the abbreviation

$$\rho(\alpha, \beta) = \left\{\gamma_j(x/\alpha, y/\beta)\right\}_{j \in J}.$$

$\rho$ will be also regarded as a function mapping equations to sets of formulas, and called an *equation-formula translation*.

A logic $\mathrm{L} = (\mathbf{For}(\mathcal{L}), \vdash_{\mathrm{L}})$ is said to be *algebraisable* with *equivalent algebraic semantics* $\mathbb{K}$ (where $\mathbb{K}$ is a class of algebras of the same language as $\mathbf{For}(\mathcal{L})$) iff there exist a formula-equation translation $\tau$, and an equation-formula translation $\rho$, such that the following conditions hold for any $\alpha, \beta \in For(\mathcal{L})$, for any $\Gamma \subseteq For(\mathcal{L})$ and for any $E \subseteq For(\mathcal{L})^2$:

AL1: $\Gamma \vdash_{\mathrm{L}} \alpha$ iff $\tau(\Gamma) \vdash_{Eq(\mathbb{K})} \tau(\alpha)$;
AL2: $E \vdash_{Eq(\mathbb{K})} \alpha \approx \beta$ iff $\rho(E) \vdash_{\mathrm{L}} \rho(\alpha, \beta)$;
AL3: $\alpha \dashv\vdash_{\mathrm{L}} \rho(\tau(\alpha))$;
AL4: $\alpha \approx \beta \dashv\vdash_{Eq(\mathbb{K})} \tau(\rho(\alpha, \beta))$.

The mappings $\tau(x)$ and $\rho(x, y)$ are respectively called a *system of defining equations* and a *system of equivalence formulas* for L and $\mathbb{K}$. A logic L is algebraisable (tout court) iff, for some $\mathbb{K}$, it is algebraisable with equivalent algebraic semantics $\mathbb{K}$. By virtue of [9,

[6] If $\Gamma, \Delta$ are sets of formulas, $\Gamma \vdash_{\mathrm{L}} \Delta$ means $\Gamma \vdash_{\mathrm{L}} \alpha$ for all $\alpha \in \Delta$; if $E, E'$ are sets of equations, $E \vdash_{Eq(\mathbb{K})} E'$ means $E \vdash_{Eq(\mathbb{K})} \epsilon$ for all $\epsilon \in E'$.

Corollary 2.11], at least in the finitary case, we may assume without loss of generality that $\mathbb{K}$ is a quasivariety; in other words, $\mathbb{K}$ is an *equivalent quasivariety semantics* for L.

## 2.3 Relatively $\tau$-regular quasivarieties

If $\mathbb{Q}$ is a quasivariety of algebras and $\mathbf{A} \in \mathbb{Q}$, a $\mathbb{Q}$-*congruence* on $\mathbf{A}$ is a congruence $\theta$ on $\mathbf{A}$ such that $\mathbf{A}/\theta \in \mathbb{Q}$.

In their article on assertionally equivalent quasivarieties [11], Blok and Raftery introduce a notion of $\tau$-*class* which relativises the usual notion of congruence class to a given formula-equation translation $\tau$. If $\mathbb{Q}$ is a quasivariety, $\mathbf{A} \in \mathbb{Q}$, $\theta \subseteq A^2$ and $\tau = \{\delta_i(x) \approx \epsilon_i(x) : i \leq n\}$ is a formula-equation translation in the language of $\mathbb{Q}$, the $\tau$-*class* of $\theta$ in $\mathbf{A}$ — in symbols $\tau^{\mathbf{A}}/\theta$ — is defined as

$$\tau^{\mathbf{A}}/\theta = \left\{a \in \mathbf{A} : \left(\delta_i^{\mathbf{A}}(a), \epsilon_i^{\mathbf{A}}(a)\right) \in \theta \text{ for every } i \leq n\right\}.$$

A quasivariety $\mathbb{Q}$ is *relatively $\tau$-regular* iff for any $\mathbb{Q}$-congruences $\theta, \varphi$ on any $\mathbf{A} \in \mathbb{Q}$, $\tau^{\mathbf{A}}/\theta = \tau^{\mathbf{A}}/\varphi$ implies $\theta = \varphi$; if $\tau = \{x \approx 1\}$, we get as a special case the standard notion of relatively 1-regular quasivariety. The following theorem holds [10, 5, 11]:

*Theorem* 4. For $\mathbb{Q}$ a quasivariety of algebras of language $\mathcal{L}$, the following are equivalent:

1. $\mathbb{Q}$ is relatively $\tau$-regular;
2. The $\tau$-assertional logic of $\mathbb{Q}$ is algebraisable with $\mathbb{Q}$ as equivalent quasivariety semantics;
3. There are binary formulas $\alpha_1, ..., \alpha_n$ in $For(\mathcal{L})$ such that for all $i \leq n$, $x \approx y \dashv\vdash_{Eq(\mathbb{K})} \tau(\alpha_i(x, y))$.

The $\tau$-assertional logic L of a relatively $\tau$-regular quasivariety $\mathbb{Q}$ can be effectively axiomatised provided an axiomatisation of $\mathbb{Q}$ and a system of equivalence formulas for L and $\mathbb{Q}$ are both known [11, Thm. 8.0.9]:

*Theorem* 5. Let $\mathbb{Q}$ be a relatively $\tau$-regular quasivariety of language $\mathcal{L}$, let L be its $\tau$-assertional logic, and let $\rho(x, y)$ be a system of equivalence formulas for L and $\mathbb{Q}$. Then L is axiomatised by the following axioms and rules:

A1 $\vdash_{\mathrm{L}} \rho(\alpha, \alpha)$;
A2 $\alpha, \rho(\alpha, \beta) \vdash_{\mathrm{L}} \beta$;
A3 $\rho(\alpha, \beta) \vdash_{\mathrm{L}} \rho(\beta, \alpha)$;
A4 For each $k$-ary $\mathcal{L}$-connective $c^k$, $\bigcup_{i\leq k} \rho(\alpha_i, \beta_i) \vdash_{\mathrm{L}} \rho\left(c^k\left(\vec{\alpha}\right), c^k\left(\vec{\beta}\right)\right)$;
A5 $\alpha \dashv\vdash_{\mathrm{L}} \rho(\tau(\alpha))$;
A6 For each axiom of $\mathbb{Q}$ $\alpha_1 \approx \beta_1 \& \cdots \& \alpha_n \approx \beta_n \Rightarrow \alpha \approx \beta$,

$$\bigcup_{i\leq n} \rho(\alpha_i, \beta_i) \vdash_{\mathrm{L}} \rho(\alpha, \beta).$$

### 2.4 Primal algebras

Recall that a finite algebra **A** is *primal* if every function on $A$ is a term operation of **A**. The most prominent example of a primal algebra is of course the 2-element Boolean algebra **2**; $n$-element Post algebras and rings (with unit) of integers modulo prime numbers make other notable instances of the concept. Also, recall that the *discriminator function* on a set $A$ is the ternary function

$$t(a,b,c) = \begin{cases} c & \text{if } a = b; \\ a, & \text{otherwise.} \end{cases}$$

The following theorem, due to Foster and Pixley, characterises primal algebras [15, Cor. IV.10.8].

*Theorem* 6. An algebra **A** is primal if and ony if the following conditions jointly hold:

1. **A** has no nontrivial subalgebra;
2. The only automorphism on **A** is the identity;
3. The discriminator function on $A$ is term definable in **A**.

For later use, we also remind the reader that a variety $\mathbb{V}$ of language $\mathcal{L}$ is a *discriminator variety* if there is a ternary term $t(x,y,z)$ of language $\mathcal{L}$ that realises the discriminator function on any subdirectly irreducible member of $\mathbb{V}$. Clearly, varieties generated by primal algebras are discriminator varieties.

## 3 Axiomatising the quasivariety generated by 4

Hereafter, $\mathcal{L}$ will denote the language $(2,2,2,2,1,0)$ containing the connectives $\wedge$, $\vee$, $\otimes$, $\oplus$, $\sqrt{\neg}$, $\top$, while $\mathcal{L}^c$ will denote the language $(2,2,1,0,0)$ containing the connectives $\sqcap, \sqcup, \sim, 1, 0$. Clearly, both classical logic and Boolean algebras can be respectively defined as a logic, and as a class of algebras, of language $\mathcal{L}^c$. Formally, classical sentential logic CPC will be identified with the pair CPC $= (\mathbf{For}(\mathcal{L}^c), \vdash_{\mathrm{CPC}})$, where $\vdash_{\mathrm{CPC}}$ is the derivability relation of some Hilbert system for classical logic in the language $\mathcal{L}^c$. Given a formula $\alpha$ of language $\mathcal{L}$, $\mathcal{L}(\alpha)$ will denote the set of connectives contained in $\alpha$.

### 3.1 The main result

A *square root algebra* is an algebra $\mathbf{A} = (A, \wedge, \vee, \otimes, \oplus, \sqrt{\neg}, \top)$ of language $\mathcal{L}$ such that, upon defining:

$$\begin{aligned} \neg x &= \sqrt{\neg}\sqrt{\neg}x; \\ \bot &= \neg\top; \\ t &= \sqrt{\neg}\top; \\ f &= \neg t, \end{aligned}$$

the following conditions are satisfied:

SQ1 $(A, \wedge, \vee, \otimes, \oplus)$ is a distributive bilattice;
SQ2 $(A, \wedge, \vee, \neg, t, f)$ and $(A, \otimes, \oplus, \neg, \top, \bot)$ are Boolean algebras;
SQ3 For all $a, b \in A$,

$$\begin{aligned}
\sqrt{\neg}\,(a \otimes b) &= \sqrt{\neg}a \wedge \sqrt{\neg}b;\\
\sqrt{\neg}\,(a \oplus b) &= \sqrt{\neg}a \vee \sqrt{\neg}b;\\
\sqrt{\neg}\,(a \wedge b) &= \sqrt{\neg}a \oplus \sqrt{\neg}b;\\
\sqrt{\neg}\,(a \vee b) &= \sqrt{\neg}a \otimes \sqrt{\neg}b.
\end{aligned}$$

Square root algebras form a variety, henceforth noted as $\mathbb{SQ}$. It is not hard to check that **4** is an example of square root algebra. Also, it can be checked that the map sending $a$ to $\sqrt{\neg}a$, for all $a \in A$, yields an isomorphism between the Boolean algebras $(A, \otimes, \oplus, \neg, \top, \bot)$ and $(A, \wedge, \vee, \neg, t, f)$.

*Lemma* 7. Let **A** be a square root algebra, and let $a, b \in A$. Then:

$$\begin{aligned}
a \otimes b = a \wedge b &\text{ iff } \sqrt{\neg}a \otimes \sqrt{\neg}b = \sqrt{\neg}a \vee \sqrt{\neg}b;\\
a \otimes b = a \vee b &\text{ iff } \sqrt{\neg}a \otimes \sqrt{\neg}b = \sqrt{\neg}a \wedge \sqrt{\neg}b.
\end{aligned}$$

**Proof.** If $a \otimes b = a \wedge b$, then $\sqrt{\neg}a \wedge \sqrt{\neg}b = \sqrt{\neg}\,(a \otimes b) = \sqrt{\neg}\,(a \wedge b) = \sqrt{\neg}a \oplus \sqrt{\neg}b$, and so, by Lemma 3, $\sqrt{\neg}a \otimes \sqrt{\neg}b = \sqrt{\neg}a \vee \sqrt{\neg}b$. Conversely, if $\sqrt{\neg}a \otimes \sqrt{\neg}b = \sqrt{\neg}a \vee \sqrt{\neg}b$, then $\sqrt{\neg}\,(a \vee b) = \sqrt{\neg}\,(a \oplus b)$, whence $a \vee b = a \oplus b$. Again, by Lemma 3, $a \otimes b = a \wedge b$. The remaining quasiequations are proved similarly. ■

Let now $\mathbf{B} = (B, \sqcap, \sqcup, \sim, 1, 0)$ be a Boolean algebra. The *pair algebra* $\mathbf{B} \boxdot \mathbf{B}$ is the algebra

$$\left(B^2, \wedge, \vee, \otimes, \oplus, \sqrt{\neg}, \top\right)$$

of language $\mathcal{L}$ such that:

- $\left(B^2, \wedge, \vee, \otimes, \oplus\right)$ is the product bilattice whose factors are both equal to the distributive lattice reduct $(B, \sqcap, \sqcup)$ of **B**;
- for all $a, b \in B$, $\sqrt{\neg}\,(a, b) = (b, \sim a)$;
- $\top = (1, 1)$.

It is not hard to check that **4** is isomorphic to $\mathbf{2} \boxdot \mathbf{2}$, where **2** is the 2-element Boolean algebra.

*Lemma* 8. If $\mathbf{B} = (B, \sqcap, \sqcup, \sim, 1, 0)$ is a Boolean algebra, the pair algebra $\mathbf{B} \boxdot \mathbf{B}$ is a square root algebra.

**Proof.** By the results mentioned in Section 2.1, $(B, \sqcap, \sqcup) \odot (B, \sqcap, \sqcup)$ is a distributive bilattice. By way of example, we check some conditions required to establish that $\left(B^2, \wedge, \vee, \neg, t, f\right)$ and $\left(B^2, \otimes, \oplus, \neg, \top, \bot\right)$ are Boolean algebras (and leave the remaining ones to the reader). If $b, c \in B$,

$$\begin{aligned}
(b, c) \oplus (0, 0) &= (b \sqcup 0, c \sqcup 0) = (b, c);\\
(b, c) \wedge (1, 0) &= (b \sqcap 1, c \sqcup 0) = (b, c);\\
(b, c) \otimes \neg\,(b, c) &= (b \sqcap \sim b, c \sqcap \sim c) = (0, 0) = \bot;\\
(b, c) \vee \neg\,(b, c) &= (b \sqcup \sim b, c \sqcap \sim c) = (1, 0) = t.
\end{aligned}$$

Finally, we check some of the conditions in SQ3 above.

$$\begin{aligned}\sqrt{\neg}\,((b_1,c_1)\otimes(b_2,c_2)) &= \sqrt{\neg}\,(b_1\sqcap b_2, c_1\sqcap c_2)\\ &= (c_1\sqcap c_2, {\sim}b_1\sqcup{\sim}b_2)\\ &= (c_1,{\sim}b_1)\wedge(c_2,{\sim}b_2)\\ &= \sqrt{\neg}\,(b_1,c_1)\wedge\sqrt{\neg}\,(b_2,c_2)\,.\end{aligned}$$

$$\begin{aligned}\sqrt{\neg}\,((b_1,c_1)\wedge(b_2,c_2)) &= \sqrt{\neg}\,(b_1\sqcap b_2, c_1\sqcup c_2)\\ &= (c_1\sqcup c_2, {\sim}b_1\sqcup{\sim}b_2)\\ &= (c_1,{\sim}b_1)\oplus(c_2,{\sim}b_2)\\ &= \sqrt{\neg}\,(b_1,c_1)\oplus\sqrt{\neg}\,(b_2,c_2)\,.\end{aligned}$$

■

The proof of the next result has been partly inspired by Bou and Rivieccio's proof of the representation theorem for distributive bilattices [13, Prop. 3.8], as well as by the representation theorem for Cartesian square root quasi-MV algebras [23, Thm. 36].

*Theorem* 9. Every square root algebra $\mathbf{A}$ is isomorphic to the pair algebra $\mathbf{B}\boxdot\mathbf{B}$, for some Boolean algebra $\mathbf{B}$.

**Proof.** Let $\mathbf{A} = (A, \wedge, \vee, \otimes, \oplus, \sqrt{\neg}, \top)$ be a square root algebra, and let

$$\theta = \{(a,b)\in A : a\otimes b = a\vee b\}\,.$$

By [13, Prop. 3.8], $\theta$ is a congruence on the bilattice reduct $(A, \wedge, \vee, \otimes, \oplus)$ of $\mathbf{A}$. We now show that $\theta$ is also a congruence on the Boolean algebra term reduct $(A, \otimes, \oplus, \neg, \top, \bot)$ of $\mathbf{A}$. In other words, we have to prove that $(a,b)\in\theta$ implies $(\neg a, \neg b)\in\theta$. In fact, $a\otimes b = a\vee b$ implies $\sqrt{\neg}a\wedge\sqrt{\neg}b = \sqrt{\neg}\,(a\otimes b) = \sqrt{\neg}\,(a\vee b) = \sqrt{\neg}a\otimes\sqrt{\neg}b$. Thus, by Lemma 3, $\sqrt{\neg}a\oplus\sqrt{\neg}b = \sqrt{\neg}a\vee\sqrt{\neg}b$, whence

$$\begin{aligned}\neg a\vee\neg b = \sqrt{\neg}\sqrt{\neg}a\vee\sqrt{\neg}\sqrt{\neg}b = \sqrt{\neg}\left(\sqrt{\neg}a\oplus\sqrt{\neg}b\right) =\\ \sqrt{\neg}\left(\sqrt{\neg}a\vee\sqrt{\neg}b\right) = \sqrt{\neg}\sqrt{\neg}a\otimes\sqrt{\neg}\sqrt{\neg}b = \neg a\otimes\neg b.\end{aligned}$$

Thus, $\mathbf{A}^* = (A, \otimes, \oplus, \neg, \top, \bot)\,/\theta$ is a Boolean algebra. Similarly, it can be shown that

$$\varphi = \{(a,b)\in A : a\otimes b = a\wedge b\}$$

is a congruence on $(A, \otimes, \oplus, \neg, \top, \bot)$. We know from [13, Prop. 3.8] that the map $f(a) = (a/\theta, a/\varphi)$ yields a bilattice isomorphism between $(A, \wedge, \vee, \otimes, \oplus)$ and $(A, \otimes, \oplus)\,/\theta\odot(A, \otimes, \oplus)\,/\varphi$. By Lemma 7, however, we have that $(a,b)\in\theta$ iff $(\sqrt{\neg}a, \sqrt{\neg}b)\in\varphi$, and $(a,b)\in\varphi$ iff $(\sqrt{\neg}a, \sqrt{\neg}b)\in\theta$. It follows that the map sending $a/\varphi$ to $\sqrt{\neg}a/\theta$ is an isomorphism, whence the map $g(a) = (a/\theta, \sqrt{\neg}a/\theta)$ yields a bilattice isomorphism between the bilattice reducts of $\mathbf{A}$ and of the pair algebra $\mathbf{A}^*\boxdot\mathbf{A}^*$. Moreover,

$$\begin{aligned}g\left(\sqrt{\neg}^{\mathbf{A}}a\right) &= \left(\sqrt{\neg}^{\mathbf{A}}a/\theta, \neg^{\mathbf{A}}a/\theta\right)\\ &= \left(\sqrt{\neg}^{\mathbf{A}}a/\theta, \neg^{\mathbf{A}^*}(a/\theta)\right)\\ &= \sqrt{\neg}^{\mathbf{A}^*\boxdot\mathbf{A}^*}\left(a/\theta, \sqrt{\neg}^{\mathbf{A}}a/\theta\right)\\ &= \sqrt{\neg}^{\mathbf{A}^*\boxdot\mathbf{A}^*}g(a)\,,\end{aligned}$$

and

$$\begin{aligned} g\left(\top^{\mathbf{A}}\right) &= \left(\top^{\mathbf{A}}/\theta, \sqrt{\neg}^{\mathbf{A}}\top/\theta\right) \\ &= \left(\top^{\mathbf{A}}/\theta, t^{\mathbf{A}}/\theta\right) \\ &= \left(\top^{\mathbf{A}}/\theta, \top^{\mathbf{A}}/\theta\right) \\ &= \left(\top^{\mathbf{A}^*}, \top^{\mathbf{A}^*}\right) = \top^{\mathbf{A}^* \boxdot \mathbf{A}^*}, \end{aligned}$$

where the third equality holds because, in $\mathbf{A}$, $\top$ and $t$ are the top elements of the Boolean algebras $(A, \otimes, \oplus, \neg, \top, \bot)$ and $(A, \wedge, \vee, \neg, t, f)$, respectively, whence $\top \otimes t = t = \top \vee t$. ■

If $\mathbb{K}$ is a class of Boolean algebras, by $\wp(\mathbb{K})$ we denote the class of all square root algebras of the form $\mathbf{B} \boxdot \mathbf{B}$, for $\mathbf{B} \in \mathbb{K}$. As usual, for $\mathbb{J}$ a class of similar algebras, $I(\mathbb{J})$, $S(\mathbb{J})$, and $P(\mathbb{J})$ denote the classes of all isomorphic images, subalgebras, and products of algebras in $\mathbb{J}$, respectively.

*Lemma* 10. For any class $\mathbb{K}$ of Boolean algebras: (i) $\wp I(\mathbb{K}) \subseteq I\wp(\mathbb{K})$; (ii)$\wp S(\mathbb{K}) \subseteq S\wp(\mathbb{K})$; (iii) $\wp P(\mathbb{K}) \subseteq P\wp(\mathbb{K})$.

**Proof.** We confine ourselves to the proof of (iii), the foregoing items being immediate. Actually, we will work out the proof for the case of binary products only. Let then $\mathbf{B}_1, \mathbf{B}_2 \in \mathbb{K}$. It will suffice to show that $\mathbf{C} = (\mathbf{B}_1 \times \mathbf{B}_2) \boxdot (\mathbf{B}_1 \times \mathbf{B}_2)$ is isomorphic to $\mathbf{D} = (\mathbf{B}_1 \boxdot \mathbf{B}_1) \times (\mathbf{B}_2 \boxdot \mathbf{B}_2)$. For $b_i, b_i' \in B_i$ $(i \leq 2)$, let then

$$f\left((b_1, b_2), (b_1', b_2')\right) = \left((b_1, b_1'), (b_2, b_2')\right).$$

Clearly, $f$ is a bijection. We now check that it respects some of the operations.

$$\begin{aligned} f\left(\sqrt{\neg}^{\mathbf{C}}\left((b_1, b_2), (b_1', b_2')\right)\right) &= f\left((b_1', b_2'), \sim^{\mathbf{B}_1 \times \mathbf{B}_2}(b_1, b_2)\right) \\ &= f\left((b_1', b_2'), \left(\sim^{\mathbf{B}_1} b_1, \sim^{\mathbf{B}_2} b_2\right)\right) \\ &= \left(\left(b_1', \sim^{\mathbf{B}_1} b_1\right), \left(b_2', \sim^{\mathbf{B}_2} b_2\right)\right) \\ &= \left(\sqrt{\neg}^{\mathbf{B}_1 \boxdot \mathbf{B}_1}(b_1, b_1'), \sqrt{\neg}^{\mathbf{B}_2 \boxdot \mathbf{B}_2}(b_2, b_2')\right) \\ &= \sqrt{\neg}^{\mathbf{D}}\left((b_1, b_1'), (b_2, b_2')\right) \\ &= \sqrt{\neg}^{\mathbf{D}} f\left((b_1, b_2), (b_1', b_2')\right). \end{aligned}$$

For $\wedge$,

$$\begin{aligned} & f\left(\left((b_1, b_2), (b_1', b_2')\right) \wedge^{\mathbf{C}} \left((c_1, c_2), (c_1', c_2')\right)\right) \\ &= f\left((b_1, b_2) \sqcap^{\mathbf{B}_1 \times \mathbf{B}_2} (c_1, c_2), (b_1', b_2') \sqcup^{\mathbf{B}_1 \times \mathbf{B}_2} (c_1', c_2')\right) \\ &= f\left(\left(b_1 \sqcap^{\mathbf{B}_1} c_1, b_2 \sqcap^{\mathbf{B}_2} c_2\right), \left(b_1' \sqcup^{\mathbf{B}_1} c_1', b_2' \sqcup^{\mathbf{B}_2} c_2'\right)\right) \\ &= \left(b_1 \sqcap^{\mathbf{B}_1} c_1, b_1' \sqcup^{\mathbf{B}_1} c_1'\right), \left(b_2 \sqcap^{\mathbf{B}_2} c_2, b_2' \sqcup^{\mathbf{B}_2} c_2'\right) \\ &= \left((b_1, b_1') \wedge^{\mathbf{B}_1 \boxdot \mathbf{B}_1} (c_1, c_1'), (b_2, b_2') \wedge^{\mathbf{B}_2 \boxdot \mathbf{B}_2} (c_2, c_2')\right) \\ &= \left((b_1, b_1'), (b_2, b_2')\right) \wedge^{\mathbf{D}} \left((c_1, c_1'), (c_2, c_2')\right) \\ &= f\left((b_1, b_2), (b_1', b_2')\right) \wedge^{\mathbf{D}} f\left((c_1, c_2), (c_1', c_2')\right). \end{aligned}$$

For $\otimes$,

$$\begin{aligned}
&f\left(((b_1,b_2),(b_1',b_2'))\otimes^{\mathbf{C}}((c_1,c_2),(c_1',c_2'))\right)\\
&= f\left((b_1,b_2)\sqcap^{\mathbf{B}_1\times\mathbf{B}_2}(c_1,c_2),(b_1',b_2')\sqcap^{\mathbf{B}_1\times\mathbf{B}_2}(c_1',c_2')\right)\\
&= f\left(\left(b_1\sqcap^{\mathbf{B}_1}c_1,b_2\sqcap^{\mathbf{B}_2}c_2\right),\left(b_1'\sqcap^{\mathbf{B}_1}c_1',b_2'\sqcap^{\mathbf{B}_2}c_2'\right)\right)\\
&= \left(b_1\sqcap^{\mathbf{B}_1}c_1,b_1'\sqcap^{\mathbf{B}_1}c_1'\right),\left(b_2\sqcap^{\mathbf{B}_2}c_2,b_2'\sqcap^{\mathbf{B}_2}c_2'\right)\\
&= \left((b_1,b_1')\otimes^{\mathbf{B}_1\boxdot\mathbf{B}_1}(c_1,c_1'),(b_2,b_2')\otimes^{\mathbf{B}_2\boxdot\mathbf{B}_2}(c_2,c_2')\right)\\
&= ((b_1,b_1'),(b_2,b_2'))\otimes^{\mathbf{D}}((c_1,c_1'),(c_2,c_2'))\\
&= f((b_1,b_2),(b_1',b_2'))\otimes^{\mathbf{D}}f((c_1,c_2),(c_1',c_2')).
\end{aligned}$$

The remaining operations are left to the reader. ■

*Theorem* 11. **4** generates $\mathbb{SQ}$ as a quasivariety.

**Proof.** Since **4** is a single finite algebra, by the Grätzer–Lakser theorem [24] it suffices to show that every $\mathbf{A}\in\mathbb{SQ}$ belongs to $ISP(\{\mathbf{4}\})$. Thus, let $\mathbf{A}\in\mathbb{SQ}$. By Theorem 9, $\mathbf{A}\in I\wp(\mathbb{BA}) = I\wp(ISP(\{\mathbf{2}\}))$. Thus, by Lemma 10, $\mathbf{A}\in IISP\wp(\{\mathbf{2}\}) = ISP(\{\mathbf{4}\})$. ■

## 3.2 Strengthenings and alternative proofs

The results in the previous subsection can be either strengthened or given alternative proof in light of the considerations that follow.

In the first place, we directly derived Theorem 11 from Theorem 9 by appealing to concrete properties of the pair algebra construction. The same result could have been obtained within the more abstract approach by Cabrer and Priestley [16], who prove a very general product representation theorem for varieties of bilattices equipped with additional operations having a certain form, out of factors belonging to a given base variety. The base variety (in our case, Boolean algebras) and the represented variety (in our case, square root algebras) are ensured to be categorically equivalent by some features of this construction.

On a different tack, we also observe that:

*Theorem* 12. The algebra **4** is primal.

**Proof.** It suffices to show that **4** satisfies the conditions (1)-(3) in Theorem 6. For a start, any subalgebra of **4** must contain $\top$ and hence also $t = \sqrt{\neg}\top, \bot = \neg\top, f = \neg\sqrt{\neg}\top$. Therefore, **4** has no proper subalgebra. Moreover, given an automorphism $\varphi$ of **4**,

$$\varphi(t) = \varphi\left(\sqrt{\neg}\top\right) = \sqrt{\neg}\varphi(\top) = \sqrt{\neg}\top = t,$$

and similarly for the other elements. It follows that the sole automorphism of **4** is the identity mapping.

Finally, consider the binary term

$$\rho(x,y) = (\neg x\vee y)\wedge(\neg y\vee x)\wedge\left(\neg\sqrt{\neg}x\vee\sqrt{\neg}y\right)\wedge\left(\neg\sqrt{\neg}y\vee\sqrt{\neg}x\right).$$

It can be checked that, for all $a,b\in\{t,f,\bot,\top\}, \rho^{\mathbf{4}}(a,b) = t$ in case $x = y$, and $\rho^{\mathbf{4}}(a,b) = f$ otherwise. Therefore the ternary term

$$t(x, y, z) = (\rho(x, y) \vee x) \wedge (\neg\rho(x, y) \vee z)$$

realises the discriminator function on $\{t, f, \bot, \top\}$. ■

As a consequence, we obtain a strengthened version of Theorem 11: the variety $\mathbb{SQ}$ is a discriminator variety and every square root algebra is a *Boolean power* of the algebra **4**.

Lastly, we show that square root algebras are nothing but bounded classical implicative bilattices in disguise. Although, for the sake of convenience, we *presuppose* Theorem 11 in this result, a direct proof of this fact would yield yet another different proof of Theorem 11.

*Theorem* 13. The variety $\mathbb{SQ}$ is term equivalent to the variety of bounded classical implicative bilattices.

**Proof.** If $\mathbf{A} = (A, \wedge, \vee, \otimes, \oplus, \sqrt{\neg}, \top)$ is a square root algebra, let

$$\mathbf{A}^* = \left(A, \wedge, \vee, \otimes, \oplus, \supset^{A^*}, -^{A^*}, \top, \bot, t, f\right),$$

where

$$a \supset^{A^*} b = \left((\neg a \wedge \top) \otimes \left(\neg\sqrt{\neg}a \vee \top\right)\right) \vee b \text{ and } -^{A^*} a = \left(\sqrt{\neg}a \wedge \top\right) \otimes \left(\neg\sqrt{\neg}a \vee \top\right).$$

Conversely, given a bounded classical implicative bilattice

$$\mathbf{B} = (B, \wedge, \vee, \otimes, \oplus, \supset, -, \top, \bot, t, f),$$

let

$$\mathbf{B}^+ = \left(B, \wedge, \vee, \otimes, \oplus, \sqrt{\neg}^{B^+}, \top^{B^+}\right),$$

where

$$\sqrt{\neg}^{B^+} a = -((a \supset \bot) \wedge (-a \supset \top)).$$

$\mathbf{A}^*$ is a bounded classical implicative bilattice. In fact, the mapping that sends $\mathbb{K}$ to $\{\mathbf{A}^* : \mathbf{A} \in \mathbb{K}\}$ is easily seen to commute with the class operators $I, S, P$, whence it is sufficient to check that $\mathbf{4}^*$ is just $\mathbf{B4}^{\supset}$. Similarly, it can be verified that $\mathbf{B}^+$ is a square root algebra.

It remains to be seen that $\mathbf{A} = \mathbf{A}^{*+}$ and that $\mathbf{B} = \mathbf{B}^{+*}$. We confine ourselves to the first verification. Ultimately, we must show that the equation

$$\sqrt{\neg}x \approx \left(\sqrt{\neg}(f(x) \wedge g(x)) \wedge \top\right) \otimes \left(\neg\sqrt{\neg}(f(x) \wedge g(x)) \vee \top\right),$$

where

$$f(x) = \left((\neg x \wedge \top) \otimes \left(\neg\sqrt{\neg}x \vee \top\right)\right) \vee \bot,$$
$$g(x) = \left((\neg h(x) \wedge \top) \otimes \left(\neg\sqrt{\neg}h(x) \vee \top\right)\right) \vee \top,$$
$$h(x) = \left(\sqrt{\neg}x \wedge \top\right) \otimes \left(\neg\sqrt{\neg}x \vee \top\right),$$

is satisfied in every square root algebra, and of course it suffices to check that it is satisfied in the algebra **4**, a task that is safely left to the reader. ■

## 4 Axiomatising the logic D

If the logic L is the $\tau$-assertional logic of the quasivariety $\mathbb{Q}$, the formula-equation translation $\tau$ encodes in equational form conditions that have to be met by elements of algebras in $\mathbb{Q}$ in order to represent "true values" in the perspective of L. Since D is the logic arising from the matrix $(\mathbf{4}, \{\top, t\})$, the "true values" in $\mathbf{4}$ are (in the perspective of D) exactly those that are greater or equal than $t$ in the information ordering $\leq_i$ of $\mathbf{4}$. For an element of $\{\top, t, f, \bot\}$, then, a necessary and sufficient condition to be a designated value of this matrix is satisfying the equation $t \otimes x \approx t$. Throughout this section, let $\tau$ be the formula-equation translation $\{t \otimes x \approx t\}$. Our remarks so far imply that D is the $\tau$-assertional logic of the class $\{\mathbf{4}\}$. Moreover, since D is finitary (in fact, it is defined by a single finite matrix), by Theorem 11 it is also the $\tau$-assertional logic of the variety $\mathbb{SQ}$.

*Theorem* 14. (i) $\mathbb{SQ}$ is a $\tau$-regular variety;

(ii) D is algebraisable with $\mathbb{SQ}$ as an equivalent variety semantics;

(iii) the set

$$\rho = \left\{\neg x \vee y, \neg y \vee x, \neg \sqrt{\neg} x \vee \sqrt{\neg} y, \neg \sqrt{\neg} y \vee \sqrt{\neg} x\right\}$$

is a system of equivalence formulas for D and $\mathbb{SQ}$.

**Proof.** (i) We use Theorems 4.(3) and 11 and verify by inspection that

$$x \approx y \dashv\vdash_{Eq(\{\mathbf{4}\})} t \otimes (\neg x \vee y) \approx t, t \otimes (\neg y \vee x) \approx t,$$
$$t \otimes \left(\neg \sqrt{\neg} x \vee \sqrt{\neg} y\right) \approx t, t \otimes \left(\neg \sqrt{\neg} y \vee \sqrt{\neg} x\right) \approx t.$$

(ii) By (i) and Theorem 4.(2).

(iii) It suffices to check AL4 in the definition of algebraisable logic. In light of Theorem 11, this condition boils down to

$$\alpha \approx \beta \dashv\vdash_{Eq(\{\mathbf{4}\})} \tau(\rho(\alpha, \beta)),$$

namely,

$$\alpha \approx \beta \dashv\vdash_{Eq(\{\mathbf{4}\})} t \otimes (\neg \alpha \vee \beta) \approx t, t \otimes (\neg \beta \vee \alpha) \approx t,$$
$$t \otimes \left(\neg \sqrt{\neg} \alpha \vee \sqrt{\neg} \beta\right) \approx t, t \otimes \left(\neg \sqrt{\neg} \beta \vee \sqrt{\neg} \alpha\right) \approx t.$$

This follows from the proof of (i). ■

Theorem 14 may assuage any remaining worries about the lack of congruentiality of $\sqrt{\neg}$ in D. In fact, using the terminology of [29, § 3.3], $\sqrt{\neg}$ actually happens to be $\leftrightarrow$-congruential with respect to the *generalised* equivalence connective $\rho$: whenever we have all of

$$\vdash_{\mathrm{D}} \neg \alpha \vee \beta, \vdash_{\mathrm{D}} \neg \beta \vee \alpha, \vdash_{\mathrm{D}} \neg \sqrt{\neg} \alpha \vee \sqrt{\neg} \beta, \vdash_{\mathrm{D}} \neg \sqrt{\neg} \beta \vee \sqrt{\neg} \alpha,$$

indeed, we also get

$$\vdash_{\mathrm{D}} \neg \sqrt{\neg} \alpha \vee \sqrt{\neg} \beta, \vdash_{\mathrm{D}} \neg \sqrt{\neg} \beta \vee \sqrt{\neg} \alpha, \vdash_{\mathrm{D}} \neg\neg \alpha \vee \neg \beta, \vdash_{\mathrm{D}} \neg\neg \beta \vee \neg \alpha.$$

We now introduce the following abbreviations:

$$\alpha \to \beta =: \neg \alpha \vee \beta;$$
$$\alpha \Rightarrow \beta =: \neg \alpha \oplus \beta;$$

$$\alpha \leftrightarrow \beta =: \{\alpha \to \beta, \beta \to \alpha\};$$
$$\alpha \Leftrightarrow \beta =: \{\alpha \Rightarrow \beta, \beta \Rightarrow \alpha\};$$
$$\Gamma(\alpha_1, \alpha_2, \beta_1, \beta_2) =: \{\alpha_1 \leftrightarrow \beta_1, \sqrt{\neg}\alpha_1 \leftrightarrow \sqrt{\neg}\beta_1, \alpha_2 \leftrightarrow \beta_2, \sqrt{\neg}\alpha_2 \leftrightarrow \sqrt{\neg}\beta_2\};$$
$$\Delta(\alpha_1, \alpha_2, \beta_1, \beta_2) =: \{(\alpha_1 \circ \alpha_2) \leftrightarrow (\beta_1 \circ \beta_2), \sqrt{\neg}(\alpha_1 \circ \alpha_2) \leftrightarrow \sqrt{\neg}(\beta_1 \circ \beta_2) : \circ \in \{\wedge, \vee, \otimes, \oplus\}\}.$$

We stipulate that $\to, \Rightarrow$ bind less strongly than the primitive connectives of $\mathcal{L}$.

*Corollary* 15. D is axiomatised by the following axioms and rules:

D1 $\vdash_{\mathrm{D}} \alpha \to \alpha$;
D2 $\alpha, \alpha \to \beta, \beta \to \alpha, \sqrt{\neg}\alpha \to \sqrt{\neg}\beta, \sqrt{\neg}\beta \to \sqrt{\neg}\alpha \vdash_{\mathrm{D}} \beta$;
D3 $\bigcup \Gamma(\alpha_1, \alpha_2, \beta_1, \beta_2) \vdash_{\mathrm{D}} \bigcup \Delta(\alpha_1, \alpha_2, \beta_1, \beta_2)$;
D4 $\alpha \dashv\vdash_{\mathrm{D}} t \otimes \alpha \to t, t \to t \otimes \alpha, \sqrt{\neg}(t \otimes \alpha) \to \sqrt{\neg}t, \sqrt{\neg}t \to \sqrt{\neg}(t \otimes \alpha)$;
D5 For all axioms $\alpha \approx \beta$ of $\mathbb{SQ}$:
(i) $\vdash_{\mathrm{D}} \alpha \to \beta$;
(ii) $\vdash_{\mathrm{D}} \beta \to \alpha$;
(iii) $\vdash_{\mathrm{D}} \sqrt{\neg}\alpha \to \sqrt{\neg}\beta$;
(iv) $\vdash_{\mathrm{D}} \sqrt{\neg}\beta \to \sqrt{\neg}\alpha$.

**Proof.** By Theorems 5 and 14, discarding trivially redundant items. ■

Corollary 15 solves the problem we posed at the outset, but the devised axiomatisation is far from pleasing from an aesthetic point of view. With an eye to doing better than this, we introduce a partial function $t^c : For(\mathcal{L}) \to For(\mathcal{L}^c)$, inductively defined as follows:

$$t^c(x) = x, \text{ for all sentential variables } x;$$
$$t^c(\top) = 1;$$
$$t^c\left(\sqrt{\neg}\alpha\right) = \text{ undefined, if either } \alpha = x \text{ or } \alpha = \beta \circ \gamma, \text{ for } \circ \in \{\wedge, \vee, \otimes, \oplus\};$$
$$t^c\left(\sqrt{\neg}\alpha\right) = {\sim}t^c(\beta), \text{ if } \alpha = \sqrt{\neg}\beta;$$
$$t^c\left(\sqrt{\neg}\alpha\right) = 1, \text{ if } \alpha = \top;$$
$$t^c(\alpha \wedge \beta) = t^c(\alpha \otimes \beta) = t^c(\alpha) \sqcap t^c(\beta), \text{ if } t^c(\alpha), t^c(\beta) \text{ are defined};$$
$$t^c(\alpha \vee \beta) = t^c(\alpha \oplus \beta) = t^c(\alpha) \sqcup t^c(\beta), \text{ if } t^c(\alpha), t^c(\beta) \text{ are defined}.$$

Let D′ be the sentential logic whose formula algebra is **For** ($\mathcal{L}$) and whose consequence relation $\vdash_{\mathrm{D}'}$ is the derivability relation of the Hilbert calculus axiomatised by the following axioms and rules:

D'1 $\vdash_{\mathrm{D}'} \sigma(\alpha)$, for every substitution $\sigma$ and for every $\alpha$ such that $\mathcal{L}(\alpha) \subseteq \{\neg, \otimes, \oplus, \top\}$ and $\vdash_{\mathrm{CPC}} t^c(\alpha)$;
D'2 $\vdash_{\mathrm{D}'} \sigma(\alpha)$, for every substitution $\sigma$ and for every $\alpha$ such that $\mathcal{L}(\alpha) \subseteq \{\neg, \wedge, \vee, t\}$ and $\vdash_{\mathrm{CPC}} t^c(\alpha)$;
D'3 if $\vdash_{\mathrm{D}'} \sigma(\alpha)$, then $\vdash_{\mathrm{D}'} \sqrt{\neg}\sigma(\alpha)$, for every substitution $\sigma$ and for every $\alpha$ such that $\mathcal{L}(\alpha) \subseteq \{\neg, \otimes, \oplus, \top\}$;
D'4 $\alpha, \alpha \to \beta \vdash_{\mathrm{D}'} \beta$;
D'5 $\vdash_{\mathrm{D}'} (\alpha \to \beta) \leftrightarrow (\alpha \Rightarrow \beta)$;
D'6 $\vdash_{\mathrm{D}'} \sqrt{\neg}(\alpha \otimes \beta) \leftrightarrow \sqrt{\neg}\alpha \wedge \sqrt{\neg}\beta$;
D'7 $\vdash_{\mathrm{D}'} \sqrt{\neg}(\alpha \oplus \beta) \leftrightarrow \sqrt{\neg}\alpha \vee \sqrt{\neg}\beta$;

D'8 $\vdash_{D'} \sqrt{\neg}(\alpha \wedge \beta) \leftrightarrow \sqrt{\neg}\alpha \oplus \sqrt{\neg}\beta$;
D'9 $\vdash_{D'} \sqrt{\neg}(\alpha \vee \beta) \leftrightarrow \sqrt{\neg}\alpha \otimes \sqrt{\neg}\beta$.

*Lemma* 16. Let $\gamma_1, ..., \gamma_n, \alpha$ be formulas in $For(\mathcal{L})$ such that either $\mathcal{L}(\gamma_1) \cup ... \cup \mathcal{L}(\gamma_n) \cup \mathcal{L}(\alpha) \subseteq \{\neg, \otimes, \oplus, \top\}$ or $\mathcal{L}(\gamma_1) \cup ... \cup \mathcal{L}(\gamma_n) \cup \mathcal{L}(\alpha) \subseteq \{\neg, \wedge, \vee, t\}$. If $t^c(\gamma_1), ..., t^c(\gamma_n) \vdash_{CPC} t^c(\alpha)$, then for every substitution $\sigma$ on $\mathbf{For}(\mathcal{L})$,

$$\sigma(\gamma_1), ..., \sigma(\gamma_n) \vdash_{D'} \sigma(\alpha).$$

**Proof.** If $\mathcal{L}(\gamma_1) \cup ... \cup \mathcal{L}(\gamma_n) \cup \mathcal{L}(\alpha) \subseteq \{\neg, \wedge, \vee, t\}$ our conclusion is immediate from D'2 and D'4. Otherwise, just observe that by D'4 and D'5, the rule

$$\alpha, \alpha \Rightarrow \beta \vdash_{D'} \beta$$

is a derivable rule of the system. ■

*Theorem* 17. D and D′ are deductively equivalent.

**Proof.** All the axioms and rules of D′ are $\vdash_D$-sound, whence, by Corollary 15, $\vdash_{D'} \subseteq \vdash_D$. For the converse direction, we need to check that all of D1-D5 are derivable from D'1-D'9.

(D1) follows from D'2.

(D2) Assume $\alpha, \alpha \leftrightarrow \beta$, $\sqrt{\neg}\alpha \leftrightarrow \sqrt{\neg}\beta$. Then, by D'4, we derive $\beta$.

(D3) Assume $\alpha_1 \leftrightarrow \beta_1$, $\sqrt{\neg}\alpha_1 \leftrightarrow \sqrt{\neg}\beta_1, \alpha_2 \leftrightarrow \beta_2$, $\sqrt{\neg}\alpha_2 \leftrightarrow \sqrt{\neg}\beta_2$. By Lemma 16 we can derive $(\alpha_1 \circ \alpha_2) \leftrightarrow (\beta_1 \circ \beta_2)$, for $\circ \in \{\wedge, \vee, \otimes, \oplus\}$. By way of example, we now derive $\sqrt{\neg}(\alpha_1 \otimes \alpha_2) \leftrightarrow \sqrt{\neg}(\beta_1 \otimes \beta_2)$:

| | | |
|---|---|---|
| 1. | $\sqrt{\neg}\alpha_1 \rightarrow \sqrt{\neg}\beta_1$ | Ass. |
| 2. | $\sqrt{\neg}\alpha_2 \rightarrow \sqrt{\neg}\beta_2$ | Ass. |
| 3. | $\sqrt{\neg}\alpha_1 \wedge \sqrt{\neg}\alpha_2 \rightarrow \sqrt{\neg}\beta_1 \wedge \sqrt{\neg}\beta_2$ | Lm 16: 1,2 |
| 4. | $\sqrt{\neg}(\alpha_1 \otimes \alpha_2) \rightarrow \sqrt{\neg}\alpha_1 \wedge \sqrt{\neg}\alpha_2$ | D'6 |
| 5. | $\sqrt{\neg}\beta_1 \wedge \sqrt{\neg}\beta_2 \rightarrow \sqrt{\neg}(\beta_1 \otimes \beta_2)$ | D'6 |
| 6. | $\sqrt{\neg}(\alpha_1 \otimes \alpha_2) \rightarrow \sqrt{\neg}(\beta_1 \otimes \beta_2)$ | Lm 16: 3,4,5 |

Similarly, the converse implication is established and the rest of D3 is proved along analogous lines.

(D4) From left to right, we have to prove all of the following:

(D4.1) $\alpha \vdash_{D'} \neg(t \otimes \alpha) \vee t$;
(D4.2) $\alpha \vdash_{D'} f \vee (t \otimes \alpha)$;
(D4.3) $\alpha \vdash_{D'} \neg\sqrt{\neg}(t \otimes \alpha) \vee \bot$;
(D4.4) $\alpha \vdash_{D'} \top \vee \sqrt{\neg}(t \otimes \alpha)$.

For D4.1 observe that by Lemma 16, $\alpha \vdash_{D'} t$ and $t \vdash_{D'} \neg(t \otimes \alpha) \vee t$, whence our conclusion follows. Similarly, $\alpha \vdash_{D'} t$ and $\alpha \vdash_{D'} \alpha$ imply $\alpha \vdash_{D'} t \otimes \alpha$, once more by Lemma 16. For the same reason, $t \otimes \alpha \vdash_{D'} f \vee (t \otimes \alpha)$, which is enough for D4.2. As regards D4.3, observe that by Lemma 16

$$\vdash_{D'} \neg\beta \oplus \neg\alpha \oplus \beta \text{ and } \neg\beta \oplus \neg\alpha \oplus \beta \dashv\vdash_{D'} \neg(\beta \otimes \alpha) \oplus \beta,$$

whence we obtain that $\vdash_{D'} \neg(\beta \otimes \alpha) \oplus \beta$ and, letting $\beta = t$, by D'3, $\vdash_{D'} \sqrt{\neg}(\neg(t \otimes \alpha) \oplus t)$. Moreover, by D'7,

$$\sqrt{\neg}(\neg(t \otimes \alpha) \oplus t) \vdash_{D'} \neg\sqrt{\neg}(t \otimes \alpha) \vee \bot,$$

and our conclusion readily follows.

Finally, for D4.4, by D'1 we have $\vdash_{D'} \top$ and by Lemma 16 $\top \vdash_{D'} \top \vee \sqrt{\neg}(t \otimes \alpha)$, and once again we can establish our claim.

For the converse direction of D4, Lemma 16 entails that $f \vdash_{D'} \alpha$ and $t \otimes \alpha \vdash_{D'} \alpha$, whence, once more in virtue of the same lemma, $f \vee (t \otimes \alpha) \vdash_{D'} \alpha$.

(D5) We illustrate the case of the axioms in SQ1 and SQ2 with the simple examples $\alpha \wedge \beta \approx \beta \wedge \alpha$ and $\alpha \otimes \beta \approx \beta \otimes \alpha$. Start with the former. By D'2, $\vdash_{D'} \alpha \wedge \beta \to \beta \wedge \alpha$, and further

1. $\sqrt{\neg}(\alpha \wedge \beta) \to \sqrt{\neg}\alpha \oplus \sqrt{\neg}\beta$ D'8
2. $\sqrt{\neg}\alpha \oplus \sqrt{\neg}\beta \to \sqrt{\neg}\beta \oplus \sqrt{\neg}\alpha$ Lm 16
3. $\sqrt{\neg}\beta \oplus \sqrt{\neg}\alpha \to \sqrt{\neg}(\beta \wedge \alpha)$ D'8
4. $\sqrt{\neg}(\alpha \wedge \beta) \to \sqrt{\neg}(\beta \wedge \alpha)$ Lm 16 1,2,3

As to the latter, the strategy is similar but with a crucial recourse to D'5. For SQ3, consider by way of illustration the axiom

$$\sqrt{\neg}(\alpha \otimes \beta) \approx \sqrt{\neg}\alpha \wedge \sqrt{\neg}\beta.$$

We are committed to showing that

$$\vdash_{D'} \begin{array}{c} \sqrt{\neg}(\alpha \otimes \beta) \to \sqrt{\neg}\alpha \wedge \sqrt{\neg}\beta,\ \sqrt{\neg}\alpha \wedge \sqrt{\neg}\beta \to \sqrt{\neg}(\alpha \otimes \beta), \\ \neg(\alpha \otimes \beta) \to \sqrt{\neg}(\sqrt{\neg}\alpha \wedge \sqrt{\neg}\beta),\ \sqrt{\neg}(\sqrt{\neg}\alpha \wedge \sqrt{\neg}\beta) \to \neg(\alpha \otimes \beta). \end{array}$$

By D'6, $\vdash_{D'} \sqrt{\neg}(\alpha \otimes \beta) \to \sqrt{\neg}\alpha \wedge \sqrt{\neg}\beta,\ \sqrt{\neg}\alpha \wedge \sqrt{\neg}\beta \to \sqrt{\neg}(\alpha \otimes \beta)$. Moreover:

1. $\neg(\alpha \otimes \beta) \to \neg\alpha \oplus \neg\beta$ D'1, D'5
2. $\neg\alpha \oplus \neg\beta \to \sqrt{\neg}(\sqrt{\neg}\alpha \wedge \sqrt{\neg}\beta)$ D'8
3. $\neg(\alpha \otimes \beta) \to \sqrt{\neg}(\sqrt{\neg}\alpha \wedge \sqrt{\neg}\beta)$ Lm 16 1,2

The remaining item is established similarly, and the reader can easily provide the missing details for the verification of the rest of D5. ■

ADDED IN PROOF. After this paper had been submitted and accepted, we became aware of some important literature items that had been regrettably ignored while writing it. The PhD thesis by Carolina Blasio [8] has an explicit mention of demi-negation in Belnap-Dunn logic. Some articles by Hitoshi Omori and collaborators [19, 30, 33, 34] investigate other deductively equivalent versions of the logic D and study functional completeness in Belnap-Dunn logic. An appraisal of the relationships between these contributions and the present paper will be left for another occasion. We warmly thank João Marcos and Hitoshi Omori for the appropriate pointers.

**Acknowledgement** The author gratefully acknowledges the support of RAS – Regione Autonoma della Sardegna (Project CRP-78705, "Argomentazione e Metafora. Effetti della comunicazione persuasiva nel territorio sardo"). Some suggestions by Bogdan Dicher, Antonio Ledda, Lloyd Humberstone, Umberto Rivieccio and two anonymous referees were extremely important in shaping the final version of this paper.

## References


1. Anderson A.R., Belnap N.D., *Entailment*, Princeton University Press, Princeton, NJ, 1975 (vol. 1), and 1992, (vol. 2, with J.M. Dunn).
2. Arieli O., Avron A., "Reasoning with logical bilattices", *Journal of Logic, Language and Information*, 5, 1, 1996, pp. 25-63.
3. Arieli O., Avron A., "The value of the four values", *Artificial Intelligence*, 102, 1, 1998, pp. 97-141.
4. Avron A., "The structure of interlaced bilattices", *Mathematical Structures in Computer Science*, 6, 3, 1996, pp. 287-299.
5. Barbour G.D., Raftery J.G., "Quasivarieties of logic, regularity conditions and parameterized algebraization", *Studia Logica*, 74, 2003, pp. 99-152.
6. Belnap N.D., "How a computer should think", in G. Ryle (Ed.), *Contemporary Aspects of Philosophy*, Oriel Press, Boston, 1976, pp. 30-56.
7. Berto F., "Adynaton and material exclusion", *Australasian Journal of Philosophy*, 86, 2, 2008, pp. 165-190.
8. Blasio C., *Sobre Noçẽs de Consequência Generalizadas e Lógicas Plurivalentes*, PhD Thesis, Unicamp, 2017.
9. Blok W.J., Pigozzi D., *Algebraizable Logics*, Memoirs of the AMS, number 396, American Mathematical Society, Providence, RI, 1989.
10. Blok W.J., Raftery J.G., "Ideals in quasivarieties of algebras", in X. Caicedo and C.H. Montenegro (Eds.), *Models, Algebras and Proofs*, Dekker, New York, 1999, pp. 167-186.
11. Blok W.J., Raftery J.G., "Assertionally equivalent quasivarieties", *International Journal of Algebra and Computation*, 18, 4, 2008, pp. 589-681.
12. Bou F., Jansana R., Rivieccio U., "Varieties of interlaced bilattices", *Algebra Universalis*, 66, 1, 2011, pp. 115-141.
13. Bou F., Rivieccio U., "The logic of distributive bilattices", *Logic Journal of the IGPL*, 19, 1, 2011, pp. 183-216.
14. Bou F., Rivieccio U., "Bilattices with implications", *Studia Logica*, 101, 4, 2013, pp. 651-675.
15. Burris S., Sankappanavar H.P., *A Course in Universal Algebra*, Springer, Berlin, 1981.
16. Cabrer L., Priestley H., "A general framework for product representations: Bilattices and beyond", *Logic Journal of IGPL*, 23, 2015, pp. 816-841.
17. Chiffi D., Schang F., "The logical burdens of proof", *Logic and Logical Philosophy*, 2017, DOI: 10.12775/LLP.2017.006.
18. Dalla Chiara M.L., Giuntini R., Greechie R., *Reasoning in Quantum Theory*, Kluwer, Dordrecht, 2004.
19. De M., Omori, H., Classical negation and expansions of Belnap-Dunn logic, *Studia Logica*, 103, 4, 2015, pp. 825-851.
20. Dunn J.M., "Intuitive semantics for first-degree entailment and 'coupled trees'", *Philosophical Studies*, 29, 3, 1976, pp. 149–168.
21. Fitting M., "Bilattices and the semantics of logic programming", *Journal of Logic Programming*, 11, 1991, pp. 91-116.
22. Ginsberg M.L., "Multivalued logics: A uniform approach to inference in artificial intelligence", *Computational Intelligence*, 4, 3, 1988, pp. 265-316.
23. Giuntini R., Ledda A., Paoli F., "Expanding quasi-MV algebras by a quantum operator", *Studia Logica*, 87, 1, 2007, pp. 99-128.
24. Grätzer G., Lakser H., "A note on the implicational class generated by a class of structures", *Canadian Mathematical Bulletin*, 16, 4, 1973, pp. 603-605.
25. Gurevich Y., "Intuitionistic logic with strong negation", *Studia Logica*, 36, 1-2, 1977, pp. 49–59.
26. Humberstone L., "Negation by Iteration", *Theoria*, 61, 1, 1995, pp. 1–24.
27. Humberstone L., "Contraclassical logics", *Australasian Journal of Philosophy*, 78, 4, 2000, pp. 438–474.
28. Humberstone L., "Sentence connectives in formal logic", *Stanford Encyclopaedia of Philosophy*, 2010, http://plato.stanford.edu/archives/sum2010/entries/connectives-logic/
29. Humberstone L., *The Connectives*, MIT Press, Cambridge-London, 2011.
30. Kamide N., Omori H., An extended first-order Belnap-Dunn Logic with classical negation, in *Proceedings of LORI 2017*, Springer, Berlin, 2017, pp. 79-93.
31. Mares E., *Relevant Logic*, Cambridge University Press, Cambridge, 2004.

32. Mizraji E., "Vector logic: A natural algebraic representation of the fundamental logical gates", *Journal of Logic and Computation*, 18, 1, 2008, pp. 97-121.
33. Omori H., Sano K., Generalizing functional completeness in Belnap-Dunn Logic, *Studia Logica*, 103, 5, 2015, pp. 883-917.
34. Omori H., Wansing, H., "On Contra-classical variants of Nelson logic N4 and its classical extension," *Review of Symbolic Logic*, 11, 4, 2018, pp. 805-820.
35. Price H., "Why 'Not'?", *Mind*, 99, 394, 1990, pp. 221-238.
36. Priest G., *In Contradiction: A Study of the Transconsistent*, Oxford University Press, Oxford, $2006^2$.
37. Restall G., "Multiple Conclusions", in P. Hájek et al. (Eds.), *Logic, Methodology and Philosophy of Science: Proceedings of the Twelfth International Congress*, Kings' College Publications, 2005, pp. 189–205.
38. Ripley D., "Paradoxes and failures of cut", *Australasian Journal of Philosophy*, 91, 1, 2013, pp. 139-164.
39. Ruet P., "Complete set of connectives and complete sequent calculus for Belnap's logic", *Logic Colloquium 96*, 1996.
40. Rumfitt I., "Yes and no", *Mind*, 109, 436, 2000, pp. 781-823.
41. Shramko Y., Dunn J. M., Takenaka T., "The trilattice of constructive truth values", *Journal of Logic and Computation*, 11, 6, 2001, pp. 761-788.
42. Shramko Y., Wansing H., *Truth and Falsehood*, Springer, Berlin, 2011.
43. Smiley T.J., "Rejection", *Analysis*, 56, 1, 1996, pp. 1-9.
44. Wansing H., *The Logic of Information Structures*, Springer, Berlin, 1993.
45. Wansing H., "Proofs, disproofs, and their duals", in V. Goranko, L. Beklemishev, V. Shehtman (Eds.), *Advances in Modal Logic* Volume 8, College Publications, London, 2010, pp. 483-505.

# Consistency, Completeness, and Classicality

Adam Přenosil

**Abstract** We study the expansion of the four-valued Belnap–Dunn logic by a pair of constants 0 and 1 which express respectively the weakest inconsistent proposition and the strongest complete proposition. We then further expand this logic by the intuitionistic implication and use this expansion to introduce a logic which is a conservative extension of both classical and intuitionistic logic. The key idea behind this way of combining classical and intuitionistic logic is that classical negation is nothing but the De Morgan negation of the Belnap–Dunn logic restricted to classical contexts, i.e. contexts in which completeness and consistency is assumed. Finally, sequent calculi are provided for the logics introduced in the paper.

**Keywords** Non-classical logic • Consistency • Completeness • Combination of logics • intuitionistic logic • Belnap–Dunn logic • Kleene logic • Logic of Paradox

## 1 Introduction

It has long been acknowledged that if we seek a logic which captures certain important aspects of our practice of asserting and denying propositions, we need to abandon the classical assumption that all inconsistent propositions are equivalent. The four-valued Belnap–Dunn logic [6, 3, 4] achieves this goal by treating truth and falsity as independent properties of each proposition: a proposition may be true (and not false), false (and not true), both true and false, or neither. This allows us to avoid the undesirable equivalence of all classical contradictions by treating consistent and inconsistent sets of propositions on a par.

It may be argued, however, that this is a step too far. After all, consistent and inconsistent sets of propositions are *not* on a par: the former may be entirely true, whereas the latter may not (in a classical world). Although we may accept that in asserting an inconsistent set of propositions, we do not thereby assert every proposition, still we ought not assert inconsistent sets of propositions.

Adam Přenosil
Department of Logic, Faculty of Arts, Charles University, Prague, Czech Republic
e-mail: `adam.prenosil@gmail.com`

 
H. Omori, H. Wansing (eds.), *New Essays on Belnap-Dunn Logic*,
Synthese Library 418, https://doi.org/10.1007/978-3-030-31136-0_15

Likewise, we may want to distinguish between complete and incomplete sets of propositions. While an inconsistent set of propositions is bound to contain a false proposition (in a classical world), a complete set of propositions is bound to contain a true proposition (in a classical world). Just as inconsistent sets of propositions are things we should avoid (in the sense of not asserting all propositions in the set), complete sets of propositions are things we should aim for (in the sense of asserting some proposition in the set).

Although completeness in the above sense is a somewhat less familiar notion than inconsistency, and for this reason it may seem less fundamental at first sight, in this paper we commit ourselves to treating the two notions as equally fundamental. This requires adopting a multiple-conclusion framework. Consistency then applies to sets of propositions interpreted conjunctively as sets of premises, while completeness applies to sets of propositions interpreted disjunctively as sets of conclusions.

It is precisely the tension between consistency and completeness which makes the whole business of asserting things about the world a non-trivial enterpise. The requirement of consistency tells us that if $\Gamma$ is an inconsistent set of propositions, then we ought not assert all of $\Gamma$. The requirement of completeness, on the other hand, tells us that if $\Delta$ is a complete set of propositions, then we ought to assert some of $\Delta$. If we disregard consistency, it would suffice to assert everything, while if we disregard completeness, it would suffice to assert nothing.

Consistency and completeness, of course, should not be interpreted as absolute requirements. Idealizing our situation somewhat, we may imagine that we start out asserting nothing and fulfilling the demands of consistency completely while also completely violating the demands of completeness. Now surely it is unreasonable to demand that out of each pair of proposition A and −A we assert one. What we claim is much less controversial, namely that the closer we approach these goals the better. Asserting more is a strict improvement only if doing so does not lead us further away from consistency, while asserting less is a strict improvement only if doing so does not lead us further away from completeness.

Now if we accept that consistency and completeness are fundamental properties of sets of propositions, it is surely desirable to have the expressive means in one's logic to differentiate between consistent and inconsistent or complete and incomplete sets of propositions. In the multiple-conclusion formulation of classical logic, the *verum* and *falsum* constants (or equivalently, the empty set of premises and the empty set of conclusions) play this role: classical inconsistency means classically implying the *falsum*, while classical completeness means being classically implied by the *verum*.

In the multiple-conclusion formulation of the Belnap–Dunn logic, the empty set of premises and the empty set of conclusions do not play this role: the proposition $p \wedge -p$ is inconsistent (i.e. it is bound to be false in a classical world), but does not imply $\emptyset$, and likewise the proposition $p \vee -p$ is complete (i.e. it is bound to be true in a classical world), but it is not implied by $\emptyset$. We therefore expand the Belnap–Dunn logic by a pair of constants 0 and 1 interpreted respectively as the weakest inconsistent proposition and the strongest complete proposition.

The notions of inconsistency and completeness may of course also be captured by simply specifying which sets of formulas are inconsistent and which are complete. We choose the approach based on constants because it involves a smaller departure from the standard view of logic as a consequence relation, allowing us to essentially internalize the notions of completeness and inconsistency as part of the consequence relation. It also brings other expressive benefits which will be discussed later.

Note that designating certain sets of propositions as complete or inconsistent provides us with information which goes beyond what the consequence relation tells us. Under the reading of $-A$ as the negation of A, the proposition $A \wedge -A$ is surely inconsistent. However, if we fix a possible world $w$ and interpret $-A$ as "it is not the case that A holds at $w$", then there is *a priori* no reason why $A \wedge -A$ should be inconsistent. Crucially, there need not be anything in the consequence relation itself which forces us decide between the two interpretations: both interpretations are consistent with the consequence relation of the Belnap–Dunn logic. Similarly, there is nothing in the consequence relation of minimal intuitionistic logic (which replaces the *falsum* constant of intuitionistic logic with a constant 0 not required to satisfy any particular condition) which forces us to claim that $A \wedge (A \to 0)$ is an inconsistent proposition: 0 may consistently be claimed to represent a perfectly innocuous proposition.

Not only does the presence of the constants 0 and 1 help us pin down the interpretation of the De Morgan negation, but it also enables us to differentiate between several kinds of contexts in which formulas may occur, namely

- arbitrary contexts $\Gamma \vDash \Delta$,
- complete contexts $1, \Gamma \vDash \Delta$,
- consistent contexts $\Gamma \vDash \Delta, 0$,
- classical contexts $1, \Gamma \vDash \Delta, 0$.

Classical logic is then the restriction of the Belnap–Dunn logic to a classical context, while restricting to complete contexts yields Priest's Logic of Paradox [11] and restricting to consistent contexts yields the strong three-valued Kleene logic.

The above observation is straightforward to make, but it becomes substantially more interesting if we note that, although combining intuitionistic and classical logic within a single system has been considered to be a non-trivial task, combining intuitionistic logic with the Belnap–Dunn logic is not difficult at all. Restricting in such a combined system to a classical context then yields a logic which is a conservative extension of both classical and intuitionistic logic. Moreover, we shall see that the semantics of this combined system is in fact strikingly simple.

One of the main goals of this paper will therefore be to convince the reader along the way that, in addition to classical logic being the expansion of intuitionistic logic by the law of the excluded middle, classical and intuitionistic logic may also naturally be viewed as *two fragments of a single logic*.

At this point it needs to be said that our proposal for combining classical and intuitionistic logic is certainly not the first. We are aware of two previous proposals in this direction, namely those due to Liang & Miller [9] and to del Cerro & Herzig [5]. Ours is close in spirit to the former approach and essentially unrelated to the latter. However, a proper comparison between the three approaches will be postponed until after we have presented the current proposal.

This paper is also a continuation of a line of research into expansions of the Belnap–Dunn logic which internalize the notion of inconsistency, either in the form of a predicate expressing that a proposition is inconsistent [14], or in the form of an expansion by a second negation operator expressing *reductio ad contradictionem* [13]. The logic introduced here is more expressive than both of these expansions.

Let us now outline the structure of the paper. In Section 2 we sketch the proof-theoretic reasons why the presence of constants expressing completeness and inconsistency allows us to avoid the collapse of the intuitionistic implication into the material implication when combining classical and intuitionistic logic. Section 3 then provides a simple relational semantics for this combined system. The fact that this semantics yields a conservative extension of classical and intuitionistic logic when we restrict to truth preservation in classical worlds is proved in Sections 4 and 5. The current approach to combining classical and intuitionistic logic is then compared with the existing proposals in Section 6. Finally, in Sections 7 and 8 we provide a sequent calculus for the full combined system, as well as a cut-free sequent calculus for a particular fragment of this system, namely the Belnap–Dunn logic expanded by the completeness and inconsistency constants.

## 2 The classical–intuitionistic collapse

We start by outlining our strategy for combining classical and intuitionistic logics within a single system. Combining the two logics in a naïve way is known to lead to the collapse of intuitionistic implication into classical implication, as we shall see below. Our strategy for avoiding this collapse is to consider a system which is a conservative extension of intuitionistic but not classical logic and then to restrict to consequence within a classical context in this system.

We have both a proof-theoretic and a semantic story to tell, and we shall start on the proof-theoretic side. Let us first recall why combining the usual Gentzen rules for classical and intuitionistic connectives within a single system results in the intuitionistic implication collapsing into the classical one.

In most of this paper, with the exception of Section 8, we adopt a framework where sequents are pairs of finite sets of formulas, written as $\Gamma \rhd \Delta$, and where the rule of Weakening

$$\frac{\Gamma \rhd \Delta}{\mathrm{A}, \Gamma \rhd \Delta} \qquad \frac{\Gamma \rhd \Delta}{\Gamma \rhd \Delta, \mathrm{A}}$$

is valid. Although some of our points may also be made in the absence of certain structural rules, we will not do so to avoid muddying the waters unnecessarily.

Now the intuitionistic implication can easily be proved to be logically stronger than the material implication in the system which naïvely combines the standard sequent calculi for classical and intuitionistic logic:

$$\dfrac{\dfrac{\dfrac{\mathrm{A} \rhd \mathrm{A} \qquad \mathrm{B} \rhd \mathrm{B}}{\mathrm{A}, \mathrm{A} \to \mathrm{B} \rhd \mathrm{B}}}{\mathrm{A} \to \mathrm{B} \rhd -\mathrm{A}, \mathrm{B}}}{\mathrm{A} \to \mathrm{B} \rhd -\mathrm{A} \vee \mathrm{B}}$$

We do not contest this: intuitionistic implication should indeed express a stronger connection between the antecedent and the consequent than the mere falsity of one or truth of the other.

We do, however, contest the opposite implication. In particular, the mere falsity of the antecedent or truth of the consequent should not be sufficient to establish that an intuitionistic implication obtains between the two. The problem, then, is how to block the following proof in our system:

$$\dfrac{\dfrac{\dfrac{\mathrm{B} \triangleright \mathrm{B}}{\mathrm{A}, \mathrm{B} \triangleright \mathrm{B}}}{\mathrm{B} \triangleright \mathrm{A} \to \mathrm{B}} \qquad \dfrac{\dfrac{\dfrac{\mathrm{A} \triangleright \mathrm{A}}{-\mathrm{A}, \mathrm{A} \triangleright \emptyset}}{-\mathrm{A}, \mathrm{A} \triangleright \mathrm{B}}}{-\mathrm{A} \triangleright \mathrm{A} \to \mathrm{B}}}{-\mathrm{A} \vee \mathrm{B} \triangleright \mathrm{A} \to \mathrm{B}}$$

The left branch of the proof, or at any rate the provability of its conclusion, is incontestable if our system is to be a conservative extension of intuitionistic logic. Likewise, the provability of the sequent $-\mathrm{A}, \mathrm{A} \triangleright \mathrm{B}$ is incontestable if our system is to be a conservative extension of classical logic.

The last step of the proof, i.e. disjunction introduction on the left, is also incontestable (due to the classical validity of the sequent $\mathrm{A} \vee \mathrm{B} \triangleright \mathrm{A}, \mathrm{B}$) if we want our system is to be (i) a conservative extension of multiple-conclusion classical logic where (ii) the multiple-conclusion form of Cut

$$\dfrac{\Gamma \triangleright \Delta, \mathrm{A} \qquad \mathrm{A}, \Gamma \triangleright \Delta}{\Gamma \triangleright \Delta}$$

is derivable, and which is (iii) substitution invariant, i.e. where the derivability of $\Gamma \triangleright \Delta$ implies the derivability of $\sigma[\Gamma] \triangleright \sigma[\Delta]$ for each substitution $\sigma$.

The only step that we can question in the above proof is therefore the last step of the right branch of the proof:

$$\dfrac{-\mathrm{A}, \mathrm{A} \triangleright \mathrm{B}}{-\mathrm{A} \triangleright \mathrm{A} \to \mathrm{B}}$$

Now on what grounds can we possibly question the validity of an inference step as fundamental to intuitionistic implication as the right introduction rule? The simple answer is that we do not, not really. What we claim, instead, is that the system which we present as a natural combination of classical and intuitionistic logic arises as the restriction of another system, in which all the rules of intuitionistic (but not classical) logic are valid, to *classical contexts*.

What do we mean by a classical context, proof-theoretically? We mean that we assume completeness and consistency. Assuming completeness amounts to assuming that there is a tacit proposition 1 in the antecedent from which we may infer all instances of the excluded middle $\mathrm{A} \vee -\mathrm{A}$. Assuming consistency amounts to assuming that there is a tacit proposition 0 in the consequent which may be inferred from any instance of $\mathrm{A} \wedge -\mathrm{A}$. Making these tacit propositions explicit makes it clear why the right introduction rule for intuitionistic implication fails in general: what we are trying to do is to draw the inference

$$\dfrac{-\mathrm{A}, \mathrm{A} \triangleright \mathrm{B}, 0}{-\mathrm{A} \triangleright \mathrm{A} \to \mathrm{B}, 0}$$

which is of course not an intuitionistically valid inferential step.

With the above remarks in mind, we may transform the proofs above into proofs (of slightly different sequents) which take the presence of 1 and 0 explicit. That is, instead of assuming the provability of the sequents

$$\emptyset \triangleright \mathrm{A}, -\mathrm{A} \qquad\qquad -\mathrm{A}, \mathrm{A} \triangleright \emptyset$$

which are in the presence of multiple-conclusion Cut equivalent to the rules

$$\dfrac{\mathrm{A}, \Gamma \triangleright \Delta}{\Gamma \triangleright \Delta, -\mathrm{A}} \qquad \dfrac{\Gamma \triangleright \Delta, \mathrm{A}}{-\mathrm{A}, \Gamma \triangleright \Delta}$$

we are assuming the provability of the sequents

$$1 \rhd \mathrm{A}, -\mathrm{A} \qquad\qquad -\mathrm{A}, \mathrm{A} \rhd 0$$

which are in the presence of multiple-conclusion Cut equivalent to the rules

$$\frac{1, \mathrm{A}, \Gamma \rhd \Delta}{1, \Gamma \rhd \Delta, -\mathrm{A}} \qquad \frac{\Gamma \rhd \Delta, \mathrm{A}, 0}{-\mathrm{A}, \Gamma \rhd \Delta, 0}$$

and which imply the standard introduction rules for negation in classical contexts, i.e. in sequents of the form $1, \Gamma \rhd \Delta, 0$.

Let us now modify the above proof of $-\mathrm{A} \vee \mathrm{B} \rhd \mathrm{A} \to \mathrm{B}$, which we rejected, to account for the presence of 1 and 0 in the context. The inclusion of the constant 1 among the premises transforms the first proof into:

$$\dfrac{\dfrac{\dfrac{\dfrac{\mathrm{A} \rhd \mathrm{A} \qquad \mathrm{B} \rhd \mathrm{B}}{\mathrm{A}, \mathrm{A} \to \mathrm{B} \rhd \mathrm{B}}}{1, \mathrm{A}, \mathrm{A} \to \mathrm{B} \rhd \mathrm{B}}}{1, \mathrm{A} \to \mathrm{B} \rhd -\mathrm{A}, \mathrm{B}}}{1, \mathrm{A} \to \mathrm{B} \rhd -\mathrm{A} \vee \mathrm{B}}$$

In other words, this proof therefore remains valid if we assume that the 1 in the antecedent was tacit in the original proof. On the other hand, the inclusion of the constant 0 among the conclusions forces us to modify the proof more substantially and to include the constant 0 explicitly in the conclusion:

$$\dfrac{\dfrac{\dfrac{\dfrac{\mathrm{B} \rhd \mathrm{B}}{\mathrm{A}, \mathrm{B} \rhd \mathrm{B}, 0}}{\mathrm{A}, \mathrm{B} \rhd \mathrm{B} \vee 0}}{\mathrm{B} \rhd \mathrm{A} \to (\mathrm{B} \vee 0)} \qquad \dfrac{\dfrac{\dfrac{\dfrac{\mathrm{A} \rhd \mathrm{A}}{-\mathrm{A}, \mathrm{A} \rhd 0}}{-\mathrm{A}, \mathrm{A} \rhd \mathrm{B}, 0}}{-\mathrm{A}, \mathrm{A} \rhd \mathrm{B} \vee 0}}{-\mathrm{A} \rhd \mathrm{A} \to (\mathrm{B} \vee 0)}}{-\mathrm{A} \vee \mathrm{B} \rhd \mathrm{A} \to (\mathrm{B} \vee 0)}$$

We see that the naïve derivation of the sequent $-\mathrm{A} \vee \mathrm{B} \rhd \mathrm{A} \to \mathrm{B}$ is blocked by the presence of the constant 0 in the consequent. In order to show that this sequent is in fact not derivable, we now turn to the semantic side of the story.

## 3 Expressing completeness and inconsistency

The key semantic idea underlying our combined system for classical and intuitionistic logic is that classical logic is simply the four-valued Belnap–Dunn logic restricted to classical worlds, and that combining intuitionistic logic with the Belnap–Dunn logic is in fact straightforward. Our recipe for the semantics is to therefore combine these two logics, to specify what the classical worlds are, and then to define the consequence relation via truth preservation in classical worlds.

The relational (Kripke) semantics of intuitionistic logic is, of course, given simply by partially ordered sets, where propositions are interpreted as upward closed sets (*upsets* for short). That is, we fix an infinite set of propositional atoms, interpret each of these by an upset, and then extend this interpretation to all formulas by means of the semantic clauses

$$u \Vdash A \wedge B \iff u \Vdash A \text{ and } u \Vdash B,$$
$$u \Vdash A \vee B \iff u \Vdash A \text{ or } u \Vdash B,$$
$$u \Vdash A \to B \iff v \Vdash A \text{ implies } v \Vdash B \text{ for all } v \geq u.$$

The *verum* and *falsum* constants, if we wish to include them in the signature of the logic, are covered by the semantic clauses

$$u \Vdash \top \text{ for all } u,$$
$$u \nVdash \bot \text{ for all } u.$$

The consequence relation of intuitionistic logic, $\Gamma \vDash_{\mathcal{IL}} B$, is then defined as follows: for each interpretation on each partially ordered set and each point $u$ of this set

$$\text{either } u \nVdash A \text{ for some } A \in \Gamma \text{ or } u \Vdash B.$$

It is well known that it suffices to restrict to cases where $u$ is the smallest element of a partially ordered set.

The relational semantics of the Belnap–Dunn logic is given by partially ordered sets equipped with an order-inverting involution $\partial$. That is, $u \leq v$ implies $\partial v \leq \partial u$, and moreover $u = \partial\partial u$. Such structures will simply be called *frames* here. We prefer to call elements of such frames *states* rather than worlds, reserving the term *world* for classical worlds as defined below. The state $\partial u$ will be dual to the state $u$ in the sense that it asserts precisely those things that $u$ does not deny, or conversely it denies precisely those things that $u$ does not assert.

Again, we fix an infinite set of propositional atoms, interpret each of these by an upset, and extend this interpretation to all formulas using the semantic clauses

$$u \Vdash -A \iff \partial u \nVdash A,$$
$$u \Vdash A \wedge B \iff u \Vdash A \text{ and } u \Vdash B,$$
$$u \Vdash A \vee B \iff u \Vdash A \text{ or } u \Vdash B.$$

The *verum* and *falsum* constants can also be included in the signature of the logic using the same semantic clauses as above. The multiple-conclusion consequence relation of the Belnap–Dunn logic, $\Gamma \vDash_{\mathcal{BD}} \Delta$, is then defined as follows: for each interpretation on each frame and each state $u$ of this frame

$$\text{either } u \nVdash A \text{ for some } A \in \Gamma \text{ or } u \Vdash B \text{ for some } B \in \Delta.$$

A relational semantics for the combination of intuitionistic logic and the Belnap–Dunn logic is now immediately obtained by simply combining all of the semantic clauses above. This yields a logic that is sometimes called *symmetric intuitionistic logic*, which was introduced by Monteiro [10].[1] Note that the *verum* and *falsum* constants are definable in this logic as $\top := p \to p$ and $\bot := -(p \to p)$.

The dual of intuitionistic implication, usually called *co-implication*, is also definable in symmetric intuitionistic logic as $A \prec B := -(-B \to -A)$. As the reader may easily verify, this defined connective is governed by the semantic clause

[1] The name "symmetric intuitionistic logic" is sometimes also used for what we call bi-intuitionistic logic, i.e. the extension of intuitionistic logic by co-implication.

$$u \Vdash A \prec B \iff \text{there is some } v \leq u \text{ such that } v \Vdash A \text{ and } v \nVdash B.$$

Symmetric intuitionistic logic is thus a conservative extension of bi-intuitionistic logic, which extends intuitionistic logic by co-implication.

Observe in particular that negated implication $-(A \to B)$ is *not* equivalent to $A \wedge -B$, as is the case in logics with Nelson's strong negation. Rather, the negated implication is equivalent to the contraposed co-implication $-B \prec -A$. (For a comparison of different interpretations negated implications, see [16].)

It now remains to specify the complete, consistent, and classical worlds in each frame. Let us call a state *complete* if for each interpretation either $u \Vdash A$ or $u \Vdash -A$ for each proposition A, *consistent* if for each interpretation either $u \nVdash A$ or $u \nVdash -A$ for each proposition A, and *classical* if it is both complete and consistent. Note that quantifying over propositional atoms rather than over all propositions would yield an equivalent definition. The following observation now tells us how to actually pick out the complete and consistent worlds in a given frame.

**Fact 1.** *A state u is consistent if and only if $u \leq \partial u$, complete if and only if $\partial u \leq u$, and classical if and only if $u = \partial u$.*

*Proof.* If $u \leq \partial u$ and $u \Vdash A$, then $\partial u \Vdash A$, thus $u \nVdash -A$. If $\partial u \leq u$ and $u \nVdash A$, then $\partial u \nVdash A$, thus $u \Vdash -A$. Conversely, if $u \nleq \partial u$, let A be a propositional atom and interpret A as the set of all worlds $w$ with $w \geq u$. Then $u \Vdash A$ and $\partial u \nVdash A$, thus $u \Vdash -A$. Likewise, if $\partial u \nleq u$, interpret the atom A as the set of all worlds $w$ with $w \geq \partial u$. Then $u \nVdash A$ and $\delta u \Vdash A$, thus $u \nVdash -A$.

With this lemma in hand, we can now provide the appropriate semantic clauses for the constants 0 and 1:

$$u \Vdash 0 \iff u \nleq \delta u,$$
$$u \Vdash 1 \iff \delta u \leq u.$$

Note that we represent inconsistency rather than consistency by a proposition because the consistent worlds form a downset, and downsets are not propositions in general. (Moreover, it is inconsistency rather than consistency which is the dual of completeness if we exchange the roles of the premises and the conclusions.) Notice also that 1 and $-0$ are in fact the same proposition.

Using propositional quantification, the constants 0 and 1 can be thought of as $\exists X\,(X \wedge -X)$ and $\forall X\,(X \vee -X)$, where the existential quantification is interpreted as disjunction over all propositions of the form $X \wedge -X$ and the universal quantification is interpreted as the conjunction over all propositions of the form $X \vee -X$, where X ranges over all propositions, i.e. over all upsets of a frame.

Let us denote the multiple-conclusion consequence relation given by the above semantics simply as $\Gamma \vDash \Delta$. Here $\Gamma$ and $\Delta$ are sets of formulas in the signature $\{\wedge, \vee, \to, -, 0, 1\}$. As we foreshadowed at the beginning of this section, our strategy will now be to restrict to truth preservation in classical worlds only. Since those are precisely the worlds where 1 holds but 0 does not, it is straightforward to express truth preservation in classical worlds in terms of the relation $\Gamma \vDash \Delta$ as $1, \Gamma \vDash \Delta, 0$. Truth preservation in complete and in consistent states can similarly be expressed respectively as $1, \Gamma \vDash \Delta$ and $\Gamma \vDash \Delta, 0$.

**Fact 2.** *The multiple-conclusion consequence* $1, \Gamma \vDash \Delta, 0$ *holds if and only if for each interpretation on each frame and each* classical world *u of the frame*

$$\textit{either } u \nVdash A \textit{ for some } A \in \Gamma \textit{ or } u \Vdash B \textit{ for some } B \in \Delta.$$

We are now finally in a position to show that the sequent $1, -A \vee B \rhd A \rightarrow B, 0$ is in general not provable in the system that we sketched in Section 2, which will be introduced more formally in Section 7. Take the formulas A and B to be propositional atoms and consider the three-element frame $u < v < w$, where necessarily $\partial u = w$, $\partial v = v$, and $\partial w = u$. Take an interpretation where $v \nVdash A$, $w \Vdash A$, and $w \nVdash B$. Then $v \Vdash -A \vee B$ but $v \nVdash A \rightarrow B$ because $w \Vdash A$ and $w \nVdash B$. This shows that the material implication is not equivalent to the intuitionistic implication in our logic of classical worlds, therefore the classical–intuitionistic collapse is avoided. The reader is encouraged to work out why the material implication is in fact strictly weaker than the intuitionistic implication, i.e. why the consequence $1, A \rightarrow B \vDash -A \vee B, 0$ (and even $1, A \rightarrow B \vDash -A \vee B$) holds.

Observe that it is the presence of inconsistent states which blocks the classical–intuitionistic collapse. The connectives of classical logic behave locally and therefore classically in classical worlds, but the truth value of an intuitionistic implication $A \rightarrow B$ at a classical world $u$, unless it is falsified classically already at $u$, depends on the truth values of A and B in states strictly above the classical world $u$, which are necessarily inconsistent. Accepting inconsistent states is therefore crucial to our strategy for reconciling classical and intuitionistic logic.

Moreover, due to the presence of inconsistent states, the classical worlds lie "in the middle" of a frame rather than "at the top", as we are accustomed to thinking of them in intuitionistic logic. This is reflected by the fact that classical worlds do not necessarily validate the law of the excluded middle in the form $A \vee (A \rightarrow \bot)$, but they satisfy the law in both of the forms $A \vee -A$ and $A \vee {\sim}A$, where ${\sim}A := A \rightarrow 0$. The inferential negation ${\sim}A$, which expresses that $A$ cannot consistently be assumed, is in general weaker than the De Morgan negation $-A$. As we shall see in the following section, ${\sim}A$ can be identified with the negation of minimal intuitionistic logic.

Although we believe that the semantic framework introduced above provides a particularly simple and natural way of reconciling classical and intuitionistic logic, we should also be clear about the price to be paid for restricting to truth preservation in a certain subset of states: interderivable formulas may no longer be substituted for each other inside of formulas while preserving interderivability. In technical terms, the system is not *selfextensional*. That is, it is in general not the case that

$$\text{if } 1, A \vDash B, 0 \text{ and } 1, B \vDash A, 0, \text{ then } 1, C(A) \vDash C(B), 0.$$

For example, we have

$$1, A \wedge -A \vDash B \wedge -B, 0 \qquad \text{and} \qquad 1, B \wedge -B \vDash A \wedge -A, 0$$

but substituting $A \wedge -A$ and $B \wedge -B$ for X in $C \rightarrow X$ does not yield interderivable formulas. That is,

$$1, C \rightarrow (A \wedge -A) \nvDash C \rightarrow (B \wedge -B), 0.$$

We encourage the reader to work out why this inference is not valid in our framework. The counter-example again relies on the presence of inconsistent states.

Note that, unlike e.g. intuitionistic logic with Nelson's strong negation, the above semantics yields a selfextensional logic if we consider truth preservation in *all* states.

## 4 Four-in-one conservativity

In the previous section, we introduced a relational semantics for a logic claimed to be a conservative extension of both classical and intuitionistic logic (if we restrict to truth preservation in classical worlds). In this section, we investigate the fragment $\{\wedge, \vee, \top, \bot, 1, 0, -\}$ of this logic, proving that not only the consequence relation of classical logic $\mathcal{CL}$, but also the consequence relations of the Logic of Paradox $\mathcal{LP}$ and the strong three-valued Kleene logic $\mathcal{K}$ can be expressed in this system by restricting to truth preservation in appropriate sets of states.

In order to do so, it will be convenient to consider a somewhat simpler semantics for the fragment $\{\wedge, \vee, \top, \bot, 1, 0, -\}$, namely (a simple extension of) the relational semantics of Dunn [6] based on what we call truth relations. The two semantics will turn out to be essentially equivalent on this fragment.

A *truth relation* specifies for each formula A whether it is true (denoted $u \Vdash^+ A$) and whether it is false (denoted $u \Vdash^- A$), allowing for the possibility that A may be neither true nor false (i.e. $u \nVdash^+ A$ and $u \nVdash^- A$) or both true and false (i.e. $u \Vdash^+ A$ and $u \Vdash^- A$). This specification is arbitrary for propositional atoms, and it extends to all formulas by means of the semantic clauses

$$u \Vdash^+ -A \iff u \Vdash^- A,$$
$$u \Vdash^- -A \iff u \Vdash^+ A,$$

$$u \Vdash^+ A \wedge B \iff u \Vdash^+ A \text{ and } u \Vdash^+ B,$$
$$u \Vdash^- A \wedge B \iff u \Vdash^- A \text{ or } u \Vdash^+ B,$$

$$u \Vdash^+ A \vee B \iff u \Vdash^+ A \text{ or } u \Vdash^+ B,$$
$$u \Vdash^- A \vee B \iff u \Vdash^- A \text{ and } u \Vdash^- B,$$

$$u \Vdash^+ \top, u \nVdash^- \top,$$
$$u \nVdash^+ \bot, u \Vdash^- \bot.$$

Truth relations provide a semantics for the Belnap–Dunn logic in the following sense: $\Gamma \vDash_{\mathcal{BD}} \Delta$ if and only if for each truth relation $u$

$$\text{either } u \nVdash^+ \gamma \text{ for some } \gamma \in \Gamma \text{ or } u \Vdash^+ \delta \text{ for some } \delta \in \Delta.$$

That is, $\mathcal{BD}$ is defined by truth preservation in all truth relations.

Moreover, restricting to certain subclasses of truth relations yields a semantics for the strong Kleene logic $\mathcal{K}$ and for the Logic of Paradox $\mathcal{LP}$. Let us call a truth relation *consistent* if no proposition (or equivalently, no propositional atom) is both true and false, *complete* if each proposition (or equivalently, each propositional atom) is either true or false, and *classical* if it is both consistent and complete.

The logic $\mathcal{K}$ is then defined by restricting to truth preservation in consistent truth relations, and $\mathcal{LP}$ by restricting to truth preservation in complete truth relations. Classical truth relations, of course, are essentially nothing but ordinary two-valued classical valuations. (The reader looking for a more thorough introduction to the logics $\mathcal{K}$ and $\mathcal{LP}$ is advised to consult Priest's textbook on non-classical logic [12].)

The most natural way of extending this semantics to account for the constants 0 and 1 is to use the expected semantic clauses

$$\begin{aligned} u \Vdash^+ 0 &\iff u \text{ is inconsistent,} \\ u \Vdash^- 0 &\iff u \text{ is complete,} \\ u \Vdash^+ 1 &\iff u \text{ is complete,} \\ u \Vdash^- 1 &\iff u \text{ is inconsistent.} \end{aligned}$$

This yields what we call *standard* truth relations. On the other hand, we may also relax those semantic clauses and merely require that the following implications hold:

$$\begin{aligned} u \Vdash^+ 0 &\Longleftarrow u \text{ is inconsistent,} \\ u \nVdash^+ 1 &\Longleftarrow u \text{ is incomplete,} \end{aligned}$$

supplemented by the equivalences

$$\begin{aligned} u \Vdash^- 0 &\iff u \Vdash^+ 1, \\ u \Vdash^- 1 &\iff u \Vdash^+ 0. \end{aligned}$$

This yields what we simply call *truth relations* (for the signature $\{\wedge, \vee, \top, \bot, 1, 0, -\}$). Allowing for non-standard truth relations will be useful for technical purposes.

The semantics of arbitrary and standard truth relations differ only marginally. The semantics of arbitrary truth relations is finitary, in the sense that

$$\Gamma \vDash \Delta \implies \Gamma' \vDash \Delta' \text{ for some finite } \Gamma' \subseteq \Gamma \text{ and } \Delta' \subseteq \Delta.$$

The same holds for the semantics of frames. (Given truth relations $u_{\Gamma',\Delta'}$ witnessing that $\Gamma' \nvDash \Delta'$, take an appropriate ultraproduct to obtain a truth relation $u$ witnessing that $\Gamma \nvDash \Delta$, and likewise for interpretations on frames. This standard argument works because truth relations and frames are objects defined by a set of first-order conditions.)

On the other hand, the semantics of standard truth relations is not finitary, since

$$\text{At}, 0 \vDash -[\text{At}] \text{ in the semantics of standard truth relations,}$$

but for each finite set of atoms $\Gamma$ we have

$$\Gamma, 0 \nvDash -[\Gamma] \text{ in the semantics of standard truth relations.}$$

However, for all practical purposes (in particular, whenever $\Gamma$ and $\Delta$ are finite), the two semantics may be treated as equivalent.

We now show that the semantics of standard truth relations and the semantics of frames are equivalent in case $\Gamma$ and $\Delta$ are finite, using the semantics of arbitrary truth relations as an intermediate step.

**Theorem 1 (Equivalence between frames and truth relations).** *Let $\Gamma$ and $\Delta$ be a pair of finite sets of formulas in the signature $\{\wedge, \vee, \top, \bot, 1, 0, -\}$. Then the following are equivalent:*

*(i)* $\Gamma \vDash \Delta$ *in the semantics of frames,*
*(ii)* $\Gamma \vDash \Delta$ *in the semantics of truth relations,*
*(iii)* $\Gamma \vDash \Delta$ *in the semantics of standard truth relations.*

*Proof.* Firstly, given an interpretation on a frame and an element $w$ of that frame, we may define a truth relation $u$ as

$$u \Vdash^{+} \mathrm{A} \iff w \Vdash \mathrm{A},$$
$$u \Vdash^{-} \mathrm{A} \iff w \Vdash -\mathrm{A}.$$

We now need to verify that this indeed yields a truth relation. We only verify the appropriate semantic clauses for 1 and 0. It suffices to consider $u \Vdash^{+} 0$ and $u \Vdash^{+} 1$:

$$\begin{aligned} u \text{ is inconsistent} &\iff \text{there is some A such that } u \Vdash^{+} \mathrm{A} \text{ and } u \Vdash^{-} \mathrm{A} \\ &\iff \text{there is some A such that } w \Vdash \mathrm{A} \text{ and } w \Vdash -\mathrm{A} \\ &\implies w \Vdash 0 \\ &\implies u \Vdash^{+} 0, \end{aligned}$$

$$\begin{aligned} u \text{ is incomplete} &\iff \text{there is some A such that } u \nVdash^{+} \mathrm{A} \text{ and } u \nVdash^{-} \mathrm{A} \\ &\iff \text{there is some A such that } w \nVdash \mathrm{A} \text{ and } w \nVdash -\mathrm{A} \\ &\implies w \nVdash 1 \\ &\implies u \nVdash^{+} 1. \end{aligned}$$

Secondly, given a truth relation $u$ which witnesses that $\Gamma \nvDash \Delta$, we may define a standard truth relation which witnesses that $\Gamma \nvDash \Delta$. Namely, if $u$ is consistent but $u \Vdash^{+} 0$, take an atom which does not occur in $\Gamma$ and $\Delta$ and take it to be both true and false in $v$. (Here we use the assumption that $\Gamma$ and $\Delta$ are finite.) Similarly, if $u$ is complete but $u \nVdash^{+} 1$, take another such atom and take it to be neither true nor false in $v$. This yields a standard truth relation.

Finally, given a truth relation $u$ (standard or not) where $u \Vdash^{+} 0$ and $u \nVdash^{+} 1$, we may define an interpretation on a frame consisting of two incomparable states $w_{+}$ and $w_{-}$ (with $\partial w_{+} = w_{-}$ and $\partial w_{-} = w_{+}$) as

$$w_{+} \Vdash \mathrm{A} \iff u \Vdash^{+} \mathrm{A},$$
$$w_{-} \Vdash \mathrm{A} \iff u \nVdash^{-} \mathrm{A}.$$

Moreover, if $u \nVdash^{+} 0$ and $u \nVdash^{+} 1$, we take $w_{+} < w_{-}$, if $u \Vdash^{+} 0$ and $u \Vdash^{+} 1$, we take $w_{-} < w_{+}$, and if $u \nVdash^{+} 0$ and $u \Vdash^{+} 1$, we take $w_{+} = w_{-}$. It is routine to verify that this indeed yields an interpretation on a two-element or one-element frame.

The following theorem is now immediate given the definitions of $\mathcal{LP}$, $\mathcal{K}$, and $\mathcal{BD}$. It shows that we can recover the multiple-conclusion consequence relations of $\mathcal{CL}$, $\mathcal{LP}$, and $\mathcal{K}$ from the multiple-conclusion consequence relation of $\mathcal{BD}$ expanded by the inconsistency and completeness constants 0 and 1 in a particularly simple way.

**Theorem 2 (Four-in-one conservativity).** *Let* $\Gamma$ *and* $\Delta$ *be a pair of finite sets of formulas in the signature* $\{\wedge, \vee, \top, \bot, -\}$. *Then:*

*(i)* $\Gamma \vDash_{\mathcal{CL}} \Delta$ *if and only if* $1, \Gamma \vDash \Delta, 0$.

(ii) $\Gamma \vDash_{\mathcal{LP}} \Delta$ *if and only if* $1, \Gamma \vDash \Delta$.
(iii) $\Gamma \vDash_{\mathcal{K}} \Delta$ *if and only if* $\Gamma \vDash \Delta, 0$.
(iv) $\Gamma \vDash_{\mathcal{BD}} \Delta$ *if and only if* $\Gamma \vDash \Delta$.

We can also interpret the signature of classical logic as $\{\wedge, \vee, 1, 0, -\}$ rather than $\{\wedge, \vee, \top, \bot, -\}$, with 1 being the *verum* constant and 0 being the *falsum* constant. In that case we obtain the following theorem.

**Theorem 3 (Classical conservativity).** *Let* $\Gamma$ *and* $\Delta$ *be set of formulas in the signature* $\{\wedge, \vee, 1, 0, -\}$. *Then* $\Gamma \vDash_{\mathcal{CL}} \Delta$ *if and only if* $1, \Gamma \vDash \Delta, 0$.

These results are closely related to a theorem of Beall [1, 2], who was concerned with expressing the multiple-conclusion consequence relation of classical logic using the multiple-conclusion consequence relations of $\mathcal{BD}$, $\mathcal{LP}$, and $\mathcal{K}$. (Beall in fact proved the theorem in the context of first-order logic in [2].) In the following, let At($\Gamma$) denote the set of all propositional atoms which occur in $\Gamma$ (more precisely, which are subformulas of some formula in $\Gamma$).

**Theorem 4 (Beall's theorem).** *Let* $\Gamma$ *and* $\Delta$ *be sets of formulas in the signature* $\{\wedge, \vee, \top, \bot, -\}$. *Then:*

(i) $\Gamma \vDash_{\mathcal{CL}} \Delta$ *if and only if* $\Gamma \vDash_{\mathcal{LP}} \Delta \cup \{A \wedge -A \mid A \in \mathrm{At}(\Gamma)\}$.
(ii) $\Gamma \vDash_{\mathcal{CL}} \Delta$ *if and only if* $\{A \vee -A \mid A \in \mathrm{At}(\Delta)\} \cup \Gamma \vDash_{\mathcal{K}} \Delta$.
(iii) $\Gamma \vDash_{\mathcal{CL}} \Delta$ *if and only if* $\{A \vee -A \mid A \in \mathrm{At}(\Delta)\} \cup \Gamma \vDash_{\mathcal{BD}} \Delta \cup \{A \wedge -A \mid A \in \mathrm{At}(\Gamma)\}$.

As the reader can see, Beall's theorem for $\mathcal{BD}$ is essentially a refinement of our conservativity theorem for $\mathcal{CL}$. Our result is blunt (but compact), whereas Beall's result pinpoints precisely which inconsistent or complete propositions need to be added at the appropriate places. Indeed, we can formulate analogous refinements of the conservativity theorems for $\mathcal{LP}$ and $\mathcal{K}$ proved above.

**Theorem 5.** *Let* $\Gamma$ *and* $\Delta$ *be sets of formulas in the signature* $\{\wedge, \vee, \top, \bot, -\}$. *Then:*

(i) $\Gamma \vDash_{\mathcal{LP}} \Delta$ *if and only if* $\{A \vee -A \mid A \in \mathrm{At}(\Delta)\} \cup \Gamma \vDash_{\mathcal{BD}} \Delta$.
(ii) $\Gamma \vDash_{\mathcal{K}} \Delta$ *if and only if* $\Gamma \vDash_{\mathcal{BD}} \Delta \cup \{A \wedge -A \mid A \in \mathrm{At}(\Gamma)\}$.
(iii) $\Gamma \vDash_{\mathcal{CL}} \Delta$ *if and only if* $\{A \vee -A \mid A \in \mathrm{At}(\Delta)\} \cup \Gamma \vDash_{\mathcal{BD}} \Delta \cup \{A \wedge -A \mid A \in \mathrm{At}(\Gamma)\}$.

*Proof.* The proof of this theorem does not essentially differ from the proof of Beall's theorem (and indeed coincides with it in the case of $\mathcal{CL}$). We therefore only sketch the proof of one implication, namely the left-to-right implication of (ii).

Suppose that $\Gamma \nvDash_{\mathcal{BD}} \Delta \cup \{A \wedge -A \mid A \in \mathrm{At}(\Gamma)\}$. Then there is by definition a truth relation $u$ which validates all of $\Gamma$ but invalidates all of $\Delta$ and which is consistent on all atoms occuring in $\Gamma$. Now consider the truth relation $v$ which agrees with $u$ everywhere except it assigns the value "neither true nor false" to all atoms which do not occur in $\Gamma$. Then $v$ is a consistent truth relation which still validates all of $\Gamma$. Moreover, we can prove by induction over the complexity of formulas that

$$v \Vdash^{+} A \implies u \Vdash^{+} A,$$
$$v \Vdash^{-} A \implies u \Vdash^{-} A,$$

therefore $v$ invalidates all of $\Delta$. It follows that $\Gamma \nvDash_{\mathcal{K}} \Delta$.

## 5 Intuitionistic conservativity

We now consider fragments of the signature $\{\wedge, \vee, \top, \bot, \rightarrow, -, 0, 1\}$ related to intuitionistic logic. Proving conservativity with respect to intuitionistic logic itself is straightforward. Let us, however, remark that we do not know whether our system is a conservative extension of bi-intuitionistic logic.

**Theorem 6 (Intuitionistic conservativity).** *Let $\Gamma$ and $\Delta$ be sets of formulas in the signature $\{\wedge, \vee, \rightarrow, \top, \bot\}$. Then $\Gamma \vDash_{\mathcal{IL}} \Delta$ if and only if $1, \Gamma \vDash \Delta, 0$.*

*Proof.* The left-to-right direction is trivial. Conversely, suppose that $\Gamma \nvDash_{\mathcal{IL}} \Delta$. Then this consequence fails in some *rooted* Kripke frame for intuitionistic logic, i.e. there is a partially ordered set $\langle W, \leq \rangle$ with a smallest state $u$ and an intuitionistic interpretation such that $u \Vdash \gamma$ for each $\gamma \in \Gamma$ but $u \nVdash \delta$ for each $\delta \in \Delta$.

Take $U$ to be the ordinal sum of the partially ordered set $\langle W, \geq \rangle$ and its order dual $\langle W, \leq \rangle$ with the two instances of $u$ identified. That is, each element of $U$ is either above $u$ (and such elements are the elements of $W$ ordered by $\leq$) or below $u$ (and such elements are the elements of $W$ ordered by $\geq$). The elements above $u$ will be denoted $w_+$ and the elements below $u$ will be denoted $w_-$ for some $w \in W$. In particular $u_+ = u = u_-$. Then we may define an order-inverting involution $\partial$ such that $\partial w_+ = w_-$ and $\partial w_- = w_+$ for each $w \in W$, in particular $\partial u = u$. Now any interpretation on this frame which agrees with the original intuitionistic interpretation on the partially ordered set $\langle W, \leq \rangle$ witnesses that $1, \Gamma \nvDash \Delta, 0$.

We can also prove that the logic of all states (rather than the logic of classical worlds which features in the previous theorem) is a conservative extension of the minimal intuitionistic logic of Johansson [8]. One could easily throw in the explosive *falsum* constant $\bot$ into the signature of this logic in addition to the non-explosive minimal intuitionistic constant 0, but we avoid this for the sake of simplicity.

Recall that the relational semantics of minimal intuitionistic logic, denoted $\mathcal{JL}$ here, is essentially the same as the semantics of intuitionistic logic, the constant 0 being interpreted by an arbitrary upset.

**Theorem 7 (Minimal intuitionistic conservativity).** *Let $\Gamma$ and $\Delta$ be sets of formulas in the signature $\{\wedge, \vee, \rightarrow, 0\}$. Then $\Gamma \vDash_{\mathcal{JL}} \Delta$ if and only if $\Gamma \vDash \Delta$.*

*Proof.* The left-to-right direction is again trivial. Conversely, suppose that $\Gamma \nvDash_{\mathcal{JL}} \Delta$. Then this consequence fails for some interpretation in some state $u$ of some partially ordered set $\langle W, \leq \rangle$ where the constant 0 is interpreted by an upset $I$.

Now take $U = W \times \{+, -\}$, i.e. two copies of $W$ with the elements of one copy being denoted $w_+$ and elements of the other copy being denoted $w_-$ for $w \in W$. The partial order $\sqsubseteq$ on $U$ is defined as follows for $u, v \in W$:

$$
\begin{aligned}
u_+ \sqsubseteq v_+ &\iff u \leq v, \\
u_- \sqsubseteq v_- &\iff v \leq u, \\
u_+ \sqsubseteq v_- &\iff \text{there is some } w \notin I \text{ such that } u \leq w \geq v, \\
u_- \sqsubseteq v_+ &\text{ never holds.}
\end{aligned}
$$

The involution $\partial \colon U \rightarrow U$ is defined as $\partial w_+ = w_-$ and $\partial w_- = w_+$. This involution is easily seen to be order-inverting with respect to the above order.

This amounts to having two disjoint copies of $W$, as shown in Figure 1, one with the original order and the other with the dual order, and only making points in distinct copies comparable if the comparability serves to witness the fact that $w_+ \sqsubseteq \partial w_+$ for $w \in W \setminus I$. If $u, v, w \notin I$ are related as in Figure 1, then we need $w_+$ to be below $w_- = \partial w_+$, which forces $u_+ \sqsubseteq v_-$ and $v_+ \sqsubseteq u_+$ (and also $u_+ \sqsubseteq u_-$ and $v_+ \sqsubseteq v_+$, although these lines are not depicted in the figure).

Observe that (for $w \in W$) $w \in I$ if and only if $w_+ \not\sqsubseteq \partial w_+$, and moreover $w_- \not\sqsubseteq \partial w_-$. It follows that $w_+ \Vdash 0$ if and only if $w \notin I$, and moreover $w_- \Vdash 0$ for each $w \in W$. Now consider an interpretation on $U$ which agrees with the original interpretation on $W \times \{+\}$ and validates each atom (hence each formula) on $W \times \{-\}$. Then, by a straightforward induction on the complexity of $A$ we have $w_+ \Vdash A$ in the new frame if and only if $w \Vdash A$ in the original frame. The state $u$ therefore witnesses that $\Gamma \nvDash \Delta$ in this interpretation.

**Fig. 1** Frame constructed in the proof of Theorem 7

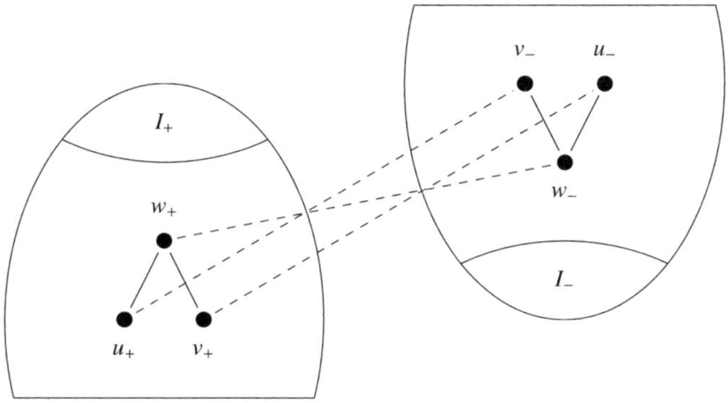


Finally, let us consider the fragment $\{\wedge, \vee, \top, \bot, -, \sim\}$, where $\sim A := A \rightarrow 0$. This fragment consists of the lattice connectives and two different negations, namely the De Morgan negation $-A$ and what we might call the inferential negation $\sim A$. While it is not our goal to provide and defend a full-fledged analysis of these two notions, let us try to sketch where the distiction between two negations lies.

De Morgan negation can be thought of as the linguistic counterpart of denial. According to this account, asserting the De Morgan negation $-A$ of $A$ amounts to denying $A$. We are, roughly speaking, justified in doing so whenever we can falsify $A$ (in the extreme case, whenever we can disprove $A$). Here falsification should be thought of as the same kind of activity as verification, except it aims to establish the falsity rather than the truth of some proposition.

This kind of direct falsification, which differs from verification only in its polarity (so to speak), is to be contrasted with showing that we can derive a contradiction from the assumption that the proposition is true. (This is simply the dual form of the distinction between verifying a proposition directly and establishing it by means of *reductio ad contradictionem*.) Asserting the inferential negation $\sim A$ of $A$ then amounts to asserting that we can reduce the assumption $A$ to a contradiction. For a more thorough discussion of the distinction between direct falsification and reduction to absurdity, we advise the reader to consult Wansing's discussion [16] of proofs, disproofs, and what he calls their duals.[2]

To get a better feel for the distinction between these two negations, we encourage the reader to work out why the inference

$$-A \vee C, -B \vee -C \vDash {\sim}(A \wedge B)$$

is valid on our semantics, even though

$$-A \vee C, -B \vee -C \nvDash -(A \wedge B),$$

and to consider whether this should or should not be the case given our informal understanding of the two negations. The example above can informally be phrased as follows: if we are able to directly verify the premises, then we are able to reduce $A \wedge B$ to a contradiction, although we need not be able to directly falsify $A \wedge B$. The presence of both the inferential negation and the De Morgan negation in the signature of a logic thus allows us to express fine-grained differences such as these in the status of a proposition.

An algebraic semantics for this fragment was studied by the present author in [14], where among other things it was proved that the algebras corresponding to this logic inherit many of the pleasant properties of De Morgan algebras and distributive pseudo-complemented lattices. In particular, they are locally finite, which in logical terms implies that given a finite set of propositional atoms there are only finitely many formulas in this fragment up to logical equivalence (interderivability).

## 6 Comparison with existing approaches

We are now finally in a position to compare our proposal for combining classical and intuitionistic logic with previous attempts. We aware of two existing approaches to the problem, namely those due to Liang & Miller [9] and del Cerro & Herzig [5].

Our approach is in fact similar in spirit to the former. Liang & Miller obtain their system of *polarized intuitionistic logic* by expanding the Kripke semantics for intuitionistic logic by a set of worlds which are designated to be classical. The signature of their logic is somewhat baroque and involves distinguishing between so-called "red" and "green" connectives. In particular, their logic has two versions of conjunction and disjunction. The red connectives are defined in all worlds, whereas the green ones are only defined in classical worlds and said to be valid in a state $u$ if they hold in all classical worlds above $u$.

The basic idea of the two approaches – namely, to designate a subset of classical worlds – is ultimately the same. In particular, Liang & Miller were the first to realize that the presence of inconsistent states (which they call "imaginary worlds") above classical states is crucial for reconciling the two logics semantically. However, we claim that our approach has the advantage of being simpler and perhaps more natural, requiring no separation between classical and non-classical connectives.

Abstracting from the fact that Liang & Miller study first-order versions of the logics, whereas we only study their propositional versions, there are at least three main differences between the two approaches. Firstly, Liang & Miller take the set of classical worlds to be

[2] Wansing talks about a slightly different kind of *reductio*, namely about reducing propositions to non-truth, i.e. to absurdity, rather than to mere contradiction.

primitive, whereas we define it in terms of an order-inverting involution. Secondly, the two systems have different expressive power. The system of Liang & Miler does not contain the De Morgan negation but it does contain what they take to be the De Morgan duals of intuitionistic connectives. In particular, they take the De Morgan dual of intuitionistic implication to be a non-local upward-looking existential connective (whereas we take the De Morgan dual of implication to be the co-implication, i.e. a non-local downward-looking existential connective). Thirdly, related to the previous point, Liang & Miller provide a cut-free sequent calculus for their system, whereas we do not, due to the problems connected with cut elimination in bi-intuitionistic logic (see [7]).

The proposal advanced by del Cerro & Herzig [5] relies on a different distinction, namely between persistent and non-persistent formulas. The classical negation ¬A of a formula A is interpreted classically at each world, i.e.

$$u \Vdash \neg A \iff u \nVdash A.$$

In particular, although propositional atoms are always interpreted by upward closed sets, complex formulas need not be. As a consequence, their system is not closed under uniform substitutions. In particular, certain instances of intuitionistically valid schemas fail, such as $\neg A \to (B \to \neg A)$ if A and B are atoms.

In contrast to the two proposals outlined above, our solution to the problem of combining classical and intuitionistic logic involves neither distinguishing between classical and non-classical connectives, nor between persistent and non-persistent formulas. However, there is a different price to be paid, as we mentioned at the end of Section 3. Namely, selfextensionality fails, i.e. logically equivalent (interderivable) formulas may no longer be substituted for each while preserving logical equivalence.

Finally, let us remark that having the constants 0 and 1 in the language is of course strictly speaking not necessary in order for our strategy to work. There is nothing about the semantics which forces us to add these constants to the language. However, we believe that doing so yields some expressive benefits, such as being able to express the inferential negation $A \to 0$. Their presence also makes the task of formulating sequent calculi somewhat easier. Providing a sequent calculus for the logic of classical worlds in the absence of the constants would presumably involve considering non-standard types of sequents, denoted e.g.

$$\Gamma \rhd_0 \Delta, \qquad \Gamma\, {}_1{\rhd}\, \Delta, \qquad \Gamma\, {}_1{\rhd}_0\, \Delta,$$

corresponding respectively to

$$\Gamma \vDash \Delta, 0, \qquad 1, \Gamma \vDash \Delta, \qquad 1, \Gamma \vDash \Delta, 0.$$

In such a framework, the rules of the sequent calculus would be formulated e.g. as:

$$\frac{\Gamma \rhd_0 \Delta, A}{-A, \Gamma \rhd_0 \Delta} \qquad \frac{\Gamma\, {}_1{\rhd}_0\, \Delta, A}{-A, \Gamma\, {}_1{\rhd}_0\, \Delta}$$

## 7 Sequent calculus for the full system

The rest of the paper is dedicated to providing sequent calculi for the logic (of all states) introduced above and for its fragment without the intuitionistic implication. In this section, we provide a sequent calculus for the logic of all states in the full signature, i.e. in the signature $\{\wedge, \vee, \top, \bot, 1, 0, -, \to\}$.

The rules and axioms of this system are shown in Figure 2, where $p$ stands for a propositional atom. Its sequents are pairs of finite sets of formulas, written as $\Gamma \rhd \Delta$.

**Fig. 2** Sequent calculus for the full system

**Identity and Cut**

$$p \rhd p \qquad 0 \rhd 0 \qquad 1 \rhd 1 \qquad 1 \rhd -0 \qquad -0 \rhd 1$$

$$\frac{\Gamma \rhd \Delta, \mathrm{A} \qquad \mathrm{A}, \Gamma \rhd \Delta}{\Gamma \rhd \Delta}$$

**Contraposition and Double Negation Elimination**

$$\frac{\Gamma \rhd \Delta}{-\Delta \rhd -\Gamma} \qquad \frac{\Gamma \rhd \Delta, --\mathrm{A}}{\Gamma \rhd \Delta, \mathrm{A}} \qquad \frac{--\mathrm{A}, \Gamma \rhd \Delta}{\mathrm{A}, \Gamma \rhd \Delta}$$

**Completeness and Consistency rules**

$$\frac{1, \mathrm{A}, \Gamma \rhd \Delta}{1, \Gamma \rhd \Delta, -\mathrm{A}} \qquad \frac{\Gamma \rhd \Delta, \mathrm{A}, 0}{-\mathrm{A}, \Gamma \rhd \Delta, 0}$$

**Introduction rules**

$$\frac{\mathrm{A}, \mathrm{B}, \Gamma \rhd \Delta}{\mathrm{A} \wedge \mathrm{B}, \Gamma \rhd \Delta} \qquad \frac{\Gamma \rhd \Delta, \mathrm{A} \qquad \Gamma \rhd \Delta, \mathrm{B}}{\Gamma \rhd \Delta, \mathrm{A} \wedge \mathrm{B}}$$

$$\frac{\Gamma \rhd \Delta, \mathrm{A}, \mathrm{B}}{\Gamma \rhd \Delta, \mathrm{A} \vee \mathrm{B}} \qquad \frac{\mathrm{A}, \Gamma \rhd \Delta \qquad \mathrm{B}, \Gamma \rhd \Delta}{\mathrm{A} \vee \mathrm{B}, \Gamma \rhd \Delta}$$

$$\frac{\Gamma, \mathrm{A} \rhd \mathrm{B}}{\Gamma \rhd \mathrm{A} \to \mathrm{B}} \qquad \frac{\Gamma \rhd \Delta, \mathrm{A} \qquad \mathrm{B}, \Gamma' \rhd \Delta'}{\mathrm{A} \to \mathrm{B}, \Gamma, \Gamma' \rhd \Delta, \Delta'}$$

The proof of completeness splits into two parts. We first provide a generalized semantics for the logic which allows for non-standard frames (reminiscent of the non-standard truth relations of Section 4) and prove it to be equivalent to the standard semantics. Then we prove completeness with respect to this generalized semantics in a routine way using a canonical model construction.

Let us first outline the generalized semantics. By a *generalized frame* we mean a frame equipped with an upset $I \subseteq W$ such that

$$u \nleq \partial u \implies u \in I.$$

A generalized frame is called *standard* if $u \nleq \partial u$ for each $u \in I$.

*Generalized interpretations* work like ordinary interpretations, except the constants 0 and 1 are governed by the semantic clauses

$$u \Vdash 0 \iff u \in I,$$
$$u \Vdash 1 \iff \partial u \notin I.$$

Generalized interpretations on standard frames are precisely the same thing as interpretations on frames in the original sense.

**Theorem 8 (Non-standard semantics).** *$\Gamma \vDash \Delta$ holds in all (standard) interpretations if and only if $\Gamma \vDash \Delta$ holds in all generalized interpretations.*

*Proof.* The right-to-left implication is trivial. Conversely, suppose that $\Gamma \vDash \Delta$ fails at some state $u$ of a generalized frame in some generalized interpretation. We define the set of worlds $X := I \cap \partial[I]$. The set of worlds $U$ will consist of one copy of each state in $W \setminus X$ and two copies of each state in $X$. That is,

$$U = W \times \{+\} \sqcup X \times \{-\} = (W \setminus X) \times \{+\} \sqcup X \times \{+, -\}.$$

We use the notation $u_+$ and $u_-$ for elements of the two sets.

We define a partial order on $U$ as follows:

$$u_+ \sqsubseteq v_+ \iff u \leq v \text{ for } u, v \in W,$$
$$u_+ \sqsubseteq v_- \iff u \leq v \text{ and } u \notin X \text{ for } u \in W, v \in X,$$
$$u_- \sqsubseteq v_+ \iff u \leq v \text{ and } v \notin X \text{ for } u \in X, v \in W,$$
$$u_- \sqsubseteq v_- \iff u \leq v \text{ for } u, v \in X.$$

This is indeed a partial order because the set $X$ is *convex* (due to being an intersection of an upset and a downset): if $u \leq v \leq w$ for $u, w \in X$, then $v \in X$. Informally speaking, the partial order contains an extra copy of $X$ which is related to $W \setminus X$ as $X$ was, but which is unrelated to the original copy of $X$.

We also define the map $\overline{\partial}\colon U \to U$ as:

$$\overline{\partial} u_+ = (\partial u)_+ \text{ for } u \in W \setminus X$$
$$\overline{\partial} u_+ = (\partial u)_- \text{ for } u \in X$$
$$\overline{\partial} u_- = (\partial u)_+ \text{ for } u \in X$$

This map is an order-inverting involution with respect to the partial order above.

To obtain a generalized frame, it remains to specify a subset $J$ of $U$. We take $J \subseteq U$ as $J := I \times \{+\} \cup X \times \{-\}$. That is, $u_- \in J$ for each $u \in X$, and $u_+ \in J$ if and only if $u \in I$. This yields a generalized frame $\langle U, \sqsubseteq, \overline{\partial}, J \rangle$: it contains all elements of the form $u_-$ and if $u \notin I$, then $u_+ \sqsubseteq \overline{\partial} u_+$ because $\langle U, \sqsubseteq, \overline{\partial}, J \rangle$ and $u \notin X$.

The generalized frame $\langle U, \sqsubseteq, \overline{\partial}, J\rangle$ is standard: if $u \in X$, then $\partial u \in X$, hence $u_{\mp} \not\sqsubseteq \overline{\partial} u_{\mp} = (\partial u)_{\pm}$. On the other hand, if $u \in I$ and $u \notin X$, then $\partial u \notin I$ and $u \nleq \partial u$ (because $I$ is an upset), hence $u_+ \not\sqsubseteq \overline{\partial} u_+ = (\partial u)_+$.

Now consider the following relation between formulas and points of the standard generalized frame $\langle U, \sqsubseteq, \overline{\partial}, J\rangle$:

$$u_+ \mathrel{\overline{\Vdash}} \mathrm{A} \iff u \Vdash \mathrm{A} \text{ for } u \in W,$$
$$u_- \mathrel{\overline{\Vdash}} \mathrm{A} \iff u \Vdash \mathrm{A} \text{ for } u \in X.$$

Clearly $u_+ \mathrel{\overline{\Vdash}} \mathrm{A}$ for each $\mathrm{A} \in \Gamma$ while $u_+ \mathrel{\overline{\Vdash}} \mathrm{B}$ for each $\mathrm{B} \in \Delta$. It only remains to show that this relation is a generalized interpretation. But this is an easy induction over the complexity of the formula A. (We can use the fact that the map $f\colon U \to W$ such that $f(u_{\pm}) = u$ is a p-morphism of intuitionistic frames commuting with $\partial$.)

**Theorem 9 (Completeness).** *$\Gamma \vDash \Delta$ for finite $\Gamma$ and $\Delta$ if and only if the sequent $\Gamma \rhd \Delta$ is provable in the above sequent calculus.*

*Proof.* Soundness is easy to verify. To prove completeness, let us first introduce some terminology. We call $\langle \Gamma, \Delta\rangle$ a *consistent pair* of formulas if $\Gamma' \rhd \Delta'$ is not provable for any pair of finite subsets $\Gamma' \subseteq \Gamma$ and $\Delta' \subseteq \Delta$. A *maximal pair* is a pair which is maximal among consistent pairs (with respect to componentwise inclusion). By Cut and Identity (which extends to arbitrary formulas) each formula belongs to exactly one of the sets in each maximal pair.

Now suppose that the sequent $\Gamma \rhd \Delta$ is not provable. Then $\langle \Gamma, \Delta\rangle$ is a consistent pair and by Zorn's lemma it extends to a maximal pair $\langle \Gamma', \Delta'\rangle$. We now construct a *generalized* frame whose states will be precisely the maximal pairs and find an interpretation on this generalized frame such that $\langle \Gamma, \Delta\rangle \Vdash \mathrm{A}$ if and only if $\mathrm{A} \in \Gamma$. To construct this generalized frame, we take the partial order

$$\langle \Gamma, \Delta\rangle \leq \langle \Pi, \Sigma\rangle \iff \Gamma \subseteq \Pi \text{ (or equivalently } \Sigma \subseteq \Delta),$$

and define $\partial\langle \Gamma, \Delta\rangle$ as the pair $\langle \Pi, \Sigma\rangle$ such that

$$\mathrm{A} \in \Pi \iff -\mathrm{A} \in \Delta,$$
$$\mathrm{A} \in \Sigma \iff -\mathrm{A} \in \Gamma.$$

Finally, we define the set of maximal pairs $I$ as follows:

$$\langle \Gamma, \Delta\rangle \in I \iff 0 \in \Gamma.$$

To show that this yields a generalized frame, we need to prove that $\partial\langle \Gamma, \Delta\rangle$ is indeed a maximal pair, $\partial$ is an order-inverting involution, and $\langle \Gamma, \Delta\rangle \in I$ whenever $\langle \Gamma, \Delta\rangle \nleq \partial\langle \Gamma, \Delta\rangle$.

If $\langle \Gamma, \Delta\rangle$ is a consistent pair, then so is $\partial\langle \Gamma, \Delta\rangle = \langle \Pi, \Sigma\rangle$: if $\Pi \rhd \Sigma$ is provable, then so is $-\Sigma \rhd -\Pi$ by contraposition, but $-\Sigma \subseteq \Gamma$ and $-\Pi \subseteq \Delta$, where $-\Sigma = \{-\mathrm{A} \mid \mathrm{A} \in \Sigma\}$ and $-\Delta = \{-\mathrm{A} \mid \mathrm{A} \in \Delta\}$. If the pair $\langle \Gamma, \Delta\rangle$ is moreover maximal, then so is $\langle \Pi, \Sigma\rangle$: $\mathrm{A} \notin \Pi \iff -\mathrm{A} \notin \Delta \iff -\mathrm{A} \in \Gamma \iff \mathrm{A} \in \Sigma$.

The map $\partial$ is order-inverting: if $\langle \Gamma, \Delta\rangle \leq \langle \Pi, \Sigma\rangle$, then $\Sigma \subseteq \Delta$, in particular $-\mathrm{A} \in \Sigma$ implies $-\mathrm{A} \in \Delta$. But this amounts to $\partial\langle \Pi, \Sigma\rangle \leq \partial\langle \Gamma, \Delta\rangle$. Moreover, the map $\partial$ is an involution: if $\partial\partial\langle \Gamma, \Delta\rangle = \langle \Pi, \Sigma\rangle$, then $\mathrm{A} \in \Pi \iff --\mathrm{A} \in \Gamma$, but $--\mathrm{A} \in \Gamma \iff \mathrm{A} \in \Gamma$ because $\mathrm{A} \rhd --\mathrm{A}$ and $--\mathrm{A} \rhd \mathrm{A}$ are provable sequents.

Finally, suppose that $\langle \Gamma, \Delta \rangle \not\leq \partial\langle \Gamma, \Delta \rangle$. Then there is some $A \in \Gamma$ such that $-A \notin \Delta$, hence $-A \in \Gamma$. But the sequent $A, -A \rhd 0$ is provable, therefore $0 \notin \Delta$ and $0 \in \Gamma$, i.e. $\langle \Gamma, \Delta \rangle \in I$.

Now consider the relation between maximal pairs and formulas defined as

$$\langle \Gamma, \Delta \rangle \Vdash A \iff A \in \Gamma.$$

It only remains to prove that this relation is a generalized interpretation. We do so by induction over the complexity of A. The steps for the intuitionistic connectives are well known. The step for 0 holds by the definition of $I$. Finally, the step for De Morgan negation holds directly by the definition of the map $\partial$.

We do not claim that this calculus admits cut elimination, and indeed we have no reason to believe that it does. The task of finding a cut-free calculus for our logic runs into the same difficulties as the task of finding a cut-free calculus for bi-intuitionistic logic (see e.g. [7]). Namely, the restriction to a singleton set of conclusions in the right rule for intuitionistic implication

$$\frac{\Gamma, A \rhd B}{\Gamma \rhd A \to B}$$

clashes with the multiple-conclusion setting, which, however, is essentially forced on us by the Completeness and Consistency rules and by Contraposition. Specifically, for cuts like

$$\frac{\Pi \rhd \Sigma, C \qquad \dfrac{C, \Gamma, A \rhd B}{C, \Gamma \rhd A \to B}}{\Pi, \Gamma \rhd \Sigma, A \to B}$$

the naïve reduction

$$\dfrac{\dfrac{\Pi \rhd \Sigma, C \qquad C, \Gamma, A \rhd B}{\Pi, \Gamma, A, \rhd \Sigma, B}}{\Pi, \Gamma \rhd \Sigma, A \to B}$$

does not work, since the last step is not an instance of the right rule for intuitionistic implication.

## 8 Sequent calculus for $\mathcal{BD}$ with 0 and 1

Finally, we provide a cut-free calculus for the fragment $\{\wedge, \vee, \top, \bot, 1, 0, -\}$ of our system, i.e. for the extension of the Belnap–Dunn logic by 0 and 1.

The calculus in question is shown in Figure 3. Sequents in this calculus are pairs of finite *multisets* of *signed* formulas, which are pairs consisting of a formula and a sign + or −, written as $A_+$ or $A_-$. In this context the Cut rule has the form:

$$\frac{\Gamma \rhd \Delta, A_\pm \qquad A_\pm, \Gamma' \rhd \Delta'}{\Gamma, \Gamma' \rhd \Delta, \Delta'}$$

Weakening has the form:

$$\frac{\Gamma \rhd \Delta}{A_\pm, \Gamma \rhd \Delta} \qquad \frac{\Gamma \rhd \Delta}{\Gamma \rhd \Delta, A_\pm}$$

**Fig. 3** Sequent calculus for $\mathcal{BD}$ with 0 and 1

**Identity axioms**

$$p_\pm, \Gamma \rhd \Delta, p_\pm$$

$$0_\pm, \Gamma \rhd \Delta, 0_\pm \qquad 1_\pm, \Gamma \rhd \Delta, 1_\pm \qquad 1_\pm, \Gamma \rhd \Delta, 0_\mp \qquad 0_\mp, \Gamma \rhd \Delta, 1_\pm$$

**Completeness and Consistency axioms**

$$1_+, \Gamma \rhd p_+, p_- \qquad 0_-, \Gamma \rhd p_+, p_- \qquad p_+, p_-, \Gamma \rhd \Delta, 0_+ \qquad p_+, p_-, \Gamma \rhd \Delta, 1_-$$

**Introduction rules**

$$\frac{A_+, \Gamma \rhd \Delta}{(-A)_-, \Gamma \rhd \Delta} \qquad \frac{A_-, \Gamma \rhd \Delta}{(-A)_+, \Gamma \rhd \Delta} \qquad \frac{\Gamma \rhd \Delta, A_+}{\Gamma \rhd \Delta, (-A)_-} \qquad \frac{\Gamma \rhd \Delta, A_-}{\Gamma \rhd \Delta, (-A)_+}$$

$$\frac{A_+, B_+, \Gamma \rhd \Delta}{(A \wedge B)_+, \Gamma \rhd \Delta} \qquad \frac{\Gamma \rhd \Delta, A_+ \qquad \Gamma \rhd \Delta, B_+}{\Gamma \rhd \Delta, (A \wedge B)_+}$$

$$\frac{\Gamma \rhd \Delta, A_-, B_-}{\Gamma \rhd \Delta, (A \wedge B)_-} \qquad \frac{A_-, \Gamma \rhd \Delta \qquad B_-, \Gamma \rhd \Delta}{(A \wedge B)_-, \Gamma \rhd \Delta}$$

$$\frac{\Gamma \rhd \Delta, A_+, B_+}{\Gamma \rhd \Delta, (A \vee B)_+} \qquad \frac{A_+, \Gamma \rhd \Delta \qquad B_+, \Gamma \rhd \Delta}{(A \vee B)_+, \Gamma \rhd \Delta}$$

$$\frac{A_-, B_-, \Gamma \rhd \Delta}{(A \vee B)_-, \Gamma \rhd \Delta} \qquad \frac{\Gamma \rhd \Delta, A_- \qquad \Gamma \rhd \Delta, B_-}{\Gamma \rhd \Delta, (A \vee B)_-}$$

and Contraction has the form:

$$\frac{A_\pm, A_\pm, \Gamma \rhd \Delta}{A_\pm, \Gamma \rhd \Delta} \qquad \frac{\Gamma \rhd \Delta, A_\pm, A_\pm}{\Gamma \rhd \Delta, A_\pm}$$

Recall that a rule is called *invertible* if its premise is provable whenever its conclusion is, and it is called *height-preserving admissible* if its conclusion has a proof of height at most $n$ whenever its premise has a proof of height at most $n$. The proofs of the following lemmas are standard, therefore we omit them. (The reader may consult e.g. [15, section 3.5] for the details.)

**Lemma 1 (Inversion Lemma).** *The introduction rules are invertible.*

**Lemma 2 (Admissibility of Weakening and Contraction).** *The rules of Weakening and Contraction are height-preserving admissible.*

**Lemma 3.** *The Identity, Completeness, and Consistency axioms are derivable for arbitrary signed formulas* $\mathrm{A}_{\pm}$ *in place of* $p_{\pm}$.

*Proof.* We prove the claim by induction over the complexity of the formula A. The induction steps for the Identity axioms are standard. Now consider the axiom $\mathrm{A}_{+}, \mathrm{A}_{-}, \Gamma \rhd \Delta, 0_{+}$. The induction step for conjunction is:

$$\dfrac{\dfrac{\mathrm{A}_{+}, \mathrm{B}_{+}, \mathrm{A}_{-}, \Gamma \rhd \Delta, 0_{+} \qquad \mathrm{A}_{+}, \mathrm{B}_{+}, \mathrm{B}_{-}, \Gamma \rhd \Delta, 0_{+}}{\mathrm{A}_{+}, \mathrm{B}_{+}, (\mathrm{A} \wedge \mathrm{B})_{-}, \Gamma \rhd \Delta, 0_{+}}}{(\mathrm{A} \wedge \mathrm{B})_{+}, (\mathrm{A} \wedge \mathrm{B})_{-}, \Gamma \rhd \Delta, 0_{+}}$$

The induction step for disjunction is:

$$\dfrac{\dfrac{\mathrm{A}_{+}, \mathrm{A}_{-}, \mathrm{B}_{-}, \Gamma \rhd \Delta, 0_{+} \qquad \mathrm{B}_{+}, \mathrm{A}_{-}, \mathrm{B}_{-}, \Gamma \rhd \Delta, 0_{+}}{(\mathrm{A} \vee \mathrm{B})_{+}, \mathrm{A}_{-}, \mathrm{B}_{-}, \Gamma \rhd \Delta, 0_{+}}}{(\mathrm{A} \vee \mathrm{B})_{+}, (\mathrm{A} \vee \mathrm{B})_{-}, \Gamma \rhd \Delta, 0_{+}}$$

Finally, the induction step for negation is:

$$\dfrac{\dfrac{\mathrm{A}_{-}, \mathrm{A}_{+}, \Gamma \rhd \Delta, 0_{+}}{(-\mathrm{A})_{+}, \mathrm{A}_{+}, \Gamma \rhd \Delta, 0_{+}}}{(-\mathrm{A})_{+}, (-\mathrm{A})_{-}, \Gamma \rhd \Delta, 0_{+}}$$

The induction steps for the other three Completeness and Consistency axioms are entirely analogous.

**Theorem 10 (Admissibility of Cut).** *The Cut rule is admissible.*

*Proof.* The argument again follows the standard proof of the admissibility of Cut (see e.g. [15, section 4.1]), which relies on the Inversion Lemma and the height-preserving admissibility of Contraction and Weakening to reduce all instances of the Cut rule to instances where *both* of the premises are axioms.

Cutting on an Identity axiom then either amounts to Weakening or results in an instance of the Identity axiom. Suppose therefore that both of the premises are Consistency or Completeness axioms. If the cut formula is one of the formulas in $\Gamma$ or $\Delta$ in the left (right) premise of the Cut rule, then such a cut results in an instance of the same Consistency or Completeness axiom that the left (right) premise is an instance of. The cut formula moreover cannot have the form $0_{\pm}$ or $1_{\pm}$, because these formulas occur only on the right or only on the left in Consistency and Completeness axioms. Finally, if the cut formula is $p_{\pm}$, then the cut results in an instance of the Identity axiom with $p_{\mp}$ on both sides of the sequent.

**Theorem 11 (Completeness).** *Let* $\Gamma$ *and* $\Delta$ *be finite sets of formulas in the signature* $\{\wedge, \vee, \top, \bot, 1, 0, -\}$. *Then* $\Gamma \vDash \Delta$ *if and only if the sequent* $\Gamma \rhd \Delta$ *is provable in the sequent calculus above.*

*Proof.* Soundness is easy to verify. To prove completeness, suppose that $\Gamma \rhd \Delta$ is not provable. Then as in the proof of Theorem 9 the pair $\langle \Gamma, \Delta \rangle$ extends to a maximal consistent pair (of signed formulas) $\langle \Gamma', \Delta' \rangle$ and it is straightforward to verify, using the admissibility of Cut, that defining

$$u \Vdash^{+} \mathrm{A} \iff \mathrm{A}_{+} \in \Gamma',$$
$$u \Vdash^{-} \mathrm{A} \iff \mathrm{A}_{-} \in \Gamma',$$

yields a (possibly non-standard) truth relation for $\mathcal{BD}$ with 0 and 1.

**Acknowledgments.** This research was supported by the project 16-07954J SEGA "From shared attitudes to group agency" of the Czech Science Foundation and DFG. The author is also grateful to the two anonymous referees for a thorough reading of the manuscript and many helpful suggestions which substantially improved the paper.

## References

1. Jc Beall. Multiple-conclusion LP and default classicality. *The Review of Symbolic Logic*, 4(2):326–336, 2011.
2. Jc Beall. LP+, K3+, FDE+, and their 'classical collapse'. *The Review of Symbolic Logic*, 6(4):742–754, 2013.
3. Nuel D. Belnap. How a computer should think. In *Contemporary Aspects of Philosophy*, pages 30–56. Oriel Press Ltd., 1977.
4. Nuel D. Belnap. A useful four-valued logic. In J. Michael Dunn and George Epstein, editors, *Modern uses of multiple-valued logic*, volume 2 of *Episteme*, pages 5–37. Springer Netherlands, 1977.
5. Luis Fariñas del Cerro and Andreas Herzig. Combining classical and intuitionistic logic. In Franz Baader and Klaus Schulz, editors, *Frontiers in Combining Systems*, volume 3 of *Applied Logic Series*, pages 93–102. Kluwer Academic Publishers, 1996.
6. J. Michael Dunn. Intuitive semantics for first-degree entailments and 'coupled trees'. *Philosophical Studies*, 29(3):149–168, 1976.
7. Rajeev Goré, Linda Postniece, and Alwen Tiu. Cut-elimination and proof search for bi-intuitionistic tense logic. In Lev Beklemishev, Valentin Goranko, and Valentin Shehtman, editors, *Advances in Modal Logic*, 2010, pages 156–177. 2010.
8. Ingebrigt Johansson. Der Minimalkalkül, ein reduzierter intuitionistischer Formalismus. *Compositio Mathematica*, 4:119–136, 1937.
9. Chuck Liang and Dale Miller. Kripke semantics and proof systems for combining intuitionistic logic and classical logic. *Annals of Pure and Applied Logic*, 164(2):86–111, 2013.
10. António Monteiro. Sur les algèbres de Heyting symétriques. *Portugaliae mathematica*, 39(1–4):1–237, 1980.
11. Graham Priest. The logic of paradox. *Journal of Philosophical Logic*, 8(1):219–241, 1979.
12. Graham Priest. *An Introduction to Non-Classical Logic: From If to Is*. Cambridge University Press, 2nd edition, 2008.
13. Adam Přenosil. Reductio ad contradictionem: an algebraic perspective. *Studia Logica*, 104(3):389–415, 2016.
14. Adam Přenosil. The inconsistency predicate on De Morgan lattices. *Acta Universitatis Carolinae Philosophica et Historica 2/2017*, pages 45–61, 2017.
15. Anne S. Troelstra and Helmut Schwichtenberg. *Basic Proof Theory*. Cambridge University Press, New York, NY, USA, 2nd edition, 2000.
16. Heinrich Wansing. Proofs, disproofs, and their duals. In L. Beklemishev, V. Goranko, and V. Shehtman, editors, *Advances in Modal Logic, Volume 8*, pages 483–505, London, 2010. College Publications.

# Natural Deduction Systems for Logics in the *FDE* Family

Graham Priest

**Abstract** There is a small family of many-valued logics associated with the logic of First Degree Entailment. These may be called the *FDE* family. The purpose of the present paper is to provide natural deduction systems for these logics. This can be done in a quite systematic fashion. An appendix to the paper deals with a closely related system which is not in the family, "Paraconsistent Weak Kleene".

**Keywords** Natural Deduction Systems • First Degree Entailment • FDEe • LP • $K_3$ • Paraconsistent Weak Kleene

## 1 Introduction

There is a small family of many-valued logics associated with the logic of First Degree Entailment, *FDE*. Elsewhere,[1] I have called them the *FDE* family. The purpose of the present paper is to provide natural deduction systems for these logics. This can be done, as we shall see, in a quite systematic way. I will start by defining the family of logics. I will then provide a natural deduction system for the weakest of the logics in the family, *FDEe*. Soundness and Completeness will be proved. It will then shown how to modify matters to accommodate the stronger systems in the family. An appendix deals with a closely related system that is not in the family, "Paraconsistent Weak Kleene", *PWK*.

## 2 The FDE Family

We will be dealing with propositional logics with connectives $\wedge$, $\vee$, and $\neg$. $\supset$ may be defined in the usual way. The logics in our family are all many-valued logics, and the weakest member, *FDEe*, has five values: $t, f, b, n, e$. The first four of these are naturally thought of as: *true only*, *false only*, *both true and false*, and *neither true nor false*. Their behaviour can be captured in the familiar Hasse diagram:

Graham Priest
CUNY Graduate Center, New York, NY, USA

The University of Melbourne, Melbourne, Vic., Australia
e-mail: priest.graham@gmail.com

[1] Priest (2014b).

 
H. Omori, H. Wansing (eds.), *New Essays on Belnap-Dunn Logic*,
Synthese Library 418, https://doi.org/10.1007/978-3-030-31136-0_16

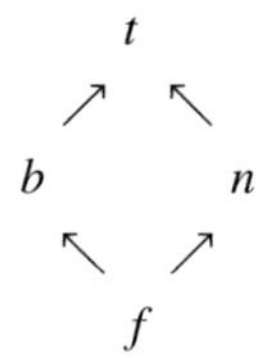


Negation maps $t$ to $f$ and vice versa; and $b$ and $n$ are fixed points for negation. Conjunction is the greatest lower bound; and disjunction is the least upper bound.

The fifth value has can be interpreted in various ways. One is as *meaningless* or defectiveness of some other kind.[2] Another is as ineffability.[3] The salient formal feature of $e$ is that any truth function has $e$ as an output iff some input is $e$.

In a many-valued logic of this kind, there is a set of designated values, $D$, and an inference is valid if any evaluation in which all the premises are designated, the conclusion is designated. In $FDEe$, $D = \{t, b\}$. So $\Sigma \models A$ iff for every evaluation: if $\nu(B) = t$ or $b$ for all $B \in \Sigma$, then $\nu(A) = t$ or $b$.

The semantic conditions for the connectives can be depicted in a more usual form by the following matrices:

| $\neg$ | |
|---|---|
| $t$ | $f$ |
| $b$ | $b$ |
| $n$ | $n$ |
| $f$ | $t$ |
| $e$ | $e$ |

| $\vee$ | $t$ | $b$ | $n$ | $f$ | $e$ |
|---|---|---|---|---|---|
| $t$ | $t$ | $t$ | $t$ | $t$ | $e$ |
| $b$ | $t$ | $b$ | $t$ | $b$ | $e$ |
| $n$ | $t$ | $t$ | $n$ | $n$ | $e$ |
| $f$ | $t$ | $b$ | $n$ | $f$ | $e$ |
| $e$ | $e$ | $e$ | $e$ | $e$ | $e$ |

| $\wedge$ | $t$ | $b$ | $n$ | $f$ | $e$ |
|---|---|---|---|---|---|
| $t$ | $t$ | $b$ | $n$ | $f$ | $e$ |
| $b$ | $b$ | $b$ | $f$ | $f$ | $e$ |
| $n$ | $n$ | $f$ | $n$ | $f$ | $e$ |
| $f$ | $f$ | $f$ | $f$ | $f$ | $e$ |
| $e$ | $e$ | $e$ | $e$ | $e$ | $e$ |

If we write the value of $A$ as $|A|$ then, as is not difficult to check, the tables can be expressed in a linear form as follows:

*Negation*

- $|\neg A| = e$ iff $|A| = e$
- $|\neg A| = t$ iff $|A| = f$
- $|\neg A| = b$ iff $|A| = b$
- $|\neg A| = n$ iff $|A| = n$
- $|\neg A| = f$ iff $|A| = t$

*Disjunction*

- $|A \vee B| = e$ iff $|A| = e$ or $|B| = e$
- $|A \vee B| = t$ iff ($|A| = t$ and $|B| \neq e$) or ($|B| = t$ and $|A| \neq e$) or ($|A| = b$ and $|B| = n$) or ($|A| = n$ and $|B| = b$)
- $|A \vee B| = b$ iff ($|A| = b$ and $|B| = b$) or ($|A| = b$ and $|B| = f$) or ($|A| = f$ and $|B| = b$)
- $|A \vee B| = n$ iff ($|A| = n$ and $|B| = n$) or ($|A| = n$ and $|B| = f$) or ($|A| = f$ and $|B| = n$)
- $|A \vee B| = f$ iff $|A| = f$ and $|B| = f$

*Conjunction*

- $|A \wedge B| = e$ iff $|A| = e$ or $|B| = e$

[2] See Goddard and Routley (1973), and the discussion of Bochvar logic in Haack (1996).

[3] See Priest (2014a).

- $|A \wedge B| = t$ iff ($|A| = t$ and $|B| = t$)
- $|A \wedge B| = n$ iff ($|A| = n$ and $|B| = n$) or ($|A| = t$ and $|B| = n$) or ($|A| = n$ and $|B| = t$)
- $|A \wedge B| = b$ iff ($|A| = b$ and $|B| = b$) or ($|A| = b$ and $|B| = t$) or ($|A| = t$ and $|B| = b$)
- $|A \wedge B| = f$ iff ($|A| = f$ and $|B| \neq e$) or ($|B| = f$ and $|A| \neq e$) or ($|A| = b$ and $|B| = n$) or ($|A| = n$ and $|B| = f$)

Let us call $t$ and $f$ the *classical values*. Then note that deleting the row and column for any non-classical value in the matrices deletes all mention of that value. In other words, the matrices define perfectly good many-valued logics when the row and column for any number of the non-classical values are deleted—and if $b$ is deleted, $D$ is simply $\{t\}$. All the logics in our family have the classical values; and we may characterise the logics in terms of which *other* values they have, as follows:[4]

- $\emptyset$: classical logic, *CL*.
- $e$: Bochvar logic (also known as weak Kleene 3-valued logic), $B_3$.
- $n$: strong Kleene 3-valued logic, $K_3$.
- $b$: logic of paradox, *LP*.
- $en$: a logic not previously formulated.
- $eb$: *AL*.
- $bn$: first degree entailment, *FDE*.
- $bne$: *FDEe*.

Note, finally, and again as is easy to check, in any of these logics, the matrix for $A \wedge B$ is the same as that for $\neg(\neg A \vee \neg B)$, and the matrix for $A \vee B$ is the same as that for $\neg(\neg A \wedge \neg B)$. For theoretical purposes, we may therefore take one of these to be defined in terms of the other. In what follows, we take conjunction to be defined in terms of disjunction.

## 3 Natural Deduction Rules

The natural deduction systems for our family of logics will be given in the style of Prawitz (1965). A basic deduction is anything of the form: $A$. $A$ is both an undischarged assumption and the conclusion. A deduction is anything that can be generated from basic deductions by applying rules from the appropriate set. $\Sigma \vdash A$ iff there is a deduction whose undischarged assumptions are in $\Sigma$, and whose conclusion is $A$.

The rules for *FDEe* are as follows. A bar over an assumption means that it is discharged by the rule. A double line means that the rule works in both directions. If $A$ is any formula, we will write $A^\dagger$ for any formula which contains all the propositional parameters in $A$.

- $\wedge I$

$$\frac{A \quad B}{A \wedge B}$$

- $\wedge E$

$$\frac{A \wedge B}{A} \qquad \frac{A \wedge B}{B}$$

[4] For *FDEe*, see Priest (2014a), where it is called *FDE*$\varphi$. See also Daniels (1990). For *FDE*, see Priest (2008), ch. 8. For $K_3$ and *LP*, see Priest (2008), ch. 7. For $B_3$, see Haack (1996), pp. 169-70. (Bochvar augments this logic with further connectives). For *AL* see Oller (1999).

- $\vee E$

$$\cfrac{A \vee B \quad \begin{matrix}\overline{A} & \overline{B} \\ \vdots & \vdots \\ C & C\end{matrix}}{C}$$

- Weak $\vee I$

$$\frac{A \quad B^{\dagger}}{A \vee B} \qquad \frac{A^{\dagger} \quad B}{A \vee B}$$

- DeM

$$\frac{\neg(A \vee B)}{\neg A \wedge \neg B} \qquad \frac{\neg(A \wedge B)}{\neg A \vee \neg B}$$

(double lines)

DN

$$\frac{A}{\neg\neg A}$$

(double line)

In due course, and for future reference, four other rules will concern us:

- Excluded Middle (EM)

$$\overline{B \vee \neg B}$$

- Explosion

$$\frac{A \wedge \neg A}{B}$$

- $\vee$I

$$\frac{A}{A \vee B} \qquad \frac{B}{A \vee B}$$

- Weak EM

$$\frac{A^{\dagger}}{A \vee \neg A}$$

# 4 Soundness and Completeness for *FDEe*

## 4.1 Soundness for *FDEe*

In this section we show that in *FDEe* (= *bne*), if $\Sigma \vdash A$ then $\Sigma \models A$.

The proof is by recursion over the construction of deductions. The basis case is trivial. There is a recursion case for each rule. These are straightforward, and mostly left to the reader. I do two casess as examples.

The first is $\vee E$. Suppose that we have proofs, $\pi_1$, $\pi_2$, and $\pi_3$, such that the undischarged assumptions and conclusions of the three deductions are, respectively:

- $\Pi_1, A \vee B$
- $\Pi_2 \cup \{A\}, C$
- $\Pi_3 \cup \{B\}, C$

The application of $\vee E$ gives us a proof whose undischarged assumptions are $\Pi_1 \cup \Pi_2 \cup \Pi_3$, and whose conclusion is $C$. Suppose that $\Sigma \supseteq \Pi_1 \cup \Pi_2 \cup \Pi_3$. Then by Recursion Hypothesis (RH), $\Sigma \models A \vee B$. Hence $\Sigma \models A$ or $\Sigma \models B$. By RH, $\Sigma \cup \{A\} \models C$ and $\Sigma \cup \{B\} \models C$. So in either case, $\Sigma \models C$, as required.

The second example is Weak $\vee I$. Suppose that we have a proofs, $\pi_1$ and $\pi_2$, whose undischarged assumptions and conclusions are, respectively:

- $\Pi_1, A$
- $\Pi_2, B^\dagger$

Suppose that $\Sigma \supseteq \Pi_1 \cup \Pi_2$. Then by RH, $\Sigma \models A$ and $\Sigma \models B^\dagger$. Let $\nu$ be any evaluation that designates all members of $\Sigma$ (i.e., gives them either the value $t$ or the value $b$). Then $\nu$ designates $A$ and $B^\dagger$. Hence it assigns no propositional parameter in $B$ the value $e$, and so it does not assign $B$ the value $e$. Hence it designates $A \vee B$. That is, $\Sigma \models A \vee B$.

## 4.2 Henkin Lemma

In this section, in preparation for proving completeness, we prove the *Henkin Lemma*:

If $\Sigma \nvdash A$, there is a $\Pi \supseteq \Sigma$ such that:

- $\Pi \nvdash A$
- $\Pi$ is deductively closed
- $\Pi$ is prime (that is, if $A \vee B \in \Pi$, $A \in \Pi$ or $B \in \Pi$)

*Proof*: Enumerate the formulas of the language: $\langle B_i : i \in \omega \rangle$. Define by recursion:

- $\Pi_0 = \Sigma$
- $\Pi_{n+1} = \Pi_n \cup \{B_n\}$ if $\Pi_n \cup \{B_n\} \nvdash A$; $\Pi_{n+1} = \Pi_n$ otherwise
- $\Pi = \bigcup_{n \in \omega} \Pi_n$

Clearly, $\Pi \supseteq \Sigma$.

We now show that the conditions hold.

- $\Pi \nvdash A$. By induction on $n$, each $\Pi_n$ is such that $\Pi_n \nvdash A$. By compactness, $\Pi \nvdash A$.
- $\Pi$ is deductively closed. Suppose for *reductio* that $\Pi \vdash B_n$ and $B_n \notin \Pi$. Then $\Pi_n \cup \{B_n\} \vdash A$. So $\Pi \vdash A$. Contradiction.
- $\Pi$ is prime. Suppose for *reductio* that $\Pi \vdash B_n \vee B_m$, but that $B_n \notin \Pi$ and $B_m \notin \Pi$. By construction $\Pi_n \cup \{B_n\} \vdash A$ and $\Pi_m \cup \{B_m\} \vdash A$. So $\Pi \vdash A$ (by $\vee$E). Contradiction.

## 4.3 Completeness for *FDEe*

In this section we show that in *FDEe*, if $\Sigma \models A$ then $\Sigma \vdash A$. We prove the contrapositive. Suppose that $\Sigma \nvdash A$. Construct $\Pi$ as in the previous subsection. Define $|B|$ as follows:

- $|B| = t$ if $B \in \Pi$ and $\neg B \notin \Pi$

- $|B| = f$ if $\neg B \in \Pi$ and $B \notin \Pi$
- $|B| = b$ if $B \in \Pi$ and $\neg B \in \Pi$
- $|B| = n$ if $B \notin \Pi$, $\neg B \notin \Pi$, and there is some $B^\dagger \in \Pi$
- $|B| = e$ if $B \notin \Pi$, $\neg B \notin \Pi$, and there is no $B^\dagger \in \Pi$

The conditions are obviously exclusive and exhaustive. Note that the last condition simplifies to:

- $|B| = e$ if there is no $B^\dagger \in \Pi$

If we can show that $|B|$ really is an evaluation, the result follows, since the evaluation makes all members of $\Pi$ (and so $\Sigma$) designated, but not $A$.

To show this, we show that the definition satisfies the (linear) conditions of Section 2. We may take conjunction to be defined in terms of negation and conjunction. So we need consider only these. For each of them, there are five cases—one for each value.

For negation, here are two of the cases. The others are left to the reader.

For $t$, we need to show that:

- $|\neg B| = t$ iff $|B| = f$

That is:

- $|\neg B| \in \Pi$ and $|\neg\neg B| \notin \Pi$ iff $|\neg B| \in \Pi$ and $|B| \notin \Pi$

This follows from deductive closure and DN.

For $e$, we need to show that:

- $|\neg B| = e$ iff $|B| = e$

That is:

- there is no $(\neg B)^\dagger \in \Pi$ iff there is no $B^\dagger \in \Pi$

This holds since any $B^\dagger$ is a $(\neg B)^\dagger$, and vice versa.

Before we do the cases for disjunction, here are a couple of useful facts, which follow simply from the rules and deductive closure:

- $\neg(A \vee B) \in \Pi$ iff $\neg A \wedge \neg B \in \Pi$ iff $\neg A \in \Pi$ and $\neg B \in \Pi$

So:

- $\neg(A \vee B) \notin \Pi$ iff $\neg A \notin \Pi$ or $\neg B \notin \Pi$

We now check the five cases for $\vee$.

For $e$, we need to show that:

- $|A \vee B| = e$ iff $|A| = e$ or $|B| = e$

That is:

- there is no $(A \vee B)^\dagger \in \Pi$ iff there is no $A^\dagger \in \Pi$ or there is no $B^\dagger \in \Pi$

That is, contraposing:

- there is some $(A \vee B)^\dagger \in \Pi$ iff there is some $A^\dagger \in \Pi$ and there is some $B^\dagger \in \Pi$

⇒: This is obvious.
⇐: if $A^\dagger \in \Pi$ and $B^\dagger \in \Pi$ then $A^\dagger \wedge B^\dagger \in \Pi$. $A^\dagger \wedge B^\dagger$ is $(A \vee B)^\dagger$.

For $f$, we have to show that :

- $|A \vee B| = f$ iff $|A| = f$ and $|B| = f$

That is:

- $A \vee B \notin \Pi$ and $\neg(A \vee B) \in \Pi$ iff $A \notin \Pi$ and $\neg A \in \Pi$ and $B \notin \Pi$ and $\neg B \in \Pi$

⇒: $\neg(A \vee B) \in \Pi$, so $\neg A \in \Pi$ and $\neg B \in \Pi$. Suppose that $A \in \Pi$. Then since $\neg(A \vee B) = B^\dagger \in \Pi$ we have $A \vee B \in \Pi$. Contradiction. The argument for $B \in \Pi$ is similar.
⇐: $\neg A \in \Pi$ and $\neg B \in \Pi$. So $\neg(A \vee B) \in \Pi$. Suppose that $A \vee B \in \Pi$. Then $A \in \Pi$ or $B \in \Pi$. Contradiction.

For $t$ we need to show that:

- $|A \vee B| = t$ iff ($|A| = t$ and $|B| \neq e$) or ($|B| = t$ and $|A| \neq e$) or ($|A| = b$ and $|B| = n$) or ($|A| = n$ and $|B| = b$)

That is:

- $A \vee B \in \Pi$ and $\neg(A \vee B) \notin \Pi$ iff the disjunction of:
  - $[\alpha]$ $A \in \Pi$ and $\neg A \notin \Pi$ and some $B^\dagger \in \Pi$
  - $[\beta]$ $B \in \Pi$ and $\neg B \notin \Pi$ and some $A^\dagger \in \Pi$
  - $[\gamma]$ $A \in \Pi$ and $\neg A \in \Pi$ and $B \notin \Pi$ and $\neg B \notin \Pi$ and some $B^\dagger \in \Pi$
  - $[\delta]$ $B \in \Pi$ and $\neg B \in \Pi$ and $A \notin \Pi$ and $\neg A \notin \Pi$ and some $A^\dagger \in \Pi$

⇐: In case $[\alpha]$, $A \in \Pi$ and some $B^\dagger \in \Pi$, so $A \vee B \in \Pi$ and $\neg A \notin \Pi$, so $\neg(A \vee B) \notin \Pi$. The other three cases are similar.
⇒: Since $A \vee B \in \Pi$, some $A^\dagger \in \Pi$ and some $B^\dagger \in \Pi$. So we need only consider the other clauses. Since $A \vee B \in \Pi$, $A \in \Pi$ or $B \in \Pi$. Since $\neg(A \vee B) \notin \Pi$ $\neg A \notin \Pi$ or $\neg B \notin \Pi$. This gives us four cases. If $A \in \Pi$ and $\neg A \notin \Pi$, then we are in case $[\alpha]$. If $B \in \Pi$ and $\neg B \notin \Pi$, we are in case $[\beta]$. So suppose that $A \in \Pi$ and $\neg B \notin \Pi$. (If $A$ and $B$ are reversed, the case is similar.) Either $\neg A \in \Pi$ or $\neg A \notin \Pi$. In the second case, we are in case $[\alpha]$. In the first case, we again argue by cases. Either $B \in \Pi$ or $B \notin \Pi$. In the first case, we are in case $[\beta]$. In the second case, we are in case $[\gamma]$.

For $b$ we have to show that:

- $|A \vee B| = b$ iff ($|A| = b$ and $|B| = b$) or ($|A| = b$ and $|B| = f$) or ($|A| = f$ and $|B| = b$)

That is:

- $A \vee B \in \Pi$ and $\neg(A \vee B) \in \Pi$ iff the disjunction of:
  - $[\alpha]$ $A \in \Pi$ and $\neg A \in \Pi$ and $B \in \Pi$ and $\neg B \in \Pi$
  - $[\beta]$ $A \in \Pi$ and $\neg A \in \Pi$ and $B \notin \Pi$ and $\neg B \in \Pi$
  - $[\gamma]$$B \in \Pi$ and $\neg B \in \Pi$ and $A \notin \Pi$ and $\neg A \in \Pi$

⇐: In case $[\alpha]$, $A \in \Pi$, so $A \vee B \in \Pi$, and $\neg A \in \Pi$ and $\neg B \in \Pi$, so $\neg(A \vee B) \in \Pi$. The other two cases are similar.

⇒: Since $\neg(A \vee B) \in \Pi$, $\neg A \in \Pi$ and $\neg B \in \Pi$. So we need address only the situation with respect to $A$ and $B$. Since $A \vee B \in \Pi$, $A \in \Pi$ or $B \in \Pi$. Take the first case. The other is similar. Either $B \in \Pi$ or $B \notin \Pi$. In the first case, we are in case [$\alpha$]. in the second case, we are in case [$\beta$].

For $n$, we have to show that:

- $|A \vee B| = n$ iff ($|A| = n$ and $|B| = n$) or ($|A| = n$ and $|B| = f$) or ($|A| = f$ and $|B| = n$)

That is:

- $A \vee B \notin \Pi$ and $\neg(A \vee B) \notin \Pi$ and some $(A \vee B)^\dagger \in \Pi$ iff the disjunction of:
  - [$\alpha$] $A \notin \Pi$ and $\neg A \notin \Pi$ and some $A^\dagger \in \Pi$ and $B \notin \Pi$ and $\neg B \notin \Pi$ and some $B^\dagger \in \Pi$
  - [$\beta$] $A \notin \Pi$ and $\neg A \notin \Pi$ and some $A^\dagger \in \Pi$ and $B \notin \Pi$ and $\neg B \in \Pi$
  - [$\gamma$] $B \notin \Pi$ and $\neg B \notin \Pi$ and some $B^\dagger \in \Pi$ and $A \notin \Pi$ and $\neg A \in \Pi$

⇐: In each case, $A \notin \Pi$ and $B \notin \Pi$, so $A \vee B \notin \Pi$. In each case $\neg A \notin \Pi$ or $\neg B \notin \Pi$, so $\neg(A \vee B) \notin \Pi$. And in each case, some $A^\dagger \in \Pi$ and some $B^\dagger \in \Pi$. (For any $C$, $\neg C$ is a $C^\dagger$.) So $A^\dagger \wedge B^\dagger = (A \vee B)^\dagger \in \Pi$.

⇒: First note that some $(A \vee B)^\dagger \in \Pi$, so some $A^\dagger \in \Pi$ and some $B^\dagger \in \Pi$. We therefore need to consider only the other clauses. Now, suppose that $A \in \Pi$; then since some $B^\dagger \in \Pi$, $A \vee B \in \Pi$. Contradiction. So $A \notin \Pi$. Similarly, $B \notin \Pi$. Now, either $\neg A \notin \Pi$ or $\neg B \notin \Pi$. Suppose $\neg A \notin \Pi$. (The other case is similar.) Then either $\neg B \in \Pi$ or $\neg B \notin \Pi$. In the first possibility, we are in case [$\beta$] . In the second, we are in case [$\alpha$].

## 5 Subtracting Values

We next consider removing each of the non-classical values from *bne*. In each case, we add a new rule of inference, which is sound in virtue of the value deleted, and the completeness proof simplifies appropriately.

### 5.1 *ne*

First, remove $b$ to give *ne*. The rules of *bne* are augmented by Explosion. All the old rules are still sound. (If there was no counter-example before, there is no counter-example now.) So only Explosion needs to be checked, and this is straightforward. For completeness, in the canonical model the case where $B \in \Pi$ and $\neg B \in \Pi$ is ruled out by Explosion. In the argument by cases, the case for $b$ disappears. The only other case that mentions $b$ is that for $t$. This becomes the following. We need to show that:

- $|A \vee B| = t$ iff ($|A| = t$ and $|B| \neq e$) or ($|B| = t$ and $|A| \neq e$)

That is:

- $A \vee B \in \Pi$ and $\neg(A \vee B) \notin \Pi$ iff the disjunction of:

- [$\alpha$] $A \in \Pi$ and $\neg A \notin \Pi$ and some $B^\dagger \in \Pi$
- [$\beta$] $B \in \Pi$ and $\neg B \notin \Pi$ and some $A^\dagger \in \Pi$

$\Leftarrow$: In case [$\alpha$], $A \in \Pi$, so $A \vee B \in \Pi$; and $\neg A \notin \Pi$, so $\neg(A \vee B) \notin \Pi$. The case for [$\beta$] similar.

$\Rightarrow$: Since $A \vee B \in \Pi$, some $A^\dagger \in \Pi$ and some $B^\dagger \in \Pi$. So we need only consider the other clauses. Since $A \vee B \in \Pi$, $A \in \Pi$ or $B \in \Pi$. Since $\neg(A \vee B) \notin \Pi$, $\neg A \notin \Pi$ or $\neg B \notin \Pi$. This gives us four cases. If $A \in \Pi$ and $\neg A \notin \Pi$, then we are in case [$\alpha$]. If $B \in \Pi$ and $\neg B \notin \Pi$, we are in case [$\beta$]. So suppose that $A \in \Pi$ and $\neg B \notin \Pi$. (If $A$ and $B$ are reversed, the case is similar.) Since $A \in \Pi, \neg A \notin \Pi$, by Explosion. So we are in case [$\alpha$].

## 5.2 *be*

Next, remove *n* to give *be*. In the proof theory, we add Weak EM. Soundness is easily established. In the canonical model, the clause for *n* is removed, and the clause for *e* becomes:

- $|B| = e$ if $B \notin \Pi$ and $\neg B \notin \Pi$

The four cases are exclusive and exhaustive.

In checking the cases, the case for *e* becomes as follows. We need to show that:

- $|A \vee B| = e$ iff $|A| = e$ or $|B| = e$

That is:

- $A \vee B \notin \Pi$ and $\neg(A \vee B) \notin \Pi$ iff the disjunction of:
  - [$\alpha$] $A \notin \Pi$ and $\neg A \notin \Pi$
  - [$\beta$] $B \notin \Pi$ and $\neg B \notin \Pi$

$\Rightarrow$: This is obvious.

$\Leftarrow$: Here is the case for [$\alpha$]. The case for [$\beta$] is similar. Since $\neg A \notin \Pi$, $\neg(A \vee B) \notin \Pi$. Suppose that $A \vee B \in \Pi$, then $A \in \Pi$ or $\neg A \in \Pi$, by Weak EM. Contradiction.

There is no case for *n* to consider. The only other case that mentions *e* or *n* is that for *t*, which is now as follows. We need to show that:

- $|A \vee B| = t$ iff ($|A| = t$ and $|B| \neq e$) or ($|B| = t$ and $|A| \neq e$)

that is:

- $A \vee B \in \Pi$ and $\neg(A \vee B) \notin \Pi$ iff the disjunction of:
  - [$\alpha$] $A \in \Pi$ and $\neg A \notin \Pi$ and ($B \in \Pi$ or $\neg B \in \Pi$)
  - [$\beta$] $B \in \Pi$ and $\neg B \notin \Pi$ and ($A \in \Pi$ or $\neg A \in \Pi$)

$\Leftarrow$: In case [$\alpha$], $A \in \Pi$ so $A \vee B \in \Pi$; and $\neg A \notin \Pi$, so $\neg(A \vee B) \notin \Pi$. Case [$\beta$] is similar.

$\Rightarrow$: $A \vee B \in \Pi$, so $A \in \Pi$ or $B \in \Pi$. Suppose the first. The second is similar. Since $\neg(A \vee B) \notin \Pi$, $\neg A \notin \Pi$ or $\neg B \notin \Pi$. Suppose the first. Since $A \vee B \in \Pi$ then $B \vee \neg B$, by Weak EM. So either $B \in \Pi$ or $\neg B \in \Pi$. In either case we are in case [$\alpha$]. In the second case $B \in \Pi$, again by Weak EM. So we are in case [$\beta$].

### 5.3 *bn (FDE)*

Finally, remove $e$ to give $be$ (that is, $FDE$). For this, Weak $\vee I$ is replaced by $\vee I$. Soundness is easily checked. In the canonical model, the clause for $n$ becomes:

- $|B| = n$ if $B \notin \Pi$ and $\neg B \notin \Pi$

The four cases are exclusive and exhaustive.

There is no case for $e$ to check. The clauses that mention $n$ are those for $t$ and $n$. The case for $n$ is as follows. We must show that:

- $|A \vee B| = n$ iff ($|A| = n$ and $|B| = n$) or ($|A| = n$ and $|B| = f$) or ($|A| = f$ and $|B| = n$)

That is:

- $A \vee B \notin \Pi$ and $\neg(A \vee B) \notin \Pi$ iff the disjunction of:
  - [$\alpha$] $A \notin \Pi$ and $\neg A \notin \Pi$ and $B \notin \Pi$ and $\neg B \notin \Pi$
  - [$\beta$] $A \notin \Pi$ and $\neg A \notin \Pi$ and $B \notin \Pi$ and $\neg B \in \Pi$
  - [$\gamma$] $B \notin \Pi$ and $\neg B \notin \Pi$ and $A \notin \Pi$ and $\neg A \in \Pi$

$\Leftarrow$: In each case, $A \notin \Pi$ and $B \notin \Pi$, so $A \vee B \notin \Pi$. In each case $\neg A \notin \Pi$ or $\neg B \notin \Pi$, so $\neg(A \vee B) \notin \Pi$.

$\Rightarrow$: Since $A \vee B \notin \Pi$, $A \notin \Pi$ and $B \notin \Pi$, by $\vee$I. Now, either $\neg A \notin \Pi$ or $\neg$B$\notin \Pi$. Suppose that $\neg A \notin \Pi$. (The other case is similar.) Then either $\neg B \in \Pi$ or $\neg B \notin \Pi$. In the first possibility, we are in case [$\beta$] . In the second, we are in case [$\alpha$].

In the case for $t$, we must now show that:

- $|A \vee B| = t$ iff $|A| = t$ or $|B| = t$ or ($|A| = b$ and $|B| = n$) or ($|A| = n$ and $|B| = b$)

That is:

- $A \vee B \in \Pi$ and $\neg(A \vee B) \notin \Pi$ iff the disjunction of:
  - [$\alpha$] $A \in \Pi$ and $\neg A \notin \Pi$
  - [$\beta$] $B \in \Pi$ and $\neg B \notin \Pi$
  - [$\gamma$] $A \in \Pi$ and $\neg A \in \Pi$ and $B \notin \Pi$ and $\neg B \notin \Pi$
  - [$\delta$] $B \in \Pi$ and $\neg B \in \Pi$ and $A \notin \Pi$ and $\neg A \notin \Pi$

$\Leftarrow$: In case [$\alpha$] $A \in \Pi$ so $A \vee B \in \Pi$; and $\neg A \notin \Pi$, so $\neg(A \vee B) \notin \Pi$. The other three cases are similar.

$\Rightarrow$: Since $A \vee B \in \Pi$, $A \in \Pi$ or $B \in \Pi$. Since $\neg(A \vee B) \notin \Pi$, $\neg A \notin \Pi$ or $\neg B \notin \Pi$. This gives us four cases. If $A \in \Pi$ and $\neg A \notin \Pi$, then we are in case [$\alpha$]. If $B \in \Pi$ and $\neg B \notin \Pi$, we are in case [$\beta$]. So suppose that $A \in \Pi$ and $\neg B \notin \Pi$. (If $A$ and $B$ are reversed, the case is similar.) Either $\neg A \in \Pi$ or $\neg A \notin \Pi$. In the second case, we are in case [$\alpha$]. In the first case, we again argue by cases. Either $B \in \Pi$ or $B \notin \Pi$. In the first case, we are in case [$\beta$]. In the second case, we are in case [$\gamma$].

## 6 Subtracting Multiple Values

Modifying matters to omit multiple values is straightforward. Essentially, we just combine the modifications for each of the values—though there is a small wrinkle in the case for $b$, as I shall note. Detailed checking is left to the reader.

If we remove $b$ and $e$, we get $n$ (i.e., $K_3$). The proof theory adds Explosion and replaces Weak $\vee$I with $\vee$I. Soundness is easy to check, and the canonical model simplifies appropriately. (The only cases to check are for $t$, $f$, and $n$. We simply remove all mention of $b$ from the argument for $bn$.)

If we remove $b$ and $n$, we get $e$ (i.e., $B_3$). The proof theory adds Explosion and Weak EM.[5] Soundness is easy to check, and in the canonical model simplifies appropriately. The only cases to check are for $t$, $f$, and $e$. We simply remove all mention of $b$ from the argument for $be$.

If we remove $n$ and $e$, we get $b$ (i.e., *LP*). One might think that the proof theory would replace Weak $\vee$I with $\vee$I, and add Weak EM; but that doesn't quite work. The rules are certainly sound, but they are not complete. The reason is that *LP* has logical truths (such as $A \vee \neg A$), but as a simple proof establishes, any deduction with just these rules has undischarged assumptions, so no logical truth is established. The trick is to add not Weak EM, but EM. Soundness is easily established, and the canonical model simplifies appropriately. (There only cases to check are those for $t$, $f$, and $b$—EM implies that they are exhaustive. We simply remove all mention of $n$ from the argument for $bn$.)

Finally, if we remove $b$, $n$, and $e$, we get $\emptyset$, that is *CL*. For this, we replace Weak $\vee$I with $\vee$I, and add Explosion and EM. Soundness is easily verified, and the canonical model argument reduces to the standard Henkin completeness proof for classical logic.[6]

## 7 Appendix: PWK

The system "Paraconsistent Weak Kleene (*PWK*)" is an interesting, but relatively little known system invented by Sören Halldén and Arthur Prior.[7] The system is the same as

[5] I note that these additions make Weak $\vee$I redundant, since this can now be derived as follows:

$$\dfrac{\dfrac{A \quad B^{\dagger}}{A \wedge B^{\dagger}[= (B \vee A)^{\dagger}]}}{(A \vee B) \vee \neg(A \vee B)}$$

We now reason by $\vee$E. In the second case, we have:

$$\dfrac{\dfrac{\dfrac{\dfrac{\neg(A \vee B)}{\neg A \wedge \neg B}}{\neg A} \quad A}{A \wedge \neg A}}{A \vee B}$$

[6] Thanks go to Roberto Ciuni and two anonymous referees of this volume for helpful comments.

[7] Halldén (1949) and Prior (1957). For discussion, see Ciuni and Carrara (2016), and Bonzo, Gil-Férez and Paoli (2017).

$B_3$, except that $e$ is designated, so it is not in the $FDE$ family as I have characterised it.[8] But the techniques of this paper extend to it naturally. In this appendix, I provide a natural deduction system for PWK.

## 7.1 Semantics

The matrices for $PWK$ are as follows:

| $\neg$ | |
|---|---|
| $t$ | $f$ |
| $f$ | $t$ |
| $e$ | $e$ |

| $\vee$ | $t$ | $f$ | $e$ |
|---|---|---|---|
| $t$ | $t$ | $t$ | $e$ |
| $f$ | $t$ | $f$ | $e$ |
| $e$ | $e$ | $e$ | $e$ |

| $\wedge$ | $t$ | $f$ | $e$ |
|---|---|---|---|
| $t$ | $t$ | $f$ | $e$ |
| $f$ | $f$ | $f$ | $e$ |
| $e$ | $e$ | $e$ | $e$ |

$t$ and $e$ being designated. In linear form, the truth conditions are:

*Negation*

- $|\neg A| = e$ iff $|A| = e$
- $|\neg A| = f$ iff $|A| = t$
- $|\neg A| = t$ iff $|A| = f$

*Disjunction*

- $|A \vee B| = e$ iff $|A| = e$ or $|B| = e$
- $|A \vee B| = f$ iff $|A| = f$ and $|B| = f$
- $|A \vee B| = t$ iff ($|A| = t$ and $|B| \neq e$) or ($|B| = t$ and $|A| \neq e$)

*Conjunction*

- $|A \wedge B| = e$ iff $|A| = e$ or $|B| = e$
- $|A \wedge B| = f$ iff ($|A| = f$ and $|B| \neq e$) or ($|B| = f$ and $|A| \neq e$)
- $|A \wedge B| = t$ iff ($|A| = t$ and $|B| = t$)

As before, we may take $A \wedge B$ to be defined as $\neg(\neg A \vee \neg B)$.

## 7.2 Natural Deduction Rules

The natural deduction rules for PWK are: $\vee$I, $\vee$E, DeM, DN, EM, $\wedge$I, plus:

- Weak Explosion

$$\frac{A \quad \neg A}{A^{\dagger}}$$

- Weak $\wedge$E

$$\frac{A \wedge B}{A \vee B^{\dagger}} \qquad \frac{A \wedge B}{A^{\dagger} \vee B}$$

The proof of soundness is left as an exercise.

[8] An obvious question is why one might designate a value that behaves in this way. I confess that I know of no very plausible reason.

## 7.3 Completeness

The proof of the Henkin Lemma of 4.2 is unaffected. The proof of completeness is the same as in 4.3, except that canonical model is defined as follows:

- $|B| = t$ if $B \in \Pi$ and $\neg B \notin \Pi$
- $|B| = f$ if $\neg B \in \Pi$ and $B \notin \Pi$
- $|B| = e$ if $B \in \Pi$ and $\neg B \in \Pi$

EM ensures that we are in one of these cases.

We have to prove that this is a PWK evaluation. The cases for negation are straightforward, and left as an exercise. The cases for disjunction are as follows.

For $e$, we have to prove that:

- $|A \vee B| = e$ iff $|A| = e$ or $|B| = e$

That is:

- $A \vee B \in \Pi$ and $\neg(A \vee B) \in \Pi$ iff the disjunction of:
  - $[\alpha]$ $A \in \Pi$ and $\neg A \in \Pi$
  - $[\beta]$ $B \in \Pi$ and $\neg B \in \Pi$

$\Leftarrow$: In each case, the result follows from Weak Explosion.
$\Rightarrow$: Since $A \vee B \in \Pi$, $A \in \Pi$ or $B \in \Pi$. Suppose the first of these. The second is similar. Since $\neg(A \vee B) \in \Pi$, $\neg A \wedge \neg B \in \Pi$. By weak $\wedge$E, $\neg A \in \Pi$ or for any $B^\dagger$, $\neg B^\dagger \in \Pi$. In the first case, we are in case $[\alpha]$. In the second case, we are in case $[\beta]$ (since both $B$ and $\neg B$ are of the form).

For $f$ we have to prove that:

- $|A \vee B| = f$ iff $|A| = f$ and $|B| = f$

That is:

- $A \vee B \notin \Pi$ and $\neg(A \vee B) \in \Pi$ iff $A \notin \Pi$ and $\neg A \in \Pi$ and $B \notin \Pi$ and $\neg B \in \Pi$.

$\Rightarrow$: Since $A \vee B \notin \Pi$, $A \notin \Pi$ and $B \notin \Pi$. Since $\neg(A \vee B) \in \Pi$, $\neg A \wedge \neg B \in \Pi$. By Weak $\wedge$E, for any $B^\dagger$, $\neg A \vee B^\dagger \in \Pi$; so $\neg A \in \Pi$ or $B^\dagger \in \Pi$. The second case is impossible, since $B \notin \Pi$. Hence $\neg A \in \Pi$. The case for $\neg B$ is similar.
$\Leftarrow$: Since $A \notin \Pi$ and $B \notin \Pi$, $A \vee B \notin \Pi$. Since $\neg A \in \Pi$ and $\neg B \in \Pi$, $\neg A \wedge \neg B \in \Pi$, so $\neg(A \vee B) \in \Pi$.

For $t$, we have to prove that:

- $|A \vee B| = t$ iff ($|A| = t$ and $|B| \neq e$) or ($|B| = t$ and $|A| \neq e$)

That is:

- $A \vee B \in \Pi$ and $\neg(A \vee B) \notin \Pi$ iff the disjunction of:
  - $[\alpha]$ ($A \in \Pi$ and $\neg A \notin \Pi$) and ($B \notin \Pi$ or $\neg B \notin \Pi$)
  - $[\beta]$ ($B \in \Pi$ and $\neg B \notin \Pi$) and ($A \notin \Pi$ or $\neg A \notin \Pi$)

$\Leftarrow$: Here is the case for [$\alpha$] the case for [$\beta$] is similar. Since $A \in \Pi$, $A \vee B \in \Pi$. Suppose $\neg(A \vee B) \in \Pi$ then $\neg A \wedge \neg B \in \Pi$. So for any $B^{\dagger}$, $\neg A \vee (\neg B)^{\dagger} \in \Pi$, and so either $\neg A \in \Pi$ or every $(\neg B)^{\dagger} \in \Pi$. The first case contradicts the first conjunct; the second case contradicts the second.

$\Rightarrow$: Since $A \vee B \in \Pi$, $A \in \Pi$ or $B \in \Pi$. Since $\neg(A \vee B) \notin \Pi$, $\neg A \notin \Pi$ or $\neg B \notin \Pi$. This gives us four cases. Suppose $A \in \Pi$ and $\neg A \notin \Pi$. (The case where $B \in \Pi$ and $\neg B \notin \Pi$ is the same.) If we could show that $B \notin \Pi$ or $\neg B \notin \Pi$, we would be in case [$\alpha$]. So suppose that $B \in \Pi$ and $\neg B \in \Pi$. Then, by Weak Explosion, every $B^{\dagger} \in \Pi$. Hence $\neg(A \vee B) \in \Pi$. Contradiction. Suppose that $A \in \Pi$ and $\neg B \notin \Pi$. (The case where $A \in \Pi$ and $\neg B \notin \Pi$ is the same.) Either $\neg A \notin \Pi$ or $\neg A \in \Pi$. In the first case, we are in case [$\alpha$]. In the second case, $A \in \Pi$ and $\neg A \in \Pi$. Hence, every $A^{\dagger} \in \Pi$. So $\neg(A \vee B) \in \Pi$, which is impossible.

## References


1. Bonzo, S., Gil-Férez, J., Paoli, F. (2017), 'On Paraconsistent Weak Kleene Logic', *Studia Logica* 105: 253-98.
2. Ciuni, R., and Carrara, M. (2016), 'Characterizing Logical Consequence in Paraconsistent Weak Kleene', pp. 165-176 of L. Felline, F. Paoli, and E. Rossanese (eds.), *New Developments in Logic and Philosophy of Science*, London: College Publications.
3. Daniels., D. (1990), 'A Note on Negation', *Erkenntnis* 32: 423-9.
4. Goddard, L. and Routley, R. (1973) *The Logic of Significance and Context*, New York, NY: Halsted Press.
5. Haack, S. (1996), *Deviant Logic, Fuzzy Logic: Beyond the Formalism*, Chicago, IL: University of Chicago Press.
6. Halldén, S. (1949) *The Logic of Nonsense*, Uppsala: Lundequista Bokhandeln
7. Oller, C. (1999), 'Paraconsistency and Analyticity', *Logic and Logical Philosophy* 7: 91-9.
8. Prawitz, D. (1965), *Natural Deduction: a Proof-Theoretical Study*, Stockholm: Almqvist & Wiksell.
9. Priest, G. (2008), *Introduction to Non-Classical Logics: From If to Is*, Cambridge: Cambridge University Press.
10. Priest, G. (2010), 'The Logic of the Catuṣkoṭi', *Comparative Philosophy* 1: 32-54.
11. Priest, G. (2014a), 'Speaking of the Ineffable. . . ', ch. 7 of J. Lee and D. Berger (eds.), *Nothingness in Asian Philosophy*, London: Routledge.
12. Priest, G. (2014b), 'Plurivalent Logic', *Australasian Journal of Logic* 11, article 1. http://ojs.victoria.ac.nz/ajl/article/view/1830.
13. Prior, A., (1957), *Time and Modality*, Oxford: Oxford University Press.

# Modelling Sources of Inconsistent Information in Paraconsistent Modal Logic

Igor Sedlár and Ondrej Majer

**Abstract** Epistemic logics based on normal modal logic are notoriously bad at handling inconsistent and yet non-trivial information. This fact motivates epistemic logics based on paraconsistent logic, examples of which can be traced back at least to the 1980s. These logics handle inconsistent and non-trivial information, but they usually do not articulate sources of the inconsistency. Yet, making the origin of an inconsistency present in a body of information explicit is important to assess the body—can we trace the mutually conflicting pieces of information to sources of information relevant to the body or is the inconsistency a result of an error unrelated to any outside sources? Is the inconsistency derived from various equally trustworthy sources or from a single inconsistent source? In this article we show that a paraconsistent modal logic, namely, the logic BK introduced by Odintsov and Wansing, is a first step toward a formalism capable of making these distinctions explicit. We interpret the accessibility relation between states in a model as a source relation—states accessible from a given state are seen as sources of potential justification of the information contained in the original state. This interpretation also motivates the study of a number of extensions of BK. We focus here on extensions of BK able to articulate the relation of compatibility between bodies of information and extensions working with labels explicitly differentiating between bodies of information. In the case of compatibility-based extensions a more detailed technical study including a completeness proof is provided; technical features of the simpler case of label-based extensions, on the other hand, are discussed without going into details.

**Keywords** Compatibility • Ex falso quodlibet • First degree entailment • Inconsistent information • Modal logic • Paraconsistent logic • Paraconsistent modal logic • Sources of information

---

Igor Sedlár
The Czech Academy of Sciences, Institute of Computer Science, Prague, Czech Republic
e-mail: sedlar@cs.cas.cz

Ondrej Majer
The Czech Academy of Sciences, Institute of Philosophy, Prague, Czech Republic
e-mail: majer@flu.cas.cz

 
H. Omori, H. Wansing (eds.), *New Essays on Belnap-Dunn Logic*,
Synthese Library 418, https://doi.org/10.1007/978-3-030-31136-0_17

## 1 Introduction

It is well known that epistemic logics[1] based on normal modal logic [12, 11] treat all *inconsistent* information as being equivalent. This is hard-wired into how information is treated within normal epistemic logic. Bodies of information, as represented by normal epistemic logics, are closed under all classically valid inference rules including *Ex falso quodlibet*

$$\frac{\varphi \quad {\sim}\varphi}{\psi}$$

This means that if a body of information contains (supports) a contradictory pair $\varphi, {\sim}\varphi$, then it contains (supports) every $\psi$. Consequently, all inconsistent bodies of information contain (support) *every* piece of information whatsoever and so they cannot be distinguished from each other. A related semantic point is that, on the *information-as-range* methodology adopted within normal epistemic logic [24], a body of information is represented by the set of possible worlds compatible with the body; however, the set of possible worlds compatible with each and every inconsistent body of information is the same, namely, the empty set. In many contexts, however, it is crucial to distinguish between inconsistent bodies of information and model logically the fact that different conclusions may be drawn from different inconsistent assumptions.

*Paraconsistent logics* invalidate *Ex falso* and so they allow us to differentiate between inconsistent bodies of information by drawing different conclusions from them. These logics are therefore a suitable basis for epistemic logic dealing with distinct inconsistent and yet non-trivial bodies of information.

Levesque [14] develops an epistemic logic based on the four-valued paraconsistent logic FDE of Belnap and Dunn [3, 4, 8, 9]. The idea is to replace possible worlds by more general "states" that allow formulas to be both true and false, or neither true nor false, while keeping most of the other aspects of epistemic logic intact. Similarly to the information-as-range approach, a body of information is represented by the set of states supporting all the information in the body. In a paraconsistent setting, however, the set of states may vary depending on the specific contradiction at hand.

Levesque's approach allows to distinguish between inconsistent bodies of information, but it fails to articulate *sources* of the inconsistency. Yet, making the origin of an inconsistency present in a body of information explicit is important to assess the body—can we trace the mutually conflicting pieces of information to sources of information relevant to the body or is the inconsistency a result of an error unrelated to any outside sources? Is the inconsistency derived from various equally trustworthy sources or from a single inconsistent source?

**Example 1.1.** In a conversation with a friend, Alice expresses contradictory beliefs about Carl, a prominent politician: she claims that Carl wants to grant asylum to a significant number of refugees but later in the discussion she mentions that Carl is xenophobic and does not want to allow any foreigners to enter the country. Assume that Alice's belief about Carl's xenophobia is based on the opinions of her trusted friend Cathy and the report about refugees is taken from an article in her favorite newspaper. Alice does not realize that these two are in mutual contradiction.

[1] To avoid complications unnecessary with respect to our motivation, we do not distinguish between "epistemic" and "doxastic" in this article and we use "epistemic" in a sense covering all relevant information-relative adjectives. This approach is taken in some literature on epistemic logic; see [15, xi], for example.

In an unrelated situation, Bob, being questioned by the police, claims that he was not in town at the time of the crime, but admits to being home by himself later during the interrogation. Assume that neither side of the inconsistency in Bob's testimony can be substantiated in any reliable way; Bob cannot produce any evidence supporting either side of his inconsistent statement.

It seems natural to say that Bob's testimony is somewhat worse off than Alice's beliefs about Carl. Imagine further that Alice's friend Cathy is in fact the author of the article about refugees and that Alice knows this. This is again relevant to our assessment of the situation—Alice's source itself is inconsistent.

Making justice to the distinctions pointed out in our example requires us to make explicit the fact that, while assessing a body of information, one may *turn to other bodies of information*, understood as *sources* for the original body—sometimes an inconsistency is a result of following these sources too closely without reflection; sometimes no such excuse is available. This requires to make explicit use of *relations between bodies of information*, something not commonly present in epistemic logic, normal or paraconsistent.

This article shows that the "four-valued" modal logic BK introduced by Odintsov and Wansing [19] (and studied subsequently in [16, 17, 18, 20], for example) is a suitable starting point. We may see states in a four-valued Kripke model as potentially inconsistent (and incomplete) bodies of information and the accessibility relation as articulating a "source" relation between these bodies of information. On this reading, $Rxy$ means that $y$ is a source for the information contained in $x$. We can also say that sources for $x$ serve as *potential justifications* for the information contained in $x$.[2] For instance, take a news article, $x$, and $R(x) = \{y \mid Rxy\}$ consisting of all the materials, interviews etc. from which the information conveyed by the article derives. It is clear that the relation of the article to the sources may be strong, but also very loose. Some articles build on their sources very closely, some add information that cannot be traced to any sources whatsoever. As another example, take $x$ consisting of a student's beliefs about the implications of the Special Theory of Relativity and assume that the student is required to justify these beliefs, for instance during an exam. In typical cases, the student can provide a number of sources (references, calculations etc.) and it is the job of the examiner to asses the relation of these sources to the student's beliefs. Or consider an empirical scientist writing an article, $x$, based on her experiments and experimental and theoretical work done by other scientists, $R(x)$. Bílková et al. [5] adopt a similar reading of the accessibility relation in models for substructural modal logics.

As shown in this article, the source-interpretation of the semantics for BK gives us the necessary tools for articulating the distinctions pointed out in Example 1.1 and it also motivates interesting extensions of BK. We focus here on extensions of BK able to articulate the relation of compatibility between bodies of information and extensions working with labels explicitly differentiating between bodies of information.

The article is structured as follows. Section 2 provides the basic information about BK, the source interpretation of modal accessibility and shows how the framework articulates different ways inconsistencies may be related to sources. In Section 3 we add to models for BK a compatibility relation and show that specific formulas of our language define interesting classes of models in the standard modal correspondence sense. In Section 4 we introduce a new negation connective based on the compatibility relation (in the style of [10, 21]) and prove a completeness result for the extension of BK with the negation.

[2] Justification logics [2] make the relation between information and justification explicit, but the setting is one where justifications are represented syntactically. In our framework, bodies of information and sources are *semantic* objects of the same kind.

Interesting technical and philosophical features of the logic are pointed out; for example, the logic is not closed under uniform substitution and there is a certain tension between the properties of the compatibility-based negation on one hand and a natural requirement concerning the compatibility relation on the other. In Section 5 we return to BK and we show how relative reliability of sources can be expressed using special formulas called labels.

## 2 The Belnapian modal logic BK

In this section we outline the paraconsistent modal logic BK (Section 2.1) and we show how it articulates the distinctions analogous to those pointed out in Example 1.1 (Section 2.2).

### *2.1* BK

This subsection outlines the technical features of BK, introduced by Odintsov and Wansing [19] and later studied, for example, in [16, 17, 18, 20]. We build mainly on the presentation in [19]. The logic BK is an FDE-based modal logic; it extends the Belnap–Dunn logic [3, 4, 8, 9] with "weak implication" $\rightarrow$ in the sense of [1], the falsum constant $\bot$ and normal modal operators $\Box, \Diamond$. BK can be seen as a conservative extension of the smallest normal modal logic K with a "strong" negation $\sim$ allowing truth-value gaps and gluts.

Formally, the language $\mathcal{L}$ (over a countable set of atomic formulas $AT$) contains a nullary connective $\bot$, unary connectives $\sim, \Box, \Diamond$ and binary connectives $\wedge, \vee, \rightarrow$. A second negation $\neg\varphi$ is defined as $\varphi \rightarrow \bot$, $\top$ is defined as $\neg\bot$ and $\varphi \leftrightarrow \psi$ is defined as $(\varphi \rightarrow \psi) \wedge (\psi \rightarrow \varphi)$. "Strong implication" $\varphi \Rightarrow \psi$ is defined as $(\varphi \rightarrow \psi) \wedge ({\sim}\psi \rightarrow {\sim}\varphi)$ and "strong equivalence" $\varphi \Leftrightarrow \psi$ is defined as $(\varphi \Rightarrow \psi) \wedge (\psi \Rightarrow \varphi)$. We use "$\mathcal{L}$" also to denote the set of formulas of the language.

A basic $\mathcal{L}$-frame ("frame" for short) is a standard Kripke frame $F = \langle S, R \rangle$, that is, a non-empty set (of "states") $S$ with a binary ("accessibility") relation $R$ on $S$. A basic $\mathcal{L}$-model $M$ ("model" for short) based on $F$ adds to $F$ two functions $V^+$ and $V^-$ from $AT$ to subsets of $S$ ("the positive and the negative valuation"). [3]

For every $M$, we define two relations $\models^+_M, \models^-_M \subseteq \mathscr{P}(S \times \mathcal{L})$ as follows (we usually omit the subscript "$M$"):

1. $x \models^+ p$ iff $x \in V^+(p)$;
   $x \models^- p$ iff $x \in V^-(p)$
2. $x \models^+ \bot$ for no $x$;
   $x \models^- \bot$ for all $x$
3. $x \models^+ {\sim}\varphi$ iff $x \models^- \varphi$;
   $x \models^- {\sim}\varphi$ iff $x \models^+ \varphi$
4. $x \models^+ \varphi \wedge \psi$ iff $x \models^+ \varphi$ and $x \models^+ \psi$;
   $x \models^- \varphi \wedge \psi$ iff $x \models^- \varphi$ or $x \models^- \psi$

[3] Alternatively, equivalent semantics for BK can be formulated using four-valued valuations, i.e. functions from $AT \times S$ to $\{True, False, Neither, Both\}$ or to subsets of $\{T, F\}$—cf. [19].

5. $x \models^+ \varphi \vee \psi$ iff $x \models^+ \varphi$ or $x \models^+ \psi$;
   $x \models^- \varphi \vee \psi$ iff $x \models^- \varphi$ and $x \models^- \psi$
6. $x \models^+ \varphi \rightarrow \psi$ iff $x \not\models^+ \varphi$ or $x \models^+ \psi$;
   $x \models^- \varphi \rightarrow \psi$ iff $x \models^+ \varphi$ and $x \models^- \psi$
7. $x \models^+ \Box\varphi$ iff for all $y$, if $Rxy$, then $y \models^+ \varphi$
   $x \models^- \Box\varphi$ iff there is $y$ such that $Rxy$ and $y \models^- \varphi$
8. $x \models^+ \Diamond\varphi$ iff there is $y$ such that $Rxy$ and $y \models^+ \varphi$
   $x \models^- \Diamond\varphi$ iff for all $y$, if $Rxy$, then $y \models^- \varphi$

The positive and negative extension of a formula $\varphi$ is defined as $[\![\varphi]\!]^+_M = \{x \mid x \models^+_M \varphi\}$ and $[\![\varphi]\!]^-_M = \{x \mid x \models^-_M \varphi\}$, respectively. Formula $\varphi$ is valid in a model $M$ ($M$-valid) iff $[\![\varphi]\!]^+_M = S_M$; $\varphi$ is valid in a frame $F$ iff it is $M$-valid for all $M$ based on $F$. A set of formulas $\Gamma$ entails $\varphi$ in $M$ ($\Gamma \models_M \varphi$) iff $\bigcap_{\psi \in \Gamma} [\![\psi]\!]^+_M \subseteq [\![\varphi]\!]^+_M$; $\Gamma$ entails $\varphi$ in frame $F$ ($\Gamma \models_F \varphi$) iff $\Gamma \models_M \varphi$ for all $M$ based on $F$.

The relation $\models^+$ is seen as a *verification* relation, i.e. $x \models^+ \varphi$ means that state $x$ verifies $\varphi$. We shall sometimes write $x \models \varphi$ instead of $x \models^+ \varphi$. Similarly, $\models^-$ is seen as the *falsification* relation, thus $x \models^- \varphi$ means that $x$ falsifies $\varphi$. The crucial feature of the semantics is that a formula can be *both* verified and falsified in a state (the semantics allows truth value gluts) and, similarly, a formula can be *neither* verified nor falsified in a state (the semantics allows truth value gaps). A quick look at the verification/falsification conditions specified above reveals that this is, in fact, the only difference between the present semantics and the Kripke semantics for normal modal logics; the clauses for the connectives in our language mimic the Kripke-style clauses (the negation of a formula is verified iff the negated formula is falsified and it is falsified iff the negated formula is verified, a conjunction is verified iff both conjuncts are verified and it is falsified iff at least one conjunct is falsified etc.).

The logic BK is the least set of formulas containing the axioms of classical propositional logic in the language $\{\wedge, \vee, \rightarrow, \bot\}$, the strong negation axioms

$$\sim\sim p \leftrightarrow p \qquad \sim(p \vee q) \leftrightarrow (\sim p \wedge \sim q) \qquad \sim(p \rightarrow q) \leftrightarrow (p \wedge \sim q)$$
$$\sim\bot \qquad \sim(p \wedge q) \leftrightarrow (\sim p \vee \sim q)$$

the K axioms

$$\Box(p \rightarrow q) \rightarrow (\Box p \rightarrow \Box q) \qquad \Box\top$$

and the modal interaction axioms

$$\neg\Box p \leftrightarrow \Diamond\neg p \qquad \Box p \Leftrightarrow \sim\Diamond\sim p$$
$$\neg\Diamond p \leftrightarrow \Box\neg p \qquad \Diamond p \Leftrightarrow \sim\Box\sim p,$$

closed under uniform substitution, modus ponens and the monotonicity rules

$$\frac{\varphi \rightarrow \psi}{\Box\varphi \rightarrow \Box\psi} \qquad \frac{\varphi \rightarrow \psi}{\Diamond\varphi \rightarrow \Diamond\psi}$$

Formula $\varphi$ is *derivable* in BK from the set of assumptions $\Gamma$, in symbols $\Gamma \vdash_{\mathsf{BK}} \varphi$, if $\varphi$ can be obtained from $\Gamma$ and BK by means of modus ponens.

**Theorem 2.1** (Odintsov and Wansing [19]). *$\Gamma \vdash_{\mathsf{BK}} \varphi$ iff $\Gamma \models_F \varphi$ for each frame $F$.*

It is clear that *Ex falso quodlibet* fails in BK, i.e. $p, \sim p \nvdash_{\mathsf{BK}} q$. Another remarkable feature of BK is that it is not closed under the Replacement rule

$$\frac{\varphi \leftrightarrow \psi}{\chi(p/\varphi) \leftrightarrow \chi(p/\psi)}$$

For example, $\sim(p \to q) \leftrightarrow (p \wedge \sim q)$ is in BK, but $\sim\sim(p \to q) \leftrightarrow \sim(p \wedge \sim q)$ is not.[4] Nevertheless, BK is closed under the Positive replacement rule

$$\frac{\varphi \leftrightarrow \psi}{\chi(p/\varphi) \leftrightarrow \chi(p/\psi)} \quad \chi \text{ is } \sim\text{-free}$$

and the Weak replacement rule

$$\frac{\varphi \Leftrightarrow \psi}{\chi(p/\varphi) \leftrightarrow \chi(p/\psi)}$$

It is easy to check that $x \models \neg\varphi$ iff $x \not\models \varphi$ and that $\varphi \models_M \psi$ iff $\models_M \varphi \to \psi$. The first fact can be seen as stating that $\neg\varphi$ is the Boolean negation of $\varphi$ but we need to be careful here. For example, if $p$ is both verified and falsified, then $\neg\neg p$ is verified but not falsified and if $p$ is neither verified nor falsified, then $\neg\neg p$ is only falsified. Hence, the "value" of $\neg\neg p$ is not necessarily identical to the "value" of $p$.

## 2.2 Sources and contradictions

We now show that the source-related interpretation of the semantics for BK can be used to articulate the different ways an inconsistency present in a body of information may be related to the sources for the body of information.

Take a frame $F = \langle S, R \rangle$. Recall that, on our interpretation outlined in the Introduction, $S$ is a set of bodies of information (or as we shall equivalently say, "information states") and $R$ is the source relation ($Rxy$ means that $y$ is a source for $x$). The relations $\models^+$ and $\models^-$ are seen as relations of support and rejection, respectively, between information states (represented by states of a model $M$) and pieces of information (represented by formulas of $\mathcal{L}$).

Our main motivation is to express the relation between information supported by a state $x$ and the information supported by the sources for $x$, that is, states in $R(x)$. The goal is to differentiate between situations where a contradiction in $x$ cannot be traced to sources for $x$, situations where a contradiction in $x$ derives from two mutually inconsistent sources, and situations where the inconsistency derives from a single inconsistent source for $x$ (recall Example 1.1).

On our interpretation of the semantics, $x \models^+ \varphi \wedge \Diamond\varphi$ iff $\varphi$ is supported by the body of information $x$ and also by some source for $x$. This is all we need to differentiate between the three situations mentioned above. Figure 1 depicts these three situations. Let $x$ be the bottom state in all three models. In the model (a), $x$ supports contradictory information about $p$, but the contradiction is not related to the sources for $x$ in any way as no source for $x$ supports either $p$ or $\sim p$. In the model (b), the contradictory information supported by $x$ is supported also by the sources for $x$: one of them gives positive information about $p$ and the other one gives negative information. In the model (c), the contradiction in $x$ is supported by a single contradictory source for $x$. To simplify notation, we define

[4] If it were, then we could show that $(p \to q) \leftrightarrow (\sim p \vee q)$ is in BK using De Morgan laws and the Double negation law.

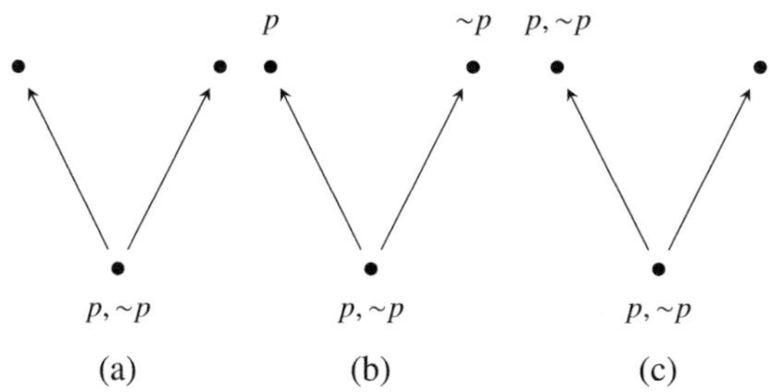


**Fig. 1** Three kinds of contradictory information.

$$\Diamonddot\varphi := \varphi \wedge \Diamond\varphi$$

($\Diamonddot\varphi$ may be read as "$\varphi$ is supported (by the present state) and justified by a source (of the present state)" or "$\varphi$ is supported and justifiable", for example.) Formally,

$$\begin{array}{ll} \text{(a)} & x \models \neg(\Diamonddot p \wedge \Diamonddot{\sim}p) \wedge \neg\Diamonddot(p \wedge {\sim}p) \\ \text{(b)} & x \models (\Diamonddot p \wedge \Diamonddot{\sim}p) \wedge \neg\Diamonddot(p \wedge {\sim}p) \\ \text{(c)} & x \models \Diamonddot(p \wedge {\sim}p) \end{array}$$

We can see $\Diamonddot$ as a kind of epistemic operator; it is thus natural to look at familiar properties of epistemic operators in normal epistemic logic and to compare them to the properties of $\Diamonddot$ in our setting. This is the content of the following proposition.

**Proposition 2.2.** *The following are not valid in* BK*, resp. do not preserve validity in* BK*:*

*(a-)* $\varphi/\Diamonddot\varphi$ *($\Diamonddot$-necessitation)*
*(b-)* $(\Diamonddot\varphi \wedge \Diamonddot\psi) \to \Diamonddot(\varphi \wedge \psi)$ *($\Diamonddot$-regularity)*
*(c-)* $(\Diamonddot\varphi \wedge \Diamonddot{\sim}\varphi) \to \Diamonddot\psi$ *($\Diamonddot$-explosion 1)*
*(d-)* $\Diamonddot(\varphi \wedge {\sim}\varphi) \to \Diamonddot\psi$ *($\Diamonddot$-explosion 2)*
*(e-)* $\Diamonddot(\varphi \vee \psi) \to (\Diamonddot\varphi \vee \Diamonddot\psi)$ *($\Diamonddot$-primeness)*
*(f-)* $\Diamonddot\varphi \to \Diamonddot\Diamonddot\varphi$ *(positive $\Diamonddot$-introspection)*
*(g-)* ${\sim}\Diamonddot\varphi \to \Diamonddot{\sim}\Diamonddot\varphi$ *(negative $\Diamonddot$-introspection)*

*The following are valid in* BK*:*

*(a+)* $\Diamonddot(\varphi \wedge \psi) \to (\Diamonddot\varphi \wedge \Diamonddot\psi)$ *($\Diamonddot$-monotonicity)*
*(b+)* $\Diamonddot\varphi \to \varphi$ *($\Diamonddot$-factivity)*

*Proof.* (a-) Take a model $M$ where $V^+(p) = S$ and $R(x) = \emptyset$ for some $x$. Then $x \not\models \Diamond p$. (b-) Let $S = \{x, y, z\}$, $V^+(p) = \{x, y\}$, $V^+(q) = \{x, z\}$ and $R = \{\langle x, y\rangle, \langle x, z\rangle\}$. Then $x \models \Diamond p \wedge \Diamond q$, but $x \not\models \Diamond(p \wedge q)$. (c-) Let $S = \{x, y, z\}$, $V^+(p) = \{x, y\}$, $V^-(p) = \{x, z\}$, $V^+(q) = \emptyset$ and $R = \{\langle x, y\rangle, \langle x, z\rangle\}$. Then $x \models \Diamonddot p \wedge \Diamonddot{\sim}p$, but $x \not\models \Diamonddot q$. (d-) Similar to (c-); let $V^+(p) = V^-(p)$. (e-) Let $S = \{x, y\}$, $R = \{\langle x, y\rangle\}$, $V^+(p) = \{x\}$ and $V^+(q) = \{y\}$. Then $x \models (p \vee q) \wedge \Diamond(p \vee q)$, but $x \not\models \Diamonddot p \vee \Diamonddot q$. The rest follows from the $\models^+$-condition for $\vee$. (f-) and (g-) are easy, it is sufficient to note that they boil down to

$$(\varphi \wedge \Diamond\varphi) \to (\varphi \wedge \Diamond\varphi \wedge \Diamond(\varphi \wedge \Diamond\varphi))$$

and

$$\sim(\varphi \wedge \Diamond\varphi) \rightarrow (\sim(\varphi \wedge \Diamond\varphi) \wedge \Diamond\sim(\varphi \wedge \Diamond\varphi))$$

(a+) follows from the $\models^+$-condition for conjunction, (b+) is trivial.

Proposition 2.2 shows that $\diamondplus$ is quite weak. Let us discuss the points of the proposition in turn: (a-) a valid formula may not be justifiable since we may have a body of information without any sources; (b-) justifiable support may not be closed under conjunctions as there may be distinct sources justifying the conjuncts, but no single one justifying the conjunction; (a+) the converse, however, holds as support is closed under conjunction elimination; (c-) two distinct sources may provide conflicting information about a formula $\varphi$ without any of them providing positive information about some completely unrelated formula $\psi$; and similarly for one inconsistent source (d-).

We note that an axiom analogous to (e-) was a source of some concern about the framework of [5]. In our setting, there is even a model $M$ such that $R_M$ is a function and (e-) is not valid in $M$. The converse of (e-) is valid in every $M$.

It is not hard to observe, however, that (a-) preserves validity in every model with a serial $R$ (i.e., $\forall x \exists y(Rxy)$); (b-) is valid in every model where $R$ is a function (i.e., $\forall xyz(Rxy \wedge Rxz \rightarrow y = z)$; and (c-) is valid in every model where every state is consistent (i.e., $V^+(p) \cap V^-(p) = \emptyset$ for all $p$) (though trivially so, as $[\![\varphi \wedge \sim\varphi]\!]^+ = \emptyset$ in such models for all $\varphi$). The formulas usually characterizing positive and negative introspection have a slightly different meaning in our source based approach—they are related to "second-order confirmation". For example, (f-) says that if $\varphi$ is supported by $x$ and justified by a source $y$, then there is a source $z$ supporting the information that $\varphi$ is supported and justified.

We note that similar facts would hold if we defined $\diamondplus$ within the normal modal logic $\mathsf{K}$ and not within $\mathsf{BK}$. Notable exceptions, of course, are the explosion axioms (d) and (e) of Proposition 2.2.

## 3 Compatibility

Admittedly, our requirements concerning sources were quite weak so far. One way to put more flesh on these bones is to relate sources to the notions of compatibility. One may require, for example, that sources be self-compatible; that sources for a state be compatible with that state; or that sources for a given state be mutually compatible.

But, one may ask, what does it take for two information states to be mutually compatible? A necessary condition for two states $x$ and $y$ being compatible, we submit, is

(NC) There is no $\varphi$ such that $x$ supports $\varphi$ and $y$ supports $\sim\varphi$.

Our semantics can be made to comply with (NC) as follows. For non-modal $\varphi$, (NC) is enforced by requiring that, for all $p \in AT$, if $x$ supports (rejects) $p$, then $y$ does not reject (support) $p$, and vice versa. For modal formulas, however, an additional condition is needed. The easiest way to enforce the condition is to add a binary *compatibility* relation on states of the frame. In what follows, (when $M$ is clear form the context) let $V^\pm(x) = \{p \mid x \in V^\pm(p)\}$ and $[\![x]\!]^\pm = \{\varphi \mid x \in [\![\varphi]\!]^\pm\}$, for $\pm \in \{+, -\}$.

**Definition 3.1.** A *compatibility frame* is a tuple $F = \langle S, R, C \rangle$ where $S \neq \emptyset$ and $R, C$ are binary relations on $S$ such that

$$Cxy \implies Cyx \tag{1}$$

$$Cxy \text{ and } Rxz \implies (\exists w)(Ryw \ \& \ Czw) \tag{2}$$

A *compatibility $\mathcal{L}$-model $M$* based on a compatibility frame $F$ is a tuple $\langle F, V^+, V^- \rangle$, where $V^+, V^-$ are functions from $AT$ to $\mathscr{P}(S)$ such that

$$Cxy \implies V^+(x) \cap V^-(y) = \emptyset \tag{3}$$

Validity of formulas in compatibility models is defined as usual; $\varphi$ is valid in a compatibility frame iff it is valid in each compatibility model based on the frame. Entailment in a model and a class of frames is defined as usual.

It is clear, that since $Cxy$ implies $Cyx$, $Cxy$ implies $V^-(x) \cap V^+(y) = \emptyset$.

**Proposition 3.2.** *If $Cxy$, then $[\![x]\!]^+ \cap [\![y]\!]^- = \emptyset$ and $[\![x]\!]^- \cap [\![y]\!]^+ = \emptyset$.*

*Proof.* The first claim is established by induction on the complexity of formulas; the second claim follows from the first one and (1).

The base case follows immediately from Definition 3.1 and the clauses for non-modal connectives are established by easy induction.

Consider modal formulas. Assume $Cxy$. $x \models \Diamond\varphi$ iff there is $z \in R(x)$ such that $z \models \varphi$. By (2), there is $w \in R(y)$ such that $Czw$. By the induction hypothesis, $w \not\models {\sim}\varphi$, so $y \not\models {\sim}\Diamond\varphi$. The fact that $x \models {\sim}\Diamond\varphi$ implies $y \not\models \Diamond\varphi$ is established similarly, using (1) in addition to (2). The case for $\Box$ is analogous.

Condition (1) corresponds to the intuitive assumption that any reasonable compatibility relation should be symmetric. Conditions (2)–(3) establish (NC), as witnessed by Proposition 3.2.

As mentioned earlier, intuitive considerations may lead us to adopting additional frame conditions. Fig. 2 lists some of these conditions. The class **SC** ('*s*ources are *c*ompatible')

| *Frame condition* | *Frame class label* |
|---|---|
| (SC) $Rxy \implies Cxy$ | **SC** |
| (SSC) $Rxy \implies Cyy$ | **SSC** |
| (SMC) $Rxy$ and $Rxz \implies Cyz$ | **SMC** |

**Fig. 2** Some intuitive frame classes.

requires every source for a state to be compatible with that state. As a result, the information supported by a state never contradicts the sources for the state. The class **SSC** ('*s*ources are *s*elf-*c*ompatible') requires every source to be self-compatible. As a result, no reliable source supports an explicit contradiction $\varphi \wedge {\sim}\varphi$. The class **SMC** ('*s*ources are *m*utually *c*ompatible') requires every pair of sources for a given state to be mutually compatible. As a result, contradictions are never implicitly supported. Note that **SMC** $\subset$ **SSC**.

We say that a formula *defines a class of frames* **F** in case $F \in$ **F** iff the formula is valid in $F$.

**Proposition 3.3.**

*(a)* **SC** *is defined by* (SC) $\Diamond{\sim}p \to \neg p$
*(b)* **SSC** *is defined by* (SSC) $\neg\Diamond(p \wedge {\sim}p)$
*(c)* **SMC** *is defined by* (SMC) $\neg(\Diamond p \wedge \Diamond{\sim}p)$

*Proof.* It is easy to check that the formulas are valid in the respective frame classes. We give a detailed proof of the converse implication in case of (a) only, the other cases are established similarly. Assume that we have a frame $F$ such that $Rxy$ and not $Cxy$ for some $x, y$. Define a model $M$ on $F$ as follows: $V^-(p) = \{y\}$ and $V^+(p) = \{x\}$ (this is possible since not $Cxy$). It is clear that $x \models^+_M \Diamond{\sim}p$ but also $x \models^+_M p$.

Proposition 3.3 has a number of interesting consequences. For instance, the following schemes are valid in **SC**:

$$(\Diamond\varphi \lor \Diamond{\sim}\varphi) \to \neg(\varphi \land {\sim}\varphi) \tag{4}$$

$$(\Diamond\varphi \land \Diamond{\sim}\varphi) \to \neg(\varphi \lor {\sim}\varphi) \tag{5}$$

As a consequence of (4), if a contradiction is supported by a state, then it cannot be supported by its sources in any way; if a state supports $\varphi \land {\sim}\varphi$, then the conflicting conjuncts are not supported by sources.

**Remark 3.4.** States $x$ in our model are seen as bodies of information and $R(x)$ as a set of sources that can be used to justify the information contained in $x$. Informally, we may also think about $x$ as being the information believed by (available to, entertained by...) an agent. The question is what is the formal counterpart, on this interpretation, of the information state of the agent. Is it $x$, $\{x\} \cup R(x)$, $\{x\} \cup R(x) \cup R^2(x)$...? In other words, are sources (sources of sources, ...) a part of the information state? We do not decide the issue here; we just note that the "distances" between a state supporting $\varphi$ and a source justifying $\varphi$ can be formalized using iterated $\Diamond$. Generally, $\varphi \land \Diamond^k\varphi$ says that $\varphi$ is supported and justified by a source of ... a source ($k$ times). Of course, our basic language cannot express the fact that there is a finite path from the present state to a source of ... a source of the present state. To this end, we would need to employ the reflexive transitive closure modality $\Diamond^*$ (the semantic conditions associated with this modality involve the reflexive transitive closure $R^*$ of the accessibility relation $R$). Such a modality is studied in the context of paraconsistent Propositional Dynamic Logic in [22].

Let BK+SC be the least set of formulas containing BK and the formula SC closed under uniform substitution, modus ponens and the monotonicity rules; BK+SSC and BK+SMC are defined similarly. It is relatively straightforward to show that $\Gamma \models_{\mathbf{SC}}$ iff $\varphi$ is derivable from $\Gamma$ and BK+SC by means of modus ponens; and similarly for **SSC** and **SMC**. We omit the details.

## 4 Compatibility negation

We have seen in the previous section that the compatibility relation $C$ is useful in articulating interesting relations between sources. However, the relation usually plays additional roles as well. In the relational semantics for substructural logics, $C$ is used to interpret a negation connective [10, 21]. It is interesting to look at the interaction between such a connective and the Belnapian negation $\sim$.

Let $\mathcal{L}^\frown$ be $\mathcal{L}$ extended with a new unary connective '$\frown$', called the *compatibility negation*. Intuitively, $\frown\varphi$ says that $\varphi$ is incompatible with the present information. In other words, no state compatible with the present state supports $\varphi$ (every state that supports $\varphi$ is incompatible with the present state). We will interpret the language in compatibility models. The new verification and falsification conditions are (the verification condition is standard in the literature on substructural logics [10, 21]):

- $x \models^+ \frown\varphi$ iff for all $y$, if $Cxy$, then $y \not\models^+ \varphi$
- $x \models^- \frown\varphi$ iff there is $y$ such that $Cxy$ and $x \models^+ \varphi$

Validity and entailment in compatibility models and classes of frames are defined as usual.

An $\mathcal{L}^\frown$-substitution is any map $AT \to \mathcal{L}^\frown$. An $\mathcal{L}^\frown$-substitution instance of a formula $\varphi$ is the result of replacing each occurrence of $p$ in $\varphi$ (for all $p \in AT$) by $\varsigma(p)$ (where $\varsigma$ is an $\mathcal{L}^\frown$-substitution).

The logic $\mathsf{BK}^\frown$ is the smallest set of $\mathcal{L}^\frown$-formulas containing

1. all $\mathcal{L}^\frown$-substitution instances of classical tautologies in the language $\{\wedge, \vee, \to, \bot\}$,
2. the strong negation axioms
$$\sim\sim\varphi \leftrightarrow \varphi \qquad \sim(\varphi \vee \psi) \leftrightarrow (\sim\varphi \wedge \sim\psi) \qquad \sim(\varphi \to \psi) \leftrightarrow (\varphi \wedge \sim\psi)$$
$$\sim\bot \qquad \sim(\varphi \wedge \psi) \leftrightarrow (\sim\varphi \vee \sim\psi)$$
3. the $\mathsf{K}$ axioms
$$\Box(\varphi \to \psi) \to (\Box\varphi \to \Box\psi) \qquad \Box\top$$
4. the modal interaction axioms
$$\neg\Box\varphi \leftrightarrow \Diamond\neg\varphi \qquad \Box\varphi \Leftrightarrow \sim\Diamond\sim\varphi$$
$$\neg\Diamond\varphi \leftrightarrow \Box\neg\varphi \qquad \Diamond\varphi \Leftrightarrow \sim\Box\sim\varphi,$$
5. the compatibility negation axioms
$$\varphi \to \frown\frown\varphi \qquad \chi \to \frown\sim\chi \qquad \chi \text{ is } \frown\text{-free}$$
$$(\frown\varphi \wedge \frown\psi) \to \frown(\varphi \vee \psi) \qquad \neg\frown\varphi \to \sim\frown\varphi$$
$$\Diamond\frown\varphi \to \frown\Box\varphi \qquad \sim\frown\varphi \to \neg\frown\varphi$$

and closed under modus ponens, the monotonicity rules

$$\frac{\varphi \to \psi}{\Box\varphi \to \Box\psi} \qquad \frac{\varphi \to \psi}{\Diamond\varphi \to \Diamond\psi}$$

and the transposition rule

$$\frac{\varphi \to \psi}{\frown\psi \to \frown\varphi}$$

(where $\varphi, \psi$ range over $\mathcal{L}^\frown$ formulas).

The derivability relation $\vdash_{\mathsf{BK}^\frown}$ is defined as expected: $\Gamma \vdash_{\mathsf{BK}^\frown} \varphi$ iff $\varphi$ can be obtained from $\Gamma$ and $\mathsf{BK}^\frown$ by means of modus ponens.

It is a matter of easy checking to establish the validity of each $\varphi \in \mathsf{BK}^\frown$ in all compatibility frames. An axiom schema that deserves special attention, however, is $\chi \to \frown\sim\chi$. Proposition 3.2 entails that each instance of the schema where $\chi$ is $\frown$-free is valid in each compatibility frame. The restriction is essential, as shown by the following example.

**Example 4.1.** Take a frame where $S = \{x, y, z\}$, $R = \emptyset$ and $C$ is the symmetric closure of $\{\langle x, y\rangle, \langle y, z\rangle\}$. Let $V^+(p) = \{z\}$, $V^-(p) = \emptyset$ and $V^+(q) = V^-(q) = \emptyset$ for all other $q \in AT$. This is obviously a compatibility model. It is clear that $x \models^+ \frown p$ and $y \not\models^+ \frown p$. The latter (together with the falsification condition for $\frown$) entails that $y \models^- \frown p$ and so $y \models^+ \sim\frown p$. This means that $x \not\models^+ \frown\sim\frown p$. Hence, $\frown p \to \frown\sim\frown p$ is not valid in all compatibility frames.

Example 4.1 shows that the set of formulas valid in a compatibility frame is not in general closed under uniform $\mathcal{L}^\frown$-substitution. This observation motivates our definition of $\mathsf{BK}^\frown$ in terms of schemes.

We will prove now that $\Gamma \vdash_{\mathsf{BK}^\frown} \varphi$ iff $\Gamma \vDash_F \varphi$ for each compatibility frame $F$. We use the techniques of [21] and [19].

A *prime* $\mathsf{BK}^\frown$*-theory* is a subset of $\mathcal{L}^\frown$ that contains $\mathsf{BK}^\frown$, does not contain $\bot$, is closed under $\vdash_{\mathsf{BK}^\frown}$ (i.e. $\Delta$ is a prime $\mathsf{BK}^\frown$-theory only if $\Gamma \subseteq \Delta$ and $\Gamma \vdash_{\mathsf{BK}^\frown} \varphi$ imply $\varphi \in \Delta$) and contains $\varphi \vee \psi$ only if it contains either $\varphi$ or $\psi$.

Let $\Delta, \Sigma \subseteq \mathcal{L}^\frown$. We say that $\langle \Delta, \Sigma \rangle$ is a $\mathsf{BK}^\frown$*-pair* iff there are no finite $\Delta' \subseteq \Delta$ and $\Sigma' \subseteq \Sigma$ such that

$$\bigwedge \Delta' \to \bigvee \Sigma' \quad \in \quad \mathsf{BK}^\frown$$

We will use 'pair' instead of '$\mathsf{BK}^\frown$-pair' and 'theory' instead of '$\mathsf{BK}^\frown$-theory' in this section.

**Lemma 4.2** (Pair Extension Theorem). *If $\langle \Delta, \Sigma \rangle$ is a pair, then there is a prime theory $\Gamma$ such that $\Delta \subseteq \Gamma$ and $\Sigma \cap \Gamma = \emptyset$.*

*Proof.* (This is a standard proof of the result based on [21, 92-95].) Let $\langle \varphi_n \mid n \in \omega \rangle$ be an enumeration of $\mathcal{L}^\frown$. Define a sequence

$$\langle \langle \Delta_n, \Sigma_n \rangle \mid n \in \omega \text{ and } \Delta_n, \Sigma_n \subseteq \mathcal{L}^\frown \rangle$$

as follows: $\langle \Delta_0, \Sigma_0 \rangle = \langle \Delta, \Sigma \rangle$ and

$$\langle \Delta_{n+1}, \Sigma_{n+1} \rangle = \begin{cases} \langle \Delta_n \cup \{\varphi_n\}, \Sigma_n \rangle & \text{if } \langle \Delta_n \cup \{\varphi_n\}, \Sigma_n \rangle \text{ is a pair} \\ \langle \Delta_n, \Sigma_n \cup \{\varphi_n\} \rangle & \text{otherwise.} \end{cases}$$

*Claim* 4.3. If $\langle \Delta_n, \Sigma_n \rangle$ is a pair, then so is $\langle \Delta_n \cup \{\varphi_n\}, \Sigma_n \rangle$ or $\langle \Delta_n, \Sigma_n \cup \{\varphi_n\} \rangle$.

If both are not pairs, then there are conjunctions $\chi, \chi'$ of some formulas in $\Delta_n$ and disjunctions $\psi, \psi'$ of some formulas in $\Sigma_n$ such that

$$\chi \wedge \varphi_n \to \psi \qquad \chi' \to \varphi_n \vee \psi'$$

are in $\mathsf{BK}^\frown$. But then $\chi \wedge \chi' \to \psi \vee \psi'$ is in $\mathsf{BK}^\frown$ (we may reason as in classical propositional logic); so $\langle \Delta_n, \Sigma_n \rangle$ is not a pair. But this is a contradiction with the assumption, so the claim is established and it follows that $\langle \Delta_n, \Sigma_n \rangle$ is a pair for each $n \in \omega$.

*Claim* 4.4. $\langle \bigcup_{n \in \omega} \Delta_n, \bigcup_{n \in \omega} \Sigma_n \rangle$ is a pair; $\bigcup_{n \in \omega} \Delta_n \cup \bigcup_{n \in \omega} \Sigma_n = \mathcal{L}^\frown$; and $\bigcup_{n \in \omega} \Delta_n$ is a prime theory.

These facts are straightforward. We show that $\bigcup_{n \in \omega} \Delta_n$ is a prime theory if the other facts hold. The set $\bigcup_{n \in \omega} \Delta_n$ does not contain $\bot$. Otherwise there would be some $\Delta_n$ containing $\bot$ and then $\langle \Delta_n, \Sigma_n \rangle$ would not be a pair, contradicting Claim 4.3. The set $\bigcup_{n \in \omega} \Delta_n$ contains $\mathsf{BK}^\frown$. If not, then there is $\chi \in \mathsf{BK}^\frown$ such that, for some $n$, $\langle \Delta_n \cup \{\chi\}, \Sigma_n \rangle$ is not a pair. This means that $\delta \wedge \chi \to \sigma$ is in $\mathsf{BK}^\frown$ (for a conjunction $\delta$ of a finite subset of $\Delta_n$ and a disjunction $\sigma$ of a finite subset of $\Sigma_n$). But then $\delta \to \sigma$ is in $\mathsf{BK}^\frown$ (since, obviously, $\delta \to \chi \vee \sigma$ is in $\mathsf{BK}^\frown$) and so $\langle \Delta_n, \Sigma_n \rangle$ is not a pair, contradicting Claim 4.3. It can be shown in a similar fashion that $\bigcup_{n \in \omega} \Delta_n$ is closed under $\vdash_{\mathsf{BK}^\frown}$. Finally, if $\varphi, \varphi' \notin \bigcup_{n \in \omega} \Delta_n$, then $\varphi, \varphi' \in \bigcup_{n \in \omega} \Sigma_n$. Therefore $\bigcup_{n \in \omega} \Delta_n \nvdash \varphi \vee \varphi'$ and $\varphi \vee \varphi' \notin \bigcup_{n \in \omega} \Delta_n$.

Finally, note that $\bigcup_{n \in \omega} \Delta_n$ contains $\Delta$ and is disjoint from $\Sigma$.

The *canonical* $\mathsf{BK}^{\frown}$*-frame* ('the canonical frame') is $F^{\vdash} = \langle S^{\vdash}, R^{\vdash}, C^{\vdash}\rangle$ where

- $S^{\vdash}$ is the set of all prime theories,
- $R^{\vdash} = \{\langle \Gamma, \Delta\rangle \mid \{\varphi \mid \Box\varphi \in \Gamma\} \subseteq \Delta\}$ and
- $C^{\vdash} = \{\langle \Gamma, \Delta\rangle \mid \{\varphi \mid \frown\varphi \in \Gamma\} \cap \Delta = \emptyset\}$.

**Proposition 4.5.** *The canonical frame is a compatibility frame.*

*Proof.* Firstly, we have to show that $C^{\vdash}$ is symmetric. Assume that $C^{\vdash}\Gamma\Delta$ and $\frown\varphi \in \Delta$. Then $\frown\frown\varphi \notin \Gamma$ and so $\varphi \notin \Gamma$. Hence, $C^{\vdash}\Delta\Gamma$.

Secondly, we have to show that if $C^{\vdash}\Gamma\Delta$ and $R^{\vdash}\Gamma\Sigma$, then there is $\Omega$ such that $R^{\vdash}\Delta\Omega$ and $C^{\vdash}\Sigma\Omega$. Take

$$\underline{\Delta} = \{\varphi \mid \Box\varphi \in \Delta\} \qquad \underline{\Sigma} = \{\psi \mid \frown\psi \in \Sigma\}$$

We show that $\langle \underline{\Delta}, \underline{\Sigma}\rangle$ is a pair. If it were not a pair, then $\bigwedge_{i\leq n}\delta_i \to \bigvee_{j\leq m}\sigma_j$ is in $\mathsf{BK}^{\frown}$ for some $\delta_i \in \underline{\Delta}$ and $\sigma_j \in \underline{\Sigma}$. But then, by $n+1$ applications of the monotonicity rule for $\Box$, $\bigwedge_{i\leq n}\Box\delta_i \to \Box(\bigvee_{j\leq m}\sigma_j)$ is in $\mathsf{BK}^{\frown}$. However, $\bigwedge_{i\leq n}\Box\delta_i \in \Delta$, so $\Box(\bigvee_{j\leq m}\sigma_j) \in \Delta$. The assumption that $C^{\vdash}\Gamma\Delta$ implies that $\frown\Box(\bigvee_{j\leq m}\sigma_j) \notin \Gamma$. It follows from the fact that each $\Diamond\frown\chi \to \frown\Box\chi \in \mathsf{BK}^{\frown}$ that $\Box\frown(\bigvee_{j\leq m}\sigma_j) \notin \Gamma$. However, $\bigwedge_{j\leq m}\frown\sigma_j \in \Sigma$, so $\frown\bigvee_{j\leq m}\sigma_j \in \Sigma$ by the fact that each $(\frown\chi_1 \wedge \frown\chi_2) \to \frown(\chi_1 \vee \chi_2)$ is in $\mathsf{BK}^{\frown}$. By the assumption that $R^{\vdash}\Gamma\Sigma$, $\Diamond\frown\bigvee_{j\leq m}\sigma_j \in \Gamma$.[5] But, using the fact that $\Diamond\frown\chi \to \frown\Box\chi \in \mathsf{BK}^{\frown}$ again, this entails a contradiction. Hence, by the Pair Extension Theorem, there is a prime theory $\Omega$ such that $\underline{\Delta} \subseteq \Omega$ and $\underline{\Sigma} \cap \Omega = \emptyset$. This means that $R^{\vdash}\Delta\Omega$ and $C^{\vdash}\Sigma\Omega$.

The *canonical* $\mathsf{BK}^{\frown}$*-model* ('canonical model') $M^{\vdash}$ is the canonical frame together with two functions from $AT$ to $\mathscr{P}(^{\vdash})$:

- $V^{\vdash+} : p \mapsto \{\Gamma \in S^{\vdash} \mid p \in \Gamma\}$
- $V^{\vdash-} : p \mapsto \{\Gamma \in S^{\vdash} \mid {\sim}p \in \Gamma\}$

**Proposition 4.6.** *The canonical model is a compatibility model.*

*Proof.* We have to prove that if $C^{\vdash}\Gamma\Delta$ and $p \in \Gamma$, then ${\sim}p \notin \Delta$. This follows from the fact that $p \to \frown{\sim}p \in \mathsf{BK}^{\frown}$.

We define the relations $\models^{+}_{M^{\vdash}}$ and $\models^{-}_{M^{\vdash}}$ in the same way as in compatibility models.

**Proposition 4.7.** *For each prime theory $\Gamma$,*

$$\Gamma \models^{+}_{M^{\vdash}} \varphi \iff \varphi \in \Gamma \quad \textit{and} \quad \Gamma \models^{-}_{M^{\vdash}} \varphi \iff {\sim}\varphi \in \Gamma$$

*Proof.* Proof by induction on the complexity of $\varphi$. The base case $\varphi \in AT$ holds by definition. The cases where the main connective of $\varphi$ is in $\{\bot, {\sim}, \wedge, \vee, \to\}$ are standard and we omit them (the proofs use the $\mathsf{BK}$ axioms and the substitution rule).

We discuss the modal cases in more detail. The cases for $\models^{+}$ are established by standard arguments; see [13, 116-119] for the case that $\Box$ is the main connective of $\varphi$ and [6, 196-199] for the case where the main connective is $\Diamond$. Now the cases for $\models^{-}$. Firstly, assume that the main connective is $\Box$. Assume that ${\sim}\Box\varphi \in \Gamma$. To prove that $\Gamma \models^{-} \Box\varphi$, we have to produce a prime theory $\Delta$ such that $R^{\vdash}\Gamma\Delta$ and ${\sim}\varphi \in \Delta$. Using the fact that $\Diamond\varphi \Leftrightarrow {\sim}\Box{\sim}\varphi$ is in $\mathsf{BK}^{\frown}$, we infer that ${\sim}\Box\varphi \leftrightarrow \Diamond{\sim}\varphi$ is in $\mathsf{BK}^{\frown}$, so $\Diamond{\sim}\varphi \in \Gamma$. Then we reason as in the standard

[5] If $R^{\vdash}\Gamma\Sigma$ and $\chi \in \Sigma$, then $\Diamond\chi \in \Gamma$. If not, then $\neg\Diamond\chi \in \Gamma$ by the fact that each $\chi \vee \neg\chi \in \mathsf{BK}^{\frown}$. But this means that $\Box\neg\chi \in \Gamma$ as $\neg\Diamond\chi \to \Box\neg\chi \in \mathsf{BK}^{\frown}$. Using the definition of $R^{\vdash}$ we may infer that $\neg\chi \in \Sigma$ and so $\chi \notin \Sigma$.

$\models^+$-case for $\Diamond$. Conversely, assume that $R^{\vdash}\Gamma\Delta$ and $\sim\varphi \in \Delta$. Reasoning as in footnote 4, we conclude that $\Diamond{\sim}\varphi \in \Gamma$. Since $\sim\Box\varphi \leftrightarrow \Diamond{\sim}\varphi$ is in $\mathsf{BK}^{\frown}$, $\sim\Box\varphi \in \Gamma$. The arguments are similar if the main connective is $\Diamond$.

To conclude the proof, we need to check the cases where $\frown$ is the main connective. If $\frown\varphi \in \Gamma$ and $C^{\vdash}\Gamma\Delta$, then $\varphi \notin \Delta$ by the definition of $C^{\vdash}$. Conversely, assume that $\frown\varphi \notin \Gamma$. We have to produce a $\Delta$ such that $C^{\vdash}\Gamma\Delta$ and $\varphi \in \Delta$. The existence of such a $\Delta$ follows from the fact that $\langle\{\varphi\}, \{\psi \mid \frown\psi \in \Gamma\}\rangle$ is a pair.[6] The cases for $\models^-$ are established in a similar fashion, using the fact that $\neg\frown\varphi \to \sim\frown\varphi$ and $\sim\frown\varphi \to \neg\frown\varphi$ are in $\mathsf{BK}^{\frown}$.

**Theorem 4.8.** *$\Gamma \vdash_{\mathsf{BK}^{\frown}} \varphi$ iff $\Gamma \models_F \varphi$ for all compatibility frames $F$.*

*Proof.* Soundness is established by induction on the length of derivations. It is a manner of easy checking that all $\mathsf{BK}^{\frown}$-axioms are verified in each state of each compatibility model and that this property is preserved by modus ponens. Completeness is also established by a standard argument. Assume that $\Gamma \nvdash_{\mathsf{BK}^{\frown}} \varphi$. The assumption implies that $\langle\Gamma, \{\varphi\}\rangle$ is a $\mathsf{BK}^{\frown}$-pair. Hence, by the Pair Extension Theorem, there is a prime theory $\Delta \supseteq \Gamma$ such that $\varphi \notin \Delta$. By Propositions 4.5, 4.6 and 4.7, we have a compatibility model $M^{\vdash}$ based on a compatibility frame where $\Gamma \not\models_{M^{\vdash}} \varphi$. Hence, $\Gamma \not\models_F \varphi$ for each compatibility frame $F$.

We have seen in Example 4.1 that (NC), our necessary condition for mutual compatibility of states, does not hold in $\mathcal{L}^{\frown}$-models. Our next observation specifies the price of repairing this. Let us call a compatibility frame a *45-frame* iff $C$ is transitive ($xCyCz \to xCz$). Note that in every 45-frame, $C$ is also euclidean ($(xCy \wedge xCz) \to yCz$).[7] 45-models are defined as expected.

**Proposition 4.9.** *In 45-models, $Cxy$ entails that $[\![x]\!]^+ \cap [\![y]\!]^- = \emptyset$ and $[\![x]\!]^- \cap [\![y]\!]^+ = \emptyset$, where $[\![\cdot]\!]^{+(-)}$ are functions from $S$ to $\mathcal{L}^{\frown}$.*

*Proof.* It is sufficient to add the induction step for $\frown$ to the proof of Prop. 3.2. Assume that the claim holds for $\varphi$. First, suppose that $x \models \frown\varphi$ and $y \models \sim\frown\varphi$. Consequently, there is $z \in C(y)$ such that $z \models \varphi$. But $C$ is transitive, so $Cxz$ in case $Cxy$. In that case, however, $x \not\models \frown\varphi$; a contradiction. The second claim follows from the fact that $C$ is euclidean.

From an intuitive viewpoint, however, 45-frames might seem to be too strong. Assume, for example, that $x$ does not support any information about $p$, $y$ supports $p$ and $z$ supports $\sim p$. Moreover, let $x, z, y$ agree on all other atomic propositions. So $xCy$ and $xCz$. By euclideanity of $C$, $yCz$ as well. But this is intuitively incorrect, as $y$ and $z$ support mutually contradictory information. Moreover, transitivity of $C$ implies that every state $x$ compatible with some state, $\exists y(xCy)$, is self-compatible ($xCy \to yCx \to xCx$) and so is every such $y$ ($xCy \to (xCy \wedge xCy) \to yCy$). Again, this might seem too strong.

This is an interesting conclusion. On the one hand, (NC) is a natural necessary condition for two states being compatible. However, (NC) is falsified once we add to the language a negation connective related to the compatibility relation in a usual way. The failure of (NC) in this case is caused by our falsification condition for the compatibility negation; yet, no other plausible candidate for the falsification condition seems available. Moreover, the class of compatibility frames that avoid the failure of (NC) embodies assumptions

[6] If this were not a pair, then we could derive a contradiction as follows. If $\varphi \to \bigvee_{i\leq n}\psi_i$ is in $\mathsf{BK}^{\frown}$, then so is $\frown(\bigvee_{i\leq n}\psi_i) \to \frown\varphi$ (use the transposition rule) and also $\bigwedge_{i\leq n}\frown\psi_i \to \frown\varphi$ (use the De Morgan axiom). But this would mean that $\frown\varphi \in \Gamma$.

[7] Assume $xCy$ and $xCz$. By symmetry $yCx$ and by transitivity $yCz$.

concerning compatibility that are obviously too strong. Does this mean that Belnapian negation $\sim$ and the compatibility negation $\frown$ are somehow "intrinsically incompatible"? We leave this issue open.

## 5 Labeling sources

When dealing with sources of information, it is natural to *categorize* them. For example, it may be useful to divide sources into witness testimonies, physical evidence, scientific theories etc. We can incorporate this idea into our framework in several ways. The present section deals with one, and perhaps the simplest, of them.

Let us from now on distinguish a finite set $L \subseteq AT$ of atomic formulas, intuitively seen as the set of *labels*. Labels represent different categories of sources. These categories might be overlapping and they might not cover the whole set $S$. The former case corresponds to the fact that

$$\bigwedge_{l \neq l'} (l \to \neg l')$$

is not valid (call this formula the "No Overlap Axiom"), the latter case corresponds to the fact that

$$\bigvee_{l_i \in L} l_i$$

is not valid (call this formula the "Exhaustion Axiom").

Labels permit us to express more fine-grained statements about *what kind* of sources support a specific piece of information. For instance, consider

$$\Diamond(l \wedge \varphi), \text{ and} \tag{6}$$

$$\Box(l \to \varphi) \tag{7}$$

(sometimes written also as $\Diamond_l\varphi$ and $\Box_l\varphi$, respectively). Formulas (6) say that some source labeled with $l$ (of category $l$) supports $\varphi$; (7) says that all sources labeled with $l$ (of category $l$) support $\varphi$. The second construction represents a notion of *strong support*. For example, let $l$ be the category "witness reports concerning the whereabouts of John Smith at the time the crime was committed". If only some witnesses place John at the scene of the crime (6), then there might be room for reasonable doubt concerning his whereabouts. If *all* witnesses do so (7), however, then the room is considerably smaller. (Of course, it might be the case that $x \models \Box(l \to \varphi)$ but also $x \models \Diamond(l \wedge {\sim}\varphi)$; in such case there is a witness that contradicts herself.)

Labels can also be used to express *relative reliability* (trustworthiness, priority) of sources. Assume $L = \{l_1, \dots, l_n\}$ where $l_1$ denotes the most reliable sources and $l_n$ the least reliable ones. (Let $O_k\varphi$ be a shorthand for $O_{l_k}\varphi$ for $O \in \{\Diamond, \Box\}$.) We may call $k$ the 'reliability degree' of sources labeled with $l_k$. This machinery allows us to express various natural notions.

**Example 5.1** (Support by the best sources)**.** The formula

$$\dot{\Diamond}_k\varphi := \varphi \wedge \Diamond_k\varphi$$

says that $\varphi$ is supported and also justified by some source of reliability degree $k$. Naturally, then, $\diamondsuit_1\varphi$ says that $\varphi$ is is supported and justified by a source of the highest reliability degree.

The formula

$$\diamondsuit_1\varphi \wedge \Box_1\varphi$$

expresses a stronger notion, namely, that $\varphi$ is supported and justified by *all* sources of the highest reliability degree.[8] The $\Diamond_1$ conjunct makes sure that this statement is not vacuous, i.e. that there is at least one source of the highest reliability degree justifying $\varphi$.

**Example 5.2** (Non-overruled justification). It is natural to assume that if $\Diamond_k\varphi$ and $\Diamond_m{\sim}\varphi$ for some $m < k$, then the justification for $\varphi$ is "overruled" by the justification for ${\sim}\varphi$. Consider

$$\varphi \wedge \bigvee_{1\leq k\leq n}\left(\Diamond_k\varphi \wedge \neg \bigvee_{m\leq k} \Diamond_m{\sim}\varphi\right)$$

This formula says that $\varphi$ has a non-overruled justification: there is a justification for $\varphi$ and there is no justification for ${\sim}\varphi$ that is at least as good as the former justification.

This "hierarchical" interpretation of labels seems to require the validity of the No Overlap Axiom—reliability degrees need to be unique. The status of the Exhaustion Axiom is not so clear; there may be sources that are 'incomparable' as to their reliability to others.

Completeness results are easy to obtain here, but we will not go into details. Consider a language with a finite set of labels $L_m = \{l_1, \ldots, l_m\}$. Consequence in the class of models where $V(l_k) \cap V(l_{k'}) = \emptyset$ for all $k \neq k'$ is axiomatized by adding the No Overlap Axiom schema (for $L_m$) to BK (the canonical model obviously satisfies the condition at hand). Similarly, consequence in the class of models where, in addition, $\bigcup_{k\leq m} V(l_k) = S$ holds is axiomatized by adding also the Exhaustion axiom (for $L_m$).

The only thing that requires care is that the use of label axioms such as the No Overlap Axiom or the Exhaustion Axiom forbids closure under uniform substitution. We may use *sorted substitution* instead, i.e. mappings $\mathcal{L} \to \mathcal{L}$ induced by any $AT \setminus L_m \to \mathcal{L}$.

Note that if the No Overlap Axiom holds in $M$, then the labels induce a preference ordering on $\bigcup_{l\in L} V(l)$ given by $x \leq^L y$ iff $l_x \leq l_y$ (where $l_x$ is *the* label of $x$). It is not hard to see that $\leq^L$ is a pre-order, not a partial order. If the Exhaustion Axiom holds as well, then the whole $S$ is ordered and $\leq^L$ is a total pre-order.

## 6 Conclusion

We have shown that the paraconsistent modal logic BK, equipped with a source-related interpretation of modal accessibility, is suitable for articulating the different ways inconsistency in a body of information may be related to the sources for that body—we may have (i) states that support $\varphi$ and ${\sim}\varphi$ without there being a source supporting either $\varphi$ or ${\sim}\varphi$; (ii) states that support $\varphi$ and ${\sim}\varphi$ while some source also supports both $\varphi$ and ${\sim}\varphi$; and (iii) states that support $\varphi$ and ${\sim}\varphi$ with sources for both $\varphi$ and ${\sim}\varphi$.

This interpretation of BK naturally motivates some of its extensions. We have studied extensions of BK with compatibility (both semantically, where a compatibility relation was added to BK models and some interesting frame classes were studied, and syntactically,

[8] Strictly speaking, justified by all "accessible" sources of the highest reliability degree.

where a modal compatibility negation was added to the language) and with source labels, propositional variables used to capture the idea of a reliability ordering among the sources.

Of course, many other extensions remain to be explored. For instance, we will look in the future on a generalization of the labeling idea using a primitive preference preorder on the set of states in the style of [7, 23, 25]. Adding preorder modalities and conditional belief modalities to our language will enable us to express the notions discussed in Examples 5.1 and 5.2 in a more general setting. It will be also interesting to look at dynamic extensions of our setting, formalizing reasoning about changes in the source relation or the source preference ordering.

**Acknowledgments** Igor Sedlár acknowledges the support of the Czech Science Foundation grant GBP202/12/G061 for the project *Institute of Theoretical Informatics* and the grant JSPS-16-08 for the project *Nonclassical Mathematics Based on Fuzzy, Paraconsistent and Substructural Logics*, provided jointly by the Czech Science Foundation and the Japanese Society for the Promotion of Science. Ondrej Majer acknowledges the support of the grant 16-07954J for the project *SEGA: From shared evidence to group attitudes*, provided jointly by the Czech Science Foundation and the German Science Foundation. Versions of the paper were presented at workshops at the Kyoto University (Kyoto Nonclassical Logic Workshop II, 2016), the University of Amsterdam (LogiCIC Workshop, 2016) and the Institute of Computer Science of the Czech Academy of Sciences (Prague Seminar on Paraconsistent Logic, 2017). We thank the audiences at these events for valuable feedback. We are grateful to the reviewers of the present paper for their suggestions.

## References


1. O. Arieli and A. Avron. Reasoning with logical bilattices. *Journal of Logic, Language and Information*, 5(1):25–63, 1996.
2. S. Artemov. The logic of justification. *The Review of Symbolic Logic*, 1(4):477–513, 2008.
3. N. Belnap. A useful four-valued logic. In J. M. Dunn and G. Epstein, editors, *Modern Uses of Multiple-Valued Logic*, pages 5–37. Springer Netherlands, Dordrecht, 1977.
4. N. Belnap. How a computer should think. In G. Ryle, editor, *Contemporary Aspects of Philosophy*. Oriel Press Ltd., 1977.
5. M. Bílková, O. Majer, and M. Peliš. Epistemic logics for sceptical agents. *Journal of Logic and Computation*, 26(6):1815–1841, 2016.
6. P. Blackburn, M. de Rijke, and Y. Venema. *Modal Logic*. Cambridge University Press, 2001.
7. C. Boutilier. Toward a logic for qualitative decision theory. In J. Doyle, E. Sandewall, and P. Torasso, editors, *Principles of Knowledge Representation and Reasoning*, pages 75–86. 1994.
8. J. M. Dunn. *The Algebra of Intensional Logics*. PhD Thesis, University of Pittsburgh, 1966.
9. J. M. Dunn. Intuitive semantics for first-degree entailments and 'coupled trees'. *Philosophical Studies*, 29(3):149–168, 1976.
10. J. M. Dunn. Star and perp: Two treatments of negation. *Philosophical Perspectives*, 7(1):331–357, 1993.
11. R. Fagin, J. Y. Halpern, Y. Moses, and M. Y. Vardi. *Reasoning about Knowledge*. MIT Press, 1995.
12. J. Hintikka. *Knowledge and Belief. An Introduction to the Logic of the Two Notions*. Cornell University Press, Ithaca, NY, 1962.
13. G. E. Hughes and M. Cresswell. *A New Introduction to Modal Logic*. Routledge, London and New York, 1996.
14. H. Levesque. A logic of implicit and explicit belief. In *Proceedings of AAAI 1984*, pages 198–202, 1984.

15. J.-J. C. Meyer and W. van der Hoek. *Epistemic Logic for AI and Computer Science*. Cambridge University Press, 1995.
16. S. P. Odintsov and E. I. Latkin. BK-lattices. Algebraic semantics for Belnapian modal logics. *Studia Logica*, 100(1):319–338, 2012.
17. S. P. Odintsov and S. O. Speranski. The lattice of Belnapian modal logics: Special extensions and counterparts. *Logic and Logical Philosophy*, 25(1):3–33, 2016.
18. S. P. Odintsov and S. O. Speranski. Belnap-Dunn modal logics: truth constants vs. truth values. Sumbitted, 2018.
19. S. P. Odintsov and H. Wansing. Modal logics with Belnapian truth values. *Journal of Applied Non-Classical Logics*, 20(3):279–301, 2010.
20. S. P. Odintsov and H. Wansing. Disentangling FDE-based paraconsistent modal logics. *Studia Logica*, 105(6):1221–1254, 2017.
21. G. Restall. *An Introduction to Substrucutral Logics*. Routledge, London, 2000.
22. I. Sedlár. Propositional dynamic logic with Belnapian truth values. In *Advances in Modal Logic. Vol. 11*, London, 2016. College Publications.
23. J. van Benthem, P. Girard, and O. Roy. Everything else being equal: A modal logic for ceteris paribus preferences. *Journal of Philosophical Logic*, 38(1):83–125, 2009.
24. J. van Benthem and M. Martinez. The stories of logic and information. In P. Adriaans and J. van Benthem, editors, *Philosophy of Information*, pages 217–280. Elsevier, 2008.
25. J. van Benthem, S. van Otterloo, and O. Roy. Preference logic, conditionals, and solution concepts in games. In H. Lagerlund, S. Lindström, and R. Sliwinski, editors, *Modality matters: Twenty-Five Essays in Honour of Krister Segerberg*. Uppsala Philosophical Studies, 2006.

# First-Degree Entailment and Structural Reasoning

Yaroslav Shramko

**Abstract** In this paper it is shown how the logic of first-degree entailment of Anderson and Belnap can be represented by a variety of deductively equivalent binary (Fmla-Fmla) consequence systems, up to a system with transitivity as the only inference rule. Some possible extensions of these systems are briefly considered as well.

**Keywords** First-degree entailment • Structural reasoning • Binary consequence system • Exactly true logic • Non-falsity logic.

## 1 Introduction: first-degree entailment, binary consequence systems, and four truth values

Let the propositional language $\mathcal{L}$ be defined as follows:

$$\varphi ::= p \mid \varphi \wedge \varphi \mid \varphi \vee \varphi \mid {\sim}\varphi.$$

A *binary consequence expression* (or simply *binary consequence*) is an expression of the form $\varphi \vdash \psi$ (to be read as "$\varphi$ has $\psi$ as a consequence" [8, p. 302]), where $\varphi, \psi \in \mathcal{L}$. A *binary consequence system* can be defined then as a proof system, which manipulates binary consequences as formal objects, cf. [9, p. 24]. A binary consequence is a particular case of a Gentzenian "sequent" where both antecedent and succedent are restricted to singletons. In this respect, a binary consequence system is a "sequent style system" [24, p. 246] constructed in a Fmla-Fmla "logical framework" (by employing terminology from [17, p. 108]).

A *logic* over the given language (in the Fmla-Fmla setting) is meant to be a set of binary consequences closed at least under the usual Tarskian conditions of *reflexivity* and *transitivity*:

(*ref*) $\varphi \vdash \varphi$

(*tr*) $\varphi \vdash \psi,\ \psi \vdash \chi \Rightarrow \varphi \vdash \chi$

as well as Łoś and Suszko's condition of *structurality* (substitution-invariance, for every uniform substitution function $s$ on $\mathcal{L}$):

(*si*) $\varphi \vdash \psi \Rightarrow s(\varphi) \vdash s(\psi)$.

Moreover, a logic is said to be *consistent* (or non-trivial) whenever $\vdash \neq \mathcal{L} \times \mathcal{L}$. A particular consequence system is consistent if the logic yield by this system (the set of all valid consequences derived in this system) is consistent.

Yaroslav Shramko
Department of Philosophy, Kryvyi Rih State Pedagogical University, Kryvyi Rih, Ukraine
e-mail: shramko@rocketmail.com

 
H. Omori, H. Wansing (eds.), *New Essays on Belnap-Dunn Logic*,
Synthese Library 418, https://doi.org/10.1007/978-3-030-31136-0_18

Clearly, the same logic can be yield by various consequence systems. If several binary consequence systems determine the same set of provable consequences, these systems are said to be *deductively equivalent*.

The present paper will focus on the logic of *first-degree entailment* by Anderson and Belnap constituted by the set of *tautological entailments* as defined, e.g., in [2]. This set can be generated in various ways, in particular by a binary consequence system defined as follows:

System $\mathbf{E}_{\mathrm{fde}}$

$(ce_1)$ $\varphi \wedge \psi \vdash \varphi$

$(ce_2)$ $\varphi \wedge \psi \vdash \psi$

$(di_1)$ $\varphi \vdash \varphi \vee \psi$

$(di_2)$ $\psi \vdash \varphi \vee \psi$

$(dis_1)$ $\varphi \wedge (\psi \vee \chi) \vdash (\varphi \wedge \psi) \vee \chi$

$(ni)$ $\varphi \vdash {\sim}{\sim}\varphi$

$(ne)$ ${\sim}{\sim}\varphi \vdash \varphi$

$(tr)$ $\varphi \vdash \psi,\ \psi \vdash \chi \,/\, \varphi \vdash \chi$

$(ci)$ $\varphi \vdash \psi,\ \varphi \vdash \chi \,/\, \varphi \vdash \psi \wedge \chi$

$(de)$ $\varphi \vdash \chi,\ \psi \vdash \chi \,/\, \varphi \vee \psi \vdash \chi$

$(con)$ $\varphi \vdash \psi \,/\, {\sim}\psi \vdash {\sim}\varphi$

It is essentially the original formulation from [3, p. 158], where this system is characterized as "a Hilbert-style formalism" with "seven axioms and four rules", conceived as "the first degree entailment fragment of the calculus **E**" (hence its name).[1] Anderson and Belnap use an "entailment connective" $\rightarrow$ instead of $\vdash$, with "first-degree entailments" considered to be formulas of the form $\varphi \rightarrow \psi$, where $\varphi$ and $\psi$ "can be truth functions of any degree but cannot contain any arrows" [3, p. 150]. Thus, strictly speaking, first-degree entailments are expressions about the consequence relation between single (Boolean, to wit, truth-functional) formulas.

Notably, every "truth function" is characterized here by a group of exactly three principles, two of which being the "direct consequences" taken as axioms: $(ce_1)$, $(ce_2)$ for conjunction elimination; $(di_1)$, $(di_2)$ for disjunction introduction; $(ni)$, $(ne)$ for negation introduction and elimination, and another one formulated as a rule of inference, saying how to obtain some consequence with certain connective from other consequence(s): $(ci)$ for conjunction introduction; $(de)$ for disjunction elimination and $(con)$ for negation contraposition.

There is also an additional axiom of distributivity $(dis_1)$ reflecting an interconnection between conjunction and disjunction, and one inference rule of transitivity $(tr)$, which involves no propositional connectives and deals purely with the consequence relation as such. Thus, by using Gentzen's terminology, $(ci)$, $(de)$ and $(con)$ are "rules for logical sym-

[1] The name $\mathbf{E}_{\mathrm{fde}}$, as well as $\mathbf{R}_{\mathrm{fde}}$ below, may look a little old-fashioned and has fallen out of favor with most people lately. It is more common nowadays to use the abbreviation FDE for the logic of first-degree entailment in general. The aim of this paper is to show how this logic can be equivalently determined by various logical systems. Thus, by employing the mentioned abbreviation (FDE) as a generic term for the logic under consideration, I will use various adscripts to it, to mark the corresponding systems, while retaining the historical labels for the two original formulations.

bols" ("Logische-Zeichen-Schlußfiguren"), see [14, p. 191],[2] and (*tr*) is the only "structural inference rule" [ibid].

An inference (proof) in $\mathbf{E}_{\mathrm{fde}}$ is a finite list of consequence expressions where every list item is either an axiom or results from preceding elements of the list by an inference rule application. It is well known that four De Morgan laws ($\sim\varphi \wedge \sim\psi \vdash \sim(\varphi \vee \psi)$, $\sim(\varphi \wedge \psi) \vdash \sim\varphi \vee \sim\psi$, $\sim\varphi \vee \sim\psi \vdash \sim(\varphi \wedge \psi)$, $\sim(\varphi \vee \psi) \vdash \sim\varphi \wedge \sim\psi$) are derivable in $\mathbf{E}_{\mathrm{fde}}$. Moreover, $\mathbf{E}_{\mathrm{fde}} + \varphi \wedge \sim\varphi \vdash \psi$ amounts to a consequence system of classical logic, cf. [25, pp. 255-256]. (Clearly, $\psi \vdash \varphi \vee \sim\varphi$ becomes then derivable by (*con*).)

$\mathbf{E}_{\mathrm{fde}}$ is a rather strong system that hardly allows interesting non-classical extensions. The only non-classical consistent extension of $\mathbf{E}_{\mathrm{fde}}$ I am aware of can be obtained by adding to it the following axiom (see, e.g., [7, p. 157 and note 7] and [13, p. 53]):

(*saf*) $\varphi \wedge \sim\varphi \vdash \psi \vee \sim\psi$.

Label (*saf*) stands for "safety", see explanations in [11, p. 443]. Safety is the characteristic principle for the first-degree entailment fragments of both "R-mingle" and "E-mingle", the Dunn–McCall systems obtained by extending systems **R** and **E** by the "mingle axiom" $\varphi \to (\varphi \to \varphi)$ and "restricted mingle axiom" $(\varphi \to \psi) \to ((\varphi \to \psi) \to (\varphi \to \psi))$ respectively.[3] I will mark the system obtained by extending $\mathbf{E}_{\mathrm{fde}}$ with (*saf*) as $\mathbf{EM}_{\mathrm{fde}}$.[4]

To make the logic of first-degree entailment more "flexible", and particularly more sensitive to possible extensions, one could try to give up some of its initial inference rules, and first of all, contraposition. In this way we obtain another well-known formulation of this logic without the contraposition rule, but with De Morgan laws taken instead as axioms. One can find this formulation, e.g. in [10, p. 12]:

System $\mathbf{R}_{\mathrm{fde}}$

| | |
|---|---|
| $(ce_1)$ | $\varphi \wedge \psi \vdash \varphi$ |
| $(ce_2)$ | $\varphi \wedge \psi \vdash \psi$ |
| $(di_1)$ | $\varphi \vdash \varphi \vee \psi$ |
| $(di_2)$ | $\psi \vdash \varphi \vee \psi$ |
| $(dis_1)$ | $\varphi \wedge (\psi \vee \chi) \vdash (\varphi \wedge \psi) \vee \chi$ |
| $(ni)$ | $\varphi \vdash \sim\sim\varphi$ |
| $(ne)$ | $\sim\sim\varphi \vdash \varphi$ |
| $(dm_1)$ | $\sim(\varphi \vee \psi) \vdash \sim\varphi \wedge \sim\psi$ |
| $(dm_2)$ | $\sim\varphi \wedge \sim\psi \vdash \sim(\varphi \vee \psi)$ |
| $(dm_3)$ | $\sim(\varphi \wedge \psi) \vdash \sim\varphi \vee \sim\psi$ |

[2] Schroeder-Heister [24, p. 247] characterizes rules of this kind as "rules for logical connectives", which is completely adequate if we stay on a propositional level or treat quantifiers as a kind of logical connectives. English translation of [14] in [15, p. 82] specifies these rules as "operational", comprising thus a hefty dose of interpretation rather than translation proper. In this paper, they will be called "logical", for brevity sake.

[3] For more explanations about mingle, and restricted mingle consult, e.g., [3, pp. 97-99]. For the definitions of systems **EM** and **RM** see, e.g., [3, pp. 340, 394]. The first appearance of system **RM** in print is [6], where also the essence of an inference of safety by using the mingle axiom is presented, see the proof of T5 in [6, p. 4] with its attribution to Meyer (cf. also [3, p. 429]).

[4] My conjecture would be that $\mathbf{EM}_{\mathrm{fde}}$ is the *maximal binary system* with respect to the classical binary consequence system, in the sense that it does not allow non-classical non-trivial extensions (i.e., adding to $\mathbf{EM}_{\mathrm{fde}}$ any binary consequence that is not provable in it gives either classical binary consequences system or is trivial). Since this conjecture deserves special consideration, it will not be examined here.

$(dm_4)$ $\sim\varphi \vee \sim\psi \vdash \sim(\varphi \wedge \psi)$

$(tr)$ $\varphi \vdash \psi,\ \psi \vdash \chi \,/\, \varphi \vdash \chi$

$(ci)$ $\varphi \vdash \psi,\ \varphi \vdash \chi \,/\, \varphi \vdash \psi \wedge \chi$

$(de)$ $\varphi \vdash \chi,\ \psi \vdash \chi \,/\, \varphi \vee \psi \vdash \chi$

Dunn uses the label $\mathbf{R}_{\mathrm{fde}}$ to highlight the fact that the first-degree entailment fragments of systems **R** and **E** are the same. He also shows that contraposition, although *not derivable* in $\mathbf{R}_{\mathrm{fde}}$, is still *admissible*, see [10, Proposition 11]. Thus, $\mathbf{E}_{\mathrm{fde}}$ and $\mathbf{R}_{\mathrm{fde}}$ are deductively equivalent.[5] Yet, the latter system is in a way weaker than the former, so that it has fewer derivable rules, and allows thus certain non-classical (and non-trivial) extensions, which are impossible with $\mathbf{E}_{\mathrm{fde}}$.

Namely, as stated in [10, Theorem 12], $\mathbf{R}_{\mathrm{fde}} + \varphi \wedge \sim\varphi \vdash \psi$ gives us the first-degree entailment fragment of Kleene's logic,[6] and $\mathbf{R}_{\mathrm{fde}} + \psi \vdash \varphi \vee \sim\varphi$ results in the first-degree entailment fragment of Priest's "logic of paradox". Clearly, the contraposition rule is no longer admissible in these extensions of $\mathbf{R}_{\mathrm{fde}}$.

Now, let us briefly run through a semantic characterization of the first-degree entailment. According to Dunn's "intuitive semantics" [7], a valuation $v$ for any $\varphi \in \mathcal{L}$ can be defined as a map from the set of propositional variables to the *subsets* of the set of classical truth-values $\{t, f\}$, with the following truth and falsity conditions for $\wedge, \vee, \sim$:

(1) $t \in v(\varphi \wedge \psi) \Leftrightarrow t \in v(\varphi)$ and $t \in v(\psi)$,
$f \in v(\varphi \wedge \psi) \Leftrightarrow f \in v(\varphi)$ or $f \in v(\psi)$;
(2) $t \in v(\varphi \vee \psi) \Leftrightarrow t \in v(\varphi)$ or $t \in v(\psi)$,
$f \in v(\varphi \vee \psi) \Leftrightarrow f \in v(\varphi)$ and $f \in v(\psi)$;
(3) $t \in v(\sim\varphi) \Leftrightarrow f \in v(\varphi)$,
$f \in v(\sim\varphi) \Leftrightarrow t \in v(\varphi)$.

Belnap's four truth values $B, T, F, N$ (cf. [4, 5]) are then explicated as follows:

$v(\varphi) = B \Leftrightarrow t \in v(\varphi)$ and $f \in v(\varphi)$,
$v(\varphi) = T \Leftrightarrow t \in v(\varphi)$ and $f \notin v(\varphi)$,
$v(\varphi) = F \Leftrightarrow t \notin v(\varphi)$ and $f \in v(\varphi)$,
$v(\varphi) = N \Leftrightarrow t \notin v(\varphi)$ and $f \notin v(\varphi)$.

Consider the following definition of an entailment relation:

*Definition* 1.1. $\varphi \models_{\mathcal{FDE}} \psi =_{df} \forall v : t \in v(\varphi) \Rightarrow t \in v(\psi)$.

Let $\varphi \vdash_{\mathrm{Efde}} \psi$ means that $\varphi \vdash \psi$ is provable in $\mathbf{E}_{\mathrm{fde}}$, and analogously for $\mathbf{R}_{\mathrm{fde}}$. We have then the following soundness and completeness result for both $\mathbf{E}_{\mathrm{fde}}$ and $\mathbf{R}_{\mathrm{fde}}$:

*Theorem* 1.2. For every $\varphi, \psi$: $\varphi \vdash_{Efde} \psi \Leftrightarrow \varphi \models_{\mathcal{FDE}} \psi \Leftrightarrow \varphi \vdash_{Rfde} \psi$.

**Proof.** See the proof of Theorem 7 in [10]. ■

*Observation* 1.3. The entailment relation $\models_{\mathcal{FDE}}$ as defined by Definition 1.1, explicitly takes $\{T, B\}$ as the set of designated truth values. That is, Definition 1.1 can be reformulated as follows: $\varphi \models_{\mathcal{FDE}} \psi =_{df} \forall v : v(\varphi) \in \{T, B\} \Rightarrow v(\psi) \in \{T, B\}$.

[5] Moreover, it is not difficult to extend Dunn's proof to the safety axiom, so that contraposition remains admissible in the system $\mathbf{RM}_{\mathrm{fde}} = \mathbf{R}_{\mathrm{fde}} + (saf)$. Thus, $\mathbf{EM}_{\mathrm{fde}}$ and $\mathbf{RM}_{\mathrm{fde}}$ are deductively equivalent as well. Dunn in [10, p. 15] observes that the logic determined by $\mathbf{RM}_{\mathrm{fde}}$ "is also the first-degree entailment fragment of Łukasiewicz's 3-valued logic, and also the logical analogue of the algebraic system of Kalman [19]".

[6] Known also as "Hao Wang's logic" as formalized by Alan Rose in [23].

A direct comparison of $\mathbf{E}_{\text{fde}}$ and $\mathbf{R}_{\text{fde}}$ shows that deleting contraposition from the list of initial inference rules, and compensating it by a set of suitable axioms (specifically, De Morgan laws), allows for some new non-classical extensions of the first-degree entailment logic. Moreover, there are only *two* non-structural inference rules left in $\mathbf{R}_{\text{fde}}$, for conjunction introduction and disjunction elimination. Could we possibly take this one step further, and obtain new interesting systems for the logic of first-degree entailment that will allow more non-trivial extensions?

## 2 First-degree entailment without the disjunction elimination and contraposition rules

Consider the following system of first-degree entailment with conjunction introduction as the only *logical* inference rule (together with the structural rule of transitivity):

System **FDE**(ci)

| | |
|---|---|
| $(ce_1)$ | $\varphi \wedge \psi \vdash \varphi$ |
| $(ce_2)$ | $\varphi \wedge \psi \vdash \psi$ |
| $(di_1)$ | $\varphi \vdash \varphi \vee \psi$ |
| $(dco)$ | $\varphi \vee \psi \vdash \psi \vee \varphi$ |
| $(did)$ | $\varphi \vee \varphi \vdash \varphi$ |
| $(das)$ | $\varphi \vee (\psi \vee \chi) \vdash (\varphi \vee \psi) \vee \chi$ |
| $(dis_2)$ | $\varphi \vee (\psi \wedge \chi) \vdash (\varphi \vee \psi) \wedge (\varphi \vee \chi)$ |
| $(dis_3)$ | $(\varphi \vee \psi) \wedge (\varphi \vee \chi) \vdash \varphi \vee (\psi \wedge \chi)$ |
| $(dni)$ | $\varphi \vee \psi \vdash {\sim\sim}\varphi \vee \psi$ |
| $(dne)$ | ${\sim\sim}\varphi \vee \psi \vdash \varphi \vee \psi$ |
| $(ddm_1)$ | ${\sim}(\varphi \vee \psi) \vee \chi \vdash ({\sim}\varphi \wedge {\sim}\psi) \vee \chi$ |
| $(ddm_2)$ | $({\sim}\varphi \wedge {\sim}\psi) \vee \chi \vdash {\sim}(\varphi \vee \psi) \vee \chi$ |
| $(ddm_3)$ | ${\sim}(\varphi \wedge \psi) \vee \chi \vdash ({\sim}\varphi \vee {\sim}\psi) \vee \chi$ |
| $(ddm_4)$ | $({\sim}\varphi \vee {\sim}\psi) \vee \chi \vdash {\sim}(\varphi \wedge \psi) \vee \chi$ |
| $(tr)$ | $\varphi \vdash \psi,\ \psi \vdash \chi \ /\ \varphi \vdash \chi$ |
| $(ci)$ | $\varphi \vdash \psi,\ \varphi \vdash \chi \ /\ \varphi \vdash \psi \wedge \chi$ |

**FDE**(ci) is obtained from $\mathbf{R}_{\text{fde}}$ by removing disjunction elimination (*de*) from the list of initial inference rules, and compensating this loss by a stock of additional axiom schemata, most crucially, "disjunctive versions" of the double negation and De Morgan laws.

It is instructive to compare this system with Font's formulation of Belnap's logic as a "Hilbert-style axiomatization" $\vdash_H$ in [12]. An important difference between **FDE**(ci) and $\vdash_H$ is that the former is a binary consequence system of FMLA-FMLA type, whereas the latter deals with SET-FMLA consequence expressions. Thus, **FDE**(ci) incorporates the principle of conjunction introduction in a form of an indirect inference rule (*ci*), as distinct from $\vdash_H$, where this principle is accepted as a direct consequence expression with two formulas in the antecedent (a two-premises direct rule of natural deduction): $\varphi, \psi \vdash \varphi \wedge \psi$.

The following lemma, which enables us to get rid of redundant disjunctions, is inspired by Proposition 3.2 from [12]:

*Lemma* 2.1. For every schema (*dni*), (*dne*), ($ddm_1$)–($ddm_4$) of the form $\varphi \vee \chi \vdash \psi \vee \chi$ the following consequences are derivable in ***FDE(ci)***:

(a) $\varphi \vdash \psi$;

(b) $\varphi \wedge \chi \vdash \psi \wedge \chi$.

**Proof.** (a) Consider any schema ($x$) from (*dni*), (*dne*), ($ddm_1$)–($ddm_4$) of the form $\varphi \vee \chi \vdash \psi \vee \chi$. We have the following schema of an inference in **FDE**(ci):

| | |
|---|---|
| 1. $\varphi \vee \psi \vdash \psi \vee \psi$ | ($x$) |
| 2. $\psi \vee \psi \vdash \psi$ | (*did*) |
| 3. $\varphi \vdash \varphi \vee \psi$ | ($di_1$) |
| 4. $\varphi \vee \psi \vdash \psi$ | 1, 2: (*tr*) |
| 5. $\varphi \vdash \psi$ | 3, 4: (*tr*) |

(b) We have the following schema of an inference in **FDE**(ci):

| | |
|---|---|
| 1. $\varphi \wedge \chi \vdash \varphi$ | ($ce_1$) |
| 2. $\varphi \vdash \psi$ | (a) |
| 3. $\varphi \wedge \chi \vdash \psi$ | 1, 2: (*tr*) |
| 4. $\varphi \wedge \chi \vdash \chi$ | ($ce_2$) |
| 5. $\varphi \wedge \chi \vdash \psi \wedge \chi$ | 3, 4: (*ci*) |

■

In what follows $\varphi \vdash_{\text{ci}} \psi$ means that $\varphi \vdash \psi$ is derivable in **FDE**(ci).

*Lemma* 2.2. The following rules (disjunction elimination and contraposition) are admissible in ***FDE(ci)***:

| | |
|---|---|
| (*de*) | If $\varphi \vdash_{ci} \chi$ and $\psi \vdash_{ci} \chi$, then $\varphi \vee \psi \vdash_{ci} \chi$; |
| (*con*) | If $\varphi \vdash_{ci} \psi$, then $\sim\psi \vdash_{ci} \sim\varphi$. |

**Proof.** (1) To prove admissibility of (*de*) first prove that the following rule is admissible in **FDE**(ci): if $\varphi \vdash_{\text{ci}} \psi$, then $\varphi \vee \chi \vdash_{\text{ci}} \psi \vee \chi$ ("Font's rule", cf. [12, Proposition 3.3]). To demonstrate this one has to show how an inference of $\varphi \vdash \psi$ in **FDE**(ci) can be transformed into the inference of $\varphi \vee \chi \vdash \psi \vee \chi$. This can be proved by induction on the length of the initial inference of $\varphi \vdash \psi$. Namely, it can be shown that replacing every step of the form $\gamma \vdash \xi$ in the given inference of $\varphi \vdash \psi$ by a derivation of the consequence $\gamma \vee \chi \vdash \xi \vee \chi$ will give an inference of $\varphi \vee \chi \vdash \psi \vee \chi$ in **FDE**(ci).

Indeed, it is not difficult to see that for every axiom schema $\gamma \vdash \xi$ of **FDE**(ci) the consequence $\gamma \vee \chi \vdash \xi \vee \chi$ is derivable in **FDE**(ci).

Next, for each inference rule (*tr*), (*ci*) it can be easily demonstrated that if one transforms every premise of the form $\gamma \vdash \varphi$ into the consequence $\gamma \vee \chi \vdash \varphi \vee \chi$, then provided that premises so transformed are derivable in **FDE**(ci), the analogous transformation $\delta \vee \chi \vdash \psi \vee \chi$ of the conclusion of the form $\delta \vdash \psi$ will be derivable as well.

Finally, the following schema of an inference in **FDE**(ci) completes the case:

| | |
|---|---|
| 1. $\varphi \vdash_{\text{ci}} \chi$ | (*assumption*) |
| 2. $\psi \vdash_{\text{ci}} \chi$ | (*assumption*) |
| 3. $\varphi \vee \psi \vdash_{\text{ci}} \chi \vee \psi$ | 1: (Font's rule) |
| 4. $\chi \vee \psi \vdash_{\text{ci}} \chi \vee \chi$ | 2: (Font's rule) |
| 5. $\chi \vee \chi \vdash_{\text{ci}} \chi$ | (*did*) |
| 6. $(\varphi \vee \psi) \vdash_{\text{ci}} \chi$ | (3, 4, 5: (*tr*), twice) |

(2) To show an admissibility of (*con*) one has to demonstrate, how an inference of $\varphi \vdash_{\mathrm{ci}} \psi$ can be transformed into the inference of $\sim\psi \vdash_{\mathrm{ci}} \sim\varphi$. This can be done by induction on the length of the initial inference of $\varphi \vdash_{\mathrm{ci}} \psi$ (cf. [10, Proposition 11]). First, observe that the contraposition of every axiom schema of **FDE**(ci) is derivable in **FDE**(ci) (one can use De Morgan laws, derivable in **FDE**(ci) by Lemma 2.1).

Moreover, for each inference rule one can show, that if the contrapositions of its premises are derivable in **FDE**(ci), so is the contraposition of the conclusion. Indeed, for (*tr*) (transitivity) it is easy to see that contraposing its premises and conclusion results again in transitivity. And for (*ci*) the admissibility of the disjunction elimination together with De Morgan laws can be applied to get the result. ■

*Lemma* 2.3. Systems $\boldsymbol{E}_{fde}$, $\boldsymbol{R}_{fde}$ and ***FDE(ci)*** are all deductively equivalent.

**Proof.** Using Theorem 7, Proposition 11 from [10], as well as Lemmas 2.1 and 2.2. ■

Finally, we have the completeness result for **FDE**(ci):

*Theorem* 2.4. For every $\varphi, \psi$: $\varphi \vdash_{ci} \psi \Leftrightarrow \varphi \vDash_{\mathcal{FDE}} \psi$.

**Proof.** By Theorem 1.2 and Lemma 2.3. ■

## 3 Nothing but the Truth in a Fmla-Fmla setting

Since the rule of disjunction elimination is not derivable in **FDE**(ci), it allows for some interesting extensions not performable on the basis of $\mathbf{R}_{\mathrm{fde}}$. In particular, a Fmla-Fmla version of "exactly true logic" by Pietz and Rivieccio can be obtained as a straightforward extension of **FDE**(ci).

Pietz[7] and Rivieccio [22] ask for a consequence relation "that preserves *truth-and-non-falsity*", i.e., takes the Belnapian truth value $T$ as *the only* designated truth value. They formalize this relation by a "Hilbert-style proof system" **ETL**, which is an extension of system $\vdash_H$ from [12] by a rule of *disjunctive syllogism*, and suggest that it could not be easy "to find a nice sequent calculus for this logic" [22, p. 129]. Remarkably, they consider a relation of Set-Fmla type, ignoring a tradition of constructing first-degree entailment as a binary Fmla-Fmla consequence system.[8]

In [27, pp. 1302-1304] we reconstructed **ETL** in the form of a Set-Fmla consequence system, and raised further question (pp. 1314-1315) on formalizing a relation determined by the following definition:

*Definition* 3.1. $\varphi \vDash_{\mathcal{T}} \psi =_{df} \forall v : v(\varphi) = T \Rightarrow v(\psi) = T$.

Now, it is natural to suppose that the corresponding Fmla-Fmla system $\mathbf{ETL}_1$ can be obtained by extending **FDE**(ci) with the following axiom (disjunctive syllogism):

$(ds) \quad \sim\varphi \wedge (\varphi \vee \psi) \vdash \psi.$

Needless to say, neither (*de*) nor (*con*) is admissible in $\mathbf{ETL}_1$ any more. Theorem 3.4 in [22] establishes the completeness of Pietz and Rivieccio's system **ETL** with respect to

[7] After the name change—Andreas Kapsner.

[8] But see also an analysis of entailment relations of the Set-Set type by Marcos in [21], based on the sets of designated truth values $\{T\}$, $\{T, B\}$ and $\{N, T, B\}$. [26] presents an analysis of various Fmla-Fmla entailment relations based on designated and antidesignated subsets from $\{N, T, B, F\}$ and interconnections between them.

the SET-FMLA version of relation $\models_{\mathcal{T}}$. Taking into account a direct analogy between $\vdash_H$ and **FDE**(ci) on the one side, and between **ETL** and **ETL**$_1$ on the other, it should be a matter of a routine exercise to adapt the proof of Theorem 3.4 from [22] to a FMLA-FMLA setting, and to obtain thus the following ($\varphi \vdash_{\text{etl}} \psi$ naturally means that $\varphi \vdash \psi$ is provable in **ETL**$_1$):

*Theorem* 3.2. For every $\varphi, \psi$: $\varphi \vdash_{etl} \psi \Leftrightarrow \varphi \models_{\mathcal{T}} \psi$.

## 4 First-degree entailment without the conjunction introduction and contraposition rules

Duality between the rules of conjunction introduction and disjunction elimination suggests construction of another version of the logic of first-degree entailment with only one logical inference rule (accompanied by transitivity), but now for disjunction elimination.

System **FDE**(de)

| | |
|---|---|
| $(di_1)$ | $\varphi \vdash \varphi \vee \psi$ |
| $(di_2)$ | $\psi \vdash \varphi \vee \psi$ |
| $(ce_1)$ | $\varphi \wedge \psi \vdash \varphi$ |
| $(cco)$ | $\varphi \wedge \psi \vdash \psi \wedge \varphi$ |
| $(cid)$ | $\varphi \vdash \varphi \wedge \varphi$ |
| $(cas)$ | $(\varphi \wedge \psi) \wedge \chi \vdash \varphi \wedge (\psi \wedge \chi)$ |
| $(dis_4)$ | $(\varphi \wedge \psi) \vee (\varphi \wedge \chi) \vdash \varphi \wedge (\psi \vee \chi)$ |
| $(dis_5)$ | $\varphi \wedge (\psi \vee \chi) \vdash (\varphi \wedge \psi) \vee (\varphi \wedge \chi)$ |
| $(cne)$ | $\sim\sim\varphi \wedge \psi \vdash \varphi \wedge \psi$ |
| $(cni)$ | $\varphi \wedge \psi \vdash \sim\sim\varphi \wedge \psi$ |
| $(cdm_4)$ | $(\sim\varphi \vee \sim\psi) \wedge \chi \vdash \sim(\varphi \wedge \psi) \wedge \chi$ |
| $(cdm_3)$ | $\sim(\varphi \wedge \psi) \wedge \chi \vdash (\sim\varphi \vee \sim\psi) \wedge \chi$ |
| $(cdm_2)$ | $(\sim\varphi \wedge \sim\psi) \wedge \chi \vdash \sim(\varphi \vee \psi) \wedge \chi$ |
| $(cdm_1)$ | $\sim(\varphi \vee \psi) \wedge \chi \vdash (\sim\varphi \wedge \sim\psi) \wedge \chi$ |
| $(tr)$ | $\varphi \vdash \psi, \psi \vdash \chi / \varphi \vdash \chi$ |
| $(de)$ | $\varphi \vdash \chi, \psi \vdash \chi / \varphi \vee \psi \vdash \chi$ |

By dualizing the proof of Lemma 2.1 one easily obtains:

*Lemma* 4.1. For every schema (*cne*), (*cni*), ($cdm_4$)–($cdm_1$) of the form $\varphi \wedge \chi \vdash \psi \wedge \chi$ the following consequences are derivable in ***FDE(de)***:

(a) $\varphi \vdash \psi$;

(b) $\varphi \vee \chi \vdash \psi \vee \chi$.

Also the following can be proved analogously to Lemma 2.2:

*Lemma* 4.2. The rules of conjunction introduction and contraposition are admissible in ***FDE(de)***:

| | |
|---|---|
| $(ci)$ | If $\varphi \vdash_{de} \psi$ and $\varphi \vdash_{de} \chi$, then $\varphi \vdash_{de} \psi \wedge \chi$; |
| $(con)$ | If $\varphi \vdash_{de} \psi$, then $\sim\psi \vdash_{de} \sim\varphi$. |

**FDE**(de) is indeed an adequate formalization of first-degree entailment, as the following lemma shows:

*Lemma* 4.3. Systems $\boldsymbol{E}_{fde}$, $\boldsymbol{R}_{fde}$, ***FDE(ci)*** and ***FDE(de)*** are all deductively equivalent.

**Proof.** Using Theorem 7, Proposition 11 from [10], together with Lemmata 2.1, 2.2, 4.1 and 4.2. ■

Hence, **FDE**(de) is sound and complete with respect to the entailment relation determined by Definition 1.1:

*Theorem* 4.4. For every $\varphi, \psi$: $\varphi \vdash_{de} \psi \Leftrightarrow \varphi \models_{\mathcal{FDE}} \psi$.

## 5 Anything but Falsehood in a Fmla-Fmla setting

The absence of conjunction introduction among the initial inference rules of **FDE**(de) enables the possibility of extending it in other way as compared with **FDE**(ci). In [27, pp. 1307-1309] we considered a "non-falsity logic" **NFL** dual to **ETL**, and also raised (pp. 1314-1315) a question of formalizing a relation determined by the following definition:

*Definition* 5.1. $\varphi \models_{\mathcal{F}} \psi =_{df} \forall v : v(\psi) = F \Rightarrow v(\varphi) = F$.

This definition introduces a Fmla-Fmla entailment relation based on the set of designated truth values $\{T, B, N\}$, preserving thus the property of having any of the four Belnapian values, except the bare falsehood $F$. In this respect, $\models_{\mathcal{F}}$ is dual to the relation determined by Definition 3.1. One obtains the corresponding binary consequence system of non-falsity logic $\mathbf{NFL}_1$ by extending **FDE**(de) with the following axiom (dual disjunctive syllogism):

$(dds)$ $\quad \varphi \vdash {\sim}\psi \vee (\psi \wedge \varphi)$.

*Theorem* 5.2. For every $\varphi, \psi$: $\varphi \vdash_{ntl} \psi \Leftrightarrow \varphi \models_{\mathcal{F}} \psi$.

**Proof.** By dualizing Theorem 3.2. ■

## 6 First-degree entailment without logical rules

Anderson and Belnap, when characterizing their system $\mathbf{E}_{\mathrm{fde}}$ as "a Hilbert-style formalism", observe: "This formulation suffers, from a proof-theoretical point of view, in having lots of rules; Hilbert would have preferred just one" [3, p. 158]. Well, both **FDE**(ci) and **FDE**(de) rectify this shortcoming significantly by reducing the number of rules to two, but still, they fail to meet "Hilbert's ideal" of *rule singularity*.

Remarkably, both systems have the same structural rule of transitivity (*tr*) in common, and each of them additionally has its own logical rule absent in the other system. This removability of logical inference rules by replacing them with appropriate axioms for the corresponding propositional connectives suggests another formalization of the first-degree entailment logic with *the only* (structural) inference rule of transitivity, obtained by a straightforward combination of **FDE**(ci) and **FDE**(de). The resulting consequence system fulfills "Hilbert's dream".

System **FDE**(-)

$(di_1)$ $\varphi \vdash \varphi \vee \psi$

$(dco)$ $\varphi \vee \psi \vdash \psi \vee \varphi$

$(did)$ $\varphi \vee \varphi \vdash \varphi$

$(das)$ $\varphi \vee (\psi \vee \chi) \vdash (\varphi \vee \psi) \vee \chi$

$(ce_1)$ $\varphi \wedge \psi \vdash \varphi$

$(cco)$ $\varphi \wedge \psi \vdash \psi \wedge \varphi$

$(cid)$ $\varphi \vdash \varphi \wedge \varphi$

$(cas)$ $(\varphi \wedge \psi) \wedge \chi \vdash \varphi \wedge (\psi \wedge \chi)$

$(dis_2)$ $\varphi \vee (\psi \wedge \chi) \vdash (\varphi \vee \psi) \wedge (\varphi \vee \chi)$

$(dis_3)$ $(\varphi \vee \psi) \wedge (\varphi \vee \chi) \vdash \varphi \vee (\psi \wedge \chi)$

$(dis_4)$ $(\varphi \wedge \psi) \vee (\varphi \wedge \chi) \vdash \varphi \wedge (\psi \vee \chi)$

$(dis_5)$ $\varphi \wedge (\psi \vee \chi) \vdash (\varphi \wedge \psi) \vee (\varphi \wedge \chi)$

$(dni)$ $\varphi \vee \psi \vdash {\sim}{\sim}\varphi \vee \psi$

$(dne)$ ${\sim}{\sim}\varphi \vee \psi \vdash \varphi \vee \psi$

$(cne)$ ${\sim}{\sim}\varphi \wedge \psi \vdash \varphi \wedge \psi$

$(cni)$ $\varphi \wedge \psi \vdash {\sim}{\sim}\varphi \wedge \psi$

$(ddm_1)$ ${\sim}(\varphi \vee \psi) \vee \chi \vdash ({\sim}\varphi \wedge {\sim}\psi) \vee \chi$

$(ddm_2)$ $({\sim}\varphi \wedge {\sim}\psi) \vee \chi \vdash {\sim}(\varphi \vee \psi) \vee \chi$

$(ddm_3)$ ${\sim}(\varphi \wedge \psi) \vee \chi \vdash ({\sim}\varphi \vee {\sim}\psi) \vee \chi$

$(ddm_4)$ $({\sim}\varphi \vee {\sim}\psi) \vee \chi \vdash {\sim}(\varphi \wedge \psi) \vee \chi$

$(cdm_4)$ $({\sim}\varphi \vee {\sim}\psi) \wedge \chi \vdash {\sim}(\varphi \wedge \psi) \wedge \chi$

$(cdm_3)$ ${\sim}(\varphi \wedge \psi) \wedge \chi \vdash ({\sim}\varphi \vee {\sim}\psi) \wedge \chi$

$(cdm_2)$ $({\sim}\varphi \wedge {\sim}\psi) \wedge \chi \vdash {\sim}(\varphi \vee \psi) \wedge \chi$

$(cdm_1)$ ${\sim}(\varphi \vee \psi) \wedge \chi \vdash ({\sim}\varphi \wedge {\sim}\psi) \wedge \chi$

$(tr)$ $\varphi \vdash \psi, \psi \vdash \chi / \varphi \vdash \chi$

This system might appear rather bulky, yet it is quite manageable, as the following lemma shows:

*Lemma* 6.1. For every of 12 schemata from (*dni*) to $(cdm_1)$ of the form $\varphi * \chi \vdash \psi * \chi$ (where $*$ is either $\vee$ or $\wedge$) the consequence $\varphi \vdash \psi$ is derivable in ***FDE(-)***.

**Proof.** Consider any schema $(x)$ from (*dni*) to $(cdm_1)$. $(x)$ is either of the form $\varphi \vee \chi \vdash \psi \vee \chi$, or of the form $\varphi \wedge \chi \vdash \psi \wedge \chi$. For the first case take the inference for (a) from Lemma 2.1, for the second case take the inference for (a) from Lemma 4.1. ■

My next task will be to establish the full admissibility of the inference rules of the first-degree entailment logic in **FDE**(-). Let $\varphi \vdash_{\text{fde}} \psi$ stands for a derivability of $\varphi \vdash \psi$ in **FDE**(-).

*Lemma* 6.2. Rules (*ci*), (*de*), and (*con*) are all admissible in ***FDE(-)***.

**Proof.** This lemma is a combination of Lemma 2.2 and Lemma 4.2. I will check some cases to make sure that the lemma holds in full generality.

(1) Consider the rule of conjunction introduction. The following dual Font's rule is admissible in **FDE**(-): if $\varphi \vdash_{\text{fde}} \psi$, then $\varphi \wedge \chi \vdash_{\text{fde}} \psi \wedge \chi$. Indeed, for every axiom of **FDE**(-) of the form $\varphi \vdash \psi$, we have $\varphi \wedge \chi \vdash_{\text{fde}} \psi \wedge \chi$. Let us check some cases. For $(di_1)$ we have:

| | |
|---|---|
| 1. $\varphi \wedge \chi \vdash (\varphi \wedge \chi) \vee (\psi \wedge \chi)$ | $(di_1)$ |
| 2. $(\varphi \wedge \chi) \vee (\psi \wedge \chi) \vdash \chi \wedge (\varphi \vee \psi)$ | $(dis_4)$ |
| 3. $\chi \wedge (\varphi \vee \psi) \vdash (\varphi \vee \psi) \wedge \chi$ | $(cco)$ |
| 4. $(\varphi \wedge \chi) \vee (\psi \wedge \chi) \vdash (\varphi \vee \psi) \wedge \chi$ | 2, 3: $(tr)$ |
| 5. $\varphi \wedge \chi \vdash (\varphi \vee \psi) \wedge \chi$ | 1, 4: $(tr)$ |

For $(dco)$ we have:

| | |
|---|---|
| 1. $(\varphi \vee \psi) \wedge \chi \vdash \chi \wedge (\varphi \vee \psi)$ | $(ce_1)$ |
| 2. $\chi \wedge (\varphi \vee \psi) \vdash (\chi \wedge \varphi) \vee (\chi \wedge \psi)$ | $(dis_5)$ |
| 3. $(\chi \wedge \varphi) \vee (\chi \wedge \psi) \vdash (\chi \wedge \psi) \vee (\chi \wedge \varphi)$ | $(dco)$ |
| 4. $(\chi \wedge \psi) \vee (\chi \wedge \varphi) \vdash \chi \wedge (\psi \vee \varphi)$ | $(dis_4)$ |
| 5. $\chi \wedge (\psi \vee \varphi) \vdash (\psi \vee \varphi) \wedge \chi$ | $(cco)$ |
| 6. $(\varphi \vee \psi) \wedge \chi \vdash (\chi \wedge \varphi) \vee (\chi \wedge \psi)$ | 1, 2: $(tr)$ |
| 7. $(\varphi \vee \psi) \wedge \chi \vdash (\chi \wedge \psi) \vee (\chi \wedge \varphi)$ | 6, 3: $(tr)$ |
| 8. $(\varphi \vee \psi) \wedge \chi \vdash \chi \wedge (\psi \vee \varphi)$ | 7, 4: $(tr)$ |
| 9. $(\varphi \vee \psi) \wedge \chi \vdash (\psi \vee \varphi) \wedge \chi$ | 8, 5: $(tr)$ |

For $(did)$ we have:

| | |
|---|---|
| 1. $(\varphi \vee \varphi) \wedge \chi \vdash \chi \wedge (\varphi \vee \varphi)$ | $(cco)$ |
| 2. $\chi \wedge (\varphi \vee \varphi) \vdash (\chi \wedge \varphi) \vee (\chi \wedge \varphi)$ | $(dis_5)$ |
| 3. $(\chi \wedge \varphi) \vee (\chi \wedge \varphi) \vdash \chi \wedge \varphi$ | $(did)$ |
| 4. $\chi \wedge \varphi \vdash \varphi \wedge \chi$ | $(cco)$ |
| 5. $(\varphi \vee \varphi) \wedge \chi \vdash (\chi \wedge \varphi) \vee (\chi \wedge \varphi)$ | 1, 2: $(tr)$ |
| 6. $(\varphi \vee \varphi) \wedge \chi \vdash \chi \wedge \varphi$ | 5, 3: $(tr)$ |
| 7. $(\varphi \vee \varphi) \wedge \chi \vdash \varphi \wedge \chi$ | 6, 4: $(tr)$ |

The cases for all other axioms are proved analogously.

Now consider the rule of inference $(tr)$, and assume that $\varphi \wedge \xi \vdash_{\text{fde}} \psi \wedge \xi$ and $\psi \wedge \xi \vdash_{\text{fde}} \chi \wedge \xi$. Then, by $(tr)$ we get $\varphi \wedge \xi \vdash_{\text{fde}} \chi \wedge \xi$.

Having admissibility of the dual Font's rule, the following schema of an inference in **FDE**(-) demonstrates admissibility of (ci):

| | |
|---|---|
| 1. $\varphi \vdash_{\text{fde}} \psi$ | *(assumption)* |
| 2. $\varphi \vdash_{\text{fde}} \chi$ | *(assumption)* |
| 3. $\varphi \wedge \varphi \vdash_{\text{fde}} \psi \wedge \varphi$ | 1: (dual Font's rule) |
| 4. $\varphi \wedge \psi \vdash_{\text{fde}} \chi \wedge \psi$ | 2: (dual Font's rule) |
| 5. $\varphi \vdash_{\text{fde}} \varphi \wedge \varphi$ | $(cid)$ |
| 6. $\varphi \vdash_{\text{fde}} \psi \wedge \chi$ | (3, 4, 5: $(tr)$, twice) |

(2) As to admissibility of $(de)$, its proof is dual to (1), and is largely presented in the proof of Lemma 2.2.

(3) To show admissibility of $(con)$ one has to demonstrate, how an inference of $\varphi \vdash_{\text{fde}} \psi$ can be transformed into the inference of $\sim\psi \vdash_{\text{fde}} \sim\varphi$. Again, this can be done by induction on the length of the initial inference of $\varphi \vdash_{\text{fde}} \psi$. First, observe that the contraposition of every axiom schema of **FDE**(-) is derivable in **FDE**(-) (one can use De Morgan laws,

derivable in **FDE**(-) by Lemma 6.1). Moreover, for the inference rule of transitivity (*tr*) it is easy to see that contraposing its premises and conclusion results again in transitivity. ■

I am now in a position to establish the main result of this paper:

*Lemma* 6.3. System ***FDE(-)*** is deductively equivalent to system $\boldsymbol{E}_{fde}$ (and hence, $\boldsymbol{R}_{fde}$, ***FDE(ci)*** and ***FDE(de)***) in the sense that they determine the same set of provable consequences.

**Proof.** Obviously, all the axioms of **FDE**(-) are provable in $\mathbf{E}_{\mathrm{fde}}$. The converse can be easily obtained by Lemma 6.1. Taking into account admissibility of all the inference rules of $\mathbf{E}_{\mathrm{fde}}$ in **FDE**(-) (Lemma 6.2), we get the result. ■

As a direct byproduct of Lemma 6.3 we get the soundness and completeness of **FDE**(-) with respect to the entailment relation determined by Definition 1.1:

*Theorem* 6.4. For every $\varphi, \psi$: $\varphi \vdash_{fde} \psi \Leftrightarrow \varphi \models_{\mathcal{FDE}} \psi$.

## 7 Concluding remarks on the inference rules and structural reasoning

**FDE**(ci), **FDE**(de) and **FDE**(-) jointly present a rather illustrative example of deductively equivalent systems which nevertheless disagree in their derivable rules of inference. "Downgrading" a rule from a derivable to merely admissible within some system may be useful, opening the door for interesting extensions of the system, which could be impossible otherwise.[9]

In particular, due to non-derivability of the disjunction elimination and contraposition rules in **FDE**(ci), it could be extended to $\mathbf{ETL}_1$, where both rules are no longer admissible. Thus, whereas adding (*ds*) to $\mathbf{E}_{\mathrm{fde}}$ collapses it into classical logic, and $\mathbf{R}_{\mathrm{fde}}$ being extended by (*ds*) just amounts to Kleene's logic, **FDE**(ci) allows more subtle distinctions, so that, e.g., the consequence $(\varphi \wedge {\sim}\varphi) \vee (\psi \wedge {\sim}\psi) \vdash \chi$ is not derivable in **ETL**, even though both $\varphi \wedge {\sim}\varphi \vdash \chi$ and $\psi \wedge {\sim}\psi \vdash \chi$ are (see, [22, p. 129]). The situation with **FDE**(de) and $\mathbf{NFL}_1$ is perfectly dual.

In this respect **FDE**(-) presents the "limiting case" among the deductively equivalent systems for the logic of first-degree entailment, permitting its most possible extensions. In particular, the system that could be obtained from **FDE**(-) by adding both (*ds*) and (*dds*) is worthy of attention. The task of investigating this and other possible extensions of **FDE**(-) is recommended for future work.

**FDE**(-) is also of interest in some other regard, being a pure *structural system* in certain precise sense of the term. This sense essentially goes back to the Gentzenian division between structural rules and rules for logical symbols. As it is well known, the standard Gentzent's sequent calculi comprise both kinds of rules "with small number (usually one) of primitive sequents" [18, p. 1297]. There is also another approach to the sequent calculi construction, which stems directly from the work of Paul Hertz (see, e.g., [16]), according

[9] Logic as a set of expressions should always be closed under the inference rules of some (underlying) proof system. Therefore, adding certain expression to the given set may result in different outcomes, depending on the system taken to be basic. It means that any accurate (syntactic) consideration of "logical extensibility" is essentially "system-dependent", so that when we speak of extensions of some "logic", we always have in mind (explicitly or implicitly) a particular *proof system* for this logic, subject to the extensions in question. For example, the comprehensive study of "super-Belnap logics" in [1] (and related works, see references therein) presents, in fact, the case for a wealth of extensions of Font's system $\vdash_H$.

to which “rules have purely structural character and all logical content is contained in primitive sequents” [18, p. 1309].

Schroeder-Heister defines *structural reasoning* as “reasoning in a sequent style system using structural rules only” [24, p. 246]. He stresses the importance of this kind of reasoning as a “subject in its own right”, particularly “in the light of modern developments such as logic programming”, and also observes that “calculi developed by Paul Hertz in the 1920s are structural systems in this independent sense” [24, p. 247].

The present paper outlines a natural way of evolving the logic of first-degree entailment towards structural reasoning—from $\mathbf{E}_{\mathrm{fde}}$ up to **FDE**(-). Namely, it shows how we can forgo logical rules due to additional axiom schemata. Starting from $\mathbf{E}_{\mathrm{fde}}$ with a special inference rule for each of the three propositional connectives of language $\mathcal{L}$, we work the way through a system without the rule for negation ($\mathbf{R}_{\mathrm{fde}}$), then to a system without the rule either for disjunction (**FDE**(ci)) or for conjunction (**FDE**(de)), and ending up with a system with no logical rules at all (**FDE**(-)). The latter is thus a purely structural sequent style system of the “Hertz-kind”.

Moreover, derivations in all these systems are constructed not as trees but are linearly ordered. An inference (proof) of a consequence is defined here as a finite *consecutive* list of (occurrences of) consequence expressions, each of which either is an axiom or comes by an inference rule from some consequence expressions preceding it in the list (cf. [20, p. 34]). **FDE**(-) deserves thus attention as a system, which represents *linear structural reasoning*, and can serve as a “proof ground” for investigating the properties of this kind of reasoning.

## References


1. Albuquerque, H., Přenosil, A., Rivieccio, U. An algebraic view of super-Belnap logics, *Studia Logica* 105 (2017), 1051-1086.
2. Anderson, A.R. and Belnap, N.D. Tautological entailments, *Philosophical Studies* 13 (1962), 9-24.
3. Anderson, A.R. and Belnap, N.D. *Entailment: The Logic of Relevance and Necessity*, vol. I, Princeton, NJ: Princeton University Press, 1975.
4. Belnap, N.D. A useful four-valued logic, in: J.M. Dunn and G. Epstein (eds.), *Modern Uses of Multiple-Valued Logic*, Dordrecht: D. Reidel Publishing Company, 1977, 8-37.
5. Belnap, N.D. How a computer should think, in: G. Ryle (ed.), *Contemporary Aspects of Philosophy*, Oriel Press, 1977, 30-55.
6. Dunn, J.M. Algebraic completeness results for R-mingle and its extensions, *Journal of Symbolic Logic* 35 (1970), 1-13.
7. Dunn, J.M. Intuitive semantics for first-degree entailment and coupled trees, *Philosophical Studies* 29 (1976), 149-168.
8. Dunn, J.M. Positive modal logic, *Studia Logica*, 55 (1995), 301-317.
9. Dunn, J.M. A comparative study of various model-theoretic treatments of negation: a history of formal negation, in: D.M. Gabbay and H. Wansing (eds.), *What is Negation?*, Dordrecht: Kluwer Academic Publishers, 1999, 23-51.
10. Dunn, J.M. Partiality and its dual, *Studia Logica* 66 (2000), 5-40.
11. Dunn, J.M. Contradictory information: too much of a good thing, *Journal of Philosophical Logic* 39 (2010), 425-452.
12. Font, J.M. Belnap's four-valued logic and De Morgan lattices, *Logic Journal of the IGPL* 5 (1997), 413-440.
13. van Fraassen, B.C. Gentlemen's wagers: Relevant logic and probability, *Philosophical Studies* 43 (1983), 47-61.
14. Gentzen G. Untersuchungen über das logische Schließen. I, *Mathematische Zeitschrift* 39 (1935), 176-210.

15. Gentzen G. *The Collected Papers of Gerhard Gentzen*, Amsterdam: North-Holland Pub. Co., 1969.
16. Hertz, P. Über Axiomensysteme für beliebige Satzsysteme, *Mathematische Annalen* 101 (1929), 457-514.
17. Humberstone, L. *The Connectives*, Cambridge, MA: MIT Press, 2011.
18. Indrzejczak, A. A survey of nonstandard sequent calculi, *Studia Logica* 102 (2014), 1295-1322.
19. Kalman, J.A. Lattices with involution, *Transactions of the American Mathematical Society* 87 (1958), 485-491.
20. Kleene, S.C. *Mathematical Logic*. New York: Wiley, 1967.
21. Marcos, J. The value of the two values, in: J.-Y. Beziau, M.E. Coniglio (eds.) *Logic without Frontiers: Festschrift for Walter Alexandre Carnielli on the Occasion of his 60th Birthday* (Tributes), College Publications, 2011, 277-294.
22. Pietz, A. and Rivieccio, U. Nothing but the Truth, *Journal of Philosophical Logic* 42 (2013), 125-135.
23. Rose, A. A formalisation of the propositional calculus corresponding to Wang's calculus of partial predicates, *Zeitschrift für mathematische Logik und Grundlagen der Mathematik* 9 (1963), 177-108.
24. Schroeder-Heister, P. Resolution and the origins of structural reasoning: early proof-theoretic ideas of Hertz and Gentzen, *Bulletin of Symbolic Logic* 8 (2002), 246-265.
25. Shramko, Y. A modal translation for dual-intuitionistic logic, *The Review of Symbolic Logic* 9 (2016), 251-265.
26. Shramko, Y. and Wansing, H. Entailment relations and/as truth values, *Bulletin of the Section of Logic* 36 (2007), 131-143.
27. Shramko, Y., Zaitsev, D., Belikov, A. First-degree entailment and its relatives, *Studia Logica* 105 (2017), 1291-1317.